Marine Electrochemistry

Marine Electrochemistry
A Practical Introduction

Edited by

M. Whitfield
Marine Biological Association of the United Kingdom,
Plymouth, UK

and

D. Jagner
Department of Analytical Chemistry,
University of Göteborg,
Sweden

A Wiley–Interscience Publication

JOHN WILEY & SONS
Chichester · New York · Brisbane · Toronto

British Library Cataloguing in Publication Data:

Marine electrochemistry.
  1. Chemical oceanography
  2. Electrochemistry
  I. Whitfield, M.
  II. Jagner, D.
  551.46'01   GC116   80-42023

ISBN 0 0471 27976 5

Filmset and printed in Northern Ireland at The Universities Press (Belfast) Ltd., and bound at the Pitman Press, Bath, Avon.

82 004057

To the late Klaus Grasshoff for his friendship and for his contributions to marine science.

List of Contributors

BEN-YAAKOV, S. *Department of Electrical Engineering, Ben-Gurion University, Beer-Sheva, Israel*

BROOKS, E. E. *Department of Chemistry, Howard University, Washington, D.C. 20059, USA*

CULBERSON, C. H. *College of Marine Studies, University of Delaware, Newark, Delaware 19711, USA*

†GRASSHOFF, K. *Institut für Meereskunde an der Universität Kiel, Düstenbrookerweg 20, D-23000 Kiel, Federal Republic of Germany*

JAGNER, D. *Department of Analytical Chemistry, University of Göteborg, S-41296 Göteborg, Sweden*

MARK, H. B., JR. *Department of Chemistry, University of Cincinatti, Cincinatti, Ohio 45221, USA*

TURNER, D. R. *Marine Biological Association of the UK, The Laboratory, Citadel Hill, Plymouth PL1 2PB, UK*

WHITFIELD, M. *Marine Biological Association of the UK, The Laboratory, Citadel Hill, Plymouth PL1 2PB, UK*

WILSON, T. R. S. *Institute of Oceanographic Sciences, Wormley, Godalming, Surrey, UK*

ZIRINO, A. *Naval Undersea Center, San Diego, California 92132, USA*

† Deceased 11 March 1981

Contents

Glossary of Abbreviations

AAS	Atomic-absorption spectrometry
A/D	Analog to digital
AE	Auxiliary electrode
AOU	Apparent oxygen utilization
ASV	Anodic stripping voltammetry
BCD	Binary coded decimal
CE	Counter electrode
CF diagram	Complexation field diagram
CPU	Central processing unit
DL	Double layer
DME	Dropping mercury electrode
DPASV	Differential pulse anodic stripping voltammetry
DPP	Differential pulse polarography
DSP	Deep-sea probe
EFE diagram	Electron free energy diagram
EIGE	Epoxy-impregnated graphite electrode
FM	Frequency modulation
HMDE	Hanging mercury drop electrode
HMDE (Pt)	HMDE attached to platinum wire
IHP	Inner Helmholtz plane
ISE	Ion-selective electrode
LAR diagram	Logarithmic activity ratio diagram
LC diagram	Logarithmic concentration diagram
LSASV	Linear scan anodic stripping voltammetry
MCGE	Mercury-coated graphite electrode
NHE	Normal hydrogen electrode
OHP	Outer Helmholtz plane
OS diagram	Oxidation state diagram
PA diagram	Predominance area diagram
PCM	Pulse code modulation
PGE	Pyrolytic graphite electrode
PLL	Phase locked loop
PROM	Programmable read-only memory

RDGCE	Rotating disc glassy carbon electrode
RE	Reference electrode
SCE	Saturated calomel electrode
SHE	Standard hydrogen electrode
STP	Standard temperature and pressure
TFE	Thin-film electrode
WIGE	Wax-impregnated graphite electrode
XRF	X-ray fluorescence

Preface

Electrochemical techniques are particularly well suited for studies of natural aqueous systems since they are chemically selective, non-destructive, and readily automated. They are relatively cheap, both in terms of the initial capital outlay on equipment and on running costs, and many of the sensors can be easily miniaturized and pressure compensated for work *in situ*. These characteristics have enabied electrochemical techniques to be used for detailed *in situ* studies of profiles of salinity, oxygen, and pH in the deep ocean. The incredibly complex fine structures revealed by these studies have radically altered our concepts of the physical and biological processes that are reflected in the water chemistry and have dramatically demonstrated the value of electrochemical techniques in environmental studies.

The ancillary equipment required for such measurements, in addition to the sensors themselves, is simple and compact and instrumentation for a wide range of techniques can be fabricated from a relatively small number of basic circuits. Most electrochemical procedures that are in general use are supported by a comprehensive body of theory that not only simplifies the quantitative interpretation of the information obtained but also provides a firm basis for the development of new and powerful techniques to take advantage of the very rapid progress in electronics. Indeed, it is likely that the dramatic evolution of the microprocessor will change the nature of electroanalytical chemistry by enabling 'intelligent' instruments to be developed that are capable of optimizing the analytical procedure to suit the nature of the sample. Such developments are already under way and there is a danger that they will transform the rather bewildering array of electrochemical procedures into a continuum of techniques that will be virtually incomprehensible to the non-specialists so that their potential for solving environmental problems might not be fully exploited. It seemed that, at this juncture, it would be appropriate to prepare a book that might provide an interface between the electrochemist and the environmental scientist by setting the widely used electrochemical methods in their proper context and by showing how these methods are used for environmental studies. In this way it might be possible not only to foster a wider understanding and appreciation of the virtues of electrochemical procedures but also to guide the development of these techniques along environmentally useful lines. With this aim in mind the book is divided into two sections: the first (Part A, Chapters 1–4) covers fundamentals and the second (Part B, Chapters 5–10) covers applications.

In Part A the fundamental environmental and electrochemical concepts are introduced and set in context. Individual chapters deal with the nature of sea water (Chapter 1), the classification of electrochemical techniques (Chapter 2), the design of electrochemical instrumentation (Chapter 3) and the use of computers in electrochemical measurements (Chapter 4). These chapters are introductory in nature and are not intended as comprehensive surveys, although they do provide a useful key for those unfamiliar with the literature.

The emphasis in Part B is on the practical aspects of the use of electrochemical techniques in oceanographic studies. All of the chapters are written by practitioners who have had considerable experience of the vagaries of the various electrochemical techniques both in the laboratory and in the field. Detailed treatments are provided of the conductometric determination of salinity (Chapter 5), arguably the most precise and widely used electrochemical technique, and of the amperometric determination of oxygen (Chapter 9), the most familiar of the chemically specific electroanalytical procedures. Two chapters deal with the use of potentiometric sensors both for direct measurement (including the determination of pH, Chapter 6) and as end-point detectors in titrimetric procedures (Chapter 7). In the chapter on direct potentiometry considerable attention is given to the definition and calibration of pH scales for use in sea water. Controversy still rages as to whether the operationally defined NBS scale should be accepted rather than one of the thermodynamically defined concentration scales. Whichever scale is eventually adopted, the excellent practical procedures described in Chapter 6 will ensure that results of the highest precision are obtained. Less familiar are the electrodeposition procedures (Chapter 8), which promise to provide a versatile and chemically selective means of obtaining uncontaminated samples for analysis by spectroscopic and neutron activation analysis. Voltammetry (Chapter 10) has undoubtedly witnessed the most rapid growth, both in the number of measurements reported and in the range of techniques available. Here it has also been necessary for the author to address the problems of obtaining and maintaining uncontaminated samples since this, and not the sensitivity of the electrochemical procedures, has proved to be the major stumbling block in the analysis of the trace metals. As readers progress through Part B it is our intention that they should be able to refer back to Part A for clarification of any unfamiliar concepts and from there to branch out into the wider literature. This might be particularly necessary in the study of voltammetric procedures, where so many different modulation techniques and electrode systems have been used.

The theoretical treatment has been restricted to the minimum required for the intelligent use of each technique although, on occasion, this demands some detailed considerations. The authors have largely confined their attention to the problems involved in obtaining and interpreting good environmental data. The use of electrochemical procedures for physico-chemical studies (e.g. the determination of rate constants and stability constants) has been considered only incidentally, as a properly detailed treatment of such investigations would have doubled the size of the book, and no doubt doubled its gestation period.

From this outline it is clear that this is intended as an essentially practical book, the purpose of which is to promote good electrochemical practice and to provide a logical framework for the future application of electrochemical procedures to environmental studies.

In preparing a multi-author work of this kind there is always a danger that some contributions could become outdated, either because of the conscientiousness of the author who submits his manuscript on time or because of the rapid rate of development of the subject. We have tried to minimize these effects by focusing on fundamental issues that are not so subject to the vagaries of scientific fashion and by judicious editing where significant advances have been made. Any outstanding discrepancies are clearly the Editors' responsibility.

We thank the authors for their enthusiastic cooperation in the preparation of this book and we hope that they are pleased with the final result. We hope, too, that the book will encourage a wider appreciation of the value of electrochemical procedures in studying natural systems and will go some way towards removing the mystification that sometimes surrounds their application.

M. WHITFIELD
D. JAGNER
October 1980

Part A   Fundamentals

Marine Electrochemistry
Edited by M. Whitfield and D. Jagner
© 1981 John Wiley & Sons Ltd.

M. WHITFIELD and D. R. TURNER

Marine Biological Association of the UK,
The Laboratory, Citadel Hill,
Plymouth PL1 2PB, UK

1

Sea Water as an Electrochemical Medium

GLOSSARY OF SYMBOLS

a_Y	activity of component Y (equation 3)
A_C	Carbonate alkalinity (equation 55)
c_Y	Concentration of component Y on the molar ($mol\,dm^{-3}$) scale (Section 1.3.1)
C	Coulombs (Figure 12)
$\bar{C}_p$	Relative partial molal constant-pressure heat capacity of Y (Table 2)
C_T	Total carbon dioxide content (equation 54)
d_Y	Density of Y (equation 15)
D_Y	Diffusion coefficient of component Y (equation 10)
E	Applied electrical field (equation 79)
E_j	Liquid junction potential (equation 77)
E_{jm}	Ideal liquid junction potential (equation 78)
$E_{j\gamma}$	Non-ideal liquid junction potential (equation 78)
E'_H	Formal electrode potential (equation 30)
$E_H^{\ominus}$	Standard electrode potential (equation 30)

$\bar{E}_Y$	Relative partial molal expansibility of Y (equation 13)
g'	Nernst slope term (RT/F) (Table 3A)
g	Acceleration due to gravity (equation 15)
g_Y	Mass of component Y in grams (Table 6)
G_Y	Gibbs free energy of component Y (equation 3)
$\bar{h}$	Hydrogen ion contribution to stability constant (equation 44)
h	Depth (equation 15)
$\bar{H}_Y$	Relative partial molal enthalpy of Y (Table 2)
I	Ionic strength (equation 71)
J_Y	Flux of element Y (equation 10)
k^*	Concentration rate constant (equation 9)
$k^{\ominus}$	Activity rate constant (equation 7)
k_d	Diffusion rate constant for CO_2 transport (equation 64)
k_Y	Molinity (mol kg^{-1} solution) of component Y (Section 1.3.1)
$K^{\ominus}$	Equilibrium constant (equation 5)
K^*	Stoichiometric equilibrium constant (equation 5)
K_p	Henry's law constant for CO_2 (equation 45)
K_{1C}	First ionization constant for carbonic acid (equation 43)
K_{2C}	Second ionization constant for carbonic acid (equation 44)
$K_Y(sw)$	Ocean/rock partition coefficient (Section 1.1)
l	$= c_Y/S$ (equation 86)
$m(sw)$	Molality of an element in sea water (Section 1.1)
$m(rw)$	Molality of an element in river water (Section 1.1)
m_Y	Molality of component Y (equation 3)
M_Y	Molecular weight of Y (equation 15)
n	Stoichiometric number of electrons accompanying an electrode reaction (equation 24)
p	Vapour pressure (Table 7)
pe	$= -\log_{10}a_{e-}$ (equation 27)
P_Y	Partial pressure of component Y (equation 45)
Q_{YO}	Electronegativity term (equation 2)
r	Ionic radius (Figure 13)
S	Salinity (equation 68)
S_T	Total salt content (equation 69)
t	Temperature (°C) (Table 7)
t_F	Freezing point (Table 7).
t_i	Hittorf transport number (equation 77)
t_R	Time required for single stirring revolution of the oceans (Figure 2)
$\bar{t}_Y$	Mean oceanic residence time of an element Y (equation 1)
T	Temperature (K) (Table 7)
U_i	Infinite dilution mobility of i (equation 75)
$\bar{V}_Y$	Relative partial molal volume of Y (Table 2)
x_Y	Electronegativity of element Y (Table 1)
z_i	Charge on ion i (equation 71)
$\bar{\alpha}$	Overall side-reaction coefficient (equation 90)
β_{ML_n}	Stepwise constant for ML_n (equation 89)
β_T	Buffer capacity (equation 56)
ε	Individual electrode potential (equation 25)
ε'_s	Dielectric constant of sea water (Table 7)
ε''_s	Dielectric loss of sea water (Table 7)
ε'_w	Dielectric constant of pure water (Table 7)
η_r	Relative viscosity of sea water (Table 7)

γ_Y	Molal activity coefficient of component Y (equation 3)
Γ	Activity coefficient product (equation 5)
$\bar{\kappa}_Y$	Relative partial molal compressibility of Y (Table 2)
Λ_i	Limiting equivalent ionic conductivity (equation 80)
μ_Y	Chemical potential of component Y (equation 3)
ν	Kinematic viscosity (Table 7)
π	Osmotic pressure (Table 7)
σ_{ML_n}	Side-reaction coefficient for the formation of ML_n (equation 89)
τ	Surface tension (Table 7)
ϕ_{sw}	Osmotic coefficient of sea water (Table 7)
$\ominus$	superscript, standard state properties (equation 3)

1.1. GEOCHEMICAL CONTROL OF SEA WATER COMPOSITION

To simplify matters we shall consider initially a global mean ocean (or ocean bucket, Figure 1) containing an 'average' sea water (Table 1). The reservoir of dissolved material in the oceans is augmented primarily by acid volatiles emanating from volcanically active areas and by dissolved and suspended river-borne solids resulting from rock weathering. Over many millions of years these ingredients have interacted within the oceans to yield not only the present sea water reservoir but also the atmosphere and the marine sediments. By careful geological book-keeping it can be shown (Sillén, 1965; Horn and Adams, 1966) that the reaction

$$\frac{Continental}{rock} + \frac{Acid}{volatiles} + Water \rightarrow Sea\ water + Sediment + Atmosphere$$

is almost perfectly balanced for nearly sixty elements. It is likely that sea water itself contains, in the dissolved state, all of the elements in the Periodic Table, although in a few cases their concentrations are too low for detection at present. The composition of sea water will depend predominantly on the partitioning of the elements between the rock and the solution phase, both in the rapid initial weathering stage where the solution has a relatively low pH (5–6) and ionic strength (*ca.* 10^{-3} M) and in the final ageing stage where the detrital mineral particles slowly settle through the ocean reservoir which contains a solution with a relatively high ionic strength (0.7 M) and pH (8.2). Although fairly convincing equilibrium control mechanisms can be deduced for some predominant components (Sillén, 1967), it is more likely that the concentrations of most of the elements in sea water are fixed by a dynamic balance between input and removal processes (Broecker, 1971; Whitfield, 1979a). When material enters the oceans it is able to accumulate up to the age limit of the reservoir itself unless its concentration is controlled at some intermediate level by removal processes. The mean residence time $(\bar{t}_Y)$ of an element (Y) in the oceans provides a useful guide to the ease with which it is removed from the system either by

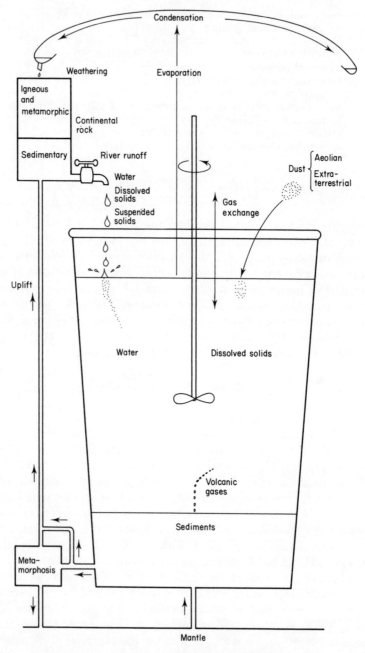

Figure 1. Schematic view of the cycling of the elements through the oceans. Reproduced with the permission of Blackwells Scientific Publications from Whitfield (1976)

the transfer of volatile components to the atmosphere (e.g. O_2, N_2, CO_2) or by the incorporation of components into the sediment directly by inclusion in the settling particles or indirectly by occlusion in the interstitial waters of the settled sediment. For an ocean at steady state, where the flux of material entering the system exactly counterbalances the flux of material leaving, $\bar{t}_Y$ can be defined as (Barth, 1952)

$$\bar{t}_Y = (\text{reservoir size})/(\text{rate of input } or \text{ removal}) \tag{1}$$

where the terms in parentheses refer to dissolved components only. The more reactive an element is in the ocean reservoir the more rapidly will it be removed and the shorter will be its residence time. The composition of sea water is therefore intimately connected with the residence times of the elements and there is a direct relationship between $\bar{t}_Y$ and the partition coefficient $[K_Y(\text{sw}) = \text{concentration of Y in sea water/concentration of Y in crustal rock}]$ which describes the distribution of the elements between the oceans and the crustal rocks (Figure 2). This implies that the more effectively a particular element can be incorporated into the solid phase, the

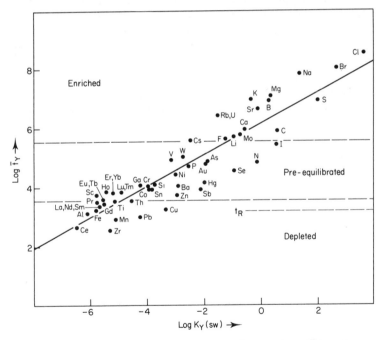

Figure 2. Relationship between mean oceanic residence time ($\bar{t}_Y$) and ocean-rock partition coefficient $[K_Y(\text{sw})]$. The various categories of the elements identified in the figure are discussed in the text. t_R is the time required for a single stirring revolution of the oceans. Reproduced with the permission of Heyden & Son, London, from Whitfield (1981)

shorter will be its stay in the ocean reservoir (Figure 2). We can see how this relates to the individual elements if we relate the concentration of a particular element in sea water, $m(sw)$, to its concentration in river water, $m(rw)$ (Whitfield, 1979a; Whitfield and Turner, 1979). Elements for which $m(sw) > 10m(rw)$ (enriched elements, Figure 2) accumulate in the oceans so effectively that they have been thoroughly mixed over geological time and their concentrations maintain a constant ratio to one another (within $\pm 10\%$), irrespective of the total salt content. The majority of elements for which $m(sw) < 0.1m(rw)$ (depleted elements, Figure 2) react so rapidly that their residence times are less than the time required for a single stirring revolution of the oceans (t_R, Figure 2). These elements consequently show an uneven distribution throughout the world oceans. The concentrations of the remaining elements for which $10m(rw) > m(sw) > 0.1m(rw)$ (pre-equilibrated elements, Figure 2) also show considerable variability in both space and time, not so much because of their rapid removal from the system but because many of them are intimately involved in an intricate web of short-term biological and geological cycles.

Even the most persistent elements (e.g., Br, Cl; Figure 2) remain in the system for no longer than a few hundred million years. This is close to the age of the oldest oceanic sediments and is only a fraction of the three billion year lifespan of the oceans themselves. It is likely, therefore, that most of the continental rocks have been round the weathering cycle several dozen times during the earth's history, enabling the partitioning of the elements between the sea water and the rocks to run close to a steady state.

Since the system has been recycling in this way for several billion years, it is tempting to look for general chemical correlations in the ocean–rock partition coefficients and hence for an inorganic rationale behind the overall composition of sea water. In general, the exchange of elements between the two phases is controlled not by solubility equilibria but by adsorption/desorption reactions at the surfaces of the suspended particles (Schindler, 1975). Since the surface sites on the mineral lattices are predominantly deprotonated hydroxyl groups, $K_Y(sw)$ can be related to the strength of the electrostatic contribution to the element–oxygen bond (Q_{YO}) by the equation (Whitfield and Turner, 1979; Turner *et al.*, 1980; see Figure 3)

$$\log K_Y(sw) = a_n Q_{YO} + b_n \tag{2}$$

where

$$Q_{YO} = (x_Y - x_O)^2$$

and x represents electronegativity (Table 1). The main sequence in the electronegativity correlation can be related to the geochemical classification of the elements first introduced by Goldschmidt (1954), suggesting that the

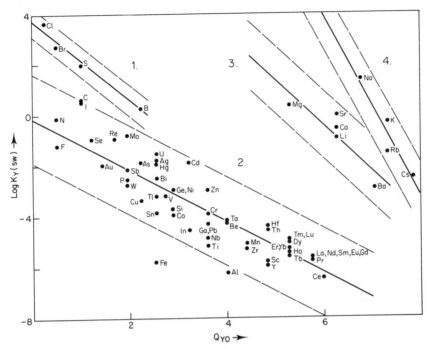

Figure 3. Relationship between the ocean-rock partition coefficient [$K_Y(sw)$] and the electronegativity function (Q_{YO}). The following groups of elements may be identified: 1, excess volatiles; 2, main sequence; 3, alkaline earth metals; and 4, alkali metals. Reproduced with the permission of Elsevier Scientific Publishing Co. from Turner *et al.*, *Marine Chemistry*, **9**, 211–218, 1980

partition coefficients themselves are controlled predominantly by the solid-state chemistry of the elements (Turner and Whitfield, 1979; Turner *et al.*, 1980). The solution chemistry of the elements exerts a significant influence on the overall control of sea water composition only when it directly affects hydroxide solubilities. Iron and aluminium fall significantly below the main correlation because their high crustal abundance and low hydroxide solubilities prevent them from attaining the partition coefficient dictated by their electronegativities. The Group IA and IIA elements (with the exception of beryllium) are far more abundant than their low electronegativities would lead us to expect. The small charge/radius ratios (or ionic potentials) of these cations give them an unusually low affinity for oxide lattices in aqueous media so that they are able to accumulate in the oceans (Turner and Whitfield, 1979). Other elements (e.g. Cl, B, S; Figure 3) also accumulate in the ocean because their concentrations are augmented by significant auxiliary inputs from volcanic emissions and they too have low affinities for oxide lattices in aqueous media.

Table 1. Characteristics of the elements (data taken from Turner *et al.*, 1980, and Whitfield 1981)

| Element | Concentration (μ mol l^{-1}) | | Mean oceanic residence time (Years) | Classification | | |
	Sea water	River water		Electro-negativity	Oceano-graphic*	Bio-logical†
Ag	4×10^{-4}	2.8×10^{-3}	4.5×10^3	1.9	P	N
Al	0.07	1.85	1.7×10^2	1.5	D	?
As	0.05	0.02	6.3×10^4	2.0	P	ET
Au	2×10^{-4}	1×10^{-5}	6.8×10^4	2.3	P	N
B‡	420	1.67	1.5×10^7	2.0	E	E
Ba	0.146	0.44	3.5×10^4	0.85	P	N
Br‡	860	0.25	1.1×10^8	2.8	E	?
C‡	2.38×10^3	100.0	7.6×10^5	2.5	E	E
Ca‡	1.05×10^4	365.0	9.1×10^5	1.0	E	E
Cd	9×10^{-4}	—	4.3×10^2	1.7	D	T
Ce	7×10^{-6}	6×10^{-4}	4.3×10^2	1.05	D	N
Cl‡	5.59×10^5	50.0	3.5×10^8	3.0	E	E
Co	9×10^{-4}	3.3×10^{-3}	1.7×10^4	1.8	P	ET
Cr	5.8×10^{-3}	0.02	1.0×10^4	1.6	P	?
Cs	3×10^{-3}	3×10^{-4}	6.6×10^5	0.7	E	N
Cu	7.8×10^{-3}	0.16	2.6×10^3	2.0	D	ET
Er	5×10^{-6}	3×10^{-5}	6.9×10^3	1.2	D	N
Eu	7×10^{-7}	7×10^{-6}	3.4×10^3	1.1	D	N
F	60	5.26	4.4×10^5	4.0	E	E
Fe	0.04	0.71	9.8×10^1	1.9	D	E
Ga	4×10^{-4}	1.3×10^{-3}	1.1×10^4	1.6	P	N
Gd	5×10^{-6}	5×10^{-5}	3.0×10^3	1.1	D	N
Hg	2×10^{-4}	3×10^{-4}	1.4×10^4	1.9	P	T
Ho	1×10^{-6}	6×10^{-6}	6.9×10^3	1.2	D	N
I‡	0.47	0.06	2.9×10^5	2.5	E	E
K‡	1.05×10^4	34.62	5.5×10^6	0.8	E	E
La	2×10^{-5}	4×10^{-4}	5.0×10^2	1.1	D	N
Li	26.1	1.71	2.0×10^6	1.0	E	N
Lu	1×10^{-6}	6×10^{-6}	6.9×10^3	1.2	D	N
Mg‡	5.5×10^4	158.3	1.1×10^7	1.2	E	E
Mn	3.6×10^{-3}	0.15	9.8×10^2	1.4	D	ET
Mo	0.10	0.01	5.6×10^5	2.1	E	ET
N	35.7	17.9	5.8×10^3	3.0	P	E
Na‡	4.79×10^5	221.7	5.6×10^7	0.9	E	E
Nd	2×10^{-5}	3×10^{-4}	2.6×10^3	1.1	D	N
Ni	0.03	0.04	1.9×10^5	1.8	P	?
P	1.94	1.29	1.0×10^5	2.1	P	E
Pb	1×10^{-4}	5×10^{-4}	3.4×10^2	1.6	D	T
Pr	4×10^{-6}	5×10^{-5}	2.9×10^3	1.1	D	N
Rb	1.40	0.02	4.0×10^6	0.8	E	?
S‡	2.89×10^4	115.6	7.9×10^6	2.5	E	E
Sb	2.0×10^{-3}	0.01	4.1×10^3	2.1	P	N
Sc	1×10^{-5}	9×10^{-5}	5.0×10^3	1.3	P	N

Table 1. (contd.)

Element	Concentration (μ mol 1^{-1})		Mean oceanic residence time (Years)	Classification		
	Sea water	River water		Electro-negativity	Oceano-graphic*	Bio-logical†
Se	2.5×10^{-3}	2.5×10^{-3}	3.4×10^{4}	2.4	P	ET
Si	71.4	193.6	3.5×10^{4}	1.8	P	E
Sm	3×10^{-6}	5×10^{-5}	2.1×10^{3}	1.1	D	N
Sn	8×10^{-5}	3×10^{-4}	8.3×10^{3}	1.9	P	?
Sr‡	90.9	0.68	3.8×10^{6}	1.0	E	?
Tb	6×10^{-7}	6×10^{-6}	3.4×10^{3}	1.2	D	N
Th	4×10^{-5}	4×10^{-4}	3.4×10^{3}	1.3	D	N
Ti	2.09×10^{-2}	0.21	1.1×10^{4}	1.6	P	?
Tm	1×10^{-6}	6×10^{-6}	6.9×10^{3}	1.2	D	N
U	1.35×10^{-2}	2×10^{-4}	2.8×10^{6}	1.9	E	?
V	4.90×10^{-2}	0.02	9.5×10^{4}	1.85	P	ET
W	5×10^{-4}	2×10^{-4}	1.1×10^{5}	2.1	P	N
Yb	5×10^{-6}	2×10^{-5}	6.9×10^{3}	1.2	D	N
Zn	7.54×10^{-2}	0.46	8.3×10^{3}	1.6	P	ET
Zr	3×10^{-4}	0.03	3.4×10^{2}	1.4	D	N

* E = enriched, concn. in sea water $\geq 10 \times$ concn. in river water; D = depleted, concn. in sea water $\leqslant 0.1 \times$ concn. in river water; P = pre-equilibrated, intermediate between E and D. (Whitfield, 1979a).
† E = essential; ET = essential trace element; N = not required; T = toxic; ? = status uncertain (Egami, 1974).
‡ Principal ionic components.

It would appear then, that in the long term the overall composition of sea water is controlled by a sequence of interlinked inorganic processes. However, when viewed on a human timescale the oceans do not present themselves as a single uniformly mixed reservoir and geologically rapid biological and geochemical processes can cause significant local variation in the concentrations of many elements. Over the short time span of our observations, seasonal and even diurnal variations in composition become significant and inhomogeneities existing over a distance of a few tens of metres can have profound biological consequences. It is precisely these details that are of most direct interest to the environmental scientist and to unravel such short-term biological and geological processes it is necessary to have an accurate understanding of the solution chemistry of the elements.

1.2. GENERAL PROPERTIES OF THE MEDIUM

1.2.1. Thermodynamic activities

Whether the chemistry of an element is controlled by equilibria or by rate processes, the key to providing a quantitative description of its behaviour in solution is the solute activity (a_Y; Whitfield 1975, 1979b) which may be defined (in molal terms) by the equation

$$a_Y = m_Y \gamma_Y = \exp{(\mu_Y - \mu_Y^\ominus)}/RT \tag{3}$$

where

$$\mu_Y = (\partial G_Y / \partial m_Y)_{T,P,m_j}$$

R is the universal gas constant ($8.314\,\text{J K}^{-1}\,\text{mol}^{-1}$, $1.987\,\text{cal K}^{-1}\,\text{mol}^{-1}$, $0.08205\,\text{l atm K}^{-1}\,\text{mol}^{-1}$) and T is the absolute temperature. G_Y, μ_Y, m_Y and γ_Y are the Gibbs free energy, chemical potential, molality and molal activity coefficient, respectively, of the component Y. $\mu_Y^\ominus$ is the chemical potential of Y in some convenient standard state where $m_Y^\ominus = \gamma_Y^\ominus = 1$. If the chemistry of Y is controlled by a series of equilibria such as

$$\text{M} + \text{Y} \rightleftharpoons \text{MY} \tag{4}$$

then the appropriate thermodynamic equilibrium constant ($K_{MY}^\ominus$) will be related to the solute activities by equations of the form

$$K_{MY}^\ominus = a_{MY}/a_M a_Y = K_{MY}^* \Gamma_{MY} \tag{5}$$

where K_{MY}^* ($= m_{MY}/m_M m_Y$) is the stoichiometric equilibrium constant and Γ_{MY} ($= \gamma_{MY}/\gamma_M \gamma_Y$) is the activity coefficient quotient.

If the forward and reverse reactions of equation 4 proceed slowly in the timescale of the observations, then we should write

$$\text{M} + \text{Y} \underset{k_+^\ominus}{\overset{k_-^\ominus}{\rightleftharpoons}} \text{MY} \tag{6}$$

where the net rate of the forward and reverse reactions may be written as

$$-\mathrm{d}a_{MY}/\mathrm{d}t = k_+^\ominus a_{MY} - k_-^\ominus a_M a_Y \tag{7}$$

At equilibrium we can write,

$$K_{MY}^\ominus = k_-^\ominus / k_+^\ominus \tag{8}$$

If equation 7 is written in terms of concentrations rather than activities we can also define stoichiometric rate constants such that

$$K_{MY}^* = k_-^* / k_+^* \tag{9}$$

where $k_+^* = k_+^\ominus$ and $k_-^* = k_-^\ominus / \Gamma_{MY}$. If significant variations in Γ_{MY} occur over the range of environmental conditions associated with the rate processes (equation 6), then the full form of the rate equation (summarized in equation 7) must be used (Eyring and Eyring, 1967).

Similarly, when the uptake or release of a component in a reaction is diffusion controlled, the full form of the diffusion equation must be used if the activity coefficient of the component is likely to vary over the concentration range which drives the diffusive process. The full form of Fick's first law may be written as (Bockris and Reddy, 1970a)

$$J_Y = -Bm_Y \, d\mu_Y/dx$$

since we are considering the transport of moles of material across a chemical potential gradient (Bockris and Reddy, 1970a).

Defining a thermodynamic diffusion coefficient $D^\ominus$ such that $D_Y^\ominus = BRT$, we can write

$$J_Y = -(D_Y^\ominus/RT)m_Y \, d\mu_Y/dx$$

or

$$J_Y = -D_Y^\ominus m_Y \, d \ln a_Y/dx \qquad (10)$$

where

$$D_Y^\ominus = D_Y^*(1 + d \ln \gamma_Y/d \ln m_Y) \qquad (11)$$

D_Y^* is the stoichiometric diffusion coefficient associated with the conventional definition of Fick's first law, *viz.*

$$J_Y = -D_Y^* \, dm_Y/dx \qquad (12)$$

Although direct measurements of ionic diffusion coefficients have been made in natural media (e.g. Krom and Berner, 1980) it is usually necessary to employ values determined in simple electrolyte solutions. The most useful sources of such data are tables of the polarographic characteristics of ions (Meites, 1965, 1963) which include values of the diffusion current constant (I), which is related to the diffusion coefficient D_Y^* by the equation

$$I = 607nD_Y^{*1/2}$$

Diffusion coefficients for a range of ionic components are listed in Appendix XIV.

Even in situations where the focus is on processes involving the rate of transport and utilization of material rather than on the rate of change of activity (e.g. electrodeposition processes), the relationships summarized in equations 7–12 are likely to be important since they will fix the values of the corresponding rate constants and diffusion coefficients. We must now consider the general properties that are likely to influence the activities of sea water components.

1.2.2. Pressure and temperature ranges

Although on the global scale the oceans appear as no more than a thin film of moisture on a ball of rock, they are sufficiently deep for considerable pressures to be generated as a result of the overburden of water. The pressure gradient within the oceans is approximately 1 bar (10^5 Nm^{-2}) for every 10 m increase in depth, so that the ambient pressures in the deepest ocean trenches are in excess of 1 kbar. A cumulative depth (or hypsometric) curve of the ocean basins (Figure 4; see Dietrich, 1963) indicates that 80% of the ocean floor is at depths in excess of 2 km and that the mean depth of the oceans is nearly 4 km. The mean volume pressure experienced within the world oceans is close to 200 bar, so that the influence of pressure on chemical processes in sea water deserves particular attention.

The temperature of sea water may range from 30°C or more in tropical surface waters to −2°C under the Arctic ice. A mean vertical temperature profile for the deep ocean (Figure 4) shows the temperature dropping away rapidly from 20°C at the surface to 2°C at a depth of 4–5 km. The region of rapid temperature change (Figure 4) is known as the thermocline. The mean volume temperature of the world ocean is 3.8°C (Dietrich, 1963). This diagram is on too large a scale to show the shallow, well mixed layer which extends to a depth of 100 m or more above the thermocline. It is clear that the mean conditions in the world oceans (200 bar, 3.8°C) are far removed from the standard physico-chemical conditions of approximately 1 bar and 25°C. The influence of pressure and temperature on solute activities and on

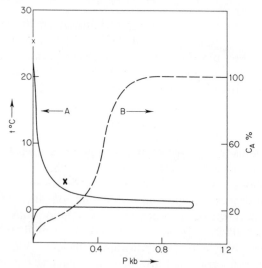

Figure 4. Relationship between temperature and pressure in the world's oceans (1 kb ≈ 10 km depth). A, Mean pressure *versus* temperature plot (data from Walton Smith, 1976). B, Hypsometric curve showing the cumulative percentage of the ocean's area (C_A) with pressures *less* than the value indicated on the abscissa (data from Dietrich, 1963). The light cross indicates the normal conditions for physico-chemical studies (25°C, 1 bar) and the heavy cross the volume mean conditions in the oceans (3–8°C, 200 bar) (1 bar = 10^5 Nm^{-2})

Table 2. Partial molal functions used to express temperature and pressure coefficients (taken with modifications from Whitfield, 1975a)[*][†]

X	$\partial X/\partial P$	$\partial^2 X/\partial P^2$	$\partial X/\partial T$	$\partial^2 X/\partial T^2$
	(equation 1)	(equation 2)	(equation 3)	(equation 4)
$-\ln a_Y$[‡]	$-\bar{V}/RT$	$\bar{\kappa}/RT$	$\bar{H}/RT^2$	$(\bar{C}_p - 2\bar{H}/T)/RT^2$
	(equation 5)	(equation 6)	(equation 7)	(equation 8)
$\ln K^{\ominus}_{MY}$	$-\Delta V^{\ominus}/RT$	$\Delta\kappa^{\ominus}/RT$	$\Delta H^{\ominus}/RT^2$	$(\Delta C^{\ominus}_p - 2\Delta H^{\ominus}/T)/RT^2$

[*] Definitions: $\bar{V}$ = relative partial molal volume ($= \bar{V}_Y - \bar{V}^{\ominus}_Y$) (*cf.* equation 3); $\bar{\kappa}$ = relative partial molal compressibility; $\bar{H}$ = relative partial molal enthalpy; $\bar{C}_p$ = relative partial molal constant pressure heat capacity. The equations for $K^{\ominus}_{MY}$ are expressed in terms of the corresponding standard partial molal property changes accompanying the reaction.
[†] Data sources:

Equation nos.   Reference
     (1), (2) Millero (1971, 1972, 1979), Leyendekkers (1976)
     (3), (4) Lewis and Randall (1961), Parker (1965), Millero (1979), Silvester and Pitzer (1978)
     (5), (6) Distèche (1975), Whitfield (1975a), Millero (1979), Hamann (1974), Sinn (1974)
     (7), (8) Helgeson (1967), Christensen and Izatt (1970), Whitfield (1975a), Millero (1979), de Bethune and Swendeman Loud (1964)
[‡] To avoid complications arising from changes in concentration as a function of T and P the equations are expressed in terms of the molal concentration scale. The same expressions can then be used for the pressure and temperature coefficients of γ_Y (equation 3).

equilibrium constants can be obtained from the standard thermodynamic relationships (Table 2). The equations for a_Y are expressed in terms of relative partial molal quantities (e.g. $\bar{V} = V_Y - V^{\ominus}_Y$) and the equations for $K^{\ominus}_{MY}$ in terms of the net change in the standard partial molal quantity (e.g. $\Delta V^{\ominus} = V^{\ominus}_{MY} - V^{\ominus}_M - V^{\ominus}_Y$).

For the most precise work it is necessary to consider also the cross-terms. The temperature derivative of $\bar{V}$ is given by

$$(\partial\bar{V}/\partial T)_p = \bar{E} \tag{13}$$

where $\bar{E}$ is known as the relative partial molal expansibility (Leyendekkers, 1976; Millero, 1974b, 1979). The pressure derivative for $\bar{H}$ is given by

$$(\partial\bar{H}/\partial P)_T = \bar{V} - T\bar{E} \tag{14}$$

Analogous equations can be defined for $\Delta V^{\ominus}$ and $\Delta H^{\ominus}$ in terms of the standard partial molal expansibility change, $\Delta E^{\ominus}$.

Since the pressure gradient within the oceans is generated over considerable depths, it is also necessary to consider the decrease in chemical potential associated with the changing influence of the earth's gravitational field as one moves down the water column. The change in chemical potential with depth (h) in an incompressible fluid must therefore be written as (MacInnes, 1939; Guggenheim, 1967)

$$\partial \mu_Y/\partial h = \partial (\ln a_Y)/\partial h = (M_Y - \bar{V}d)g \tag{15}$$

M_Y is the molecular weight of component Y, d is the density of the fluid and g is the acceleration due to gravity. In the real system, diffusive processes are too slow to even out the activity gradient that results from the turbulent convective mixing of water masses. A proper understanding of equation 15 is therefore of considerable importance in studies of biological processes that are sensitive to the activities of gases (particularly oxygen and carbon dioxide) in solution (Pytkowicz, 1968; Andrews, 1972). Since the gravitational field effects depend solely on the mass of the component involved the conservation of mass will ensure that they will cancel out when chemical reactions are considered.

The influence of pressure and temperature on chemical equilibria can either be measured directly by observing changes in $K^{\ominus}_{MY}$ over a range of temperature and pressure or indirectly by measuring the appropriate thermodynamic parameters (see Table 2). The direct approach involves the selection of an integrated form of equations 5–8 (Table 2) and the estimation of the corresponding fitting parameters from the experimental data. Such procedures are rarely as precise as the route involving the measurement of the appropriate thermodynamic parameters (Whitfield, 1975a).

The simplest and most general integrated form of equation 5 (Table 2) is (El'yanov and Hamann, 1975; El'yanov, 1975)

$$RT \ln K^{\ominus}_{MY,P}/K^{\ominus}_{MY,1} = -\Delta V^{\ominus}P/(1 + bP) \tag{16}$$

where $b = 9.2 \times 10^{-5}\,\text{bar}^{-1}$ ($9.2\,\text{m}^2\,\text{N}^{-1}$). For more precise work, b must be considered as a function of temperature (Nakahara, 1974).

For the moderate temperature range normally encountered in the marine environment we can assume that $\bar{C}_p$ (equation 4, Table 2) is not a function of temperature so that we can write

$$\bar{H} = H_1 + \bar{C}_p T \tag{17}$$

where H_1 is constant. The integrated form of equation 3 (Table 2) then becomes (Lewis and Randall, 1961; Harned and Owen, 1958),

$$R \ln a_Y = H_1/T - \bar{C}_p \log T + \text{constant} \tag{18}$$

There appears to be considerable flexibility in the choice of an integrated form for equation 7 (Table 2). The simple equation

$$-R \ln K^{\ominus} = (A/T) - B + CT \tag{19}$$

seems to be most suitable for acid–base equilibria (King, 1965), although a polynomial form expressing $\Delta \bar{H}^{\ominus}$ as a Taylor series expansion in temperature has more general application (Timimi, 1974). If the necessary data are not available for calculating the influence of pressure or temperature on particular equilibria a number of general correlations have been observed between the various parameters that might enable approximate values to be interpolated (Whitfield, 1975a; Helgeson, 1967; Millero, 1974, 1979).

Electrochemical methods are frequently used for the direct determination of $K_{MY}^{\ominus}$ at high pressures (Adams and Davis, 1973; Whitfield, 1975a; Hamann, 1974). With the increasing use of electrochemical sensors for *in situ* measurement in the deep ocean (see, for example, Chapters 3, 6 and 9), it is likely that more attention will be given to the adaptation of potentiometric and voltammetric systems for work at high pressures (Hills and Ovenden, 1966; Hills, 1969, 1972; Hamann, 1974; Distèche, 1972). It is already possible, for example to measure pH *in situ* at depths of several thousand meters (Chapter 6). A detailed understanding of the influence of pressure and temperature on the equilibrium constants controlling the ionization of carbonic and boric acids is required to interpret the measurements made (Pytkowicz, 1968; Distèche, 1975). The pressure coefficients of a number of pH buffers have been tabulated by El'yanov (1975), but these data have not yet been used to test the performance of glass electrodes on a routine basis.

The influence of temperature and pressure on the rate constant of a kinetically controlled process (equation 7) can be described by equations analogous to those used for $K_{MY}^{\ominus}$ (Table 2, equations 5–8). Using transition state theory (Hamann, 1963; Whalley, 1966; Hills, 1969) we can write, for example,

$$\partial \ln k^{\ominus}/\partial P = -(V_{\ddagger}^{\ominus} - V_{r}^{\ominus})/RT \tag{20}$$

for the influence of pressure on the formation constant (equation 6). $V_{\ddagger}^{\ominus}$ and $V_{r}^{\ominus}$ are the standard partial molal volumes of the transition state and of the reactants, respectively. Insufficient data are available for these relationships to be applied to oceanic systems.

1.2.3. Redox potential

All chemical reactions involve electron transfer and it is precisely this effect that is exploited by electrochemical methods of analysis. The progress of electron transfer reactions in solution will be controlled by the activity of electrons in the medium, and this parameter may in turn be defined in terms of the oxidation–reduction (or redox) potential.

The general equation for an oxidation–reduction (redox) reaction may be written as

$$O + R^* \rightleftharpoons R + O^* \tag{21}$$

where O and R represent the oxidized and reduced forms, respectively. The overall reaction may be divided into two individual (or 'half-cell') reactions:

$$O + ne^- \rightleftharpoons R \tag{22}$$

$$R^* \rightleftharpoons O^* + ne^- \tag{23}$$

For the first reaction (equation 22), the overall chemical potential change is related to the solute activities by the equation

$$\Delta G = \mu_R - (\mu_O + n\mu_{e^-})$$
$$= \Delta G^\ominus + RT \ln (a_R/a_O a_{e^-}^n) \tag{24}$$

where $\Delta G^\ominus = \mu_R^\ominus - (\mu_O^\ominus + n\mu_{e^-}^\ominus)$. This free energy change can be related to the potential (ε) that would be generated at equilibrium at an inert conducting electrode immersed in the solution so that (Lewis and Randall, 1961, p. 350)

$$\Delta G = -nF\varepsilon \tag{25a}$$

and

$$\Delta G^\ominus = -nF\varepsilon^\ominus = -RT \ln K^\ominus \tag{25b}$$

where F is the Faraday (96 500 C equiv^{-1}, 1 volt Faraday = 23 060 cal mol^{-1} = 1 eV). $\varepsilon^\ominus$ is the standard half-cell potential and its value is independent of the detailed solution composition. Equations 24–25b together give the Nernst equation:

$$\varepsilon = \varepsilon^\ominus + (RT/nF) \ln (a_O a_{e^-}^n/a_R) \tag{26}$$

When the system is at equilibrium $\Delta G = -nF\varepsilon = 0$ and we can rearrange equation 26 to give

$$pe = -\log a_{e^-}$$
$$= (F/RT \ln 10)\varepsilon^\ominus + n^{-1} \log (a_O/a_R) \tag{27}$$

pe corresponds to the oxidation power (or oxidizing intensity) of the half-cell reaction at equilibrium (Sillén, 1965). When $a_O = a_R = 1$, then $pe = n^{-1} \log K^\ominus$. Since it is not possible to measure the potential at an isolated electrode the redox potentials in published tables (e.g. de Bethune and Swendeman–Loud, 1964; Dobos, 1975; Sillén and Martell, 1964, 1971; Hunsberger, 1976; Milazzo and Caroli, 1978) refer to reaction 21 with the second half-cell (equation 23) written as

$$\frac{n}{2} H_2 \rightleftharpoons nH^+ + ne^- \tag{28}$$

whose potential is ε_H. The overall cell potential ($E_H = \varepsilon + \varepsilon_H$) is then given by

$$E_H = E_H^\ominus + (RT/nF) \ln (a_O a_{H_2}^{n/2}/a_R a_{H^+}^n) \tag{29}$$

where $E_H^\ominus = \varepsilon^\ominus + \varepsilon_H^\ominus$. The potentials given have usually been extrapolated to the point where $a_{H^+} = a_{H_2} = 1$ so that

$$E'_H = E_H^\ominus + (RT/nF) \ln (a_O/a_R) \tag{30}$$

The redox potential E'_H expressed relative to the Standard Hydrogen Electrode (SHE; Ives and Janz, 1961) is a *formal* potential, relevant to the particular medium in which the measurements were made. Adopting the convention that $\varepsilon_H^\ominus = 0$, equations 27 and 30 give

$$pe = FE'_H/RT \ln 10 \tag{31}$$

If the measured *formal* potentials are also extrapolated to zero ionic strength the resulting values are reversible *standard* potentials expressed on the hydrogen scale.

Since

$$\Delta G^\ominus = -RT \ln K^\ominus \tag{32}$$

the influence of pressure and temperature on $E_H^\ominus$ can be described by equations analogous to those used for the equilibrium constant in Table 2 (Lewis and Randall, 1961; de Bethune and Swendeman-Loud, 1964; MacInnes, 1939). It is clear that formal and standard redox potentials can be derived from thermodynamic data and that the inclusion of these potentials in calculations or in illustrations does not necessarily imply that a suitable electrode system exists for their direct measurement.

pe values can be used to prepare *electron free energy (EFE) diagrams* (Reilley and Murray, 1963; Gurney, 1953; Stumm and Morgan, 1970), which show directly the energy required to transfer electrons from one redox couple to another (Figure 5). When electrons are added to a system containing several redox couples (either at the cathode of an electrochemical cell or by the addition of reductants), the oxidized species will be reduced in sequence (kinetics permitting) beginning with the species with the lowest electron level. Conversely, when electrons are removed from the system they will be removed first from the reduced form with the highest electron energy level and then in sequence down the scale. The range of pe values that can be accommodated in aqueous solutions is limited by the stability of water itself with respect to the reactions

$$2H^+ + 2e^- \rightleftharpoons H_2(g) \qquad \log K^\ominus = 0 \tag{33}$$

and

$$O_2(g) + 4H^+ + 4e^- \rightleftharpoons 2H_2O \qquad \log K^\ominus = 83.1 \tag{34}$$

Components of redox systems with $0 > pe^\ominus > 20$ will therefore react with water (Figure 5). The *EFE* diagram can be used to calculate the oxidation or reduction capacity of a system of redox couples (Stumm and Morgan,

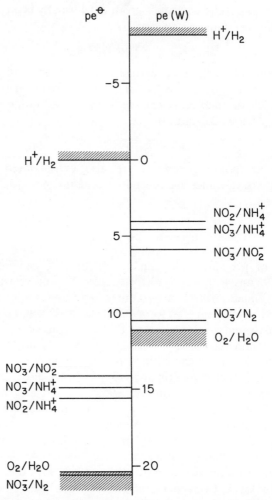

Figure 5. Electron free-energy (EFE) diagram for components of the nitrogen system (Table 5). The shaded regions indicate the thermodynamic limits of the stability of water with respect to oxidation and reduction at 25°C and 1 atm pressure. $pe^{\ominus}$ values refer to pH = 0 and $pe(W)$ values to pH = 8.3

1970) and hence to prepare 'electron titration' diagrams (Sillén, 1965, 1966), showing the variation in pe of the system in response to a continuous addition or removal of electrons. In formulating such titration curves it is useful to remember that 1 Faraday is equivalent to 1 mol of electrons.

Many elements can exist in several oxidation states and it is useful to be able to identify the thermodynamically stable states under given environmental conditions. *Oxidation state (OS)* diagrams (Phillips and Williams,

1965) provide a convenient means for summarizing the redox potential data for a particular element. In such a diagram the product of the redox potential and the algebraic value of the oxidation state (Wong, 1980; Stumm and Morgan, 1970, p. 302) is plotted against the oxidation state (Table 3, Figure 6A). The gradient of the line joining any two oxidation states will be equal to the redox potential of the couple formed by those two species. Increasingly large positive gradients indicate progressively more strongly oxidizing couples and increasingly large negative gradients indicate progressively more strongly reducing couples. *OS* diagrams can also be useful for identifying possible metastable species in solution (Figure 6A). In preparing such a diagram it is important to remember that the redox potential relevant to the natural medium (E'_H) will differ from the standard potential $(E^\ominus_H)$ often given in tables. Procedures for calculating the appropriate redox ratios for trace metals will be discussed in Section 1.4.1.

A broader view of the redox chemistry of a particular element can be obtained by using equations 31 and 32 to construct a *logarithmic activity ratio (LAR)* diagram, in which $\log(a_O/a_R)$ is plotted against pe (Figure 6B). To prepare such a diagram the reaction equations are rearranged so that they all refer to a single oxidized or reduced form (Table 3). Sillén (1966) provides a particularly well documented example of the construction of an *LAR* diagram for manganese where all redox couples are referenced to Mn^{2+} (see also Butler, 1964).

A more detailed picture of the actual composition of the solution can be obtained from the *logarithmic concentration (LC)* diagram (Figure 6C). Here an additional relationship, the mass balance, is introduced into the calculations since the total concentration of the element in question is now fixed. The distribution of species can be read directly off the diagram at any particular pe value and the stability fields of the various species are available at a glance. If a number of *LC* diagrams or *LAR* diagrams, prepared at different pH values, were assembled into a single three-dimensional diagram [with the axes pH/pe/pX and pH/pe/log (a_O/a_R), respectively] then the view from above would show the stability areas of the various species on the pH/pe plane. The resulting *predominance area (PA)* diagram can be readily constructed from the half-cell reaction equations (Table 3, Figure 6D; Butler, 1964; Stumm and Morgan, 1970). Such diagrams have been widely used for thermodynamic studies of metal corrosion (Pourbaix, 1966). They provide an overall summary of the dominant species in solution although they give no indication of the relative proportions of the various species that might co-exist at equilibrium at any particular pH or pe.

To use the various thermodynamic diagrams to describe the chemical behaviour of natural systems we must be able (i) to estimate the individual values of E'_H (equation 30) relevant to the natural system and (ii) to assess the pe levels set in a particular system by the dominant redox couples. It is

Table 3. The construction of redox diagrams

A. *Data**

No.	Equation	Log$_{10}$ K		pe		E$_H$	
		A†	B‡	A†	B‡	A†	B‡
1	$O_2 + 4H^+ + 4e^- \rightleftharpoons 2H_2O$	83.1	49.9	20.75	12.48	1.23	0.74
2	$NO_3^- + 6H^+ + 5e^- \rightleftharpoons \frac{1}{2}N_2(g) + 6H_2O$	105.3	55.5	21.05	11.09	1.25	0.66
3	$NO_3^- + 2H^+ + 2e^- \rightleftharpoons NO_2^- + H_2O$	28.3	11.7	14.15	5.85	0.84	0.35
4	$NO_3^- + 10H^+ + 8e^- \rightleftharpoons NH_4^+ + 3H_2O$	119.2	36.2	14.90	4.53	0.88	0.27
5	$NO_2^- + 8H^+ + 6e^- \rightleftharpoons NH_4^+ + 2H_2O$	90.9	24.5	15.14	4.09	0.90	0.24
6	$NO_2^- + 4H^+ + 3e^- \rightleftharpoons \frac{1}{2}N_2(g) + 2H_2O$	77.0	43.8	25.67	14.58	1.52	0.86
7	$\frac{1}{2}N_2(g) + 4H^+ + 3e^- \rightleftharpoons NH_4^+$	13.95	−19.25	4.65	−6.42	0.28	−0.38

* pe = $(1/n) \log K$, $E_H = g'$pe ln 10, where $g' = RT/F$ (Stumm and Morgan, 1970).
† Based on equilibrium stability constants, $K^\ominus$, etc.
‡ Conditional values at pH 8.3, K(W), etc.

B. *Electron free energy (EFE) diagram (Figure 5)*

Constructed simply by plotting pe$^\ominus$ or pe(W) values on an ordinate graduated in pe values or in electronvolts. ($\Delta G^\ominus = -RT \ln 10pe^\ominus \approx -0.237pe^\ominus$ (25°C); $R = 8.62 \times 10^{-5}$ eV).

C. *Oxidation state (OS) diagram (Figure 6A)*

The E_H values are all expressed relative to the production of a particular form (most conveniently the form with zero oxidation number). According to Wong (1980) the data are then expressed relative to the oxygen–water couple so that $VE'_H = N[E_H(O/R) - E_H(O_2/H_2O)]$, where N is the oxidation number of the oxidized form of the couple. For example, for NH$_4^+$, $VE'_H = -3[0.28 - 1.23] = 2.85$.

D. *Logarithmic activity ratio (LAR) diagram (Figure 6B)*

Express all reactions relative to the production of a particular form (most conveniently the form with a zero oxidation number). Write out the equation for the stability constant setting the ratio of the oxidized and reduced species as a function of pH and pe. For example, from equation 7 (Table 3A)

$$\log(a_{NH_4^+}/a_{N_2}^{1/2}) = -\log K(W) - 3pe$$
$$= -19.25 - 3pe$$

E. *Logarithmic concentration (LC) diagram (Figure 6C)*

Draw a horizontal line to represent the total concentration on a log–log plot. Calculate the system points (pe$_s$) for which $\log(O/R) = 0$ for the selected pH. For example, from equation 7, $13.95 - 4$pH $- 3$pe$_s = 0$ so that pe$_s = -19.25/3 = -6.42$ (point marked 7, Figure 6C). Draw lines from this point at the appropriate slope (e.g. -3 for NH$_4^+$ but $+6$ for N$_2$ since each molecule contains two nitrogen atoms—for the same reason the level of the N$_2$ plateau is depressed 0.3 logarithmic units below the level of the other species).

F. *Predominance area (PA) diagram (Figure 6D)*

Calculate the relationship between pH and pe when $\log(O/R) = 0$. For example,

clear that in natural systems the oxidized and reduced components in the equilibrium reaction (equation 22) will also be involved in chemical reactions with other components in solution. To estimate a_O and a_R it is necessary to account for these various reactions and in a later section (Section 1.4.1) we will consider the influence of hydrolysis and complexation side reactions on the redox potentials of individual couples. The calculation procedures are straightforward provided that the necessary data are available. It is not so simple, however, to define a pe level representative of the natural system (Parsons, 1975; Morris and Stumm, 1966; Stumm and Morgan, 1970; Stumm, 1978).

The redox potential of natural systems can rarely be measured directly by means of an inert (e.g. platinum) electrode. The potential of such an electrode is likely to be the resultant of a range of electrode processes, not all of them reversible (Morris and Stumm, 1966) and, in oxygenated conditions, noble metal electrodes are more likely to respond to changes in pH than to changes in redox potential (Whitfield, 1974). Although such measurements may be useful for the operational comparison of different environments (Zobell, 1946; Whitfield, 1969) they can be used to estimate E'_H (equation 30) only under exceptional circumstances. A more direct way of estimating the pe of a system is to assess the state of balance of dominant redox couples.

In well aerated waters the dominant redox process is likely to be the reduction of oxygen to produce water:

$$O_2 + 4H^+ + 4e^- \rightleftharpoons 2H_2O \qquad \log K_1^\ominus = 83.1 \qquad (33)$$

The overall reaction can be considered to take place in two stages:

$$O_2 + 2H^+ + 2e^- \rightleftharpoons H_2O_2 \qquad \log K_2^\ominus = 23.1 \qquad (34)$$

and

$$H_2O_2 + 2H^+ + 2e^- \rightleftharpoons 2H_2O \qquad \log K_3^\ominus = 60 \qquad (35)$$

These two separate reactions can be clearly observed when oxygen is reduced at the surface of a mercury electrode (see Section 9.3.6). Here reaction 34 proceeds rapidly but reaction 35 proceeds very slowly, suggesting that H_2O_2 may be present as a kinetically stable intermediate (Parsons, 1975). It has consequently been suggested (Breck, 1972, 1975) that reaction

---

from Equation 7 we have

$$13.95 = 4pH + 3pe$$

Therefore

$$pe = 4.65 - 1.33pH$$

Plot these lines out on a pH/pe surface.

34 is likely to set the equilibrium pe of air-saturated sea water. However, no direct measurements of hydrogen peroxide in sea water have yet been made and the available evidence suggests that the oxidation of many metallic species [e.g. Fe(II), Mn(II)] and the respiratory utilization of oxygen proceed directly to the formation of H_2O (equation 33). Oxygen is certainly a much stronger oxidizing agent when the reduction occurs as a four-electron step. We shall consequently assume that the pe of air-equilibrated sea water is given by

$$p e(\text{oxic}) = \tfrac{1}{4}[\log K_1^{\ominus} + \log P_{O_2}] - \tfrac{1}{2} \log a_{H_2O} - pH \qquad (36)$$

Under typical oceanic conditions ($\log P_{O_2} = -0.68$, $\log a_{H_2O} = -0.01$) we can write

$$p e(\text{oxic}) = 20.6 - pH \qquad (37)$$

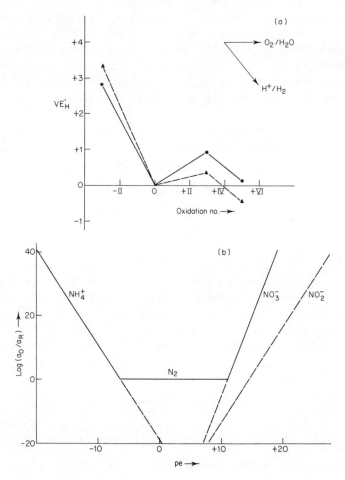

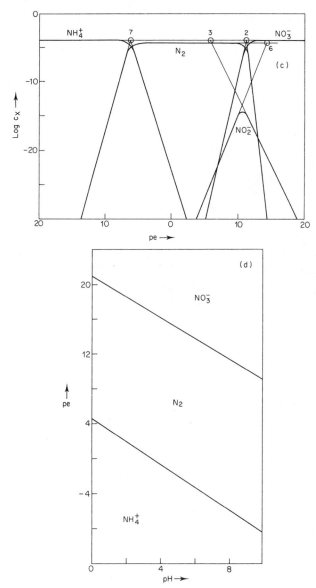

Figure 6. Redox diagrams for components of the nitrogen system. (a) Oxidation state (OS) diagrams (see Table 3, Section B). Broken lines and triangles indicate $VE'_H(W)$ values; solid lines and circles indicate VE_H values. (b) Logarithmic activity ratio (LAR) diagram (see Table 3, Section C). Dashed lines indicate unstable species. (c) Logarithmic concentration (LC) diagram (see Table 3, Section D). The system points are numbered in accordance with the equation numbers in Section A of Table 3. Feint rules indicate how the diagram is constructed. (d) Predominance area (PA) diagram (see Table 3, Section E)

In stagnant systems where the oxygen has all been utilized in the microbial degradation of organic matter the pe will be set by redox reactions involving major dissolved components such as carbon and sulphur. The relevant reactions are

$$SO_4^{2-} + 9H^+ + 8e^- \rightleftharpoons HS^- + 4H_2O \qquad \log K_4^{\ominus} = 33.04 \qquad (38)$$

$$HCO_3^- + 9H^+ + 8e^- \rightleftharpoons CH_4 + 3H_2O \qquad \log K_5^{\ominus} = 30.78 \qquad (39)$$

Typically only 10% of each element is present in the reduced form (Dyrssen and Wedborg, 1980; Reeburgh, 1972), so that we can define the likely lower limit of pe in natural waters as

$$pe(\text{anoxic}) = (34 - 9pH)/8 \qquad (40)$$

Although we may consider equations 37 and 40 as setting the background pe in natural systems, it does not necessarily follow that other redox couples will be adjusted to match this electron activity. Direct measurement of the concentration of the oxidized and reduced forms of various redox couples in solution [e.g. I^-/IO_3^-, Se(IV)/Se(VI), Cr(III)/Cr(VI)] indicate that they are neither in equilibrium with each other nor with the pe specified by the oxygen–water couple (see, for example, Stumm and Morgan, 1970). In situations where the oxidized form is a solid phase [e.g. Mn(II)/MnO_2, Fe(II)/Fe(OH)$_3$] even larger displacements from equilibrium are likely to be encountered (Parsons, 1975; Sillén, 1967). This suggests that the balance between the oxidized and reduced forms is decided by kinetic rather than by thermodynamic considerations because the equilibration rates are slow compared with the residence times of the components in natural systems. When dealing with natural waters it is essential to remember that they are active culture media for a vigorous and diverse community of organisms. When these organisms become involved in the cycling of the elements it is frequently possible to find situations arising that would be unexpected from the point of view of equilibrium thermodynamics. However, it is only by taking proper account of the thermodynamic background that such effects can be identified in an unequivocal manner.

1.2.4. pH

Although in the long term (millions of years) the pH of sea water may be buffered by ion-exchange reactions with silicate minerals (Sillén, 1961), in the short term (millenia or less) the role of buffering and pH control is largely performed by the carbon dioxide system (Pytkowicz, 1967) with minor contributions from boric acid. For our purposes it will suffice to consider only the influence of the CO_2 system (see Skirrow, 1975, for full details). The reactions involved are

$$H^+ + HCO_3^- \rightleftharpoons H_2O \cdot CO_2 \tag{41}$$

$$H^+ + CO_3^{2-} \rightleftharpoons HCO_3^- \tag{42}$$

In sea water these reactions can be described by equilibrium constants of the form

$$K_{1C} = [H_2O \cdot CO_2]/\bar{h}[HCO_3^-] \tag{43}$$

and

$$K_{2C} = [HCO_3^-]/\bar{h}[CO_3^{2-}] \tag{44}$$

where square brackets indicate total (stoichiometric) concentrations and $\bar{h}$ represents the hydrogen ion contribution on some conventional pH scale. $H_2O \cdot CO_2$ represents the un-ionized carbon dioxide whether or not it is hydrated. The partial pressure of carbon dioxide in the atmosphere (P_{CO_2}) is related to $[H_2O \cdot CO_2]$ at equilibrium by the equation

$$[H_2O \cdot CO_2] = K_p P_{CO_2} \tag{45}$$

where (Weiss, 1974)

$$\ln K_p = [-58.0931 + 90.5069(100/T) + 22.2940 \ln (T/100)]$$
$$+ [0.027766 - 0.25888(T/100) + 0.0050578(T/100)^2]S \tag{46}$$

for K_p in mol dm^{-3} atm^{-1}. S represents the salinity or salt content of the sea water (see Section 1.3.1). At a salinity of 35‰ and a temperature of 25°C, $K_p = 2.91 \times 10^{-2}$ mol dm^{-3} atm^{-1} (1 atm = 101325 Pa).

The values of the stability constants in equations 43 and 44 will depend not only on the temperature, pressure, and salinity of the sea water, but also on the scale used to define the pH (see Section 6.1.2). If the pH is determined via electrodes calibrated in sea water buffers, a total hydrogen ion concentration scale can be used (Hansson, 1973), so that

$$\bar{h} = [H^+] + [HSO_4^-] \tag{47}$$

The corresponding stoichiometric stability constants are related to the salinity and the temperature by the equations (Almgren *et al.*, 1975)

$$pK_{1C}^* = 841/T + 3.272 - 0.0101S + 0.0001S^2 \tag{48}$$

$$pK_{2C}^* = 1373/T + 4.854 - 0.01935S + 0.000135S^2 \tag{49}$$

If the electrodes are calibrated in low ionic strength National Bureau of Standards (NBS) buffers (Bates, 1973), then the pH is defined so that

$$pH_{NBS} = pH_b + (E_b - E_x)/g' \ln 10 \tag{50}$$

where $g' = RT/F$ and E_b and E_x represent the potentials of the electrode couple in the NBS buffer and in the sample, respectively. This implies that

$$\bar{h} = 10^{-\text{pH}_{\text{NBS}}} \tag{51}$$

Since $\bar{h}$ is defined in terms of an experimental procedure rather than in terms of clearly defined solution components, the corresponding values of K_{1C} and K_{2C} are known as 'apparent' stability constants. They are related to temperature and salinity by the equations (Edmond and Gieskes, 1970)

$$pK'_{1C} = 3404.71/T + 0.032786T - 14.7122 - 0.10616S^{1/3} \tag{52}$$

and

$$pK'_{2C} = 2902.39/T + 0.02379T - 6.4170 - 0.2598S^{1/3} \tag{53}$$

The different pH scales will be discussed in detail in Chapter 6. Here we shall use both scales to examine the likely influence of the carbon dioxide system on the behaviour of sea water as an electrochemical medium.

From the electrochemical point of view the most important characteristics of a pH control system are its ability to buffer the pH of the solution at a reasonably stable value and its propensity to form strong complexes with electroactive species in the solution. Complexation reactions will be considered in Section 1.4. Before we consider the buffer capacity of sea water we must define two more parameters that are characteristic of the carbon dioxide system, the carbonate alkalinity (A_C) and the total carbon dioxide content (C_T). these may be defined as

$$C_T = [H_2O \cdot CO_2] + [HCO_3^-] + [CO_3^{2-}] \tag{54}$$

and

$$A_C = [HCO_3^-] + 2[CO_3^{2-}] \tag{55}$$

The buffer capacity (β_T) of the carbon dioxide system in sea water in response to an addition (δC) of strong acid or base is given by the equation (Stumm and Morgan, 1970; Skirrow, 1975)

$$\frac{\delta C}{\delta \text{pH}} = \frac{\beta_T}{\ln 10} = [H^+] + [OH^-] + \frac{C_T K_{1C} \bar{h}(\bar{h}^2 + 4K_{2C}\bar{h} + K_{1C}K_{2C})}{(\bar{h}^2 + K_{1C}\bar{h} + K_{1C}K_{2C})^2} \tag{56}$$

The first two terms represent the buffer capacity associated with the self-ionization of water which is dominant at pH < 4 and pH > 10. The buffer capacity curve (Figure 7) is not very sensitive to the pH scale used and exhibits two maxima at around pH 6 and pH 9. In a typical sea water sample $C_T \approx 2$ mM and $A_C \approx 2.25$ mM. The maximum buffer capacity of the system is

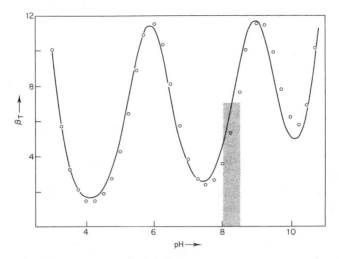

Figure 7. The buffer capacity of the carbon dioxide system in sea water. The line follows the values obtained using Hansson's constants (equations 48 and 49) in equation 56 and the circles indicate the values obtained using Lyman's constants (equations 52 and 53). The shaded area indicates the normal pH range in sea water

given approximately by

$$(\beta_T)_{max} \approx C_T \ln 10/4 \tag{57}$$

which for our typical sea water sample is *ca.* 1.15 mM. Sea water therefore has a relatively low buffer capacity and, in its normal pH range, rests uneasily on the shoulder of the pH 9 peak (Figure 7). It is for this reason that alternative buffering reactions must be active in the long term to prevent the pH of sea water from shifting too far from neutrality (Sillén, 1961).

The pH of sea water is most readily altered under normal experimental conditions by the exchange of CO_2 with the atmosphere during stirring or deoxygenation by gas bubbling. The reaction involved may be summarized by the equation

$$H_2O \cdot CO_2 + CO_3^{2-} \rightleftharpoons 2HCO_3^- \tag{58}$$

The stability constant of this reaction is given by

$$K_{12} = [HCO_3^-]^2/[CO_3^{2-}][H_2O \cdot CO_2] = K_{2C}/K_{1C} \tag{59}$$

This constant is independent of the pH scale used.

There is no change in A_C (equation 55) when CO_2 is lost or gained by the solution. To estimate the related pH change at equilibrium we need to be able to calculate the individual concentrations of the components involved in reaction 58.

Once the stability constants are known (equation 46 together with equations 52 and 53 or 48 and 49) the status of the carbon dioxide system can be precisely defined provided that at least two of the following four parameters are known: pH, P_{CO_2}, A_C and C_T. It is significant that all four parameters can be determined electrochemically (see Chapters 6 and 7) although the precision of electrochemical P_{CO_2} measurements is too low for this purpose (Dickson and Riley, 1978; Dickson, 1977). Equations detailing the relationships between these parameters, the appropriate stability constants, and the concentrations of the aqueous components of the carbon dioxide system have been summarized by Park (1969) (see also Skirrow, 1975, and Dickson, 1977).

We can look at the influence of gas exchange on the pH of the system if we alter P_{CO_2} over a solution with a fixed carbonate alkalinity (A_C, equation 55). The appropriate equations are then (Park, 1969)

$$[HCO_3^-] = [-a + (a^2 K_{12}^2 + 8A_C K_{12} a)^{1/2}]/4 \qquad (60)$$

$$[CO_3^{2-}] = (A_c - [HCO_3^-]/2 \qquad (61)$$

and

$$a = [H_2O \cdot CO_2] = K_p P_{CO_2} \qquad (62)$$

Once these concentrations have been calculated, the pH can be estimated from equation 43 or 44 using the appropriate stability constants. Using Hansson's constants (equations 48 and 49), our typical sea water would have a pH of 8.12 in equilibrium with an atmosphere containing 330 ppm of carbon dioxide. A plot of $\log P_{CO_2}$ *versus* pH (Figure 8) shows only small differences between the two pH scales and indicates that the pH of a sea water sample would be increased to pH 9.5 if the $H_2O \cdot CO_2$ content were reduced to 1% of its normal value by degassing. It is likely that calcium carbonate precipitation will be initiated before that pH value is attained since surface sea waters, at least, tend to be supersaturated with respect to both calcite and aragonite. This supersaturation may be more apparent than real because the surface of these mineral phases have high magnesium contents, giving them a higher apparent solubility (see discussion in Andersen and Malahoff, 1977). Nonetheless, a two- or three-fold increase in $[CO_3^{2-}]$ is likely to initiate some precipitation with subsequent pH buffering. Furthermore, the exchange of CO_2 between the gas and the solution phases is likely to be kinetically controlled either by the diffusion of the gas across the interface or by the rate of the dehydration reaction

$$H_2O \cdot CO_2 \rightarrow H_2O + CO_2 \qquad (63)$$

The rate of diffusion of gas across the air–water interface is given by (Skirrow, 1975)

$$J = k_d \Delta P \qquad (64)$$

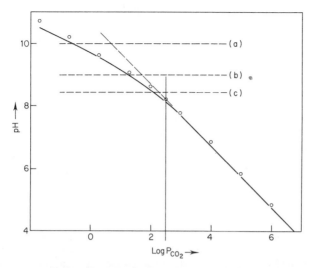

Figure 8. The equilibrium relationship between pH and P_{CO_2} for sea water at 25°C and 1 atm pressure. The line follows the values obtained using Hansson's constants (equations 48 and 49) and the circles indicate the values obtained using Lyman's constants (equations 52 and 53). Horizontal lines indicate (a) the approximate region of $Mg(OH)_2$ precipitation; (b) the pH at which $[CO_3^{2-}]$ reaches three times its normal value; and (c) the pH at which $[CO_3^{2-}]$ is twice its normal value. The vertical line indicates the normal atmospheric P_{CO_2} level

where k_d is the rate constant for transfer (mmol cm^{-2} atm^{-1} min^{-1}) and ΔP is the partial pressure difference across the interface (atm). For degassing $\Delta P = [H_2O \cdot CO_2]/K_p$ and $k_d = 0.020$ mmol cm^{-2} atm^{-1} min^{-1} (3.29 × 10^{-5} mmol N^{-1} s^{-1}) (Skirrow, 1975). The rate of change of pH can be estimated using a stepwise integration process whereby a small portion of CO_2 is released according to equation 64 and the new solution equilibrium is re-calculated from the equilibrium equations (Park, 1969) using the carbonate alkalinity (A_C) and the new value of C_T so that

$$[HCO_3^-] = \frac{C_T K_{12} - [(C_T K_{12})^2 - A_C K_{12}(K_{12}-4)(2C_T - A_C)]^{1/2}}{(K_{12}-4)} \quad (65)$$

$$[CO_3^{2-}] = (A_C - [HCO_3^-])/2 \quad (66)$$

$$[H_2O \cdot CO_2] = C_T - A_C + [CO_3^{2-}] \quad (67)$$

The new value of $[H_2O \cdot CO_2]$ is used to calculate P_{CO_2} and the calculation is repeated.

The rate of release of CO_2 will obviously depend on the rate of bubbling (i.e. on the ratio of exchange surface area to solution volume). For a ratio of 1 cm^{-1} (10^2 m^{-1}) the pH is unlikely to rise above 8.4 over a 30 min degassing period whereas for a ratio of 10 cm^{-1} (10^3 m^{-1}) the pH can

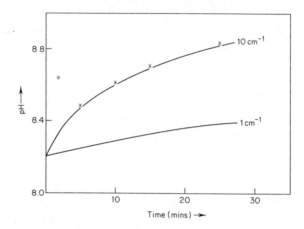

Figure 9. The rate of pH change on degassing a sea water sample. The curves show the effects of diffusion-controlled carbon dioxide release according to equation 64 with different surface to volume ratios. The crosses indicate the rate of pH change expected if the rate of carbon dioxide release is controlled by the dehydration process

increase to 8.8 (Figure 9). At this ratio the rates of diffusion and dehydration become comparable and for more vigorous degassing regimes the rate of dehydration is likely to become the rate-determining step (Figure 9). The rate of this process will be independent of the ratio of exchange surface area to solution volume.

Since gaseous CO_2 plays a key part in the pH buffering of sea water, it is important to ensure that interactions with the gas phase are treated reproducibly in electrochemical experiments. For example, if the carbonate alkalinity of sea water is varied progressively (e.g. by the addition of a strong acid), the accompanying pH changes will be different for an open system (in equilibrium with the atmosphere) and for a closed system. Ignoring kinetic effects which make the system even more unpredictable we can calculate the pH change for open systems using equations 60–62 and for closed systems using equations 65–67 (see Figure 10). Since most experimental systems are likely to be intermediate between these two equilibrium conditions, it is clear that care is needed to obtain reproducible conditions. The best procedure to adopt for voltammetric measurements at natural pH values, where degassing is essential, is to use a nitrogen (or argon)/carbon dioxide mixture of known composition so that the equilibrium P_{CO_2} can be controlled. Alternatively, a gas mixture can be prepared on-line and bubbled through a presaturator containing the sample solution. The pH of this solution can be monitored without the risk of contamination of the sample used for voltammetric analysis. The nitrogen/carbon dioxide mixture can, if necessary, be prepared simply by bubbling the nitrogen gas through a sodium bicarbonate buffer solution.

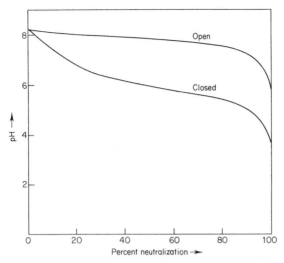

Figure 10. pH shifts expected on neutralization of the carbonate system in a sea water sample at 25°C and 1 atm pressure in open and closed systems. The curves are drawn on the Hansson pH scale

1.3. MAJOR CONSERVATIVE COMPONENTS

1.3.1. Sea water composition

Eleven components (Table 4) together account for more than 99.5% of the total dissolved solids in sea water. Since 1 kg of average sea water contains approximately 35 g of dissolved solids, the differences between the molal (moles per kilogram of water, m_i) and molar (moles per cubic decimetre of solution, c_i) concentration scales are not trivial (Table 5). Furthermore, since it is not always possible to determine the concentration of each of the eleven major components in a particular sample, it is often convenient to express concentrations on the molinity scale (moles per kilogram of solution, k_i). The concentration of the major components in natural and artificial sea waters expressed in terms of these concentration scales are shown in Table 4 and suitable conversion factors are given in Table 5. The major components are all enriched elements (Figure 3) and they have been uniformly mixed throughout the world oceans. Consequently, their concentration ratios remain essentially constant (apart from minor variations in Ca^{2+}, Sr^{2+} and HCO_3^-) so that the corresponding molalities and molinities at other salinities can be calculated by taking the appropriate proportions of the concentrations given in Table 4.

Conversions to and from the molar scale depend on the density of sea water, which may be summarized by

$$d_{sw} = d_0 + AS + BS^{3/2} \tag{68}$$

Table 4. Sea water recipes at 35‰ salinity and 25°C (recalculated from data given in Whitfield, 1979b)

Constituent	Natural sea water*			Artificial sea water†		
	Molinity	Molarity	Molality	Molinity	Molarity	Molality
Na^+	0.46818	0.47912	0.48525	0.46825	0.47926	0.48532
Mg^{2+}	0.05325	0.05449	0.05519	0.05328	0.05453	0.05522
Ca^{2+}	0.01027	0.01051	0.01064	0.01033	0.01057	0.01071
K^+	0.01021	0.01045	0.01058	0.00990	0.01013	0.01026
Sr^{2+}	0.00009	0.00009	0.00009	0.00009	0.00009	0.00009
Cl^-	0.54589	0.55865	0.56579	0.54589	0.55871	0.56579
SO_4^{2-}	0.02824	0.02890	0.02927	0.02822	0.02889	0.02925
Br^-	0.00084	0.00086	0.00087	0.00083	0.00085	0.00086
F^-	0.00006	0.00006	0.00006	0.00005	0.00005	0.00005
HCO_3^-	0.00233	0.00238	0.00241	0.00233	0.00238	0.00241
$B(OH)_3$	0.00042	0.00043	0.00044	0.00042	0.00043	0.00044
$0.5\Sigma m_i z_i^2$	0.69746	0.71374	0.72286	0.69747	0.71385	0.72289

* Millero (1974b), $a = 1.00544$ (equation 69).
† Kester *et al.* (1967), $a = 1.00471$ (equation 69).

at 1 atm pressure, where

$$d_0 = (0.99987 + 18.23 \times 10^{-3}t - 7.92 \times 10^{-6}t^2 - 55.45 \times 10^{-9}t^3 + 1.50 \times 10^{-10}t^4) \times (1 + 18.16 \times 10^{-3}t)^{-1}$$
$$A = 8.15 \times 10^{-4} - 3.92 \times 10^{-6}t + 9.20 \times 10^{-8}t^2 - 1.48 \times 10^{-9}t^3$$
$$B = -1.25 \times 10^{-6} + 2.73 \times 10^{-8}t$$
$$S = \text{salinity (see below)}$$
$$t = \text{temperature (°C)}$$

Table 5. Conversion parameters at 35‰ salinity and 25°C for concentration scales using the equation $Y = X/E$

			E	
Y	X	Definition*	Natural sea water†	Artificial sea water‡
m_i	k_i	$1 - \Sigma k_i M_i$	0.964821	0.964832
m_i	c_i§	$d_{sw} - \Sigma c_i M_i$	0.987374	0.987506
k_i	c_i§	d_{sw}	1.0234	1.0235

* d_{sw} can be calculated from equation 68. For further details see MacIntyre (1976) and Whitfield (1979b).
† Recipe of Millero (1974b).
‡ Recipe of Kester *et al.* (1969).
§ Conversions to and from the molar scale depend on d_{sw} and therefore on temperature, pressure, and salinity. Appendix I gives values of d_{sw} at regular temperature and salinity intervals.

Equation 68 is a simplified form of the density equation given by Millero and Kremling (1976) and is good to ±0.0001 unit.

The astute reader who is not familiar with the vagaries of chemical oceanography will no doubt be puzzled by the fact that the total mass of dissolved solids in both the natural and artificial sea waters (Table 4) is in excess of 35 g per kilogram of solution even though the salinity is, in both cases, given as 35‰ (35 parts per thousand). This is because salinity is a *defined* quantity which is related directly to measured properties of sea water (e.g. refractive index, conductivity) and only indirectly to the total salt content. The reasons for this are historical and we need not discuss them here (see Wallace, 1974). The total mass of dissolved solids (S_T, grams per kilogram of solution) is related to the conventional salinity (S) by the equation

$$S_T = aS \tag{69}$$

and values of a are listed in Table 4.

The conventional salinity is itself defined relative to the chlorinity (Cl) which is the chlorine equivalent of the total mass of halides (Cl, Br and I) that can be precipitated from 1 kg of sea water by the addition of silver nitrate (Section 5.3.3). The relationship is

$$S = 1.80655 Cl \tag{70}$$

In many compilations of data (Millero, 1979; Whitfield, 1979b; Walton-Smith, 1976) the physical and chemical properties of sea water are tabulated as a function of either chlorinity or the conventional salinity.

Artificial sea water can be most easily prepared from the recipe provided by Kester *et al.* (1967). All of the monovalent cations (Table 6) are added as dried crystalline material and are dissolved in 700 g of distilled water (solution A, Table 6). The divalent cations are added as concentrated standard solutions. Their molalities are checked by standard silver nitrate titration and they are made up in solution B (Table 6) with sufficient water to bring the total mass of water in A and B to 1000 g. The two solutions are then mixed slowly together with vigorous stirring to prevent the precipitation of insoluble calcium and strontium salts. The pH is stabilized by drawing air through the solution overnight (Kester *et al.*, 1967).

The ionic strength of the sea water (natural or artificial) is given by

$$I_J = 0.5 \Sigma J_i z_i^2 \tag{71}$$

where J is a general term for the concentration scale. From the data available (Table 4) we can see that

$$I_k = 0.5 \Sigma k_i z_i^2$$
$$= 1.9927 \times 10^{-2} S \tag{72}$$

Table 6. Preparation of artificial sea water by weight

Component	Molecular weight	No. of moles per kg of H_2O	No. of grams in 700 grams of H_2O	
NaCl	58.44	0.42435	17.3593	
Na_2SO_4	142.04	0.02925	2.9083	
KCl	74.56	0.00940	0.4906	Solution A
$NaHCO_3$	84.00	0.00241	0.1417	(salts dissolved
KBr	119.01	0.00086	0.07164	in 700 g of H_2O)
H_3BO_3	61.83	0.00044	0.01904	
NaF	41.99	0.00005	0.00147	
			Grams of stock solution at 23°C	
$MgCl_2$	94.94*	0.05522	59.1406 (1 M)†	Solution B
$CaCl_2$	110.99*	0.01071	11.6204 (1 M)†	(make up to 306.45 g‡
$SrCl_2$	158.53*	0.00009	0.0912 (0.1 M)†	with H_2O)

* Not allowing for water of hydration.
† Using densities from Kester *et al.* (1967).
‡ $300 \text{ g} + \Sigma g_i$.

From the conversion factors in Table 5 we can also see that

$$I_m = I_k/(1 - \Sigma k_i M_i)$$
$$= 19.927S/(1000 - 1.0051S) \tag{73}$$

and that

$$I_c = I_k d_{sw} \tag{74}$$

A number of sea water properties that are useful in electrochemical calculations are expressed as a function of salinity in Table 7. Calculated values over normal salinity and temperature ranges are given in Appendix Tables I to XI.

A number of models have been used to calculate the conventional single-ion activity coefficients of the ionic constituents of sea water (see summaries by Whitfield, 1975b, 1979b). For convenience here we shall use the values calculated via a specific interaction model (Table 8; Whitfield, 1975d) based on equations derived by Pitzer and Kim (1974) for the total single-ion activity coefficients on the molal scale (equation 3). Conversion parameters given in Table 5 can be used to calculate the corresponding values on the other concentration scales (MacIntyre, 1976).

Since sea water provides an ionic medium of constant and clearly defined composition, it is possible to consider sea water itself as the solvent and to re-adjust the reference state for the definition of activity coefficients from infinite dilution in pure water to infinite dilution in the sea water medium.

Table 7. Sea water properties as a function of salinity

No.	Property*	Relationship to sea water composition	Appendix Table	Reference
1	Dielectric constant	$\varepsilon_s' = (\varepsilon_w' - A_1 Cl)/(1 + A_1 Cl)$ $A_1 = 6.00 \times 10^{-3} + 5.33 \times 10^{-6} t$	II	Ho and Hall (1973)
	Dielectric loss	$\varepsilon_s'' = (C_0 + C_1 Cl)(\varepsilon_s' - 1)$ $C_0 = 0.282 - 6.43 \times 10^{-3} t$ $C_1 = 1.18 \times 10^{-2} + 6.70 \times 10^{-4} t$	II	
2	Relative viscosity (η_r)	$\eta_r = \eta$ (sea water)/η (pure water) $= 1 + AS_V^{1/2} + BS_V$ $A = 0.00203$ (5°C), 0.000777 (25°C) $B = 0.001526$ (5°C), 0.001891 (25°C) $S_V = S/d_{sw}$ η (pure water), $cm^{-1} g\, s^{-1} = 1.519$ $\times 10^{-2}$ (5°C), 8.904×10^{-3} (25°C)	III	Millero (1974)
3	Kinematic viscosity† (ν, $cm^2\, s^{-1}$)	$\nu = \eta$ (sea water)/d_{sw}	IV	Opekar and Beran (1976)
4	Osmotic coefficient‡ (ϕ_{sw})	$1 - \phi_{sw} = \ln 10 \cdot S' I^{1/2}(\sigma/3) + B'I + C'I^{3/2} + D'I^2$ $S' = 20.661 - 432.579/T - 3.712 \ln T + 8.638$ $\times 10^{-3} T$ $B' = -831.659 + 17022.399/T + 157.653 \ln T$ $- 0.4936T + 2.595 \times 10^{-4} T^2$ $C' = 553.906 - 11200.445/T - 105.239 \ln T$ $+ 0.3332T - 1.774 \times 10^{-4} T^2$ $D' = -0.1511$ $\sigma = (3/I^{3/2})[(1 + I^{1/2}) - (1 + I^{1/2})^{-1} - 2 \ln (1 + I^{1/2})]$	V	Millero and Leung (1976)
5	Osmotic pressure (π, bars)	$\pi = AS + BS^{3/2} + CS^2$ $A = -2.3311 \times 10^{-3} - 1.4799 \times 10^{-4} t - 7.520$ $\times 10^{-6} t^2 - 5.5185 \times 10^{-8} t^3$ $B = -1.1320 \times 10^{-5} - 8.7086 \times 10^{-6} t + 7.4936$ $\times 10^{-7} t^2 - 2.6327 \times 10^{-8} t^3$	VI	Millero and Leung (1976)
6	Vapour pressure (p, mmHg)	$p = p_0 + AS + BS^{3/2}$ $A = 0.70249 + 2.3938 \times 10^{-3} t - 3.7170$ $\times 10^{-6} t^2$ $B = -2.1601 \times 10^{-2} + 4.8460$ $\times 10^{-6} t - 1.0492 \times 10^{-6} t^2$ $C = 2.7984 \times 10^{-3} - 1.5520$ $\times 10^{-5} t - 2.7048 \times 10^{-8} t^2$	VII	Millero and Leung (1976)§
7	Freezing point (t_f, °C)	$t_f = -0.0137 - 0.05199S - 7.225$ $\times 10^{-5} S^2 - 7.58 \times 10^{-4} h$ where $h =$ depth in metres	VIII	Riley and Skirrow (1975)
8	Surface tension (τ, Nm^{-1})	$\tau \times 10^{-3} = 75.64 - 0.144t + 0.0221.S$	IX	Wilde (1971)
9	Solubility of gases (c_Y, $cm^3\, l^{-1}\, atm^{-1}$)	$\ln c_Y = a_1 + a_2 S$ $a_1 = A_1 + A_2(100/T) + A_3 \ln (T/100)$ $+ A_4(T/100)$ $a_2 = B_1 + B_2(T/100) + B_3(T/100)^2$	X	Weiss (1974)

* Conversion factors to SI units are given in the Appendix.
† Used where shear forces generated by the motion of the electrolyte relative to the electrodes are important (e.g. rotating disc electrode; Opekar and Beran, 1976).
‡ The colligative properties of the solution (freezing point depression, boiling point elevation, vapour pressure, osmotic pressure) can be calculated directly from ϕ (Robinson and Stokes, 1970; Harned and Owen, 1958; Millero and Leung, 1976). The equation is written in terms of the molal ionic strength.
§ Values of p_0 (vapour pressure of pure water at t°C) taken from Ambrose and Lawrenson (1972).

Table 8. Conventional single-ion activity coefficients on the molal scale for sea water constituents at 25°C and 1 atm pressure[*]

Ion	Ionic strength					
	0.2	0.4	0.6	0.7	0.8	1.0
H^+	0.746	0.720	0.717	0.719	0.723	0.734
Na^+	0.726	0.681	0.657	0.649	0.643	0.634
K^+	0.712	0.658	0.628	0.616	0.607	0.592
NH_4^+	0.710	0.655	0.624	0.613	0.603	0.588
Be^{2+}	0.211	0.136	0.102	0.090	0.081	0.068
Mg^{2+}	0.301	0.247	0.224	0.217	0.213	0.207
Ca^{2+}	0.289	0.232	0.207	0.199	0.193	0.185
Sr^{2+}	0.289	0.231	0.204	0.196	0.189	0.180
Ba^{2+}	0.279	0.218	0.190	0.180	0.173	0.162
Zn^{2+}	0.289	0.228	0.198	0.188	0.179	0.166
Cu^{2+}	0.279	0.219	0.191	0.182	0.175	0.165
Fe^{2+}	0.287	0.230	0.206	0.198	0.192	0.184
Co^{2+}	0.292	0.235	0.211	0.204	0.198	0.192
Ni^{2+}	0.294	0.238	0.213	0.206	0.201	0.194
Mn^{2+}	0.297	0.238	0.212	0.204	0.198	0.189
UO_2^{2+}	0.308	0.256	0.235	0.229	0.225	0.222
F^-	0.695	0.631	0.592	0.578	0.565	0.543
Cl^-	0.741	0.705	0.688	0.684	0.680	0.677
Br^-	0.750	0.720	0.710	0.708	0.708	0.711
I^-	0.763	0.743	0.741	0.743	0.747	0.757
OH^-	0.712	0.660	0.632	0.623	0.615	0.604
$H_2PO_4^-$	0.666	0.585	0.535	0.515	0.498	0.469
NO_2^-	0.693	0.631	0.595	0.582	0.571	0.553
NO_3^-	0.717	0.663	0.632	0.620	0.610	0.593
$B(OH)_4^-$	0.677	0.599	0.549	0.529	0.513	0.484
SO_4^{2-}	0.244	0.174	0.140	0.128	0.118	0.103
HPO_4^{2-}	0.243	0.165	0.126	0.113	0.102	0.085
CO_3^{2-}	0.238	0.165	0.130	0.119	0.109	0.095
PO_4^{3-}	0.046	0.021	0.012	0.010	0.008	0.006

[*] Interactions of most *cations* with Cl^-, Br^-, I^-, NO_3^-, and SO_4^{2-} have been included—a notable exception is H^+–SO_4^{2-}. With the exception of Cl^-, Br^-, I^-, NO_3^-, and SO_4^{2-} only Na^+–X and K^+–X interactions have been considered for the anions.
Reproduced from Whitfield (1975d) with the permission of Pergamon Press Ltd.

This enables the activity coefficients of minor and trace components (constituting less than a few percent of the total solute concentration) to be set equal to unity. The thermodynamic equations can then be rephrased in terms of solute concentrations rather than solute activities. This approach is valid, for example, for calculations of hydrogen ion activity in sea water and hence can provide the basis for a useful and thermodynamically meaningful pH scale (see Chapter 6).

1.3.2. Transport processes

All electrochemical techniques are significantly influenced by transport processes in solution (Bockris and Reddy, 1970a). The direct measurement of ion-transport rates provides the basis for conductometric methods (Chapter 5) and the nature of the response of all polarographic and voltammetric procedures is critically dependent on the processes which transport the electroactive components to and from the electrode surface. This is most clearly seen in the development of the oxygen electrode (Chapter 9). Although less obvious, transport processes also play an important part in the response of potentiometric systems (Buck, 1978; Koryta, 1975) since they influence both the response rate of the working electrode and the stability of the liquid junction which separates the reference electrode from the sample solution in most practical cells. Despite the importance of transport processes in electrochemical studies they have attracted relatively little attention in the oceanographic literature and the bulk of the effort that has been applied has been directed towards improving conductometric methods for the estimation of salinity (see Chapter 5 and Millero *et al.*, 1977).

The flux of a dissolved species (i) in a solution at constant temperature will depend on the degree of convective movement and on the molecular diffusion of the species in response to chemical potential gradients. Changes in the electrical field must be taken into account since they can have a marked effect on the chemical potential of ions in solution (Guggenheim, 1967). If we ignore convective transport the flux equation can be written as (Robinson and Stokes, 1970; Harned and Owen, 1958)

$$J_i = c_i U_i(-\nabla \mu_i + z_i FE) \tag{75}$$

where U_i is the infinite dilution mobility of i, E represents the applied electrical field and $\nabla = \partial/\partial x + \partial/\partial y + \partial/\partial z$. If the ions are diffusing in the absence of an applied electrical field then the term E will arise because of charge separations resulting from the different mobilities of anions and cations in solution—a phenomenon that gives rise to the liquid junction potential in electrochemical cells. If an electrical field is applied, the ions are set in motion with the cations and anions moving in opposite directions Under these circumstances two new effects arise. The electrophoretic effect arises because the migrating ions carry with them at least a part of their hydration sheaths. The resulting viscous drag will affect not only the ion in question but also ions of opposite charge moving against the induced solvent flow. The relaxation effect arises because the motion of the ions distorts the ionic atmospheres surrounding the ions in motion and hence provides an additional 'electrostatic drag' on the motion of the ions. A fuller version of equation 75 would have to take these effects into account (Robinson and Stokes, 1970; Harned and Owen, 1958).

Considering for the present molecular diffusion, the one-dimensional form of equation 75 can be rewritten in the form of Fick's law (equation 12) with a new diffusion coefficient (D_i') defined by (Katz and Ben-Yaakov, 1980)

$$D_i' = U_i RT \left[1 + \frac{\mathrm{d} \ln \gamma_i / \mathrm{d}x}{\mathrm{d} \ln c_i / \mathrm{d}x} - \frac{z_i F c_i E}{RT \, \mathrm{d} \ln c_i / \mathrm{d}x} \right] \tag{76}$$

The transport of a particular species is therefore affected by its interactions with all other components in the solution. Despite the difficulties involved, a number of theories have been developed in recent years which provide solutions to equation 75 for multi-component mixtures (see, for example, Quint and Viallard, 1978a,b; Chen, 1979a,b; Anderson and Graf, 1978, and references cited therein). These equations have not yet been applied to sea water but their use will hopefully provide a more rigorous basis for studying the influence of changes in composition on the conductivity/density relationship (see Chapter 5).

A simpler approach has been adopted by Ben-Yaakov (1973; Katz and Ben-Yaakov, 1980), who studied the diffusion of major electrolyte components from concentrated solutions into distilled water in the absence of an applied electrical field. These measurements emphasize the need to make adequate corrections for specific interactions between the ions when calculating rates of diffusion. A simple ion-pair model provides a significant improvement in the fit of theoretical predictions on the experimental data even if the diffusing solutions only contain a single electrolyte (Katz and Ben-Yaakov, 1980). An obvious practical application of such measurements is in the calculation of liquid junction potentials.

Frequently in electrochemical cells two solutions of different composition are brought into contact at a liquid–liquid boundary. Ions from both solutions will tend to migrate across such a boundary in response to the chemical potential gradient. Since the component anions and cations will have different mobilities in solution some charge separation will inevitably occur and a potential gradient will eventually be established which will oppose further net ionic migration. The potential established across such a junction (the liquid junction potential, E_j) will consequently be directly related to the transport numbers of the ions involved and to the chemical potential gradient across the interface so that at steady state the solution to equation 75 becomes (MacInnes, 1939, p. 221)

$$E_j = -g' \int_I^{II} \Sigma_i (t_i / z_i) \, \mathrm{d} \ln a_i \tag{77}$$

where t_i is the Hittorf transport number of the ion i and z_i is its charge. I and II represent the pure component solutions on either side of the liquid junction. Using the definition of a_i (equation 3), equation 77 can be split into ideal (E_{jm}) and non-ideal ($E_{j\gamma}$) components so that

$$E_j = E_{jm} + E_{j\gamma} = -g' \int_I^{II} \Sigma_i(t_i/z_i) \, d\ln m_i$$

$$-g' \int_I^{II} \Sigma_i(t_i/z_i) \, d\ln \gamma_i \tag{78}$$

To evaluate these integrals it is necessary to specify the mixing regime that characterizes the distribution of the components of the two solutions throughout the liquid junction. Equation 78 can be integrated in a straightforward manner if it is assumed that the two solutions are mixed in a linear fashion so that there is a smooth gradation from pure solution I to pure solution II. The ideal component of the liquid junction potential (E_{jm}, equation 74) is then given by (Koryta, 1975; Morf, 1977)

$$E_{jm} = -g' \left\{ \frac{\Sigma_i z_i U_i[m_i(II) - m_i(I)]}{\Sigma_i z_i^2 U_i[m_i(II) - m_i(I)]} \right\} \ln \left[\frac{\Sigma_i z_i^2 U_i m_i(II)}{\Sigma_i z_i^2 U_i m_i(I)} \right] \tag{79}$$

U_i is the absolute mobility of the ion i, which is related to the transport number, t_i by the equation

$$t_i = |z_i| \, U_i / \Sigma_i \, |z_i| \, U_i$$

and to the limiting equivalent ionic conductivity (Λ_i) by the equation (Robinson and Stokes, 1970; Koryta, 1975)

$$U_i = \Lambda_i / |z_i| \, F^2 \tag{80}$$

Values of Λ_i can be found in various electrochemical texts (e.g. Dobos, 1975; Robinson and Stokes, 1970; Kay, 1973; see also Appendix XIV). The use of this definition of U_i in equation 79 implies that the mobilities of the ions in sea water are identical with those observed at infinite dilution (Ben-Yaakov, 1972). The electrophoretic mobility of $^{22}Na^+$, $^{90}Sr^{2+}$ and $^{36}Cl^-$ ions in sea water has been measured by Kniewald and Pučar (1976) and fitted to equations of the form (see Table 9)

$$U_i' = a_1 - a_2 I^{1/3} \tag{81}$$

The electrophoretic mobility is defined by (Robinson and Stokes, 1970)

$$U_i' = \Lambda_i / F \tag{82}$$

Values of U_i' calculated from equation 82 and from equation 81 when $I = 0$ are in reasonable agreement with one another (Table 9) but are nearly twice as great as the mobilities observed in sea water at an ionic strength of 0.7 M (Table 9). Experiments on solutions of the component sea salts indicate that the lowering of the electrophoretic mobility of the cations in sea water is attributable largely to interactions with sulphate anions. Clearly a more

Table 9. Electrophoretic mobilities in electrolyte solutions ($I =$ 0.7 M)

Ion	Medium	Fitting parameters for equation 81			Calculated mobilities $(m^2 V^{-1} s^{-1}) \times 10^8$		
		$a_1 \times 10^4$	$a_2 \times 10^4$	r	A*	B†	C‡
$^{22}Na^+$	Sea water§	4.53	1.83	0.96	2.90	4.53	5.19
	NaCl	5.09	1.70	0.99			
	Na_2SO_4	3.45	1.00	0.94			
$^{90}Sr^{2+}$	Sea water§	4.53	2.28	0.94	2.51	4.53	6.16
	NaCl	4.98	2.17	0.98			
	Na_2SO_4	1.89	1.08	0.97			
$^{36}Cl^-$	Sea water§	7.51	3.31	0.99	4.56	7.51	7.91
	NaCl	7.95	2.47	0.99			
	Na_2SO_4	7.09	1.83	0.94			

* Calculated from equation 81, $I = 0.7$ M.
† Calculated from equation 81, $I = 0$ M.
‡ Calculated from equation 82, using Λ_i values from Robinson and Stokes (1970).
§ Fitted over the range $I = 0.07$–0.73 M (approx. 3.8–36.5‰ salinity, Kniewald and Pučar, 1976).

accurate interpretation of equation 77 should take into account the interactions which contribute to this discrepancy (Ben-Yaakov, 1973). Substituting from equation 80 into equation 79 we obtain

$$E_{jm} = - \frac{g'\{[\bar{C}(II) - \bar{A}(II)] - [\bar{C}(I) - \bar{A}(I)]\}}{\bar{D}(II) - \bar{D}(I)} \cdot \ln \left[\frac{\bar{D}(II)}{\bar{D}(I)} \right] \tag{83}$$

where $\bar{C} = \Sigma_c \Lambda_+ m_+$, $\bar{A} = \Sigma_a \Lambda_- m_-$ and $\bar{D} = \Sigma_i |z_i| \Lambda_i m_i$, the summations being for all cations (c), all anions (a), and all ions (i) respectively. Equations 79 and 83 give liquid junction potentials which are of the same sign as those calculated by Morf (1977) but of opposite sign to those quoted by Bates (1973, p. 38).

A more general solution of the general diffusion relationship (equation 75) was provided by Planck. Although his equations have now been set in a simpler form (Morf, 1977), the solutions for mixtures containing ions of different charge types (e.g. +1 and +2 cations) are still too complex for general use. A derivation of E_j (equation 78) has been described (Moreno and Zahradnik, 1973) which follows the assumptions made by Henderson and estimates $E_{j\gamma}$ from the Debye–Hückel theory. The resulting equations are cumbersome and not strictly applicable to the high ionic strengths normally associated with liquid junctions in electrochemical systems. A more convenient derivation described by Barry and Diamond (1970) suggests that E_j may be calculated from an equation directly analogous to equation 83 where the molality (m) is replaced by the activity (a). Ben-Yaakov (1978, 1979; Katz and Ben-Yaakov, 1980) has also discussed the use of the ion-association model to correct the Henderson equation for specific ion interactions. For simplicity we will treat only the ideal portion of the liquid junction potential and derive a convenient form of equation 83. Although there are indications that the Henderson equation is less reliable in situations where the bridge and sample solutions contain ions of different charge types (Rock, 1967; Henderson, 1907), this equation should provide a reasonable estimate of residual liquid junction potentials. To estimate junction potentials arising at reference electrodes in sea water we can treat solution I as the salt bridge and solution II as the sample. The salt bridge solution is invariably a $1:1$ electrolyte preferably with a cation and anion with almost identical mobilities. This enables us to define parameters x and y such that

$$x = \Lambda_+ - \Lambda_- \tag{84}$$

and

$$y = \Lambda_+ + \Lambda_- \tag{85}$$

Similarly for the sea water component we can define parameters X and Y such that

$$X = \Sigma_c l_c \Lambda_c - \Sigma_a l_a \Lambda_a \tag{86}$$

and

$$Y = \Sigma_i l_i \Lambda_i |z_i| \tag{87}$$

l is the ratio of the concentration of the subscripted component to the salinity in standard sea water. Values of X and Y can be readily evaluated for a particular sea water recipe (see Table 10). By combining equations 83 to 87 we obtain the simple relationship

$$E_{jm} = -g' \left(\frac{XS - xm}{YS - ym} \right) \ln \left(\frac{YS}{ym} \right) \tag{88}$$

Values of x and y for a variety of possible bridge solutions are given in Table 11. Liquid junction potentials calculated for a range of liquid junction compositions at different salinities using equation 85 are given in Table 12. For comparison, Bates (1973) estimates for liquid junction potentials in standard buffer solutions are also given with the sign convention adjusted to match equation 88. It is interesting that liquid junction potential changes observed in moving from 0.05 M potassium hydrogen phthalate solutions to 0.725 M NaCl solutions agree closely with those calculated from equation 88 (Hawley and Pytkowicz, 1973).

Table 10. Calculation of sea water parameters for the Henderson equation

Component	z_i	Λ_i	$l_i = C_i/S$	$\Lambda_i l_i$	$\Lambda_i l \, |z_i|$
Na	+1	50.10	0.01369	0.6859	0.6859
K	+1	73.50	0.00030	0.0221	0.0221
Mg	+2	53.05	0.00156	0.0828	0.1656
Ca	+2	59.50	0.00030	0.0179	0.0358
Cl	−1	76.35	0.01596	1.2185	1.2185
SO$_4$	−2	80.02	0.00083	0.0664	0.1328

$$\Sigma\Lambda_c l_c = 0.8087, \quad \Sigma\Lambda_a l_a = 1.2849, \quad X = -0.4762.$$
$$\Sigma\Lambda_c l_c z_c = 0.9094, \quad \Sigma\Lambda_a l_a \, |z_a| = 1.3513, \quad Y = 2.2607.$$

1.3.3. Electrical double layer structure

The homogeneous distribution of anions and cations is also disturbed when a solution comes into contact with a solid surface (Figure 11). The layer in immediate contact with the electrode surface consists of water molecules, hydrophobic organic molecules, and those ions (usually anions) that are not obstructed by bulky primary hydration sheaths. This strongly adsorbed layer is known as the Inner Helmholtz Plane (IHP, Figure 11). Immediately outside the IHP is a layer of hydrated ions that are attracted to the surface by the net charge that results both from the initial charge on the solid

Table 11. Parameters for the simplified version of the Henderson equation (equation 88)

Electrolyte	x	y
KCl	−2.85	149.85
NaCl	−26.25	126.45
NaNO$_3$	−21.31	121.53
KNO$_3$	2.04	144.96
NH$_4$Cl	−2.80	149.90
NH$_4$NO$_3$	2.09	145.10
RbCl	1.45	154.15
CsCl	0.95	153.65
KNO$_3$/KCl (1:1)	−0.37	147.41
KNO$_3$/KCl (4:1)	−1.87	148.72
HCl	273.47	426.17
Sea water*	−0.476	2.261

* Values are given for X (equation 86) and Y (equation 87). (use salinity rather than molarity in equation 88)

Table 12. Liquid junction potentials (mV) calculated from the Henderson equation*

Sample salinity, ‰	Bridge solution: KCl 4.16 M (sat.)	KCl 3.5 M	KCl 2 M	KCl 1 M	KNO_3–KCl 4 M (1:1)	RbCl 4 M	CsCl 4 M	NaCl 0.7 M	Sea water 35‰
5	−1.6	−1.5	−1.0	−0.2	0.2	1.4	1.1	−11.0	−10.5
10	−1.0	−0.8	−0.2	0.7	0.5	1.5	1.2	−7.2	−6.8
15	−0.1	−0.4	0.3	1.4	0.8	1.7	1.4	−5.1	−4.6
20	−0.3	−0.1	0.7	2.0	1.0	1.8	1.6	−3.5	−3.0
25	0.0	0.2	1.1	2.4	1.2	1.9	1.7	−2.3	−1.8
30	0.3	0.5	1.4	2.8	1.4	2.1	1.9	−1.4	−0.8
35	0.5	0.7	1.7	3.2	1.5	2.2	2.0	−0.5	0.0
40	0.7	0.9	2.0	3.5	1.7	2.3	2.2	0.2	0.7

* The liquid junction potentials between saturated (4.16 M) KCl and pH buffer systems are: KH phthalate (0.05 M) 2.6 mV, KH_2PO_4 (0.025 M) and Na_2HPO_4 (0.025 M) 1.9 mV, and $NaHCO_3$ (0.025 M) and Na_2CO_3 (0.025 M) 1.8 mV on this sign convention (Bates, 1973, p. 38).

surface itself and any charge contributed by the constituents of the IHP. The ions in this layer, known as the Outer Helmholtz Plane (OHP, Figure 11), are immobilized to some extent by the strength of the electrostatic interactions. From the surface to the OHP there is an approximately linear change in potential with distance. The structure of the electrical double layer bounded by the OHP is largely dictated by the characteristics of the solid surface itself. If the ions within this zone do not compensate fully for the charge generated at the solid surface, a third zone exists in which the ionic components of the solution are not uniformly distributed. This is known as the diffuse layer (DL) and it marks the final transition between the bulk solution and the solid surface. Here the ions are free to move but the net charge at the interface ensures an uneven distribution of anions and cations. The structure of the DL is largely dictated by the composition of the bulk solution and the magnitude of the uncompensated surface charge. Oldham (1975) has developed a model for the DL in sea water. Using the ion-pair model for sea water (Garrels and Thompson, 1962) to calculate the concentrations of ions of different charge types he was able to calculate (i) the concentrations of the various ionic components at the interface between the DL and the OHP and (ii) the total amount of each ion which is held within the DL by the coulombic forces. The ionic concentrations show some surprising variations with charge density within the DL (Figure 12). Notably, at high positive and negative charge densities the divalent ions became progressively more important as solution components. The composition of the electrolyte in the immediate vicinity of the electrode surface may therefore be very different to that of the bulk solution.

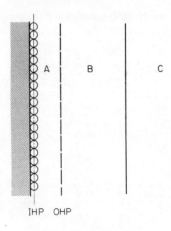

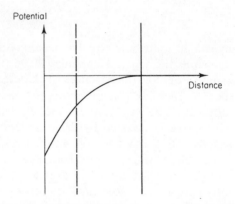

Figure 11. Schematic view of the solution structure at an electrode surface. Adsorbed molecules constitute the inner Helmholtz plane (IHP). The outer Helmholtz plane (OHP) marks the boundary of the solution layer (A) whose composition is dictated largely by the properties of the surface. The composition of the diffuse layer (B) is dictated largely by the properties of the solution. Beyond the boundary of the diffuse layer the solution is structurally homogeneous (C)

1.4. PRE-EQUILIBRATED AND DEPLETED ELEMENTS

1.4.1. Chemical periodicity of solution chemistry

The major enriched elements in sea water provide a background ionic medium of surprisingly constant composition so that it is possible, from the thermodynamic point of view, to treat sea water as a solvent in its own right. We have seen that the concentrations of the minor and trace elements in this

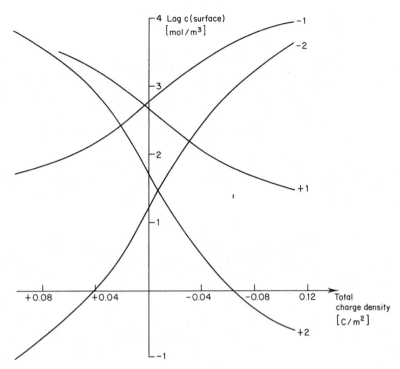

Figure 12. Ionic concentrations within the diffuse layer (region B, Figure 11) plotted as a function of surface charge density. The charges on the ions are identified on the individual curves. $0.01\,C\,m^{-2}$ is equivalent to one electron charge for each $1.5\times 10^5\,nm^2$ of surface area. Data taken from Oldham (1975)

medium are dictated largely by their solid-state chemistry. However, it is the solution chemistry of the elements that controls their role in biological cycling within the oceans. Since the most interesting changes in sea water chemistry result from biological activity we shall briefly review the solution chemistry of the elements before considering the links between biological cycling and chemical composition in the oceans.

It should also be emphasized that electrochemical methods are themselves very sensitive to the chemical form of the elements. This enables electrochemical methods to be used effectively for the determination of chemical speciation and also suggests that they may provide useful models for biological assimilation processes (Whitfield and Turner, 1979b). Since all of the elements in the Periodic Table are dissolved in sea water, there is ample scope for both ionic and covalent interactions and the chemical form (or forms) that a particular element adopts must represent a compromise between the various influences experienced. The major factors involved are

(i) the extent of hydrolysis of the element at natural pH values,

(ii) the interaction of the element (or its hydrolysed form) with dominant components in solution, and

(iii) the influence of redox potential and complexation on the stability of the various oxidation states that the element can adopt.

The extent of complexation of a particular element, with water or with other solution components, is most simply summarized by the side-reaction coefficient (σ_{ML_n}) (Ringbom, 1963; Elder, 1975). This is related to the overall stability constant for the formation of the complex (β_{ML_n}) and the concentration of the complexing agent (m_L) by the equation

$$\sigma_{ML_n} = \beta_{ML_n} m_L^n. \tag{89}$$

If the element M is involved in a number of reactions the individual σ values can be summed to give an overall side-reaction coefficient ($\bar{\alpha}_M$) defined by

$$\bar{\alpha}_M = 1 + \Sigma \sigma_{ML_n} \tag{90}$$

For a minor or trace component, M, in sea water the percentage present as the complex ML_n is given by

$$ML_n(\%) = 100 \sigma_{ML_n} / \bar{\alpha}_M \tag{91}$$

provided that (i) M forms only mononuclear complexes, (ii) interactions between trace components are negligible, and (iii) complexation does not affect the free concentration of L significantly. The percentage of uncomplexed M is given by

$$M(\%) = 100 / \bar{\alpha}_M \tag{92}$$

We shall use the side-reaction concept to consider briefly the various factors that influence the speciation of the elements in sea water.

Hydrolysis of the elements

With a concentration in excess of 50 M, water is the dominant reactant. The hydrolysis of the positive oxidation states in sea water can be considered in terms of interactions of water with the elemental cations. These interactions can range in strength from simple hydration (e.g. for Na^+, Ca^{2+}), through complexation (e.g. for Al^{3+} and Cu^{2+} to yield $Al(OH)_4^-$, $CuOH^+$) to full hydrolysis and condensation (e.g. for S^{6+} and C^{4+} to yield SO_4^{2-} and CO_3^{2-}). This range of interactions can, however, be summarized by a simple electrostatic theory (King, 1965; Millero, 1977) which relates the free energy of hydrolysis to the term z^2/r, where z is the charge on the central cation and r is its ionic radius. A plot of log $\bar{\alpha}_{M,OH}$ versus log z^2/r for conditions typical of seawater (Figure 13) shows that for the main sequence of elements there is a sharp transition from weakly hydrolysed to fully hydrolysed cations within

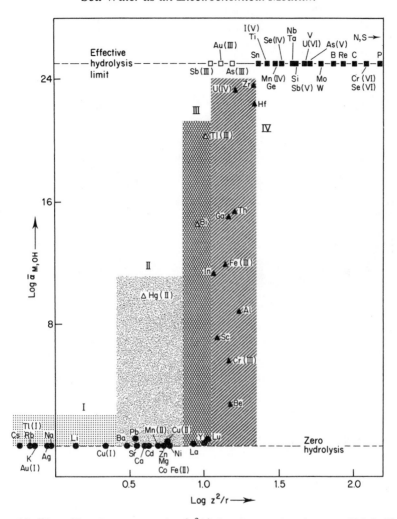

Figure 13. Plot of $\log \bar{\alpha}_{M,OH}$ *versus* $\log (z^2/r)$ for elemental cations at pH 8.2: $\square$, $\blacksquare$, fully hydrolysed; $\triangle$, $\blacktriangle$, hydrolysis dominated; $\bullet$, weakly hydrolysed. The significance of the open symbols is discussed in the text. The cations with $\log \bar{\alpha}_{M,OH} < 25$ are further divided into four groups, identified on the figure. Reproduced with the permission of Pergamon Press from Turner *et al.* (1981)

the z^2/r range of 10–20 (Turner *et al.*, 1981). The fully hydrolysed elements occupy two distinct regions of the Periodic Table (Figure 14) and form solution components with the general formula $MO_p(OH)_q$. The elements are mainly present in their group oxidation states and their chemical speciation will depend primarily on the degree of condensation experienced [i.e. on the extent to which water molecules are lost from the hydroxy complexes to

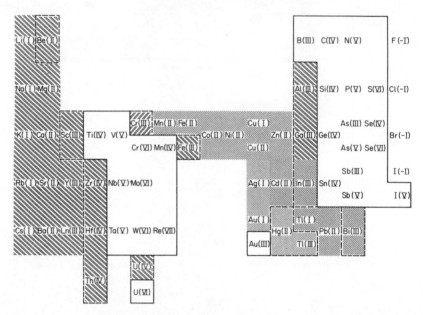

Figure 14. Periodic Table showing the classification of the elements. The symbol Ln represents the lanthanide series. ☐, Fully hydrolysed oxidation states; ⸬, oxidation states in the transition zone of the Born plot (Figure 13). Heavy hatching, (a) cations; light stippling, (b) cations; ⧷, (a)′ cations; ⧹, (b)′ cations. Reproduced with the permission of Pergamon Press from Turner *et al.* (1981)

yield oxyanions (e.g. NO_3^-) or oxocations (e.g. UO_2^{2+})]. Condensation is in general favoured by the lighter elements and the later periodic groups because of a mixture of steric and electrostatic effects. The dominant species for the fully hydrolysed elements in sea water are identified in Table 13. The chemical speciation of these elements will be further modified by protolytic reactions which are strongly pH dependent (e.g. see the discussion of the carbon dioxide system in Section 1.2.4) or by interactions between the oxyanions and the major cations. Little information is available for systems other than those involving the major oxyanions [CO_3^{2-}, SO_4^{2-}, $B(OH)_4^-$]. The only oxocation to have received any attention is the uranyl ion, which is strongly complexed with carbonate ions in sea water.

The speciation of the cationic elements that are *not* fully hydrolysed in sea water is controlled by interactions with hydroxyl ions and with the major anionic components (Cl^-, SO_4^{2-}, CO_3^{2-}, F^-).

Classification of cations

The nature and stability of the complex species formed by cations in sea water will depend largely on the extent to which they form covalent or

Table 13. Calculated speciation of fully hydrolysed elements in sea water (pH 8.2, 25°C, 1 atmosphere pressure)

Complex	%	Log $\bar{\alpha}$	Complex	%	Log $\bar{\alpha}$	Complex	%	Log $\bar{\alpha}$
$As(OH)_3$	87	0.06	$NaIO_3$	5		$CaSO_4$	7	
$As(OH)_4^-$	13		KIO_3	*				
			$MgIO_3^+$	5		$Sb(OH)_3$	100	0.00
H_3AsO_4	*	8.21	$CaIO_3^+$	1		$Sb(OH)_2^+$	*	
$H_2AsO_4^-$	1					$Sb(OH)_4^-$	*	
$HAsO_4^{2-}$	98		MoO_4^{2-}	100	0.00			
AsO_4^{3-}	1		$HMoO_4^-$	*		$Sb(OH)_5$	*	5.72
			H_2MoO_4	*		$Sb(OH)_6^-$	100	
BO_2^-	10	1.00				H_2SeO_3	*	6.09
HBO_2	81		NO_3^-	95	0.02	$HSeO_3^-$	35	
$NaBO_2$	3		$NaNO_3$	4		SeO_3^{2-}	65	
$MgBO_2^+$	3		KNO_3	*				
$CaBO_2^+$	4		$CaNO_3^-$	1		SeO_4^{2-}	100	0.00
						$HSeO_4^-$	*	
CO_3^{2-}	14	0.85	$Nb(OH)_5$	12	0.92			
$NaCO_3^-$	8		$Nb(OH)_4^+$	*		$H_2SiO_4^{2-}$	*	5.44
KCO_3^-	*		$Nb(OH)_6^-$	88		$H_3SiO_4^-$	5	
$MgCO_3$	59					H_4SiO_4	93	
$CaCO_3$	19		PO_4^{3-}	*	3.57	MgH_2SiO_4	*	
			HPO_4^{2-}	35		CaH_2SiO_4	*	
HCO_3^-	75	0.13	$MgPO_4^-$	1		$MgH_3SiO_4^+$	1	
H_2CO_3	1		$CaPO_4^-$	5		$CaH_3SiO_4^+$	*	
$NaHCO_3$	11		$H_2PO_4^-$	1				
$MgHCO_3^+$	10		$NaHPO_4^-$	40		$Ta(OH)_5$	96	0.02
$CaHCO_3^+$	2		$KHPO_4^-$	1		$Ta(OH)_4^+$	*	
			$MgHPO_4$	13		$Ta(OH)_6^-$	4	
CrO_4^{2-}	71	0.15	$CaHPO_4$	4				
$HCrO_4^-$	*		H_3PO_4	*		VO_4^{3-}	*	5.65
H_2CrO_4	*		$MgH_2PO_4^+$	*		HVO_4^{2-}	38	
$NaCrO_4^-$	28		$CaH_2PO_4^+$	*		$H_2VO_4^-$	19	
$KCrO_4^-$	1					H_3VO_4	*	
			ReO_4^-	98	0.01	VO_2^+	*	
$H_2GeO_4^{2-}$	*	4.46	$KReO_4$	2		$NaHVO_4^-$	43	
$H_3GeO_4^-$	14					$KHVO_4^-$	1	
H_4GeO_4	86		SO_4^{2-}	33	0.48			
			$NaSO_4^-$	25		WO_4^{2-}	100	0.00
IO_3^-	89	0.05	KSO_4^-	*		HWO_4^-	*	
HIO_3	*		$MgSO_4$	35		H_2WO_4	*	

* Abundance < 1%.

electrostatic bonds with the major ligands. A number of classification procedures have been developed that attempt to reflect the nature of the chemical bond formed between cation and ligand (Ahrland, 1968, 1975; Ahrland *et al.*, 1958; Pearson, 1969; Nieboer and Richardson, 1980). A simple, pragmatic approach with sea water is to use the difference in stability between the fluoro and chloro complexes of a particular element as a guide to its propensity to form covalent bonds. A term $\Delta\beta$ may be defined (Turner *et al.*, 1981), such that

$$\Delta\beta = \log \beta^0_{MF} - \log \beta^0_{MCl} \tag{93}$$

Elements for which $\Delta\beta > 2$ are known as (a)-type cations and form relatively weak complexes that are largely electrostatically bound. Elements for which $\Delta\beta < -2$ are known as (b)-type cations and form strong complexes that are largely covalently bound. The classification of the cations into (a) and (b) types was first proposed by Ahrland (1968, 1975; Ahrland *et al.*, 1958). The intermediate elements $(2 > \Delta\beta > -2)$ can be further subdivided into borderline (a)-type [or (a)′−type] cations for which $\Delta\beta$ is positive and borderline (b)-type [or (b)′-type] cations for which $\Delta\beta$ is negative. The cations in each group occupy clearly defined regions in the Periodic Table (Figure 14). If we use z^2/r and $\Delta\beta$ as the coordinates wè can lay out a grid which summarizes the major factors controlling the intensity of the interactions between anions and cations. In the resulting complexation field (CF) diagram (Figure 15;

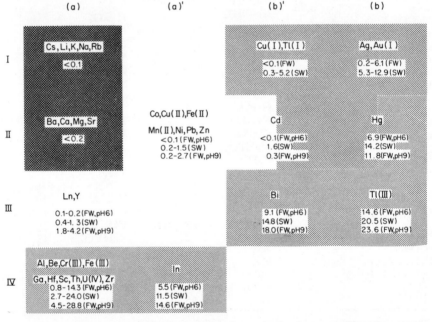

Figure 15. Complexation field diagram for the cationic elements. The elements are divided into four groups both on the horizontal axis (according to $\Delta\beta$, equation 93) and on the vertical axis (according to z^2/r, Figure 13). The figures given below each group of elements show the range of log $\bar{\alpha}_M$ values calculated for model sea water and fresh waters (Turner *et al.*, 1981). Heavy stippling, very weakly complexed cations; light stippling, hydrolysis-dominated cations; cross-hatching, chloro-dominated cations. Reproduced with the permission of Pergamon Press from Turner *et al.* (1981)

Turner *et al.*, 1981) the elements collected into the individual squares should exhibit similar solution chemistry in natural waters. In using this diagram it is important to note that the ligands themselves can be classified into *hard, soft,* and *intermediate* categories (Pearson, 1969). The hard ligands (e.g. SO_4^{2-}, F^-) form largely electrostatic bonds with the central metal ion and the stability of their complexes will increase in moving from top to bottom in the CF diagram following the increasing intensity of electrostatic interaction. There will be little shift in the stability of complexes formed with hard ligands on moving across the diagram. The soft ligands (e.g. Cl^-, S^{2-}) in contrast form largely covalent bonds with the central metal ion so that the stability of their complexes will increase markedly on moving from left to right across the diagram. The electrostatic properties of the cations will in this instance have little influence on the strength of the complexes formed. Intermediate ligands (e.g. CO_3^{2-}, OH^-) show 'hard' behaviour towards (a)-type cations and 'soft' behaviour towards (b)-type cations so that the stability of their complexes will increase on moving diagonally across the CF diagram from the top left-hand corner. The result of these influences is that in sea water four general groupings of elements can be delineated on the CF diagram: (i) (a)-type cations with a low polarizing power whose speciation is dominated by the free cation (dark stippled area, Figure 15), (ii) elements from the transition zone in the Born diagram (triangles, Figure 13) where speciation is dominated by hydrolysis products (light stippled region, Figure 15), (iii) (b)-type cations whose speciation is dominated by chloro complexes (hatched region, Figure 15), and (iv) elements intermediate between these extremes of behaviour that exhibit a complicated array of dissolved species which includes an appreciable fraction of the free metal ion (unshaded region, Figure 15). These groupings are reflected in speciation patterns of the elements in sea water (Table 14) and they have been shown to be related to the biological function of the elements (Nieboer and Richardson, 1980). From the electrochemical point of view it is the elements of intermediate character [group (iv) above] that have attracted the most attention. In particular the amalgam-forming elements Zn, Cu, Cd, and Pb have provided the focus for intensive studies by anodic stripping voltammetry (Chapter 10). Of these elements two (Zn, Cu) are biologically essential elements at low concentration and two (Cd, Pb) are considered to perform no useful biological function. All four elements become toxic at high ($>10^{-6}$ M) concentrations.

Redox chemistry

We can generalize the schematic redox reaction (equation 22) to include hydrolysis so that

$$O + mH^+ + ne^- \rightleftharpoons R + H_2O \qquad (94)$$

Table 14. Calculated speciation of the cationic elements in sea water (25°C, 1 atmosphere pressure) (Turner *et al.*, 1981) Reproduced with the permission of Pergamon Press Ltd

Ion	Free	OH^-	F^-	Cl^-	SO_4^{2-}	CO_3^{2-}	Log $\bar{\alpha}$
Ag^+	*	*	*	100	*	*	5.26
Al^{3+}	*	100	*	—†	—	*	9.22
Au^+	*	—	—	100	—	—	12.86
Au^{3+}‡	*	100	—	*	—	—	27.30
Ba^{2+}	86	*	*	9	5	*	0.07
Be^{2+}	*	99	1	*	*	*	2.74
Bi^{3+}	*	100	*	*	*	—	14.79
Cd^{2+}	3	*	*	97	*	*	1.57
Ce^{3+}	21	5	1	12	10	51	0.68
Co^{2+}	58	1	*	30	5	6	0.24
Cr^{3+}	*	100	*	*	*	—	5.82
Cs^+	93	—	—	7	—	—	0.03
Cu^+	*	—	—	100	—	—	5.18
Cu^{2+}	9	8	*	3	1	79	1.03
Dy^{3+}	11	8	1	5	6	68	0.94
Er^{3+}	8	12	1	4	4	70	1.08
Eu^{3+}	18	13	1	10	9	50	0.74
Fe^{2+}	69	2	*	20	4	5	0.16
Fe^{3+}	*	100	*	*	*	*	11.98
Ga^{3+}	*	100	*	*	—	*	15.35
Gd^{3+}	9	5	1	4	6	74	1.02
Hf^{4+}	*	100	*	*	*	—	22.77
Hg^{2+}	*	*	*	100	*	*	14.24
Ho^{3+}	10	8	1	5	5	70	0.99
In^{3+}	*	100	*	*	*	*	11.48
La^{3+}	38	5	1	18	16	22	0.42
Li^+	99	*	—	—	1	—	0.00
Lu^{3+}	5	21	1	1	1	71	1.32
Mn^{2+}	58	*	*	37	4	1	0.23
Nd^{3+}	22	9	1	10	12	46	0.66
Ni^{2+}	47	1	*	34	4	14	0.33
Pb^{2+}	3	9	*	47	1	41	1.51
Pr^{3+}	25	8	1	12	13	41	0.61
Rb^+	95	—	—	5	—	—	0.02
Sc^{3+}	*	100	*	*	*	*	7.41
Sm^{3+}	18	10	1	8	11	52	0.75
Sn^{4+}‡	*	100	—	—	—	—	32.05
Tb^{3+}	16	11	1	8	9	55	0.80
Th^{4+}	*	100	*	*	*	*	15.64
TiO^{2+}‡	*	100	—	—	*	—	11.14
Tl^+	53	*	*	45	2	—	0.28
Tl^{3+}	*	100	—	*	*	—	20.49
Tm^{3+}	11	21	1	5	6	55	0.94
U^{4+}	*	100	*	*	*	—	23.65
UO_2^{2+}‡	*	*	*	*	*	100	6.83

Table 14. (contd.)

Ion	Free	OH⁻	F⁻	Cl⁻	SO₄²⁻	CO₃²⁻	Log $\bar{\alpha}$
Y^{3+}	15	14	3	7	6	54	0.81
Yb^{3+}	5	9	1	2	3	81	1.30
Zn^{2+}	46	12	*	35	4	3	0.34
Zr^{4+}	*	100	*	*	*	—	23.96

* Ligand considered but abundance <1%.
† —, Ligand not considered.
‡ Classified as fully hydrolysed oxidation states.

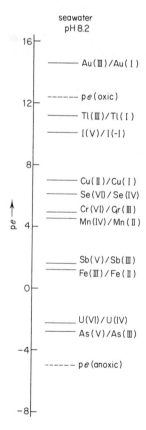

Figure 16. Electron free-energy (EFE) diagram for various redox couples in sea water at pH 8.2 (see Figure 5 and Section 1.2.3). Reproduced with the permission of Pergamon Press from Turner *et al.* (1981)

We can define an overall stability constant ($\beta_{OR}^{\ominus}$) for this reaction if we ignore activity coefficient corrections such that

$$\log \beta_{OR}^{\ominus} = \log^F m_R + m\text{pH} + n\text{p}e - \log^F m_O \tag{95}$$

where the superscript F indicates free, uncomplexed concentrations. Since $\ln \beta_{OR}^{\ominus} = -\Delta G^{\ominus}/RT$ we can use equations 25b and 90 to write

$$\log (^T m_O/^T m_R) = -16.9E^{\ominus} + m\text{pH} + n\text{p}e + \log (\bar{\alpha}_O/\bar{\alpha}_R) \tag{96}$$

where the superscript T indicates total or stoichiometric concentrations. The pe value at which the oxidized and reduced forms are present at equal concentrations is therefore given by

$$\text{p}e_{eq} = 16.9E - (m/n)\text{pH} - (1/n) \log (\bar{\alpha}_O/\bar{\alpha}_R) \tag{97}$$

At higher pe values the oxidized form (O) will dominate and at lower pe values the reduced form (R) will dominate. The side-reaction coefficients provide a particularly simple means for estimating the influence of complexation reaction on the redox potential. Using the tabulated $\bar{\alpha}$ values for sea water (e.g. Tables 13 and 14) and standard electrode potentials (e.g. Dobos, 1975) it is possible to calculate an electron-free energy diagram (*cf.* Figure 5) showing the sequence of redox couples to be expected at equilibrium in natural sea water (Figure 16). Typical limits to the pe range in natural waters (see Section 1.2.3) are included on the diagram for reference.

1.4.2. Biological cycling of the elements in the sea

General principles

In the surface layers of the ocean, microscopic algae (phytoplankton) use light energy to convert simple inorganic components into complex organic molecules by photosynthesis. The dissipation of the active wavelengths of light effectively restricts photosynthesis to the top 100 m or so, even in the clearest ocean waters. This photosynthetic (or euphotic) zone forms a thin, warm, well mixed layer which is separated from the deep cold waters of polar origin by a temperature transition zone known as the thermocline (Section 1.2.2). Photosynthesis releases oxygen from the inorganic carbon species and converts the resulting reduced carbon into particulate organic matter. In the process nitrogen and phosphorus are also assimilated in the ratio $C:N:P = 106:16:1$ (Redfield *et al.*, 1963). With the fixation of one atom of phosphorus 276 atoms of oxygen will be released. In the euphotic zone, significant changes in oxygen and carbon dioxide partial pressures and in pH will accompany the photosynthetic removal of the nutrient elements. The distribution of components involved in the photosynthetic process will exhibit a diurnal rhythm associated with the availability of light and a spatial pattern associated with the availability of the essential nutrients. In addition to organic tissue (or 'soft parts') the phytoplankton also synthesize mineral skeletons (or 'hard parts') usually from silica or calcite (Lowenstam, 1974). Into both the 'hard' and 'soft' parts the organisms incorporate a wide range of trace elements for biological or structural reasons (Banin and Navrot, 1975; Egami, 1974). It is interesting that a close correlation exists between the composition of sea water and the distribution of the bio-essential elements on the Periodic Table (Egami, 1974; Whitfield, 1981). All elements with sea water concentrations greater than 10^{-3} M (Table 1), including carbon and the major charge-carrying ions, are bio-essential. The elements with concentrations between 10^{-3} and 10^{-7} M include the major nutrient elements and the elements with concentrations between 10^{-7} and 10^{-9} M include all of the trace metals that are essential in enzyme functions. With the exception of cobalt, and possibly tin, none of the forty-five elements with concentrations less than 10^{-9} M in contemporary sea water have been shown to perform any essential biological function.

The phytoplankton cells, being slightly denser than the surrounding water, tend to sink out of the euphotic zone but the majority of them are intercepted by grazing animals which are in turn eaten by a sequence of predators. The net result is a steady rain of detrital particles (dead organisms, faecal pellets, skeletal fragments) from the euphotic zone into the thermocline and, eventually, into the deep ocean. The organic 'soft parts' of this particulate material will eventually be broken down by bacteria which gain energy from reversing the photosynthetic process to release carbon dioxide and the nutrient components with a consequent removal of oxygen from the surrounding water. The redissolution of the soft parts occurs rapidly below the surface layer and the bulk of the regeneration occurs within the thermocline with more than 90% of the organic matter being recycled in the top 400 m. Consequently, the depth profiles of the nutrient anions (e.g. NO_3^-, HPO_4^{2-}) often show a characteristic maximum in this region with a corresponding minimum in the oxygen profile. The fine structure of such profiles, and of the corresponding profiles of pH and total carbon dioxide, can reveal much useful information about the dynamics of the nutrient regeneration process and about the balance struck between biological release and utilization and physical transport. Consequently, considerable effort has been devoted to the construction of *in situ* profiling systems (Sections 3.4, 5.8, 6.3.9, and 9.7) and to the automated analysis of water samples. Trace metals incorporated in the soft parts will also be released as the tissue is degraded. The vertical profiles of zinc (Bruland *et al.*, 1978), cadmium (Boyle *et al.*, 1976), and nickel (Sclater *et al.*, 1976) consequently follow closely those of phosphate.

The solubility of the crystalline phases increases with increasing pressure so that the ocean waters became progressively more corrosive as the mineral particles descend. Elements incorporated in the mineral hard parts (e.g. Si, Ca) will therefore be released in the deep ocean so that their concentration profiles show a progressive increase with depth. This will apply not only to the structural elements themselves but also to any trace elements incorporated into the mineral structures or adsorbed into the surface (Martin, 1970). The vertical profiles of barium follow this pattern (Chan *et al.*, 1977; Gurvich *et al.*, 1978) whereas those of copper appear to indicate a significant contribution from the dissolution of both hard and soft components (Boyle *et al.*, 1977; Moore, 1978). Components released from the particulate phase by biological or physical processes below the euphotic zone will gradually be returned to the surface layers by slow oceanic circulation. The components released just below the euphotic zone will naturally be returned more rapidly than those released in the deep ocean and will consequently be more efficiently recycled.

When we look more closely inside the ocean bucket (Figure 1) we therefore find an intricate and carefully regulated feedback system interposed between the river input and the sedimentary output. It is possible to show that the sequence of particulate removal and biogenic regeneration confers some degree of stability on conditions in the euphotic zone (Broecker, 1971, 1974; Whitfield, 1981). However, the response of these feedback mechanisms is restricted. For example, large amounts of oxygen are removed during the oxidative regeneration of the inorganic nutrient components from the soft organic detritus. In the contemporary oceans this oxygen debt is repaid by the slow circulation of oxygen-rich water from the poles. However, the system is in a state of dynamic balance and it would require little more than a doubling of the surface nutrient levels (and hence the flux of detrital particles) to produce the deoxygenation of large tracts of the deep ocean (Bender and Graham, 1978). Anoxic zones of this kind are found in many areas where the circulation is restricted for topographic reasons (e.g. Black Sea, Norwegian fjords). Such zones provide very interesting sites for electrochemical studies since they are naturally deoxygenated and they contain a variety of electroactive components at high concentrations and with variable redox states.

REFERENCES

Adams, W. A., and A. R. Davis (1973). Survey of techniques to study aqueous systems at pressures of 1–3 kbar. In *Marine Electrochemistry* (Eds. J. B. Berkowitz, R. A. Horne, M. Banus, P. L. Howard, M. J. Pryor, G. C. Whitnack, and H. V. Weiss). The Electrochemical Society Inc., Princeton, N.J., pp. 53–75.

Ahrland, S. (1968). Thermodynamics of complex formation between hard and soft receptors and donors. *Struct. Bonding (Berlin)*, **5**, 118–149.

Ahrland, S. (1975). Metal complexes in sea water. In *The Nature of Sea Water* (Ed. E. D. Goldberg). Dahlem Konferenzen, Berlin, pp. 219–244.

Ahrland, S., J. Chatt, and N. R. Davies (1958). The relative affinities of ligand atoms for acceptor molecules and ions. *Quart. Rev. Chem. Soc.*, **12**, 265–276.

Almgren, T., D. Dyrssen, and M. Strandberg (1975). Determination of pH on the moles per kg seawater scale (M_w). *Deep-Sea Res.*, **22**, 635–646.

Ambrose, D., and I. J. Lawrenson (1972). The vapour pressure of water. *J. Chem. Thermodyn.*, **4**, 755–761.

Andersen, N. R., and A. Malahoff, Eds. (1972) *The Fate of Fossil Fuel CO_2*. Marine Science, Vol. 6, Plenum Press, New York.

Anderson, D. E., and D. L. Graf (1978). Ionic diffusion in naturally occurring aqueous solutions: use of activity coefficients in transition-state models. *Geochim. Cosmochim. Acta*, **42**, 251–262.

Andrews, F. C. (1972). Gravitational effects on concentrations and partial pressures in solutions: a thermodynamic analysis. *Science*, **178**, 1199–1201.

Banin, A., and J. Navrot (1975). Origin of life: clues from relations between chemical compositions of living organisms and natural environments. *Science*, **189**, 550–551.

Barry, P. H., and J. M. Diamond (1970). Junction potentials, electrode standard potentials and other problems in interpreting electrical properties of membranes. *J. Membrane Biol.*, **3**, 93–122.

Barth, T. F. W. (1952). *Theoretical Petrology*. Wiley, New York. pp. 29–34.

Bates, R. G. (1973). *Determination of pH, Theory and Practice*, 2nd edition. Wiley–Interscience, New York.

Ben-Yaakov, S. (1972). Diffusion of seawater ions into a dilute solution. *Geochim. Cosmochim. Acta*, **56**, 1396–1406.

Ben-Yaakov, S. (1973). The incremental concentration cell and its application for studying ionic diffusion in seawater. In *Marine Electrochemistry* (Eds. J. B. Berkowitz, R. A. Horne, M. Banus, P. L. Howard, M. J. Pryor, G. C. Whitnack, and H. V. Weiss). The Electrochemical Society Inc., Princeton, N.J., pp. 111–123.

Bender, M. L., and D. W. Graham (1978). Long-term constraints on the global marine carbonate system. *J. Mar. Res.*, **36**, 551–567.

Bockris, J. O'M., and A. K. N. Reddy (1970a). *Modern Electrochemistry*, Vol. 1. Macdonald, London, pp. 296–299.

Bockris, J. O'M., and A. K. N. Reddy (1970b). *Modern Electrochemistry*, Vol. 2. Macdonald, London.

Boyle, E. A., F. Sclater, and J. M. Edmond (1976). On the marine geochemistry of cadmium. *Nature*, **263**, 42–44.

Boyle, E. A., F. Sclater, and J. M. Edmond (1977). The distribution of dissolved copper in the Pacific. *Earth Planet. Sci. Lett.*, **37**, 38–54.

Bradley, D. J., and K. S. Pitzer (1979). Thermodynamics of electrolytes. 12. Dielectric properties of water and Debye–Hückel parameters to 350°C and 1 kbar. *J. Phys. Chem.*, **83**, 1599–1603.

Breck, W. G. (1972). Redox potentials by equilibration. *J. Mar. Res.*, **30**, 121–139.

Breck, W. G. (1975). Redox levels in the sea. In *The Sea*, Vol. 5 (Ed. E. D. Goldberg) Wiley, New York, pp. 153–179.

Brewer, P. G. (1975). Minor elements in sea water. In *Chemical Oceanography*, Vol. 1 (Eds. J. P. Riley and G. Skirrow). Academic Press, London, pp. 415–496.

Broecker, W. S. (1971). A kinetic model for the chemical composition of sea water. *Quaternary Res.*, **1**, 188–207.

Broecker, W. S. (1974). *Chemical Oceanography*. Harcourt, Brace and Jovanovich, New York, 214 pp.

Bruland, K. W., G. A. Knauer, and J. H. Martin (1978). Zinc in north-east Pacific water. *Nature*, **271**, 741–743.

Buck, R. P. (1978). Theory and principles of membrane electrodes. In *Ion-selective Electrodes in Analytical Chemistry*, Vol. 1 (Ed. H. Frieser). Plenum Press, New York, pp. 1–141.

Buffle, J., F.-L., Greter, G. Nembrini, J. Paul, and W. Haerdi (1976), Capabilities of voltammetric techniques for water quality control problems. *Z. Anal. Chem.*, **282**, 339–350.

Butler, J. N. (1964). *Ionic Equilibrium. A Mathematical Approach.* Addison-Wesley, Reading Mass., 547 pp.

Chan, L. H., D. Drummond, J. M. Edmond, and B. Grant (1977). On barium data from the Atlantic GEOSECS Expedition. *Deep-Sea Res.*, **24**, 613–649.

Chen, M. (1979a). Conductance equation of dilute mixed strong electrolytes. II. Hydrodynamic and osmotic terms in relaxation field. *J. Solut. Chem.*, **8**, 165–173.

Chen, M. (1979b). Conductance equation of dilute mixed strong electrolytes. III. electrophoresis. *J. Solut. Chem.*, **8**, 509–518.

Christensen, J., and R. M. Izatt (1970). *Handbook of Metal Ligand Heats.* Marcel Dekker, New York.

Cox, R. A. (1965). The physical properties of sea water. In *Chemical Oceanography* 1st edition, Vol. I (Eds. J. P. Riley and G. Skirrow). Academic Press, London, pp. 73–120.

de Bethune, A., and N. A. Swendeman Loud (1964). Standard aqueous electrode potentials and their temperature coefficients at 25°C. In *Encyclopedia of Electrochemistry* (Ed. C. A. Hampel). Reinhold, New York, pp. 414–426.

Dietrich, G. (1963). *General Oceanography.* Wiley–Interscience, New York.

Dickson, A. G. (1977). Some studies on acid–base behaviour in artificial seawater. *PhD Thesis*, University of Liverpool.

Dickson, A. G., and J. P. Riley (1978). The effect of analytical error on the evaluation of the components of the aquatic carbon dioxide system *Mar. Chem.*, **6**, 77–85.

Distèche, A. (1972). Electrochemical devices for *in situ* or simulated deep-sea measurements. In *Barobiology and the Experimental Biology of the Deep Sea*, (Ed. R. W. Brauer). North Carolina Sea Grant Program, Chapel Hill, N.C., pp. 234–265.

Distèche, A. (1975). The effect of pressure on dissociation constants and its temperature dependency. In *The Sea*, Vol. 5 (Ed. E. D. Goldberg). Wiley, New York, pp. 81–122.

Dobos, D. (1975). *Electrochemical Data.* Elsevier, Amsterdam, 339 pp.

Dyrssen, D., and M. Wedborg (1980). Major and minor elements, chemical speciation in estuaries. In *Chemistry and Biochemistry of Estuaries* (Eds. E. Olausson and I. Cato). Wiley, New York.

Edmond, J. M., and J. M. T. M. Gieskes (1970). On the calculation of the degree of saturation of sea-water with respect to calcium carbonate under *in situ* conditions. *Geochim. Cosmochim. Acta*, **34**, 1261–1291.

Egami, F. (1974). Minor elements and evolution. *J. Mol. Evol.*, **4**, 113–120.

Elder, J. F. (1975). Complexation side reactions involving trace metals in natural water systems. *Limnol. Oceanogr.*, **20**, 96–102.

El'yanov, B. S. (1975). Linear free energy relationships and some quantitative regularities of the effect of pressure on chemical processes. *Aust. J. Chem.*, **28**, 933–943.

El'yanov, B. S., and S. D. Hamann (1975). Some quantitative relationships for ionization reactions at high pressures. *Aust. J. Chem.*, **28**, 945–954.

Eyring, H., and E. M. Eyring (1967). Reaction rates in solution. In *Principles and Applications of Water Chemistry* (Eds. S. D. Faust and J. V. Hunter). Wiley, New York, pp. 1–22.

Goldberg, E. D., W. S. Broecker, M. G. Gross, and K. K. Turekian (1971). Marine chemistry. In *Radioactivity in the Marine Environment*, (Ed. A. H. Seymour) National Academy of Sciences, Washington, D.C., pp. 137–146.

Goldschmidt, V. M. (1954). *Geochemistry*, Clarendon Press, Oxford.

Guggenheim, E. A. (1967). *Thermodynamics.* North-Holland, Amsterdam, pp. 327–332.

Gurney, R. W. (1953). *Ionic Processes in Solution.* McGraw-Hill, New York.

Gurvich, Ya. G., Yu. A. Bogdanov, and A. P. Lisitsyn (1978). Behaviour of barium in recent sedimentation in the Pacific. *Geochem. Int.*, **15**, 28–43.

Hamann, S. D. (1963). Chemical kinetics. In *High Pressure Physics and Chemistry*, Vol. 2 (Eds. R. S. Bradley and D. C. Munro). Academic Press, London, pp. 161–207.

Hamann, S. D. (1974). Electrolyte solutions at high pressure. In *Modern Aspects of Electrochemistry* (Eds. B. E. Conway and J. O'M Bockris). Plenum Press, New York, pp. 47–158.

Hansson, I. (1973). A new set of pH-scales and standard buffers for seawater. *Deep-Sea Res.*, **20**, 479–491.

Harned, H. S., and B. B. Owen (1958). *The Physical Chemistry of Electrolytic Solutions.* Reinhold, New York.

Hawley, J. E., and R. M. Pytkowicz (1973). Interpretation of pH measurements in concentrated electrolyte solutions. *Mar. Chem.*, **1**, 245–250.

Helgeson, H. C. (1967). Thermodynamics of complex dissociation in aqueous solutions at elevated temperatures. *J. Phys. Chem.*, **71**, 3121–3136.

Henderson, P. (1908). Zur Thermodynamik der Flussigkeitsketten. *Z. Phys. Chem.*, **63**, 325–345.

Hills, G. J. (1969). Pressure coefficients of electrode processes. *Adv. High Press. Res.*, **2**, 226–255.

Hills, G. J. (1972). The physics and chemistry of high pressures. *Symp. Soc. Exp. Biol.*, **26**, 1–26.

Hills, G. J., and P. Ovenden (1966). Electrochemistry at high pressures. *Adv. Electrochem. Electrochem. Eng.*, **4**, 185–248.

Holland, H. D. (1978). *The Chemistry of the Atmosphere and Oceans.* Wiley, New York, 351 pp.

Horn, M. K., and J. A. S. Adams (1966). Computer derived geochemical balances and element abundances. *Geochim. Cosmochim. Acta*, **30**, 279–297.

Hunsberger, J. F. (1976). Electrochemical Series. In *Handbook of Chemistry and Physics*, 56th edition (Ed. R. C. Weast). CRC Press, Cleveland, Ohio, pp. D-141–D-146.

Ives, D. J. G., and G. J. Janz (1961). *Reference Electrodes, Theory and Practice.* Academic Press, London, 651 pp.

Katz, A., and S. Ben-Yaakov (1980). Diffusion of sea water ions. Part II. The role of activity coefficients and ion-pairing. *Mar. Chem.*, **8**, 263–280.

Kay, R. L. (1973). Ionic transport in water and mixed aqueous solvents. In *Water—A Comprehensive Treatise*, Vol. 3 (Ed. F. Franks) Plenum Press, New York, pp. 173–209.

Kester, D. R., I. W. Duedall, D. N. Connors, and R. M. Pytkowicz (1967). Preparation of artificial sea water. *Limnol. Oceanogr.*, **12**, 176–179.

King, E. J. (1965). *Acid–Base Equilibria.* Pergamon Press, Oxford.

Kniewald, Z., and Z. Pučar (1976). Electrophoretic mobilities of $^{22}Na^+$, $^{90}Sr^{2+}$ and $^{36}Cl^-$ ions in concentrated aqueous solutions of some inorganic 1:1, 2:1, 1:2 and 2:2 salts and in seawater. *J. Chem. Soc. Faraday Trans. I*, **72**, 987–995.

Koryta, J. (1975). *Ion-selective Electrodes*. Cambridge University Press, Cambridge.

Krom, M. D., and R. A. Berner (1980). The diffusion coefficients of sulphate, ammonium and phosphate ions in anoxic marine sediments. *Limnol. Oceanogr.*, **25**, 327–337.

Lewis, G. N., and M. Randall (1961). *Thermodynamics* (revised by K. S. Pitzer and L. Brewer). McGraw-Hill, New York.

Leyendekkers, J. V. (1976). *Thermodynamics of Seawater*. Marcel Dekker, New York.

Leyendekkers, J. V. (1979). The viscosity of aqueous electrolyte solutions and the TTG model. *J. Solut. Chem.*, **8**, 853–869.

Leyendekkers, J. V. (1980). Prediction of the heat capacities of sea water and other multicomponent electrolyte solutions from the Tammam–Tait–Gibson model. *Mar. Chem.*, **9**, 25–35.

Li, Y.-H., and S. Gregory (1974). Diffusion of ions in sea water and in deep-sea sediments. *Geochim. Cosmochim. Acta*, **88**, 703–714.

Lowenstam, H. A. (1974). Impact of life on chemical and physical processes. In *The Sea*, Vol. 5 (Ed. E. D. Goldberg). Wiley, New York, pp. 715–796.

MacInnes, D. A. (1939). *The Principles of Electrochemistry*. Reinhold, New York.

MacIntyre, F. (1976). Concentration scales: a plea for physico-chemical data. *Mar. Chem.*, **4**, 205–224.

Martin, J. H. (1970). The possible transport of trace metals via moulted copepod exoskeletons. *Limnol. Oceanogr.*, **15**, 756–761.

Meites, L. (1963). *Handbook of Analytical Chemistry*. McGraw-Hill, New York, pp. 5–101–5–139.

Meites, L. (1965). *Polarographic Techniques*. Wiley–Interscience, New York.

Milazzo, G., and S. Caroli (1978). *Tables of Standard Electrode Potentials*. Wiley, New York.

Millero, F. J. (1971). The molal volume of electrolytes. *Chem. Rev.*, **71**, 147–169.

Millero, F. J. (1972). The partial molal volumes of electrolytes in aqueous solutions. In *Water and Aqueous solutions* (Ed. R. A. Horne). Wiley–Interscience, New York, pp. 519–595.

Millero, F. J. (1974a). The physical chemistry of sea water. *Ann. Rev. Earth Planet. Sci.*, **2**, 101–150.

Millero, F. J. (1974b). Sea water as a multicomponent electrolyte solution. In *The Sea*, Vol. 5 (Ed. E. D. Goldberg). Wiley, New York, pp. 3–80.

Millero, F. J. (1979). Effects of pressure and temperature on activity coefficients. In *Activity Coefficients in Electrolyte Solutions*, Vol. 2 (Ed. R. M. Pytkowicz). CRC Press, Boca Raton, Florida, pp. 63–151.

Millero, F. J., and K. Kremling (1976). The densities of Baltic Sea waters. *Deep-Sea Res.*, **23**, 1129–1138.

Millero, F. J., and W. H. Leung (1976). The thermodynamics of sea water at one atmosphere. *Am. J. Sci.*, **276**, 1035–1077.

Millero, F. J., P. Chetirkin, and F. Culkin (1977). The relative conductivity and density of sea water. *Deep-Sea Res.*, **24**, 315–321.

Moore, R. M. (1978). The distribution of dissolved copper in the eastern Atlantic Ocean. *Earth Planet. Sci. Lett.*, **41**, 461–468.

Moreno, E. C., and R. T. Zahradnik (1973). Calculation of liquid junction potentials. *J. Electrochem. Soc.*, **120**, 641–643.

Morf, W. E. (1977). Calculation of liquid-junction potentials and membrane potentials on the basis of the Planck Theory. *Anal. Chem.*, **49**, 810–813.

Morris, J. C., and W. Stumm (1966). Redox equilibria and measurements of potentials in the aquatic environment. In *Equilibrium Concepts in Natural Water Systems* (Ed. W. Stumm). Am. Chem. Soc. Adv. Chem. Ser., No. 67. American Chemical Society, Washington, D.C., pp. 270–285.

Nakahara, M. (1974). Derivation of El'Yanov and Hamann's empirical formula and another new formula for pressure dependence of ionization constants. *Rev. Phys. Chem. Jap.* **44**, 57–64.

Nieboer, E., and D. H. S. Richardson (1980). The replacement of the nondescript term 'heavy metals' by a biologically and chemically significant classification of metal ions. *Environ. Pollut. (Ser. B)*, **1**, 3–26.

Oldham, K. B. (1975). Composition of the diffuse double layer in sea water and other ionic media containing ionic species of +2, +1, −1 and −2 charge types. *J. Electroanal. Chem.*, **63**, 139–156.

Opekar, F., and P. Beran (1976). Rotating disc electrodes. *J. Electroanal. Chem.*, **69**, 1–105.

Park, P. K. (1969). Oceanic CO_2 systems: an evaluation of ten methods of investigation. *Limnol. Oceanogr.*, **14**, 179–186.

Parker, V. B. (1965). Thermal properties of aqueous uni-univalent electrolytes. *Nat. Stand. Ref. Data Ser. Nat. Bur. Std. (U.S.)*, NSRDS-NBS-2. US Dept. of Commerce, Washington, D.C.

Parsons, R. (1975). The role of oxygen in redox processes in aqueous solutions. In *The Nature of Seawater* (Ed. E. D. Goldberg). Dahlem Konferenzen, Berlin, pp. 505–522.

Pearson, R. G. (1969). Hard and soft acids and bases. *Survey Progr. Chem.*, **5**, 1–52.

Phillips, C. S. G., and R. J. P. Williams (1965). *Inorganic Chemistry*, Vol. 1. Clarendon Press, Oxford, pp. 314–321.

Pitzer, K. S., and J. J. Kim (1974). Thermodynamics of electrolytes. IV. Activity and osmotic coefficients for mixed electrolytes. *J. Am. Chem. Soc.*, **96**, 5701–5707.

Pottel, R. (1973). Dielectric properties. In *Water—A Comprehensive Treatise*, Vol. 3 (Ed. F. Franks). Plenum Press, New York, pp. 401–431.

Pourbaix, M. J. N. (1966). *Atlas of Electrochemical Equilibrium in Aqueous Solutions* (translated by J. A. Franklin). Pergamon Press, Oxford.

Pytkowicz, R. M. (1967). Carbonate cycle and the buffer mechanism of recent oceans. *Geochim. Cosmochim. Acta*, **31**, 63–73.

Pytkowicz, R. M. (1968). The carbon dioxide–carbonate system at high pressures in the oceans. *Oceanogr. Mar. Biol. Ann. Rev.*, **6**, 83–135.

Pytkowicz, R. M., and D. R. Kester (1971). The physical chemistry of sea water. *Oceanogr. Mar. Biol. Ann. Rev.*, **9**, 11–60.

Quint, J., and A. Viallard (1978a). The electrophoretic effect for the case of electrolyte mixtures. *J. Solut. Chem.* **7**, 525–531.

Quint, J., and A. Viallard (1978b). Electrical conductance of electrolyte mixtures of any type. *J. Solut. Chem.*, **7**, 533–548.

Redfield, A. C., B. H. Ketchum, and F. A. Richards (1963). The influence of organisms on the composition of sea water. In *The Sea*, Vol. 2 (Ed. M. N. Hill). Wiley, New York, pp. 26–77.

Reeburgh, W. S. (1972). Processes affecting gas distributions in estuarine sediments. In *Environmental Framework of Coastal Plain Estuaries* (Ed. B. W. Nelson). The Geological Society of America, Boulder, Colorado, pp. 383–389.

Reilley, C. N., and R. W. Murray (1963). In *Treatise in Analytical Chemistry*, Part I, Vol. 4 (Eds. I. M. Kolthoff and P. J. Elving). Wiley, New York, pp. 2163–2232.

Riley, J. P., and G. Skirrow, Eds. (1975). *Chemical Oceanography*, 7 volumes. Academic Press, London.

Riley, J. P., and R. Chester (1971). *Introduction to Marine Chemistry*. Academic Press, London, 465 pp.

Ringbom, A. (1963). *Complexation in Analytical Chemistry*. Interscience, New York.

Robinson, R. A., and R. H. Stokes (1970). *Electrolyte Solutions*. Butterworths, London.

Rock, P. A. (1967). The use of the Henderson equation to estimate the EMF difference between cells with and without liquid junction. *Electrochim. Acta*, **12**, 1531–1535.

Rösler, H. J., and H. Lange (1972). *Geochemical Tables*. Elsevier, Amsterdam, 468 pp.

Schindler, P. W. (1975). Removal of trace metals from the oceans: a zero order model. *Thalassia Jugosl.*, **11**, 101–111.

Sclater, F., E. Boyle, and J. M. Edmond (1976). On the marine geochemistry of nickel. *Earth Planet. Sci. Lett.*, **31**, 119–128.

Sillén, L. G. (1959). Graphic presentation of equilibrium data. In *Treatise on Analytical Chemistry*, Part I, Vol. 1 (Eds. I. M. Kolthoff and P. J. Elving). Wiley–Interscience, New York, pp. 277–317.

Sillén, L. G. (1961). The physical chemistry of sea water. In *Oceanography* (Ed. M. Sears). Publ. 67, American Association for the Advancement of Science, New York, pp. 43–56.

Sillén, L. G. (1965). The oxidation state of Earth's ocean and atmosphere. I. A model calculation on earlier states. The myth of the 'probiotic soup'. *Ark. Kemi*, **24**, 431–456.

Sillén, L. G. (1966). Oxidation state of Earth's ocean and atmosphere. II. The behaviour of Fe, S and Mn in earlier states. Regulating mechanisms for O_2 and N_2. *Ark. Kemi*, **25**, 159–176.

Sillén, L. G. (1967). The oceans as a chemical system. *Science*, **156**, 1189–1197.

Sillén, L. G., and A. E. Martell, (1964). *Stability Constants of Metal-ion Complexes*. Spec. Publ. No. 17, Chemical Society, London, 754 pp.

Sillén, L. G., and A. E. Martell (1971). *Stability Constants of Metal-ion Complexes*. *Supplement No. 1*. Spec. Publ. No. 25, Chemical Society, London, 865 pp.

Silvester, L. F., and K. S. Pitzer (1978). Thermodynamics of electrolytes. X. Enthalpy and the effect of temperature on the activity coefficients. *J. Solut. Chem.*, **7**, 327–337.

Sinn, E. (1974). High pressure in coordination chemistry. *Coord. Chem. Rev.*, **12**, 185–220.

Skirrow, G. (1975). The dissolved gases—carbon dioxide. In *Chemical Oceanography*, Vol. 2 (Eds. J. P. Riley and G. Skirrow). Academic Press, London, pp. 1–192.

Stumm, W. (1967). Redox potential as an environmental parameter; conceptual significance and operational limitation. *Adv. Water Pollut. Res.*, **1**, 283–307.

Stumm, W. (1978). What is the pe of sea water? *Thalassia Jugosl.*, **14**, 197–208.

Stumm, W., and P. A. Brauner (1975). Chemical speciation. In *Chemical Oceanography*, Vol. 1 (Eds. J. P. Riley and G. Skirrow). Academic Press, London, pp. 173–239.

Stumm, W., and J. J. Morgan (1970). *Aquatic Chemistry*. Wiley–Interscience, New York.

Timimi, B. A. (1974). The evaluation of ΔC_p^0 for acid–base equilibria from pK measurements. *Electrochim. Acta*, **19**, 149–158.

Turner, D. R., and M. Whitfield (1979). Control of sea water composition. *Nature*, **281**, 468–469.

Turner, D. R., A. G. Dickson, and M. Whitfield (1980). Water–rock partition coefficients and the composition of natural waters—a reassessment. *Mar. Chem.*, **9**, 211–218.

Turner, D. R., M. Whitfield, and A. G. Dickson (1981). The equilibrium speciation of dissolved components in fresh water and sea water at 25°C and 1 atmosphere pressure. *Geochim. Cosmochim. Acta* (in press).

Wallace, W. J. (1974). *The Development of the Salinity Concept in Oceanography*. Elsevier, Amsterdam.

Walton Smith, F. G. (1974). *Handbook of Marine Science*, Vol. I. CRC Press, Cleveland, Ohio.

Wedepohl, K. H. (1979). *Handbook of Geochemistry*, Vol. 1. Springer-Verlag, Berlin, pp. 394–395.

Weiss, R. F. (1974). Carbon dioxide in water and sea water: the solubility of a non-ideal gas. *Mar. Chem.*, **2**, 203–215.

Whalley, E. (1966). Chemical reactions in solutions under pressure. *Ber. Bunsenges. Phys. Chem.*, **70**, 958–1112.

Whitfield, M. (1969). *Eh* as an operational parameter in estuarine studies. *Limnol. Oceanogr.*, **14**, 547–558.

Whitfield, M. (1974a). The ion-association model and the buffer capacity of the carbon dioxide system in sea water at 25°C and 1 atmosphere total pressure. *Limnol. Oceanogr.*, **19**, 235–248.

Whitfield, M. (1974b). Thermodynamic limitations on the use of the platinum electrode in *Eh* measurements. *Limnol. Oceanogr.*, **19**, 857–865.

Whitfield, M. (1975a). The effects of temperature and pressure on speciation. In *The Nature of Seawater* (Ed. E. D. Goldberg). Dahlem Konferenzen, Berlin, pp. 137–164.

Whitfield, M. (1975b). Sea water as an electrolyte solution. In *Chemical Oceanography*, Vol. 1 (Eds. J. P. Riley and G. Skirrow). Academic Press, London, pp. 43–171.

Whitfield, M. (1975c). The electroanalytical chemistry of sea water. In *Chemical Oceanography*, Vol. 4 (Eds. J. P. Riley and G. Skirrow). Academic Press, London, pp. 1–154.

Whitfield, M. (1975d). The extension of chemical models for sea water to include trace components at 25°C and 1 atmosphere total pressure. *Geochim. Cosmochim. Acta*, **39**, 1545–1557.

Whitfield, M. (1976). The evolution of the oceans and the atmosphere. In *Environmental Physiology of Animals* (Eds. J. Bligh, J. L. Cloudsley-Thompson and A. G. Macdonald). Blackwells, Oxford, pp. 30–45.

Whitfield, M. (1979a). The mean oceanic residence time (MORT) concept—a rationalisation. *Mar. Chem.*, **8**, 101–123.

Whitfield, M. (1979b). Activity coefficients in natural waters. In *Activity Coefficients in Electrolyte Solutions*, Vol. 2 (Ed. R. M. Pytkowicz). CRC Press, Boca Raton, Florida, pp. 153–299.

Whitfield, M. (1981). The world ocean—mechanism or machination? *Interdiscipl. Sci. Rev.* **6**, 20–45.

Whitfield, M., and D. R. Turner, (1979a). Water-rock partition coefficients and the composition of sea water and river water. *Nature*, **278**, 132–137.

Whitfield, M., and D. R. Turner (1979b). Critical assessment of the relationship between biological, thermodynamic and electrochemical availability. In *Chemical Modelling in Aqueous Systems* (Ed. E. A. Jenne). ACS Symposium Series, No. 93, American Chemical Society, Washington, D.C., pp. 657–680.

Wilde, P. (1971). *Formulae of Various Properties of Seawater as Functions of Temperature, Pressure and Concentration.* Paper COOE 1–5, University of California, Berkeley, Calif.

Williams, A. F. (1979). *A Theoretical Approach to Inorganic Chemistry.* Springer, Berlin.

Wong, G. T. F. (1980). The oxidation state diagram—a potential tool for studying redox chemistry in sea water. *Mar. Chem.*, **9**, 1–12.

Zobell, C. E. (1946). Studies on redox potential of marine sediments. *Bull. Am. Ass. Pet. geol.*, **30**, 477–513.

Marine Electrochemistry
Edited by M. Whitfield and D. Jagner
© 1981 John Wiley & Sons Ltd.

D. R. TURNER and M. WHITFIELD

Marine Biological Association of the UK,
The Laboratory, Citadel Hill,
Plymouth PL1 2PB, UK.

2

The Classification of Electroanalytical Techniques

GLOSSARY OF SYMBOLS

a	Potential scan rate
b	Amplitude of a.c. modulation
c	Concentration of electroactive species
E	Potential of working electrode
E_d	Deposition potential
E_{dc}	D.c. component of potential
E_{ecm}	Potential of the electrocapillary maximum
E_o	Initial potential
f	Frequency of a.c. modulation
G	Conductance
i	Current at working electrode
$i_{ac}^{(f)}$	A.c. component of current at frequency f
i_C	Capacitance current*
i_F	Faradaic current*
i_l	Limiting current

*Section 2.3; all other terms introduced in Table 1 and Figure 5.

i_p	Peak current
J_d	Electrodeposition flux
k_{diss}	Dissociation rate constant for the complex ML
K_{ML}	Stability constant for the complex ML
L^{y-}	Ligand anion
M^{z+}	Metal cation
$M(Hg)$	Metal amalgam
Q	Quantity of material electrodeposited
S	Salinity
t	Time
t_d	Time of electrodeposition
Δi	Differential current
δ_i	Charging current displacement
δ	Diffusion layer thickness
θ	Surface coverage
τ	Transition time in potentiometric stripping

2.1. INTRODUCTION

The basic unit in electrochemical measurements is a cell in which two or more conducting electrodes are brought into contact with the solution under investigation. By applying a suitable current or voltage excitation across a pair of electrodes in such a cell, it is possible to generate a signal that is related to the concentration of selected species within the solution. By varying the material and configuration of the electrodes, the excitation waveform and the method of measuring the cell response the whole range of electroanalytical techniques can be realized. In this chapter we intend to introduce a simple classification procedure which provides a logical framework for the rather bewildering array of electrochemical techniques now available. In particular we shall try to rationalize the wide range of electrode systems and excitation techniques that have been used in voltammetric studies.

2.2. THE IUPAC CLASSIFICATION SCHEME

A systematic approach to the classification and nomenclature of electroanalytical techniques has been proposed by the International Union of Pure and Applied Chemistry (IUPAC, 1975). A simplified version of the IUPAC classification scheme (Figure 1), restricted to techniques that have found application in the marine context, indicates that the major criteria involved are the nature of the electrode process and the nature of the applied excitation. A third criterion related to the manipulation of the recorded data has, as yet, made little impact on electrochemical studies of sea water (Figure 1). Details of the individual techniques and references to typical applications are given in Table 1.

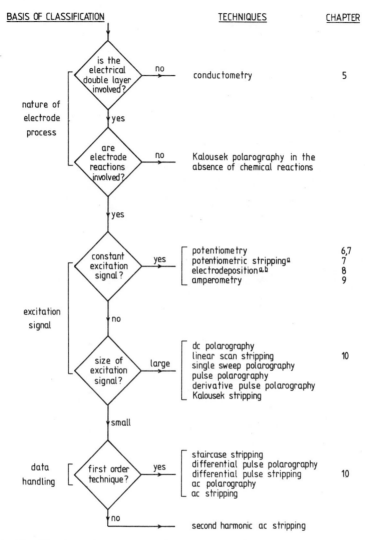

Figure 1. Classification of electroanalytical techniques (IUPAC, 1975). [a] Techniques not specifically mentioned in IUPAC scheme; [b] process preceding all stripping techniques

2.2.1. Procedures not involving specific electrode reactions

The chemical complexity of sea water (Chapter 1) would seem to suggest that only procedures with a high degree of chemical selectivity would find general application. Nonetheless, the most widely used electrochemical procedure in oceanographic studies is conductometry (Chapter 5). The

Table 1. Characteristics of electroanalytical techniques used in natural water chemistry. Presentation adapted from IUPAC (1975). Techniques are grouped according to the criteria given in Figure 1. Symbols are explained in the glossary.

Technique	Excitation signal	Measured response	Independent variable*	References
Conductometry	Alternating voltage	Conductance, G	Salinity, S	See Chapter 5
Kalousek polarography in the absence of electrode reactions			Surface coverage, θ	Ćosović and Branica (1973); Ćosović et al. (1977); Kozarac et al. (1976); Žutić et al. (1977)
Potentiometry	$i = 0$	E	Log c	See Chapter 6
Potentiometric stripping†	$i = 0$ Convective mass transport of oxidizing material to the electrode		Q	Granéli et al. (1980); Jagner (1978); Jagner and Årén (1978, 1979); Jagner and Granéli (1976); see also Chapters 4 and 7
Electrodeposition‡	$E = $ constant Convective mass transport	Quantity deposited $Q = J_d t_d$	c	Batley and Matousek (1977): Lund et al. (1977); Lund and Larsen (1974); Thomassen et al. (1976); see also Chapter 8
Amperometry	$E = $ constant Convective mass transport	i	c	Kester et al. (1973); Lambert (1974); Lambert et al. (1973); Van Landingham and Greene (1971); see also Chapters 3 and 9

Technique	Signal		Response	References
D.c. polarography	Linear voltage ramp $E = E_0 \pm at$		c	Davison (1976, 1977a,b); Luther and Meyerson (1978); Luther et al. (1978)
Linear scan stripping voltammetry†	Linear voltage ramp $E = E_0 \pm at$		Q	Allen et al. (1970); Ariel and Eisner (1963); Ariel et al. (1964); Batley and Florence (1976a,b); Batley and Gardner (1978); Bradford (1972); Branica et al. (1976); Brezonik et al. (1976); Bubić et al. (1973); Florence (1970, 1971, 1974); Florence and Farrar (1974); Fukai and Huynh-Ngoc (1976); Gardiner and Stiff (1974); Gilbert (1971); Gilbert and Hume (1973); Hoshika et al. (1977); Huynh-Ngoc (1973); Jagner and Kryger (1975); Kremling (1973); Laser and Ariel (1974); Lieberman and Zirino (1974); Lund and Salberg (1975); Macchi (1964); Matson (1968); Salim and Cooksey (1979); Schieffer and Blaedel (1978); Seelig and Blount (1979); Seitz (1970); Seitz et al. (1973); Šinko and Doležal (1970); Sipos et al. (1978); Smith and Redmond (1971); Sugai and Healy (1978);

Marine Electrochemistry

Table 1. (contd.)

Technique	Excitation signal	Measured response	Independent variable*	References
				Vydra and Nghi (1977); Wang and Ariel (1977a,b, 1978); Whitnack and Sasselli (1969); Zirino and Healy (1971, 1972); Zirino and Lieberman (1975)
Single sweep polarography	Linear voltage ramp on single drop of dropping mercury electrode $E = E_0 \pm at$	i_p (vs E)	c	Afghan et al. (1972); Whitnack (1966); Whitnack and Brophy (1969)
Pulse polarography	measure i; drop fall (E vs t)	(vs E)	c	Davison and Gabbutt (1979)
Derivative pulse polarography	measure i_1; measure i_2; drop fall; $\Delta i = i_2 - i_1$ (E vs t)	i_p (Δi vs E)	c	Abdullah and Royle (1972); Harvey and Dutton (1973); Milner et al. (1961); Petek and Branica (1969)

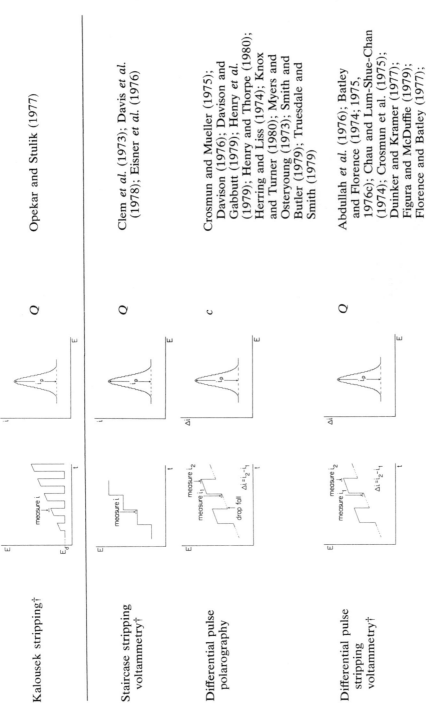

Kalousek stripping†

Opekar and Stulik (1977)

Staircase stripping voltammetry†

Clem *et al.* (1973); Davis *et al.* (1978); Eisner *et al.* (1976)

Differential pulse polarography

Crosmun and Mueller (1975); Davison (1976); Davison and Gabbutt (1979); Henry *et al.* (1979); Henry and Thorpe (1980); Herring and Liss (1974); Knox and Turner (1980); Myers and Osteryoung (1973); Smith and Butler (1979); Truesdale and Smith (1979)

Differential pulse stripping voltammetry†

Abdullah *et al.* (1976); Batley and Florence (1974; 1975, 1976c); Chau and Lum-Shue-Chan (1974); Crosmun *et al.* (1975); Duinker and Kramer (1977); Figura and McDuffie (1979); Florence and Batley (1977);

Table 1. (contd.)

Technique	Excitation signal	Measured response	Independent variable*	References
				Gillain et al. (1979); Guy et al. (1975); Kantin (1977); Lund and Onshus (1976); Magier and Branica (1977); Nürnberg (1977); Seelig and Blount (1979); Sipos et al. (1977); Valenta et al. (1977); Zirino and Lieberman (1975)
A.c. polarography	Sinusoidal modulation on linear voltage ramp $E_{dc} = E_0 \pm at$ $E = E_{dc} + b \sin(2\pi t/f)$		c	Odier and Plichon (1970)
A.c. stripping voltammetry†	$E_{dc} = E_0 \pm at$ $E = E_{dc} + b \sin(2\pi t/f)$		Q	Batley and Florence (1974); Rojahn (1972); Seitz (1970); Velghe and Claeys (1971)

2nd harmonic a.c. stripping voltammetry†	$E_{dc} = E_0 \pm at$ $E = E_{dc} + b \sin(2\pi t/f)$		Q	Batley and Florence (1974); Kopanica and Stara (1977)

* The independent variable in each case is directly proportional to the parameter indicated under 'measured response'.

† Stripping techniques immediately follow an electrodeposition process. The quantity Q is directly proportional to the concentration c of electroactive species in the sample.

‡ The quantity Q may be measured either by an electrochemical stripping technique or by non-electrochemical analysis. The references given here related to electrodeposition followed by non-electrochemical analysis.

simplest design of conductometric cell consists of two identical platinized platinum electrodes immersed in the solution, although most modern oceanographic instruments use an inductively coupled system in which there is no direct contact between the electrodes and the solution.

Despite the non-selective nature of conductometric measurements, the constancy of the relative concentrations of the major sea salt components (see Chapter 1) enables the conductivity of sea water to be related very precisely to its total salt content and hence to its density. *In situ* conductivity measurements have been developed to a high degree of sophistication and every set of observations on the chemistry, physics, or biology of the oceans is accompanied by a series of conductivity (salinity) measurements. The only other example of the use of chemically non-specific procedures in marine studies is the application of Kalousek polarography (Table 1) for the determination of surface-active organic components arising from biological activity within the oceans (Section 1.4.2) or from terrestrial run-off. The polarographic cell is described briefly in Section 2.3. All other procedures show some degree of chemical selectivity and are most usefully classified in terms of the excitation signal applied to the cell (Figure 1, Table 1).

2.2.2. Procedures involving specific electrode reactions and constant excitation signals

Potentiometric methods differ qualitatively from the other techniques utilizing electrode reactions since they are equilibrium measurements involving no net current flow. The working cell in this case consists of two electrodes: the working electrode (or ion-selective electrode, ISE) and the reference electrode (RE). The ISE is designed so that it responds directly to the activity of the selected ion in solution whereas the potential of the RE is assumed to remain constant during the period of measurement. Although in principle the cell responds to the activity of the selected ion, the thermodynamic interpretation of the cell potentials is complicated by the behaviour of the reference electrode. Problems related to the solubility of the ion-selective membranes restrict the direct use of potentiometric methods to ion concentrations in excess of 10^{-6} M in unbuffered solutions, although a theoretical response can be obtained in buffered systems down to 10^{-14} M in some instances. In general, the application of potentiometric procedures in marine chemistry has been confined to the major components (Figure 2) and to the determination of pH (Chapter 6). Frequently titrimetric procedures are required (Chapter 7) to overcome the sensitivity and selectivity problems associated with the available ion-selective electrodes.

Of the other procedures involving a constant excitation signal, amperometry is the only technique that has been widely used in oceanographic

Figure 2. Potentiometric techniques in the analysis of natural waters. References: 1, Almgren *et al.* (1975); Ben-Yaakov and Kaplan (1968, 1971); Ben-Yaakov and Ruth (1974); Ben-Yaakov *et al.* (1974); Park (1966); Takahashi *et al.* (1970); Zirino (1975). 2, *Gast and Thompson (1958)*. 3, *Almgren et al. (1977)*. 4, *Almgren and Fonselius (1976)*; Culberson *et al.* (1970); *Edmond (1970)*. 5, Ammonium: Garside *et al.* (1978); Gilbert and Clay (1973); Merks (1975); Srna *et al.* (1973). 6, *Anfält and Jagner (1971)*; Kosov *et al.* (1976); Rix *et al.* (1976); Warner (1971, 1973). 7, Garrels (1967). 8, *Lebel and Poisson (1976)*; *Whitfield et al. (1969)*. 9, *Jagner (1967)*. 10, Sulphide: Berner (1963); Conti (1972); Conti and Wilde (1972, 1973); *Green and Schitker (1976)*; Whitfield (1971). Sulphate: *Mascini (1973)*; *Ouzounian and Michard (1978)*. 11, *Jagner and Årén (1970)*; *Wilson (1975)*. 12, *Anfält and Jagner (1973)*. 13, Bradford (1968); *Dyrssen et al. (1968)*; Lebel and Poisson (1975). References in italics describe multiple standard addition or titration techniques. For a summary of conventional single-ion activity coefficients determined using ion-selective electrodes see Chapter 6, Table 23

studies. The development of membrane-covered electrodes for the amperometric determination of oxygen (Chapter 9) has led to the design of *in situ* probes (Chapters 3 and 9) capable of recording fine details in the oxygen profiles that reveal the influence of the biota on the recycling of the elements (see Section 1.4.2). Here the cell itself, usually consisting of a noble metal cathode (gold or platinum) and a silver/silver chloride anode immersed in a potassium chloride electrolyte, is separated from the solution under examination by a gas-permeable membrane. The partial pressure of oxygen in the thin film of electrolyte next to the cathode quickly equilibrates with that of the sample solution. A constant potential is applied across the cell and the current flowing is directly proportional to the concentration of oxygen in the electrolyte adjacent to the electrode surface and hence in the bulk solution. After the conductometric salinity probe and the glass pH electrode, this is the most commonly used electrochemical sensor in marine studies.

Electrodeposition techniques for the pre-concentration of trace metals from sea water prior to analysis are still largely in the developmental stage (Table 1, Chapter 8), although electrodeposition prior to electrochemical analysis (via anodic stripping voltammetry, ASV) has provided the basis for a very useful range of procedures for the determination of amalgam-forming metals (Chapter 10). These procedures employ time-varying excitation signals in the subsequent stripping process.

2.2.3. Procedures involving specific electrode reactions and time-varying excitation signals

Although the measurement of salinity, pH, and oxygen concentration still dominate the oceanographic use of electroanalytical techniques, undoubtedly the most rapid developments have been witnessed in the use of voltammetric methods for studying trace metal chemistry. Some difficulties arise in the classification of these techniques because they are distinguished only by the shape of the potential–excitation waveform and by the timing of the measurements (Table 1). As the use of microcomputer- and microprocessor-controlled instruments becomes more widespread (Chapter 4), the number of alternative excitation and measurement procedures can be expected to increase and to develop into a continuum of methods causing new problems in classification and nomenclature. This is particularly true

Table 2. Summary of main electroanalytical techniques used in natural systems

Type	Characteristic	Elements measured	Useful limit in sea water
Conductometry	Measurement of total ionic content	Salinity	$0.1‰$ *
Potentiometry	Equilibrium measurement	Major and minor elements (see Figure 2)	10^{-6} M
Direct polarographic techniques	Kinetic measurement	Minor and trace elements (see Figure 3)	10^{-8} M
Stripping techniques	2-stage kinetic measurement	Trace elements (see Figure 4)	10^{-11} M

* Application restricted by the range of validity of the salinity concept in estuarine waters.

where complex schemes can be used to apply several different excitation techniques in a single voltage scan (see, for example, Bond *et al.*, 1979). Because of these difficulties and also because of the wide range of electrode systems that have been used in voltammetric studies, these techniques will be considered in detail in the next section. The four broad categories of electroanalytical techniques that we have discussed are summarized in Table 2.

2.3. POLAROGRAPHIC AND VOLTAMMETRIC TECHNIQUES

2.3.1. Classification

IUPAC (1975) has recommended that the term polarography be reserved for the measurement of current–voltage curves at a liquid electrode whose surface is continuously renewed, while the term voltammetry is reserved for the measurement of current–voltage curves at solid or stationary electrodes. We have accordingly divided these techniques into two categories in Table 2: direct polarographic techniques (all at the dropping mercury electrode) and stripping techniques (at a variety of electrodes including the hanging mercury drop electrode). Once again the classification recommended by IUPAC is being challenged by technical development: this time by interrupted flow mercury electrodes, such as the Static Mercury Drop Electrode (Princeton Applied Research, Princeton, N.J., USA), which have many of the characteristics associated with polarographic systems but have a discontinuously renewed mercury surface. Whatever the nature of the working electrode, voltammetric procedures involve the application of a time-varying potential across the cell. In most applications a three-electrode cell is employed (Sawyer and Roberts, 1974), which consists of a working electrode (WE), a counter or auxiliary electrode (CE or AE) and a reference electrode (RE). A potentiostat maintains the required potential difference between the WE and RE by the application of a potential at the CE. The current flowing at the working electrode is then monitored as a function of the WE–RE potential difference. Different electroactive components will be discharged at different potentials and from the current *versus* potential (or time) curves it is possible not only to identify the individual components but also (after suitable calibration) to measure their concentrations. The current produced when a potential is applied across the electrode/solution interface (Section 1.3.3) includes not only the current associated with the reduction of electroactive species (the faradaic current, i_F) but also the current associated with the charging of the double-layer capacitance (the capacitance current, i_C). The purpose of most of the modulations applied to the voltage ramp (Table 1) is to reduce to a minimum the contribution of i_C to the observed current.

In a medium such as sea water the very low concentrations of many elements ensure that the detection limit, determined by the ability to discriminate faradaic from capacitance current, is the major restriction on the analytical applications of direct polarography. Although only a few elements in sea water can be determined without pre-concentration, even with modulated techniques (Table 1), applications are more widespread in estuarine, fresh, and anoxic waters where the concentrations of many components are higher (Figure 3; see also Davison and Whitfield, 1977). The other major requirement is, of course, the occurrence of a suitable electrode process within the potential window offered by the mercury electrode. In stripping analysis the elements of interest are deposited electrochemically on the working electrode at constant potential and are subsequently stripped off by the action of a voltage ramp. This pre-concentration step makes these techniques much more sensitive than direct polarography

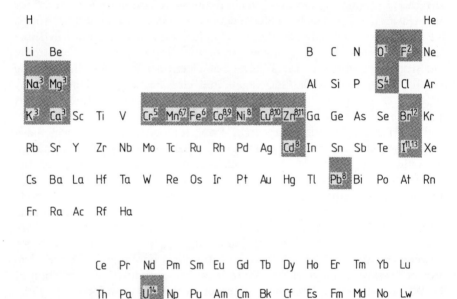

Figure 3. Polarographic techniques in the analysis of natural waters. References: 1, Davison (1977a,b); see also Chapter 9. 2, *Berge and Brugman (1972)*. 3, *Berge and Brugman (1969)*. 4, Sulphide: Davison (1977b); Davison and Gabbutt (1979). Sulphate: *Berge and Brugman (1970); Luther and Meyerson (1975); Luther et al. (1978)*. 5, Crosmun and Mueller (1975). 6, Davison (1976). 7, Knox and Turner (1980). 8, *Abdullah and Royle (1972)*. 9, *Harvey and Dutton (1973)*. 10, Odier and Plichon (1970). 11, Petek and Branica (1969). 12, *Berge and Brugman (1971)*. 13, Herring and Liss (1974); Smith and Butler (1979); Truesdale and Smith (1979). 14, *Milner et al. (1961)*. References in italics describe indirect polarographic determinations or polarography following chemical pre-concentration

so that detection limits are less of a restriction on their use but the demands on the electrode process are much more specific. Only in a very few cases is stripping from a solid (usually gold) electrode a reliable analytical process so a mercury microelectrode is normally employed (Table 3). The electrode process must therefore be a reduction which results in a product (elemental metal) with a significant solubility in mercury and the overall process must be reversible or nearly so if a well defined stripping peak is to be obtained. Within these requirements stripping techniques have been applied to the analysis of 11 metals in natural waters (Figure 4), most of which are significant as micronutrient or toxic elements.

While polarography is, by definition, restricted to a single electrode system, a variety of electrode systems have been employed in stripping analysis (Table 3). The relatively large volume and limited sensitivity of the hanging mercury drop electrode led to the development of the thin mercury film electrode, which has been adapted for use in rotating electrode and flow-through systems. In addition to the use of modulated waveforms to increase sensitivity, alternative electrode designs have been adopted, including differential systems (twin electrodes and rotating split-disc, Table 3) and 'downstream collection' systems (twin tubular electrodes and the rotating ring-disc, Table 3).

2.3.2. Chemical selectivity

The fact that each electroactive species has its own characteristic discharge potential at the mercury electrode confers a degree of chemical selectivity on polarographic techniques and, at the electrodeposition stage, on other voltammetric techniques. When the electroactive species are simple hydrated metal ions this enables the concentrations of several metals to be determined simultaneously (see Chapter 10), although problems associated with peak overlap and with the formation of intermetallic compounds impose restrictions (Section 10.5.2).

If the trace metal is also involved in complexation reactions (Section 1.4.1), then complexes that are able to dissociate to release the free metal ion within the diffusion layer will also contribute to the measured current (Figure 5). The ability of a complex to contribute in this way will depend on the diffusion layer thickness (δ, Figure 5), which sets the timescale of the electrochemical measurement (Davison, 1978), and on the stability constant (K_{ML}) and dissociation rate constant (k_{diss}) of the complex in question (Turner and Whitfield, 1979a). The proportion of the total metal in solution that is sensed by the voltammetric procedure is known as the electrochemically available fraction. For a given electrochemical procedure the size of this fraction will depend on the nature and extent of the complexation reactions in which the electroactive species is involved. A number of studies

Table 3. Working electrodes employed for anodic stripping voltammetry

Electrode system	References
Mercury drop electrodes	
Slowly dropping mercury electrode, solution stirred during deposition	Branica *et al.* (1976); Bubić *et al.* (1973); Macchi (1964); Velghe and Claeys (1972)
Hanging mercury drop electrode, solution stirred during deposition	Ariel and Eisner (1963); Ariel *et al.* (1964); Batley and Florence (1974, 1975, 1976c); Brezonik *et al.* (1976); Chau and Lum-Shue-Chan (1974); Duinker and Kramer (1977); Figura and McDuffie (1979); Florence and Batley (1977); Gillain *et al.* (1979); Guy *et al.* (1975); Jagner and Granéli (1976); Kantin (1977); Kremling (1973); Lund and Onshus (1976); Nürnberg (1977); Opekar and Stulik (1977); Rojahn (1972); Salim and Cooksey (1979); Šinko and Doležal (1970); Smith and Redmond (1971); Whitnack and Sasselli (1969)
Twin hanging mercury drop electrodes; differential measurement for background subtraction. Metals are accumulated in only one electrode	Zirino and Healy (1971, 1972)
*Mercury film electrodes in stirred solutions**	
Stationary thin mercury film electrode, solution stirred during deposition	Abdullah *et al.* (1976); Allen *et al.* (1970); Bradford (1972); Crosmun *et al.* (1975); Gardiner and Stiff (1974); Gilbert (1971); Gilbert and Hume (1973); Hoshika *et al.* (1977); Huynh-Ngoc (1973); Jagner (1978); Jagner and Årén (1979); Jagner and Kryger (1975); Matson (1968); Seelig and Blount (1979); Seitz (1970); Sugai and Healy (1978)
Stationary electrode in rotating cell	Clem *et al.* (1973); Vydra and Nghi (1977)
Stationary thin mercury film electrode with vibrating stirrer	Magjer and Branica (1977)
*Mercury film electrodes in flowing solutions**	
Stationary thin mercury film electrode in flow-through cell	Wang and Ariel (1977a)

Table 3. (contd.)

Electrode system	References
Twin stationary thin mercury film electrodes in flow-through cell, differential measurement	Wang and Ariel (1977b)
Rotating thin mercury film electrode in flow-through cell	Wang and Ariel (1978)
Tubular electrode surrounding flowing stream	Lieberman and Zirino (1974); Seitz *et al.* (1973); Zirino and Lieberman (1975)
Twin tubular electrodes surrounding flowing stream. Metal stripped from the upstream electrode and the collection current monitored at the downstream electrode	Schieffer and Blaedel (1978)
Rotating mercury film electrodes*	
Rotating disc electrode	Batley and Florence (1974; 1976a, b, c); Batley and Gardner (1978); Branica *et al.* (1976); Bubić *et al.* (1973); Eisner *et al.* (1976); Florence (1970, 1971, 1974); Florence and Farrar (1974); Florence and Batley (1977); Granéli *et al.* (1980); Jagner and Årén (1978); Kopanica and Stara (1977); Lund and Onshus (1976); Lund and Salberg (1975); Nürnberg (1977); Opekar and Stulik (1977); Valenta *et al.* (1977); Vydra and Nghi (1977)
Rotating split-disc electrode (differential measurement)	Sipos *et al.* (1978)
Rotating ring-disc electrode. Metal stripped from the disc electrode and the collection current monitored at the ring electrode	Laser and Ariel (1974)
Other electrodes	
Stationary graphite electrode (for Hg)	Fukai and Huynh-Ngoc (1976)
Rotating gold split-disc electrode (for Hg)	Sipos *et al.* (1977)
Stationary gold film electrode (for As)	Davis *et al.* (1978)

*Mercury film electrodes are generally prepared on a substrate of glassy carbon, although a range of treated graphite materials have also been used.

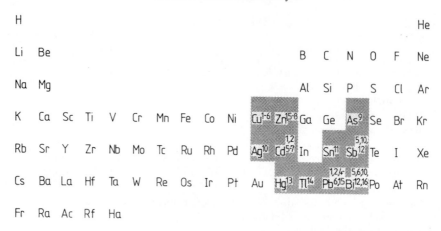

H																	He
Li	Be											B	C	N	O	F	Ne
Na	Mg											Al	Si	P	S	Cl	Ar
K	Ca	Sc	Ti	V	Cr	Mn	Fe	Co	Ni	Cu[1-6]	Zn[1,5-8]	Ga	Ge	As[9]	Se	Br	Kr
Rb	Sr	Y	Zr	Nb	Mo	Tc	Ru	Rh	Pd	Ag[10]	Cd[5-7][1,2]	In	Sn[11]	Sb[12][5,10]	Te	I	Xe
Cs	Ba	La	Hf	Ta	W	Re	Os	Ir	Pt	Au	Hg[13]	Tl[14]	Pb[6,15][1,2,4]	Bi[12,16][5,6,10]	Po	At	Rn
Fr	Ra	Ac	Rf	Ha													

Ce	Pr	Nd	Pm	Sm	Eu	Gd	Tb	Dy	Ho	Er	Tm	Yb	Lu
Th	Pa	U	Np	Pu	Am	Cm	Bk	Cf	Es	Fm	Md	No	Lw

Figure 4. Stripping voltammetry in the analysis of natural waters. References: 1, Abdullah *et al.* (1976); Allen *et al.* (1970); Ariel *et al.* (1964); Branica *et al.* (1976); Bubić *et al.* (1973); Chau and Lum-Shue-Chan (1974); Duinker and Kramer (1977); Gardiner and Stiff (1974); Huynh-Ngoc (1973); Jagner and Årén (1979); Kantin (1977); Kremling (1973); Rojahn (1972); Seitz (1970); Whitnack and Sasselli (1969); Zirino and Healy (1972); Zirino and Lieberman (1975). 2, Batley and Florence (1976a,b,c); Batley and Gardner (1978); Hoshika *et al.* (1977); Lund and Salberg (1975); Lund and Onshus (1976); Magjer and Branica (1977); Nürnberg (1977); Salim and Cooksey (1979); Sipos *et al.* (1978); Valenta *et al.* (1977). 3, Florence and Batley (1977). 4, Sugai and Healy (1978). 5, Gillain *et al.* (1979). 6, Florence (1971); Jagner and Kryger (1975). 7, Ariel and Eisner (1963). 8, Bradford (1972); Lieberman and Zirino (1974); Macchi (1964); Zirino and Healy (1971). 9, *Davis et al. (1978).* 10, Gilbert (1971). 11, *Florence and Farrar (1974).* 12, Gilbert and Hume (1973). 13, Fukai and Huynh-Ngoc (1976); Sipos *et al.* (1977). 14, *Batley and Florence (1975).* 15, Seelig and Blount (1979). 16, Florence (1974). References in italics describe ASV following chemical pre-concentration

have taken advantage of this chemical sensitivity to provide useful operational definitions of the chemical forms of the amalgam-forming trace metals in natural water samples (Table 4; see also Section 10.7). However, the difficulties of relating data obtained in this way to the chemical speciation of trace metals (i.e. to the distribution of clearly defined solution components, Section 1.4.1) become apparent on a closer perusal of Figure 5. The reversible process (unshaded area) in this scheme can be modelled in a multi-ligand system such as sea water (Turner and Whitfield, 1979b), but a great deal of data is required for the calculations. In particular, data on the

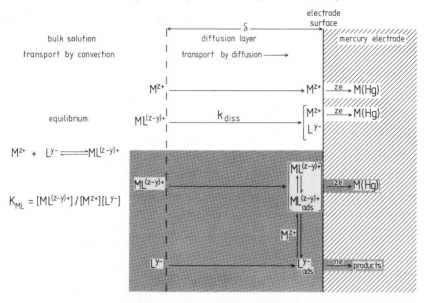

Figure 5. Processes occurring during the reduction of a metal M^{z+} in the presence of a ligand L^{y-}. Where only reversible electrode processes occur (unshaded area) the electrochemically available fraction is determined by the values of diffusion layer thickness δ, stability constant K_{ML}, complex dissociation rate constant k_{diss}, and the diffusion coefficients of the various species (Turner and Whitfield, 1979a,b). Detailed discussions of the effects of complexation on polarography and voltammetry can be found in standard texts (Heyrovsky and Kůta, 1966; Crow, 1969; Galus, 1976). The processes shown in the shaded area (irreversible reduction of complex and ligand, surface adsorption and complexation) constitute interferences in the measurement. More detailed discussions of this type of scheme can be found in the papers of Buffle and co-workers (Buffle *et al.*, 1976; Buffle, 1979; Buffle and Greter, 1979)

kinetics of complex association and dissociation are very sparse for many systems of interest in marine chemistry. More data are available in the literature for association constants, K_{ML} (Turner *et al.*, 1981), and over the last few years this has been supplemented by electrochemical measurements of K_{ML} in model solutions related to sea water. In cases where the rate constant, k_{diss}, for complex dissociation is fast on the timescale set by (Figure 5), K_{ML} can be estimated from the dependence of the reduction potential on ligand concentration (Crow, 1969). Some measurements of this type are listed in Table 5. Experiments in model solutions have also shown clearly the influence of side-reactions in determining the extent of interaction of organic ligands with trace metals in sea water (Raspor *et al.*, 1978).

Studies of this type have been carried out in model solutions partly because of the difficulties of working in a medium with many competing ligands, but also because of the interferences which may be encountered in

Table 4. Applications of polarography and stripping voltammetry to studies of chemical form in natural water analysis

Procedure	References
Measurement of pseudopolarograms* of untreated natural samples	Branica *et al.* (1976)
Measurement of electrochemically available fraction in untreated natural sample	Chau and Lum-Shue-Chan (1974); Bradford (1972); Duinker and Kramer (1977); Florence and Batley (1977); Knox and Turner (1980); Zirino and Healy (1971)
Titration of trace metal into untreated sample to measure complexation capacity†	Chau *et al.* (1974); Duinker and Kramer (1977); Florence and Batley (1977)
Study of metal distribution by combination of electrochemistry and other techniques	Abdullah *et al.* (1976); Allen *et al.* (1970); Batley and Florence (1976a,b); Batley and Gardner (1978); Salim and Cooksey (1979)
Study of distribution of oxidized and reduced species	Crosmun and Mueller (1975); Liss *et al.* (1973); Myers *et al.* (1973); Smith and Butler (1979)
Study of the dissociation rate of complexes with naturally occurring organic ligands	Shuman and Michael (1978)

* Plots of stripping peak current *versus* deposition potential (Bubić and Branica, 1973; Shuman and Cromer, 1979; Turner and Whitfield, 1979a; Zirino and Kounaves, 1977).
† O'Shea and Mancy (1976); Shuman and Woodward (1973).

Table 5. Measurements of stability constants by polarography and voltammetry

Complexes	Technique*	Medium	References
$CdCl^+$	ASV	$NaCl/NaClO_4$, 0.7 M	Bubić and Branica
$CdCl_2$	ASV	$NaCl/NaClO_4$, 2 M	(1973)
$CuCO_3$, $PbCO_3$ Cu, Pb + humic acid	DPP/ASV	KNO_3, 0.1 M	Ernst *et al.* (1975)
$PbOH^+$, $Pb(OH)_2$ $PbCO_3$, $Pb(CO_3)_2^{2-}$ $CuCO_3$, $Cu(CO_3)_2^{2-}$ $CdCO_3$, $ZnCO_3$	DPP/ASV	KNO_3, 0.1 M	Bilinski *et al.* (1976)
$PbCO_3$, $Pb(CO_3)_2^{2-}$	ASV	$NaClO_4$, 0.7 M	Sipos *et al.* (1980)

* ASV = anodic stripping voltammetry; DPP = differential pulse polarography.

untreated natural waters. Where large surface active ligands (e.g. humic and fulvic acids) are encountered it is likely that additional reactions will occur at the electrode surface (stippled region, Figure 5). The metal complex may itself be adsorbed at the electrode surface and the metal discharged from the adsorbed complex rather than directly from solution. Alternatively, the ligand may be adsorbed at the electrode surface and become involved in electron exchange reactions in its own right. Studies in natural and model systems have demonstrated the occurrence of interference by such processes (see, for example, Batley and Florence, 1976c; Zur and Ariel, 1977; Buffle and Greter, 1979; Greter *et al.*, 1979; van Leeuwen, 1979; Turner and Whitfield, 1979b).

2.4. CONCLUSIONS

This brief survey indicates the range of electrochemical techniques that have been applied to the study of marine chemistry, and the IUPAC scheme itself provides a logical framework describing the relationships between the various procedures. At this point some useful general references may be introduced that will provide more detailed discussion of the various techniques. A number of reviews have been presented on the use of potentiometric (Warner, 1975) and voltammetric (Davison and Whitfield, 1977; Buffle *et al.*, 1976; Nürnberg and Valenta, 1975; Buffle, 1979) procedures in the study of natural waters and on the general problems associated with the use of electroanalytical procedures in sea water (Whitfield, 1975). Useful practical details on the design and construction of electrochemical cells will be found in the book by Sawyer and Roberts (1974) and the theoretical basis for the polarographic and voltammetric methods has been lucidly reviewed by Galus (1976). The following chapters will consider in more detail the practical aspects of the application of electroanalytical techniques to the investigation of oceanographic problems.

REFERENCES

Abdullah, M. I., O. A. El-Rayis, and J. P. Riley, (1976). Re-assessment of chelating ion-exchange resins for trace metal analysis of seawater. *Anal. Chim. Acta*, **84,** 363–368.

Abdullah, M. I., B. Reusch Berg, and R. Klimek (1976). The determination of zinc, cadmium, lead and copper in a single sea-water sample by differential pulse anodic stripping voltammetry. *Anal. Chim. Acta*, **84,** 307–317.

Abdullah, M. I., and L. G. Royle (1972). The determination of copper, lead, cadmium, nickel, zinc and cobalt in natural waters by pulse polarography. *Anal. Chim. Acta*, **58,** 283–288.

Afghan, B. K., P. D. Goulden and J. F. Ryan (1972). Automated method for determination of nitrilotriacetic acid in natural water, detergents and sewage samples. *Anal. Chem.*, **44,** 354–359.

Allen, H. E., W. R. Matson, and K. H. Mancy (1970). Trace metal characterisation in aquatic environments by anodic stripping voltammetry. *J. Water Pollut. Control Fed.*, **42**, 573–581.

Almgren, T., and S. H. Fonselius (1976). Determination of alkalinity and total carbonate. In *Methods of Sea Water Analysis* (Ed. K. Grasshoff). Verlag Chemie, Weinheim, pp. 97–115.

Almgren, T., D. Dyrssen, and M. Strandberg (1975). Determination of pH on the moles per kg sea water scale (M_W). *Deep-Sea Res.*, **22**, 635–646.

Almgren, T., D. Dyrssen, and M. Strandberg (1977). Computerised high-precision titrations of some major constituents of sea water on board R. V. Dmitry Mendeleev. *Deep-Sea Res.*, **24**, 345–364.

Anfält, T., and D. Jagner (1971). A standard addition method for the potentiometric determination of fluoride in sea water. *Anal. Chim. Acta*, **53**, 13–22.

Anfält, T., and D. Jagner (1973). The potentiometric titration of potassium in sea water with a valinomycin electrode. *Anal. Chim. Acta*, **66**, 152–155.

Ariel, M., and U. Eisner (1963). Trace analysis by anodic stripping voltammetry. I. Trace metals in Dead Sea brine. 1. zinc and cadmium. *J. Electroanal. Chem.*, **5**, 362–374.

Ariel, M., U. Eisner, and S. Gottesfield (1964). Trace analysis by anodic stripping voltammetry. II. The method of medium exchange. *J. Electroanal. Chem.*, **7**, 307–314.

Batley, G. E., and T. M. Florence (1974). An evaluation and comparison of some techniques of anodic stripping voltammetry. *J. Electroanal. Chem.*, **55**, 23–43.

Batley, G. E., and T. M. Florence (1975). Determination of thallium in natural waters by anodic stripping voltammetry. *J. Electroanal. Chem.*, **61**, 205–211.

Batley, G. E., and T. M. Florence (1976a). A novel scheme for the classification of heavy metal species in natural waters. *Anal. Lett.*, **9**, 379–388.

Batley, G. E., and T. M. Florence (1976b). Determination of the chemical forms of dissolved cadmium, lead and copper in seawater. *Mar. Chem.*, **4**, 347–363.

Batley, G. E., and T. M. Florence (1976c). The effect of dissolved organics on the stripping voltammetry of seawater. *J. Electroanal. Chem.*, **72**, 121–126.

Batley, G. E., and D. Gardner (1978). A study of copper, lead and cadmium speciation in some estuarine and coastal marine waters. *Estuarine Coastal Mar. Sci.*, **7**, 59–70.

Batley, G. E., and J. P. Matousek (1977). Determination of heavy metals in sea-water by atomic absorption spectrometry after electrodeposition on pyrolytic graphite-coated tubes. *Anal. Chem.*, **49**, 2031–2035.

Ben-Yaakov, S., and I. R. Kaplan (1968). pH-temperature profiles in ocean and lakes using an *in situ* probe. *Limnol. Oceanogr.*, **13**, 688–693.

Ben-Yaakov, S., and I. R. Kaplan (1971). Deep-sea *in situ* calcium carbonate saturometry. *J. Geophys. Res.*, **96**, 722–731.

Ben-Yaakov, S., and E. Ruth (1974). An improved *in situ* pH sensor for oceanographic and limnological applications. *Limnol. Oceanogr.*, **19**, 144–151.

Ben-Yaakov, S., E. Ruth, and I. R. Kaplan (1974). Calcium carbonate saturation in northeastern Pacific: *in situ* determination and geochemical implications. *Deep-Sea Res.*, **21**, 229–243.

Berge, H., and L. Brugman (1969). Möglichkeiten zur polarographischen Bestimmung einiger Hauptkomponenten in Meerwasser. *Beitr. Meereskd.*, **26**, 47–57.

Berge, H., and L. Brugman (1970). Die indirekte polarographische Bestimmung von Sulfationen in Meerwasser. *Beitr. Meereskd.*, **27**, 5–13.

Berge, H., and L. Brugman (1971). Polarographische Methoden zur Bestimmung von Bromidionen in Meerwasser. *Beitr. Meereskd.*, **28**, 19–32.

Berge, H., and L. Brugman (1972). Indirekte Bestimmung von Fluoridionen in Meerwasser durch Wechselströmpolarographie. *Beitr. Meereskd.*, **29**, 115–127.

Berner, R. A. (1963). Electrode studies of hydrogen sulphide in marine sediments. *Geochim. Cosmochim. Acta*, **27**, 563–575.

Bilinski, H., R. Huston, and W. Stumm (1976). Determination of the stability constants of some hydroxo and carbonato complexes of Pb(II), Cu(II), Cd(II) and Zn(II) in dilute solutions by anodic stripping voltammetry and differential pulse polarography. *Anal. Chim. Acta*, **84**, 157–164.

Bond, A. M., B. S. Grabaric, R. D. Jones, and N. W. Rumble (1979). Theory of a.c. differential pulse polarography—a waveform for simultaneously obtaining d.c., a.c. and pulse polarograms. *J. Electroanal. Chem.*, **100**, 625–640.

Bradford, W. L. (1968). Calcium analysis in seawater by an ion sensitive electrode. *MS Thesis*, Oregon State University, Corvallis, Ore.

Bradford, W. L. (1972). A study on the chemical behaviour of zinc in Chesapeake Bay water using anodic stripping voltammetry. *Tech. Rep. Chesapeake Bay Inst.*, No. 76.

Branica, M., L. Sipos, S. Bubić, and S. Kozar (1976). Electroanalytical determination and characterisation of some heavy metals in seawater. In *Accuracy in Trace Analysis: Sampling, Sample Handling and Analysis*, NBS Special Publication, No. 422, pp. 917–928.

Brezonik, P. L., P. A. Brauner, and W. Stumm (1976). Trace metal analysis by asv: effect of sorption by natural and model organic compounds. *Water Res.*, **10**, 605–612.

Bubić, S., and M. Branica (1973). Voltammetric characterisation of the ionic state of cadmium present in seawater. *Thalassia Jugosl.*, **9**, 47–53.

Bubić, S., L. Sipos, and M. Branica (1973). Comparison of different electroanalytical techniques for the determination of heavy metals in seawater. *Thalassia Jugosl.*, **9**, 55–63.

Buffle, J. (1979). Le rôle des méthodes électrochimiques en analyse des eaux. *Cebedeau*, **426**, 165–176.

Buffle, J., F.-L. Greter, G. Nembrini, J. Paul, and W. Haerdi (1976). Capabilities of voltammetric techniques for water control problems. *Z. Anal. Chem.*, **282**, 339–350.

Buffle, J., and F.-L. Greter (1979). Voltammetric study of humic and fulvic substances. Part II. Mechanism of reactions of the Pb–fulvic complexes on the mercury electrodes. *J. Electroanal. Chem.*, **101**, 231–251.

Chau, Y. K., R. Gachter, and K. Lum-Shue-Chan (1974). Determination of the apparent complexing capacity of lake waters. *J. Fish. Res. Bd. Can.*, **31**, 1515–1519.

Chau, Y. K., and K. Lum-Shue-Chan (1974). Determination of labile and strongly bound metals in lake water. *Water Res.*, **8**, 383–388.

Clem, R. G., G. Litton, and L. D., Ornelas (1973). New cell for rapid anodic stripping analysis. *Anal. Chem.*, **45**, 1306–1317.

Conti, U. (1972). Study of an underwater environmental monitor. *PhD Thesis*, University of California, Berkeley, Calif.

Conti, U., and P. Wilde (1972). Diver-operated *in situ* electrochemical measurements. *Mar. Tech. Soc. J.*, **6**, 17–23.

Conti, U., and P. Wilde (1973). Self-contained towable underwater system for environmental monitoring. In *Oceans 73, Proc. I.E.E.E. Conference on Engineering in Ocean Environment, Seattle.*

Ćosović, B., and M. Branica (1973). Study of the adsorption of organic substances at a mercury electrode by the Kalousek commutator technique. *J. Electroanal. Chem.*, **46**, 63–69.

Ćosović, B., V. Žutić, and Z. Kozarac (1977). Surface active substances in the sea surface microlayer by electrochemical methods. *Croat. Chim. Acta*, **50**, 229–241.

Crosmun, S. T., J. A. Dean, and J. R. Stokely (1975). Pulsed anodic stripping voltammetry of zinc, cadmium and lead with a mercury-coated wax-impregnated graphite electrode. *Anal. Chim. Acta*, **75**, 421–430.

Crosmun, S. T., and T. R. Mueller (1975). The determination of chromium(VI) in natural waters by differential pulse polarography. *Anal. Chim. Acta*, **75**, 199–205.

Crow, D. R. (1969). *Polarography of Metal Complexes*, Academic Press, London.

Culberson, C. H., R. M. Pytkowicz, and J. E. Hawley (1970). Sea-water alkalinity determination by the pH method. *J. Mar. Res.*, **28**, 15–21.

Davis, P. H., G. R. Dulude, R. M. Griffin, W. R. Matson, and E. W. Zink (1978). Determination of total arsenic at the nanogram level by high-speed anodic stripping voltammetry. *Anal. Chem.*, **50**, 137–143.

Davison, W. (1976). Comparison of differential pulse and d.c. sampled polarography for the determination of ferrous and manganous ions in lake water. *J. Electroanal. Chem.*, **72**, 229–237.

Davison, W. (1977a). Sampling and handling procedures for the polarographic measurement of oxygen in hypolimnetic waters. *Freshwater Biol.*, **7**, 393–401.

Davison, W. (1977b). The polarographic measurement of O_2, Fe^{2+}, Mn^{2+} and S^{2-} in hypolimnetic water. *Limnol. Oceanogr.*, **22**, 746–753.

Davison, W. (1978). Defining the electroanalytically measured species in a natural water sample. *J. Electroanal. Chem.*, **87**, 395–404.

Davison, W., and C. D. Gabbutt (1979). Polarographic methods for measuring uncomplexed sulphide ions in natural waters. *J. Electroanal. Chem.*, **99**, 311–320.

Davison, W., and M. Whitfield (1977). Modulated polarographic and voltammetric techniques in the study of natural water chemistry. *J. Electroanal. Chem.*, **75**, 763–789.

Duinker, J. C., and C. J. M. Kramer (1977). An experimental study on the speciation of dissolved zinc, cadmium, lead and copper in River Rhine and North-Sea water by differential pulse anodic stripping voltammetry. *Mar. Chem.*, **5**, 207–228.

Dyrssen, D., D. Jagner, and H. Johansson (1968). On the potentiometric titration of calcium in sea-water using a calcium selective membrane electrode. *Reports on the Chemistry of Sea-Water*, V. Dept. of Analytical Chemistry and Marine Chemistry, University of Göteborg, Sweden.

Edmond, J. M. (1970). High precision determination of titration alkalinity and total carbon dioxide content of seawater by potentiometric titration. *Deep-Sea Res.*, **17**, 737–750.

Eisner, U., J. A. Turner, and R. A. Osteryoung (1976). Staircase voltammetric stripping at thin film mercury electrodes. *Anal. Chem.*, **48**, 1608–1610.

Ernst, R., H. E. Allen, and K. H. Mancy (1975). Characterisation of trace metal species and measurement of trace metal stability constants by electrochemical techniques. *Water Res.*, **9**, 969–979.

Figura, P., and B. McDuffie (1979). Use of chelex resin for determination of labile trace metal fractions in aqueous ligand media and comparison of the method with anodic stripping voltammetry. *Anal. Chem.*, **51**, 120–125.

Florence, T. M., (1970). Anodic stripping voltammetry with a glassy carbon electrode mercury plated *in situ*. *J. Electroanal. Chem.*, **27**, 273–281.

Florence, T. M. (1971). Determination of trace metals in marine samples by anodic stripping voltammetry. *J. Electroanal. Chem.*, **35**, 237–245.

Florence, T. M. (1974). Determination of bismuth in marine samples by anodic stripping voltammetry. *J. Electroanal. Chem.*, **49**, 255–264.

Florence, T. M., and G. E. Batley (1977). Determination of copper in sea-water by anodic stripping voltammetry. *J. Electroanal. Chem.*, **75**, 791–798.

Florence, T. M., and Y. J. Farrar (1974). Determination of tin by thin film anodic stripping voltammetry. Application to marine samples. *J. Electroanal. Chem.*, **51**, 191–200.

Fukai, R., and L. Huynh-Ngoc (1976). Direct determination of mercury in sea-water by anodic stripping voltammetry with a graphite electrode. *Anal. Chim. Acta*, **83**, 375–379.

Galus, Z. (1976). *Fundamentals of Electrochemical Analysis*, Ellis Horwood, Chichester.

Gardiner, J., and M. J. Stiff (1974). The determination of cadmium, lead, copper and zinc in ground water, estuarine water and sewage effluent by anodic stripping voltammetry. *Water Res.*, **9**, 517–523.

Garrels, R. M. (1967). Ion-sensitive electrodes and individual ion activity coefficients. In *Glass Electrodes for Hydrogen and Other Cations* (Ed. G. Eisenman). Marcel Dekker, New York.

Garside, C., G. Hull, and S. Murray (1978). Determination of submicromolar concentrations of ammonia in natural waters by a standard addition method using a gas-sensing electrode. *Limnol. Oceanogr.*, **23**, 1073–1076.

Gast, J. A., and T. G. Thompson (1958). Determination of alkalinity and borate concentration of sea-water. *Anal. Chem.*, **30**, 1549–1551.

Gilbert, T. R. (1971). Electrochemical studies of environmental trace metals. *PhD Thesis*, Massachusetts Institute of Technology.

Gilbert, T. R., and D. N. Hume (1973). Direct determination of bismuth and antimony in seawater by anodic stripping voltammetry. *Anal. Chim. Acta*, **65**, 451–459.

Gilbert, T. R., and A. M. Clay (1973). Determination of ammonia in aquaria and in seawater using the ammonia electrode. *Anal. Chem.*, **45**, 1757–1759.

Gillain, G., G. Duyckaerts, and A. Distèche (1979). Direct and simultaneous determination of Zn, Cd, Pb, Cu, Sb and Bi dissolved in seawater by differential pulse anodic stripping voltammetry with a hanging mercury drop electrode. *Anal. Chim. Acta*, **106**, 23–37.

Granéli, A., D. Jagner, and M. Josefsson (1980). Microcomputer system for potentiometric stripping analysis. *Anal. Chem.*, **52**, 2220–2223.

Green, E. J., and D. Schitker (1974). The direct titration of water soluble sulfides in estuarine muds of Montsweag Bay, Maine. *Mar. Chem.*, **2**, 111–126.

Greter, F.-L., J. Buffle, and W. Haerdi (1979). Voltammetric study of humic and fulvic substances. Part I. Study of the factors influencing the measurement of their complexing properties with lead. *J. Electroanal Chem.*, **101**, 211–229.

Guy, R. D., C. L. Chakrabarti, and L. L. Schramm (1975). The application of a simple chemical model of natural waters to metal fixation in particulate matter. *Can. J. Chem.*, **53**, 661–669.

Harvey, B. R., and J. W. R. Dutton (1973). The application of photo-oxidation to the determination of stable cobalt in seawater. *Anal. Chim. Acta*, **67**, 377–385.

Henry, F. T., T. O. Kirch, and T. M. Thorpe (1979). Determination of trace level arsenic(III), arsenic (V), and total inorganic arsenic by differential pulse polarography. *Anal. Chem.*, **51**, 215–218.

Henry, F. T., and T. M. Thorpe (1980). Determination of arsenic(III), arsenic(V), monomethylarsonate and dimethylarsonate by differential pulse polarography after separation by ion exchange chromatography. *Anal. Chem.*, **52**, 80–83.

Herring, J. R., and P. S. Liss (1974). A new method for the determination of iodine species in seawater. *Deep-Sea Res.*, **21**, 777–783.

Heyrovsky, J., and J. Kůta (1966). *Principles of Polarography*, Academic Press, New York.

Hoshika, A., O. Takimura, and T. Shiozawa (1977). Determination of cadmium, lead and copper in interstitial water by anodic stripping voltammetry. *J. Oceanogr. Soc. Jap.*, **33**, 161–164.

Huynh-Ngoc, L. (1973). L'application de l'électrode indicatrice "graphite–mercure" en redissolution anodique par voltammétrie pour la détermination simultanée du zinc, du cadmium, du plomb et du cuivre dans l'environment aquatique. *Thesis*, University of Nice.

IUPAC, 1975. Classification and nomenclature of electroanalytical techniques. *Pure Appl. Chem.*, **45**, 81–97.

Jagner, D. (1967). A potentiometric titration of magnesium in seawater.*Reports on the Chemistry of SeaWater, IV.* Dept. of Analytical Chemistry and Marine Chemistry, University of Göteborg, Sweden.

Jagner, D. (1978). Instrumental approach to potentiometric stripping analysis of some heavy metals. *Anal. Chem.*, **50**, 1924–1929.

Jagner, D., and K. Årén (1970). A rapid semi-automatic method for the determination of the total halide concentration in sea water by means of potentiometric titration. *Anal. Chim. Acta*, **52**, 491–499.

Jagner, D., and K. Årén (1978). Derivative potentiometric stripping analysis with a thin film of mercury on a glassy carbon electrode. *Anal. Chim. Acta*, **100**, 375–378.

Jagner, D., and K. Årén (1979). Potentiometric stripping analysis for zinc, cadmium, lead and copper in seawater. *Anal. Chim. Acta*, **107**, 29–35.

Jagner, D., and A. Granéli (1976). Potentiometric stripping analysis. *Anal. Chim. Acta*, **83**, 19–26.

Jagner, D., and L. Kryger (1975). Computerised electroanalysis. Part III. Multiple scanning anodic stripping and its application to seawater. *Anal. Chim. Acta*, **80**, 255–266.

Kantin, R. (1977). Utilisation de la polarographie impulsionelle à redissolution anodique pour le dosage simultané de Cu(II), Pb(II), Cd(II) et Zn(II) en milieu marin. *Téthys*, **7**, 419–425.

Kester, D. R., K. T. Crocker, and G. R. Miller (1973). Small scale oxygen variations in the thermocline. *Deep-Sea Res.*, **20**, 409–412.

Knox, S., and D. R. Turner (1980). Polarographic measurement of manganese(II) in estuarine waters. *Estuarine Coastal Mar. Sci.*, **10**, 317–324.

Kopanica, M., and V. Stara (1977). Stripping determination of traces of lead using second harmonic alternating voltammetry and rotating disc electrode. *J. Electroanal. Chem.*, **77**, 57–65.

Kosov, A. E., P. D. Novikov, O. T. Krylov, M. P. Nesterova, and A. F. Litvinova (1976). A potentiometric method for the determination of F⁻ in seawater. *Oceanologiya*, **16**, 5–15.

Kozarac, Z., B. Ćosović, and M. Branica (1976). Estimation of surfactant activity of polluted sea-water by Kalousek commutator technique. *J. Electroanal. Chem.*, **68**, 75–83.

Kremling, K. (1973). Voltammetrische Nersungen über die Verteilung von Zink, Cadmium, Blei und Kupfer in der Ostsee. *Kiel. Meeresforsch.*, **29**, 77–84.

Lambert, R. B. (1974). Small-scale dissolved oxygen variations and the dynamics of Gulf Stream eddies. *Deep-Sea Res.*, **21**, 529–546.

Lambert, R. B., D. R. Kester, M. E. Q. Pilson, and K. E. Kenyon (1973). *In situ* dissolved oxygen measurements in the north and west Atlantic Ocean. *J. Geophys. Res.*, **78**, 1479–1483.

Laser, D., and M. Ariel (1974). Anodic stripping with collection using thin mercury films. *J. Electroanal. Chem.*, **49**, 123–132.

Lebel, J., and A. Poisson (1975). Dosage potentiométrique du calcium dans les eaux de mer. *C. R. Acad. Sci. Paris, Ser. D*, **280**, 2533–2538.

Lebel, J., and A. Poisson (1976). Potentiometric determination of calcium and magnesium in sea water. *Mar. Chem.*, **4**, 321–332.

van Leeuwen, H. P. (1979). Complications in the interpretation of pulse polarographic data on complexation of heavy metals with natural polyelectrolytes. *Anal. Chem.*, **51**, 1322–1323.

Lieberman, S. H., and A. Zirino (1974). Anodic stripping voltammetry of zinc in seawater with a tubular mercury–graphite electrode. *Anal. Chem.*, **46**, 20–23.

Liss, P. S., J. R. Herring, and E. D. Goldberg (1973). The iodide/iodate system in seawater as a possible measure of redox potential. *Nature Phys. Sci.*, **242**, 108–109.

Lund, W., and B. V. Larsen (1974). The application of electrodeposition techniques to flameless atomic absorption spectroscopy. Part II. Determination of cadmium in seawater. *Anal. Chim. Acta*, **72**, 57–62.

Lund, W., and D. Onshus (1976). The determination of copper, lead and cadmium in sea water by differential pulse anodic stripping voltammetry. *Anal. Chim. Acta*, **86**, 109–122.

Lund, W., and M. Salberg (1975). Anodic stripping voltammetry with the Florence mercury film electrode. Determination of copper, lead and cadmium in seawater. *Anal. Chim. Acta*, **76**, 131–141.

Lund, W., Y. Thomassen, and P. Dovle (1977). Flame atomic absorption analysis for trace metals after electrochemical preconcentration on a wire filament. *Anal. Chim. Acta*, **93**, 53–60.

Luther, G. W., and A. L. Meyerson (1975). Polarographic analysis of sulphate ion in seawater samples. *Anal. Chem.*, **47**, 2058–2059.

Luther, G. W., A. L. Meyerson, and A. D'addio (1978). Voltammetric methods of sulphate ion analysis in natural waters. *Mar. Chem.*, **6**, 117–124.

Macchi, G. (1964). The determination of ionic zinc in seawater by anodic stripping voltammetry using ordinary capillary electrodes. *J. Electroanal. Chem.*, **9**, 290–298.

Magjer, T., and M. Branica (1977). A new electrode system with efficient mixing of electrolyte. *Croat. Chim. Acta*, **49**, L1–L5.

Mascini, M. (1973). Titration of sulphate in mineral waters and seawater using the solid-state lead electrode. *Anal. Chim. Acta*, **38**, 325–328.

Matson, W. R. (1968). Trace metals, equilibrium and kinetics of trace metal complexes in natural media. *Thesis*, Massachusetts Institute of Technology.

Merks, A. G. A. (1975). Determination of ammonia in seawater with an ion-selective electrode. *Neth. J. Sea Res.*, **9**, 371–375.

Milner, G. W. C., J. D. Wilson, G. A. Barnett, and A. A. Smales (1961). The determination of uranium in sea water by pulse polarography. *J. Electroanal. Chem.*, **2**, 25–38.

Myers, D. J., M. E. Heimbrook, J. Osteryoung, and S. M. Morrison (1973). Arsenic oxidation in the presence of microorganisms: examination by differential pulse polarography. *Environ. Lett.*, **5**, 53–61.

Myers, D. J., and J. Osteryoung (1973). Determination of arsenic(III) at the parts per billion level by differential pulse polarography. *Anal. Chem.*, **45**, 267–271.

Nürnberg, H. W., and P. Valenta (1975). Polarography and voltammetry in marine chemistry. In *The Nature of Seawater* (Ed. E. D. Goldberg) Dahlem Konferenzen, Berlin, pp. 87–136.

Nürnberg, H. W. (1977). Potentialities and applications of advanced polarographic and voltammetric methods in environmental research and surveillance of toxic metals. *Electrochim. Acta*, **22**, 935–949.

Odier, M., and V. Plichon (1970). Le cuivre en solution dans l'eau de mer: forme chimique et dosage. *J. Electroanal. Chem.*, **55**, 209–220.

Opekar, F., and K. Stulik (1977). Application of Kalousek polarography in stripping analysis. *J. Electroanal. Chem.*, **85**, 207–212.

O'Shea, T. A., and K. H. Mancy (1976). Characterisation of trace metal–organic interactions by anodic stripping voltammetry. *Anal. Chem.*, **48**, 1603–1607.

Ouzounian, G., and G. Michard (1978). Dosage des sulfates dans les eaux naturelles a l'aide d'une electrode selective au baryum. *Anal. Chim. Acta*, **96**, 405–409.

Park, K. (1966). Surface pH of the northeastern Pacific Ocean. *J. Oceanolog. Soc. Korea*, **1**, 1–6.

Petek, M., and M. Branica (1969). Hydrographical and biotical conditions in North Adriatic. III. Distribution of zinc and iodate. *Thalassia Jugosl.*, **5**, 257–261.

Raspor, B., P. Valenta, H. W. Nürnberg, and M. Branica (1978). Chelation of Cd with NTA in seawater as a model for the typical behaviour of trace heavy metal chelates in natural waters. *Sci. Total Environ.*, **9**, 87–109.

Rix, C. J., A. M. Bond, and J. D. Smith (1976). Direct determination of fluoride in seawater with a fluoride selective ion electrode by a method of standard addition. *Anal. Chem.*, **48**, 1236–1239.

Rojahn, T. (1972). Determination of copper, lead, cadmium and zinc in estuarine water by anodic stripping alternating-current voltammetry on the hanging mercury drop electrode. *Anal. Chim. Acta*, **62**, 438–441.

Salim, R., and B. G. Cooksey (1979). The analysis of river water for metal ions (lead, cadmium and copper) both in solution and adsorbed on suspended particles. *J. Electroanal. Chem.*, **105**, 127–141.

Sawyer, D. T., and J. L. Roberts, Jr. (1974). *Experimental Electrochemistry for Chemists*. Wiley, New York.

Schieffer, G. W., and W. J. Blaedel (1978). Anodic stripping voltammetry with collection at tubular electrodes for the analysis of tap water. *Anal. Chem.*, **50**, 99–102.

Seelig, P. F., and H. N. Blount (1979). Application of recursive estimation to the real time analysis of trace metal analytes by linear sweep, pulse and differential pulse anodic stripping voltammetry. *Anal. Chem.*, **51**, 1129–1134.

Seitz, W. R. (1970). Trace metal analysis in seawater by anodic stripping voltammetry. PhD Thesis, Massachusetts Institute of Technology.

Seitz, W. R., R. Jones, L. N. Klatt, and W. D. Mason (1973). Anodic stripping voltammetry at a tubular mercury-covered graphite electrode. *Anal. Chem.*, **45**, 840–844.

Shuman, M. S., and J. L. Cromer (1979). Pseudopolarograms : applied potential-anodic stripping peak current relationships. *Anal. Chem.*, **51**, 1546–1550.

Shuman, M. S., and L. C. Michael (1978). Application of the rotated disc electrode to measurement of copper complex dissociation rate constants in marine coastal samples. *Environ. Sci. Technol.*, **12**, 1069–1072.

Shuman, M. S., and G. P. Woodward (1973). Chemical constants of metal complexes from a complexometric titration followed with anodic stripping voltammetry. *Anal. Chem.*, **45**, 2032–2035.

Šinko, I., and J. Doležal (1970). Simultaneous determination of copper, cadmium, lead and zinc in water by anodic stripping polarography. *J. Electroanal. Chem.*, **25**, 299–306.

Sipos, L., S. Kozar, I. Kontušić, and M. Branica (1978). Subtractive anodic stripping voltammetry with rotating mercury coated glassy carbon electrode. *J. Electroanal. Chem.*, **87**, 347–352.

Sipos, L., P. Valenta, H. W. Nürnberg, and M. Branica (1977). Applications of polarography and voltammetry to marine and aquatic chemistry. IV. A new voltammetric method for the study of mercury traces in sea water and inland waters. *J. Electroanal. Chem.*, **77**, 263–266.

Sipos, L., P. Valenta, H. W. Nürnberg, and M. Branica (1980). Voltammetric determination of the stability constants of the predominant labile lead complexes in seawater. In *Lead–Occurrence, Fate and Pollution in the Marine Environment* (Eds. M. Branica and Z. Konrad). Pergamon Press, Oxford.

Smith, J. D., and E. C. V. Butler (1979). Speciation of dissolved iodine in estuarine waters. *Nature*, **277**, 468–469.

Smith, J. D., and J. D. Redmond (1971) Anodic stripping voltammetry applied to trace metals in seawater. *J. Electroanal. Chem.*, **33**, 169–175.

Srna, R. F., C. Epitanio, M. Hartman, G. Pruder, and A. Stubbs (1973). The use of ion specific electrodes for chemical monitoring of marine systems. I. The ammonia electrode as a sensitive water quality indicator probe for recirculating mariculture systems. *Rep. Coll. Mar. Studi.*, No. DEL-59-14-73, University of Delaware, Newark, Del.

Sugai, S. F., and M. L. Healy (1978). Voltammetric studies of the organic association of copper and lead in two Canadian inlets. *Mar. Chem.*, **6**, 291–308.

Takahashi, T., R. F. Weiss, C. H. Culberson, J. M. Edmond, D. E. Hammond, C. S. Wong, Y. H. Li, and A. E. Bainbridge (1970). A carbonate chemistry profile at the 1969 Geosecs calibration station in the eastern Pacific Ocean. *J. Geophys. Res.*, **75**, 7648–7666.

Thomassen, Y., B. V. Larsen, F. J. Langmyhr, and W. Lund (1976). The application of electrodeposition techniques to flameless atomic absorption spectrometry. Part IV. Separation and preconcentration on graphite. *Anal. Chim. Acta*, **83**, 103–110.

Truesdale, V. W., and C. J. Smith (1979). A comparative study of three methods for the determination of iodate in seawater. *Mar. Chem.*, **7**, 133–139.

Turner, D. R., and M. Whitfield (1979a). The reversible electrodeposition of trace metal ions from multi-ligand systems. Part I. Theory. *J. Electroanal. Chem.*, **103**, 43–60.

Turner, D. R., and M. Whitfield (1979b). The reversible electrodeposition of trace metal ions from multi-ligand systems. Part II. Calculations on the electrochemical availability of lead at trace levels in seawater. *J. Electroanal. Chem.*, **103**, 61–79.

Turner, D. R., M. Whitfield, and A. G. Dickson (1981). The equilibrium speciation of dissolved components in fresh water and seawater at 25°C and 1 atmosphere pressure. *Geochim. Cosmochim. Acta*, (in press).

Valenta, P., L. Mart, and H. Rützel (1977). New potentialities in ultra trace analysis with differential pulse anodic stripping voltammetry. *J. Electroanal. Chem.*, **82**, 327–343.

Van Landingham, J. W., and M. W. Greene (1971). An *in situ* molecular oxygen profiler. *Mar. Tech. Soc. J.*, **5**, 11–23.

Velghe, N., and A. Claeys (1972). The use of a slowly dropping mercury electrode in anodic stripping alternating current voltammetry. *J. Electroanal. Chem.*, **35**, 229–235.

Vydra, F., and T. V. Nghi (1977). Application of a rotating disc electrode and a rotating cell with stationary electrode in stripping voltammetry for the determination of lead and zinc. *Anal. Chim. Acta*, **91**, 335–338.

Wang, J., and M. Ariel (1977a). Anodic stripping voltammetry in a flow-through cell with a fixed mercury film glassy carbon disc electrode. Part I. *J. Electroanal. Chem.*, **83**, 217–224.

Wang, J., and M. Ariel (1977b). Anodic stripping voltammetry in a flow-through cell with fixed mercury film glassy carbon disc electrodes. Part II. The differential mode (DASV). *J. Electroanal. Chem.*, **85**, 289–297.

Wang, J., and M. Ariel (1978). The rotating disc electrode in flowing systems. Part 2. A flow system for automated anodic stripping voltammetry of discrete samples. *Anal. Chim. Acta*, **101**, 1–8.

Warner, T. B. (1971). Normal fluoride content of seawater. *Deep-Sea Res.*, **18**, 1255–1263.

Warner, T. B. (1973). Fluoride analysis in seawater and in other complex natural waters using an ion-selective electrode—techniques, potentialities and limitations. In *Chemical Analysis of the Environment and Other Modern Techniques* (Eds. S. Ahuja, E. M. Cohen, T. J. Kneip, J. L. Lambert, and G. Zweig). Plenum Press, New York.

Warner, T. B. (1975). Ion selective electrodes in thermodynamic studies. In *The Nature of Seawater* (Ed. E. D. Goldberg). Dahlem Konferenzen, Berlin, pp. 191–217.

Whitfield, M. (1971). A compact potentiometric sensor of novel design. *In situ* determination of pH, pS^{2-} and *Eh*. *Limnol. Oceanogr.*, **16**, 829–837.

Whitfield, M. (1975). The electroanalytical chemistry of seawater. In *Chemical Oceanography*, Vol. 4 (Eds. J. P. Riley and G. Skirrow). Academic Press, London pp. 1–154.

Whitfield, M., J. V. Leyendekkers, and J. D. Kerr (1969). Liquid ion-exchange electrodes as end-point detectors in compleximetric titrations. Part II. Determination of calcium and magnesium in the presence of sodium. *Anal. Chim. Acta*, **45**, 399–410.

Whitnack, G. C. (1966). Application of single-sweep polarography to the analysis of trace elements in seawater. In *Polarography 1964*, Vol. 1, (Ed. G. J. Hills). Macmillan, London, pp. 641–651.

Whitnack, G. C., and R. G. Brophy (1969). A rapid and highly sensitive single-sweep polarographic method of analysis for arsenic(III) in drinking water. *Anal. Chim. Acta*, **48**, 123–127.

Whitnack, G. C., and R. Sasselli (1969). Application of anodic stripping voltammetry to the determination of some trace elements in seawater. *Anal. Chim. Acta*, **47**, 267–274.

Wilson, T. R. S. (1971). Salinity and the major elements of seawater. In *Chemical Oceanography*, 2nd edition, Vol. 1 (Eds. J. P. Riley and G. Skirrow). Academic Press, London.

Zirino, A. (1975). Measurement of the apparent pH of sea water with a combination microelectrode. *Limnol. Oceanogr.*, **20**, 654–657.

Zirino, A., and M. L. Healy (1971). Voltammetric measurement of zinc in the northeastern tropical Pacific Ocean. *Limnol. Oceanogr.*, **16**, 773–778.

Zirino, A., and M. L. Healy (1972). pH-controlled differential voltammetry of certain trace transition elements in natural waters. *Environ. Sci. Technol.*, **6**, 243–249.

Zirino, A., and S. P. Kounaves (1977). Anodic stripping peak currents : electrolysis potential relationships for reversible systems. *Anal. Chem.*, **49**, 56–59.

Zirino, A., and S. H. Lieberman (1975). Automated anodic stripping voltammetry for the measurement of copper, zinc, cadmium and lead in seawater. In *Analytical Methods in Oceanography* (Ed. T. R. P. Gibb). Advances in Chemistry Series No. 147, American Chemical Society, Washington, D.C., pp 82–98.

Zur, C., and M. Ariel (1977). The determination of cadmium in the presence of humic acid by anodic stripping voltammetry. *Anal. Chim. Acta*, **88,** 245–251.

Žutić, V., B. Ćosović, and Z. Kozarac (1977). Electrochemical determination of surface active substances in natural waters. On the adsorption of petroleum fractions at mercury electrode/seawater interface. *J. Electroanal. Chem.*, **78,** 113–121.

Marine Electrochemistry
Edited by M. Whitfield and D. Jagner
© 1981 John Wiley & Sons Ltd.

S. BEN-YAAKOV
*Department of Electrical Engineering,
Ben-Gurion University,
Beer-Sheva, Israel*

3

Electrochemical Instrumentation

GLOSSARY OF SYMBOLS

a_Y	Ion activity (equation 1)
B	Constant (equation 7)
E	Electrode potential (equation 1)
$E^{\ominus}$	Standard electrode potential (equation 1)
F	Faraday (equation 1)
I_B	Input bias current (equation 2)
I_T	Current sensitivity of oxygen electrode (equation 5)
J	Constant (equation 5)
K_0, K_F	Constants (equations 4 and 5, respectively)
R_{in}	Amplifier input resistance (equation 2)
R_0	Constant (equation 7)
R_p	Internal resistance (Figure 2)
R_s	Variable resistance (Section 3.2.2)
V_{in}	Voltage input to amplifier (Figure 2)
V_{os}	Offset voltage (equation 2)
V_{out}	Voltage output from amplifier (Figure 2)
V_p	Polarization potential (Figure 5)
ε	Maximum acceptable voltage error (equation 3)

3.1. INTRODUCTION

Marine electrochemistry, by definition, involves the application of electrochemical methods in marine studies. As such, experimental marine electrochemistry is often based on scientific principles and methods originally developed by 'land electrochemists'. Unfortunately, however, the direct application of conventional electrochemical instrumentation is seldom possible, especially when on-site or *in situ* studies are attempted. The complexity of sea water composition, the harsh marine environment, and the constant quest for more precise and accurate data pose special instrumentation challenges to the marine electrochemist. To meet these challenges, conventional electrochemical techniques have been refined and special instrumentation systems developed. Thus, a research programme in experimental marine electrochemistry often involves a considerable instrumentation effort.

This chapter is written with two aims in mind. First, it attempts to present a coherent outline of basic instrumentation concepts and techniques applicable to the solution of marine electrochemistry problems. The discussion is mainly concerned with problems related to on-site and in particular *in situ* instrumentation. A thorough discussion of general electrochemical instrumentation problems can be found, of course, in numerous review articles and texts (e.g. Sawyer, 1974). A second objective of this chapter is to present marine electrochemistry cells and electrodes in electronic engineering terms. This may assist engineers to better understand the special problems of marine electrochemistry and enable them to contribute to their solution.

An experimental marine electrochemistry system will usually encompass at least some of the units shown schematically in Figure 1. A sensor, probing the tested solution in the laboratory, on-site or *in situ*, produces an electrical signal. The signal may be either a self-generated response, such as the

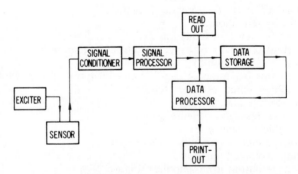

Figure 1. Schematic representation of a general configuration of a marine electrochemical instrumentation system

response of an ion-selective electrode, or an induced response dependent on some external excitation, as in the case of conductance measurements. The signal is then fed to a signal conditioner which transforms it to a high-level signal capable of driving a readout unit, a data transmitter, or a signal processor. The function of the last unit is to transform the raw signal to a more convenient form. This may involve signal amplification, linearization, or more sophisticated processing, such as automatic temperature compensation. The processed signal can then be read directly or stored for further processing. More elaborate data processing can then be done on-line or off-line by a digital computer.

In this chapter an attempt is made to review some of the instrumentation approaches taken by various investigators in the past in designing marine electrochemical studies. The discussion will follow the outline presented schematically in Figure 1. It is unfortunate, however, that many of the design details of the instrumentation used by marine electrochemists have not been reported in the scientific or technical literature. Extremely valuable experience has thus been lost and marine electrochemists often have to 're-invent the wheel' when designing an experiment. Consequently, many of the specific design details discussed in this chapter are examples of approaches taken by the author and his co-workers. An attempt is made, however, to generalize and to sort out and discuss the key problems that proved to be crucial to the success of the experiments.

3.2. SIGNAL CONDITIONERS AND EXCITERS

Apart from the electrochemical sensor itself, the signal conditioner (Figure 1) is the most important unit in determining the accuracy and precision of the electrochemical measurement. For this reason, great care must be taken when specifying the requirements for the signal conditioner. When considering the desired parameters, it is helpful to specify the parameters of the electrochemical sensor in electrical terms. This procedure is required because the parameters of commercially available signal conditioners are defined for electrical sources and loads. The relevant source parameters are signal range, internal resistance, frequency, bandwidth, and acceptable error range. The error is usually referred to the input of the signal conditioner, i.e. compared with the signal level of the sensor. Some of these considerations are discussed in conjunction with the specific examples given below.

3.2.1. D.c. amplifiers (electrometers)

Marine potentiometric studies have mainly involved ion-selective electrodes (Whitfield, 1975). These electrochemical sensors consist of two electrodes, an ion-selective electrode and a reference electrode. The signal is

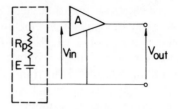

Figure 2. Schematic representation of an ion-selective electrode and signal conditioner. The electrode pair is represented by an equivalent circuit consisting of a
voltage source E and an internal resistance R_p

a potential established between the leads of the two electrodes. From an
electrical engineering point of view the electrode pair can be described as a
voltage source and an internal resistance (Figure 2). The voltage source E
represents the total potential measured between the leads of the electrodes
including the reference electrode potential and the potential of the inner
reference electrode of the ion-selective electrode. The internal resistance R_p
represents the total (d.c.) resistance seen between the electrode leads. The
ideal response of an ion-selective electrode is described by the Nernst
equation (Warner, 1972):

$$E = E^{\ominus}(P, T) + \frac{RT}{nF} \ln a_Y \tag{1}$$

where a_Y is the activity of the ion in the tested solution, P is pressure and R,
T, F, and n have their usual significance.

Assuming that the response of the ion-selective electrode is near ideal, E
(equation 1) is the open-circuit voltage of the cell. However, any amplifier
will load the sensor to some degree and, as a result, the actual voltage V_{in}
(Figure 2) at the input of the d.c. amplifier (electrometer) input will differ in
magnitude from the open-circuit voltage E. This can readily be seen by
considering the equivalent circuit of a practical amplifier (Burr Brown,
1971) shown schematically in Figure 3. As can easily be shown the actual
voltage V_{in} at the amplifier's input is

$$V_{in} = (E + V_{os} + I_B R_p) \frac{R_{in}}{R_{in} + R_p} \tag{2}$$

where V_{os} is the input offset voltage of the amplifier, R_{in} is the amplifier's
input resistance, and I_B is the input bias current.

It is thus evident that in order to minimize the errors introduced by the
amplifier its parameters should obey the relationships

$$R_{in} \gg R_p; \qquad I_B R_p < \varepsilon; \qquad V_{os} < \varepsilon \tag{3}$$

where ε is the maximum acceptable voltage error at the input.

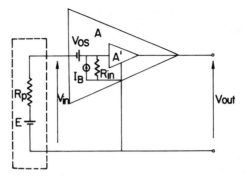

Figure 3. An equivalent circuit of the amplifier shown in Figure 2. V_{os} is the input offset voltage, I_B is the input bias current (represented by a current source), R_{in} is the input resistance, and A′ is an ideal amplifier

The internal resistances of ion-selective electrodes span a relatively large range from a few thousand megohms ($>10^9\ \Omega$) for a glass electrode (Bates, 1964) to a few ohms for solid-state electrodes (Sensorex Inc., Irvine, Calif., USA). Furthermore, the internal resistance R_p is not constant and may depend strongly on temperature. The resistance of glass membrane electrodes increases markedly at low temperature (Bates, 1964), which dictates a very low value for I_B (equation 2) if large errors are to be avoided. On the other hand, a relatively large offset voltage V_{os} could be tolerated because it is compensated for during calibration (in fact it can be considered as part of $E^{\ominus}$ in equation 1). However, the drift of V_{os} with respect to temperature and time must be kept small. Consequently, for situations where the temperature varies the conditions of equation 3 are modified to

$$R_{in} \gg R_p$$

$$I_B R_p < \varepsilon$$

$$\left| \int_{T_1}^{T_2} \frac{dV_{os}}{dT} \cdot \Delta T \right| < \varepsilon$$

$$\left| \int_{t_1}^{t_2} \frac{dV_{os}}{dt} \cdot \Delta t \right| < \varepsilon$$

where T is temperature, t is time and subscripts 1 and 2 signify the range of the experiment.

It is thus evident that the basic specifications of a d.c. amplifier (electrometer) for ion-selective electrodes are high input resistance, low input bias current, and low input offset voltage drift with temperature. Recent advances in electronic engineering have made possible the miniaturization of these types of amplifiers and have thus made practical the design of *in situ* instrumentation for ion-selective electrodes. Some of the design details of *in situ* instrumentation, relevant to the subject matter discussed here, were reported by Ben-Yaakov and Kaplan (1968a,b, 1973), Wilde and Rodgers (1970), Conti and Wilde (1972), and Ben-Yaakov and Ruth (1974).

Wilde and co-workers used a Type 132A or a Type 1302 (Zeltex Inc., Concord, Calif., USA) amplifier as a signal conditioner for ion-selective electrodes. These are FET input operational amplifiers (Burr Brown, 1971) with the following specifications; input resistance, $10^{12}\,\Omega$; input bias current, 5×10^{-11} A; and input offset voltage drift, $20\,\mu$V $^\circ$C^{-1}.

Assuming that the internal resistance of the glass electrode was $10^9\,\Omega$, the input voltage to the amplifier (V_{in} in Figures 2 and 3) is 0.99 of E (equation 1) and the internal voltage drop due to the internal resistance of the electrode and input bias current ($I_B R_p$) is 50 mV. Assuming a 20°C temperature range, the input voltage drift with temperature is 0.4 mV. The major source of error is thus the internal voltage drop due to the input bias current. As stated by Wilde and Rodgers (1970), this voltage drop could be adjusted by a balancing potentiometer. Unfortunately, however, this procedure cannot correct for errors due to a change in the input bias current or in the internal resistance of the glass electrode due to a temperature drop. As already discussed, the expected resistance change is large (Bates, 1964) and one would therefore expect relatively large (and probably unacceptable) errors if the amplifier and sensor are subjected to a variable ambient temperature during the experiment.

In the *in situ* pH sensor assembly used by Ben-Yaakov and Ruth (1974), a d.c. amplifier of novel design was used, incorporating an FET transistor as its input stage. The FET transistor Type AD 832 (Analog Devices, Norwood, Mass., USA) had an input bias current of 10^{-13} A at room temperature (lower at low temperatures) and an input offset voltage drift of $40\,\mu$V $^\circ$C^{-1}. The *in situ* glass electrode (Sensorex Inc., Irvine, Calif., USA) was fabricated from a low-resistance glass and had a large area to decrease further the internal resistance. The resistance of a typical electrode was less than $10^7\,\Omega$ at room temperature. Assuming again a 20°C temperature range and a ten-fold increase in the resistance of the glass electrode, the d.c. errors contributed by the input bias current and input offset voltage drift are $[I_B R_p]$ 0.01 mV and $[(dV_{os}/dt) \cdot \Delta t]$ 0.8 mV, respectively. The largest error was therefore contributed by the input voltage drift with temperature. Assuming an ideal response for the glass electrode, the corresponding pH error at 4°C would be approximately 0.8/50, i.e. about 0.016 pH unit. The actual experimental error was further reduced by measuring the input offset voltage of the amplifier at the *in situ* conditions (Ben-Yaakov and Kaplan, 1973). This was accomplished by adding two relays at the input of the d.c. amplifier (Figure 4), one in parallel to the input (A) and the other in series with the pH electrode lead (B). During pH measurements the contacts of relay A were open while those of relay B were closed. To measure the

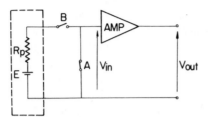

Figure 4. Relay arrangement for measuring the input offset voltage of amplifier AMP under *in situ* conditions. After Ben-Yaakov and Kaplan (1973)

amplifier's drift, the contacts of relay A were closed and those of B opened and the voltage at the amplifier's output was recorded. This voltage was later subtracted from the output voltage obtained during pH measurement (when relay A is open and B closed), resulting in an offset free value. The relays were special dry reed relays (Compac Inc., Series 10-1A-HIR) with an insulation resistance of 10^{15} Ω. This high resistance is required to prevent signal attenuation due to leakage to earth and across contacts when the relay contacts are open. These points will be further discussed below.

The specific examples given above illustrate some of the difficulties encountered when designing a signal conditioner for *in situ* ion-selective electrode studies. However, once the characteristics of the electrochemical sensor are defined in electronic engineering terms, electronic components are usually available off the shelf to meet the requirements set by the investigators. The selection of a specific amplifier to be used often involves a trade-off between a large number of parameters, such as electrical specifications, size, price, availability, reliability, power supply requirements, and power drain. Recent advances in electronic technology have made possible the manufacture of small and inexpensive high input impedance and low input bias current amplifiers. For example, modern microelectronic technology has made possible the fabrication of inexpensive monolithic operational amplifiers with a MOSFET (metal oxide field effect transistor) front end having a very low input bias current, e.g. RCA CA3130 with a typical input bias current of 5×10^{-12} A. The present price of these devices is less than one twentieth of the price of a device with similar electrical characteristics available a decade ago. It is thus evident that *in situ* marine electrochemical studies are in no way limited by the electronics. Furthermore, one can safely speculate that future *in situ* studies will be easier to design and less expensive, at least as far as electronic instrumentation is concerned.

3.2.2. *In situ* signal processing

Since the response of marine electrochemical sensors is, in general, a function of temperature and pressure, the ambient conditions must be known for a proper interpretation of *in situ* measurements. As discussed by Brown (1968a), it is possible to transmit (or record) either the raw data or data that have been compensated for pressure and temperature effects.

In the latter approach, the output signals of the pressure and temperature transducers and the electrochemical sensor are combined so as to generate an electrical signal which can be directly calibrated in the appropriate physical units for the measured parameter. For example, conductivity, temperature, and pressure signals can be combined to produce a salinity signal (Brown, 1968b; see Chapter 5). This approach not only produces a signal that can be directly calibrated in the desired units, but may also considerably simplify the accuracy requirements of the telemetry (or recording) devices. Brown (1968a) has shown that, if conductivity, temperature and pressure are transmitted or recorded separately, the following accuracies must be maintained in order to retain an accuracy of 0.02‰ in salinity: conductivity, 0.025%; temperature, 0.033%; and depth, 0.4%. This estimate assumes a mean salinity of 35‰, a temperature range of 0–35°C, and a depth range of 0–6000 m. Since the oceanic salinity range is 30–40‰, 0.02‰ accuracy represents 0.05% of full-scale. Hence, the accuracy requirements of the data link are less stringent if the combined (salinity) signal is transmitted/recorded. These considerations and, of course, the convenience of handling signals which are directly calibrated in physical units has encouraged investigators to explore various methods for *in situ* compensation. Hamon (1956), Brown and Hamon (1961), and Brown (1968b) described methods for conductivity to salinity conversion and Briggs and Viney (1964) discussed a temperature-compensated electrode for oxygen measurements. The basic approach of *in situ* compensation will be discussed here in further detail by describing the method used by the author in the design of a temperature-compensated dissolved oxygen (DO) sensor (see also Chapter 9).

Automatic temperature compensation of DO sensors

Membrane-covered polarographic sensors (Clark *et al.*, 1953; Hoare, 1968) have been successfully applied in oceanographic and limnological studies (Van Landigham and Greene, 1971; Lambert *et al.*, 1973). These sensors are usually referred to as membrane-covered polarographic electrodes, although the method is clearly amperometry rather than polarography (see Chapter 2). Atwood *et al.* (1977) found that present technology and calibration procedures are inadequate to permit the use of polarographic oxygen electrodes for the determination of dissolved oxygen in discrete samples on board ship. They pointed out, however, that *in situ* oxygen probes are clearly useful since they give a continuous profile of dissolved oxygen *versus* depth. It was suggested that absolute calibration could be accomplished by simultaneously collecting discrete samples and analysing them by Winkler titration (Carpenter, 1965).

A major source of error in the membrane-covered polarographic sensor is its extremely high sensitivity to temperature (about a 5% change in current sensitivity per degree Celsius). As discussed by Mancy *et al.* (1962), the sensor's current, I_T, as a function of concentration, c, and temperature, T (K), can be represented by

$$I_T = K_0 c e^{-J/T} \qquad (4)$$

where J and K_0 are constants for a particular sensor geometry and membrane (Pijanowski, 1975). The exponential temperature relationship is due to the temperature dependence of gas permeation through the thin plastic film which follows the activation energy relationship (Lebovits, 1966). Mancy *et al.* (1962) pointed out that a single fixed thermistor cannot offer adequate temperature compensation due to the variability in membrane permeability from one batch to another and due to possible variations in the degree of stretching during mounting. Nevertheless, in a number of applications where less accurate results are acceptable (Lambert *et al.*, 1973; Atwood *et al.*, 1977), automatic temperature compensation may be highly desirable.

Following the general diagram of Figure 1, temperature compensation can be accomplished by feeding the sensor's signal to a signal processing unit which multiplies the signal by the reciprocal of the temperature dependence function. For the case under consideration, the transfer function of the signal processing unit should follow the function $f(T)$:

$$f(T) = K_F e^{J/T} \qquad (5)$$

where K_F and J are constants. As pointed out by Briggs and Viney (1964), this correction can be accomplished by incorporating a thermistor in a feedback loop as shown in Figure 5a. The output voltage of the amplifier

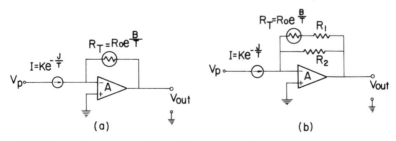

Figure 5. Schematic representation of an automatic temperature compensator for DO sensor. A is an operational amplifier, V_p is the polarization potential, and R_T is a thermistor. (a) Direct thermistor compensation; (b) improved compensator

will be (Burr Brown, 1971)

$$V_{out} = -R_T K_0 c e^{J/T} \tag{6}$$

where R_T is the thermistor resistance.

Fortunately, the temperature dependence of the thermistor's resistance is of the form (Fenwal, 1976)

$$R_T = R_0 e^{B/T} \tag{7}$$

where R_0 and B are constants. Substituting R_T in equation 6 we obtain

$$V_0 = -K_0 c e^{(B-J)/T} \tag{8}$$

If B is now chosen to be equal in magnitude to J, V_0 will be independent of temperature. However, good compensation can be achieved only if one is successful in finding a commercial thermistor with the desired B value. If such a thermistor is not commercially available one can still obtain good temperature compensation by replacing the single thermistor by a network (Figure 5). A method for selecting the resistors R_1 and R_2 in the network was described by Hamon (1956) and is based on a three-point fit. This approach is demonstrated in Figure 6, which summarizes experimental

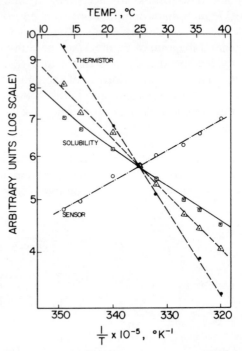

Figure 6. The response of an uncompensated DO sensor (open circles), after direct thermistor compensation (triangles), and after improved compensation (squares). The solid line represents the solubility of O_2 in water and the broken line the resistance of the thermistor (YSI 4006)

results for a specific DO sensor (Model S4022, MBK Instruments and Controls Ltd., Beer Sheva, Israel).

As pointed out by Mancy *et al.* (1962), the response of a membrane-covered polarographic sensor is sensitive to salinity. This stems from the fact that the sensor responds to the activity (i.e. to the partial pressure) of O_2 rather than to its concentration. Hence, if the sensor's output is to be calibrated in concentration units one has to take into consideration the activity coefficient of molecular oxygen in solution, that is, the salt effect on the solubility of oxygen at a constant partial pressure of oxygen. Although automatic salinity compensation is theoretically possible, it would be complicated to achieve, especially if universal compensation is attempted. The salt effect is a function of the ionic composition of the salt and hence every solution would require a special compensation scheme. Compensation for sea water and diluted sea water (estuaries) can be accomplished by taking advantage of the fixed relative chemical composition of the world oceans (Chapter 1).

Manual salinity compensation for DO measurements in sea water and diluted sea water can be easily accomplished by feeding the sensor's signal to a variable gain amplifier, as shown schematically in Figure 7. A is an operational amplifier, R_{in} and R_1 are fixed resistors and R_s is a variable resistor. The overall gain of the amplifier is

$$\frac{V_{out}}{V_{in}} = -\frac{R_s + R_1}{R_{in}}$$

The scale of R_s could now be marked in salinity units to give the appropriate overall gain. This compensation scheme (which can be found on a number of commercially available instruments) could be made accurate for a given salinity for one temperature. The salt effect, in general, is temperature dependent, whereas the compensation scheme of Figure 7 assumes a constant salinity dependence. For sea water, the solubility c [in ml(STP) dm^{-3}] can be expressed as (Weiss, 1970)

$$\text{Ln } c = A_1 + A_2(100/T) + A_3 \ln (T/100) + A_4(T/100)^2$$
$$+ S[B_1 + B_2(T/100) + B_3(T/100)^2]$$

where the As and Bs are experimentally determined constants (Appendix,

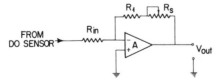

Figure 7. Salinity compensator for DO sensor. The R_s scale is marked in salinity units

Table X) and S is the salinity (‰). It is evident that, due to the cross-term factors, a temperature-independent salinity compensation is only a first-order approximation. This could be overcome by providing R_s with a different salinity scale for each temperature (Figure 7).

The purpose of the foregoing discussion was to illustrate some of the instrumentation problems that need to be handled when attempting to compensate the output of marine electrochemical sensors. High-precision electroanalytical studies should be conducted at constant temperature and pressure. Such ideal conditions may be realized in the analytical laboratory, but seldom on-site and practically never *in situ*. When variable conditions are encountered one has to weigh the pros and cons of automatic signal compensation. To date, salinometers are the only commercially available instruments which truly accomplish high-precision compensation for temperature and pressure (Brown, 1968b). This has been possible simply because modern conductivity cells are remarkably linear and stable (see Chapter 5). Indeed, the stability of the sensor is the single most important factor when considering automatic signal compensation. Efforts to develop a high-precision compensator are warranted only when it is possible to achieve a high overall accuracy. In other cases, such as the DO sensor, a less accurate compensation scheme may be acceptable because the sensor itself is subject to many errors. The constant improvement in reliability and stability of electrochemical sensors for marine studies will eventually justify comprehensive studies of various automatic compensation schemes to suit particular applications.

3.3. EARTHING, SHIELDING, AND INSULATING

Special attention must be paid to the electrical earthing of *in situ* marine electrochemical instrumentation to avoid spurious errors due to noise and electrical leakage. The problem is especially severe when line transmission is used between a submersible unit and a surface unit. As an example of the problems that may arise, consider an *in situ* pH sensor housed in a pressure housing and linked to the surface via a coaxial cable (Figure 8). In conventional designs of pH meters, the reference electrode lead (point E in Figure 8) is the ground of the electronic circuitry. To provide efficient shielding of electrical noise it is desirable to earth the case, i.e. to provide a galvanic path between the case and the electronic earth. However, direct connection of point E (Figure 8) to the case is not possible as it would short out the reference electrode (via E_R, sea water and E_H in Figure 8). An alternative approach would be to connect a relatively large capacitor (say $1\,\mu F$) between the case and point E so as to earth the case with respect to a.c. signals but to block the d.c. path. If this approach is selected then utmost caution must be exercised to avoid electrical leakage between any part of

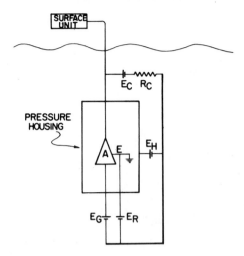

Figure 8. Electrical equivalent circuit of an *in situ* pH meter. A is an amplifier, E_G represents the glass electrode, E_R the reference electrode, E_H the casing to sea water half-cell, R_c the cable leakage to sea water, and E_c the cable to sea water half-cell. The heavy line signifies conductance via sea water

.the electrical system and sea water. Any path such as wire-to-sea water leakage (R_C in Figure 8) will load the reference electrode and may cause large errors due to polarization. The same precautions must be taken with respect to the surface unit. It is thus evident that a straightforward approach in which the electrical cable has a galvanic path to the electrochemical cell is not recommended as, sooner or later, it will cause spurious errors.

Two approaches have been used in the past to overcome the problem of possible polarization of the reference electrode due to cable leakage. One approach is based on transformer coupling and a second on the use of a differential amplifier. The latter will be discussed below in connection with the design of the UCLA Deep Sea Probe.

The transformer coupling approach has been used by Ben-Yaakov and Kaplan (1968b) in the design of an *in situ* probe which measured temperature, pH, and depth (pressure). The probe included the sensors, signal conditioners (but no signal processing), a voltage to frequency converter (to be discussed later) and a battery pack. The data were transmitted to the surface via a two-conductor cable, which was also used to relay electrical commands from the surface to the *in situ* unit and for mechanical support. The signals were time multiplexed by a stepping relay and the frequency signals were sent one at a time to the surface. The command to advance the stepping relay was sent from the surface so that the operator on board ship could select any one of the probe's channels. The operator could also send a command to short out the electrometer input and thus measure the voltage

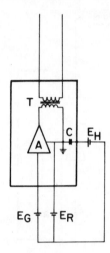

Figure 9. Earthing arrangement in the *in situ* pH probe of Ben-Yaakov and Kaplan (1968b). The reference electrode was earthed and the housing was a.c. earthed via capacitor C. The inner electronics were galvanically insulated from the cable by the transformer T

offset of the d.c. amplifier under *in situ* conditions (see Section 3.2.1). To avoid the problem of possible polarization of the reference electrode due to cable leakage, a transformer coupling was used to insulate (in the galvanic sense) the inner electronics circuitry from the cable–sea water system. As shown schematically in Figure 9, the reference electrode was earthed and the stainless steel pressure housing was a.c. earthed via capacitor C. A transformer was then used to insulate galvanically the electrical cable–sea water system from the probe's electronics. Hence, even if the cable developed a leakage to sea water (which it did), the probe's electronics and hence the reference electrode had no galvanic path to sea water, owing to the insulation between the primary and secondary windings of the transformer. It is obvious that this scheme can be used only if the signal is converted to a form which can be transmitted via transformers. In the particular design of Ben-Yaakov and Kaplan (1968b) frequency modulation was used, i.e. the analog signal was represented by a pulse train whose frequency was linearly proportional to the analog signal (Ben-Yaakov, 1968). A transformer was therefore an adequate coupling device for the converted signal. The general problems of data modulation will be discussed below in detail.

To improve further the electrical shielding and earthing in an *in situ* marine electrochemical probe, Ben-Yaakov and Ruth (1974) used a differential amplifier for the signal conditioner of a pH–reference electrode pair. In this case the electronics earth is connected directly to the pressure

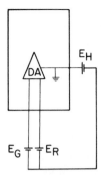

Figure 10. Earthing arrangement in the *in situ* probe described by Ben-Yaakov and Ruth (1974). DA is a differential amplifier having two high-impedance, low-current inputs

housing (Figure 10). This arrangement not only improves electrostatic shielding but also eliminates possible polarization of the reference electrode due to leakage. However, when using differential amplifiers one must pay special attention to the problem of common mode input voltage (Burr Brown, 1971). The common mode input voltage is defined as the voltage between the inputs of the differential amplifier and its earth. This voltage must be kept within a given range for proper operation. As can be seen in Figure 10, the common mode input voltage of the differential amplifier is a function of the potential of the housing–sea water half-cell (E_H). It was found that this potential will normally be below 1 V, which is acceptable for practical amplifiers. However, when the probe is not completely immersed, the galvanic path between the electrodes and the electronic earth is lost even if the electrodes are soaked in a buffer held in a beaker. This breaks the path of the amplifier's input bias current, however small, and will cause a malfunction of the differential amplifier. Hence, calibration of the pH electrode when the probe is not completely immersed could be performed only after providing a galvanic path between the probe's housing and the buffer solution. This was usually accomplished by connecting an electrical wire to the housing and immersing its bare end in the buffer.

An additional problem that requires considerable attention when designing *in situ* marine electrochemical instrumentation is the problem of d.c. insulation to prevent electrical leakage. This problem is especially important when high resistance electrodes, such as glass electrodes, are used.

In situ marine electrochemical instrumentation is prone to surface leakage problems due to the humid and salty air on board ship. Whenever the housing is opened, for battery replacement, etc., the electronics circuitry is exposed to the ambient corrosive atmosphere. Special precautions must therefore be taken in the selection of components and their layout when designing the electronics of an *in situ* probe which incorporates high-resistance electrodes. This dictates the use of PTFE-insulated wires and cables, PTFE-insulated stand-offs, guard techniques for shielded cable, and the application of silicone grease to all critical surfaces. These are standard techniques which have proved to be helpful in preventing signal attenuation due to poor insulation and surface leakage.

Another problem area is lead penetration to the high-pressure housing. To provide proper insulation, the wire-to-body resistance of the feed-throughs must be at least two (but preferably three or four) orders of magnitude higher than the internal resistance of the sensor under the worst operating conditions. For low to moderate resistance glass electrodes (*ca.* $10^8 \, \Omega$) the wire-to-body resistance of the feedthrough should be at least $10^{11} \, \Omega$. Ben-Yaakov and Kaplan (1968a) have described the design of a high-pressure, high-resistance feedthrough that was found to function satisfactorily to an ocean depth of 6000 m.

The inlet was built around a Swagelok (Crawford Fitting Co., Solon, Ohio, USA) Type 201-A-OR stainless-steel pipe fitting and consisted of a platinum wire sealed into glass and cemented with epoxy to the pipe fitting (see also Chapter 6).

3.4. *IN SITU* DATA ACQUISITION

Three methods have been used in the past for the acquisition of data from *in situ* marine electrochemical probes: direct reading by a diver, line transmission and *in situ* recording. Direct reading by a diver (Conti and Wilde, 1972) is practical only for shallow water studies and for very small volumes of data. The use of line transmission enables the on-board operator to look at the data in real time. This may have an advantage if an attempt is made to locate special features, for example when probing the sediment–water interface (Whitfield, 1971). For deep sea studies, line transmission is possible only if the research vessel is equipped with the proper equipment, i.e. an electrical cable winch. Another possible approach is the use of expendable probes which are lowered by a very thin electrical wire (Jeter *et al.*, 1972). This approach, however, is clearly applicable only to very special marine electrochemical sensors intended for routine use, which will make possible high-volume, low-cost production.

Design details of marine electrochemical *in situ* probes employing line transmission have been given by Hamon (1956) and Ben-Yaakov and Kaplan (1968b). In both cases frequency modulation—rather than direct transmission of analog signals—was employed. The latter is not practical owing to unpredictable cable attenuation which introduces large errors when cables longer than a few hundred metres are used. Also, as discussed earlier, direct transmission of analog signals may cause polarization of the reference electrode for certain earthing schemes. It is thus evident that frequency modulation has a distinct advantage over direct analog transmission.

Hamon (1956) and Hamon and Brown (1958) described an *in situ* salinometer which incorporated an FM (frequency-modulated) oscillator. The conductivity bridge and a thermistor were part of the oscillator so that the frequency of oscillation was a linear function of salinity. In subsequent studies, Brown (1968b) developed a very precise FM subcarrier oscillator whose frequency is a linear function of salinity corrected not only for temperature but also for *in situ* pressure. A different approach was taken by Ben-Yaakov and Kaplan (1968b). They first obtained an analog signal and then converted it to a proportional signal using a linear analog-to-frequency converter (Ben-Yaakov, 1968). This approach is readily applicable to electrochemical sensors such as ion-selective electrodes, whereas integration of the sensor into the oscillator circuit is adaptable only to sensors that can be incorporated into a transfer-function type network.

If several parameters are measured by an *in situ* probe, data can still be sent along a single transmission channel by using time multiplexing or frequency multiplexing. When using time multiplexing (Ben-Yaakov and Kaplan, 1968b) only one data channel is available at a time. While this simplifies the shipboard electronics it may require more time for data collection. Frequency multiplexing, on the other hand, requires more elaborate shipboard electronics but is faster as all channels are transmitted simultaneously.

A more advanced system for transmission of electrochemical data from an *in situ* probe has been described by Brown (1974). The system used pulse code modulation (PCM), which has several advantages over frequency modulation. Here the information is already coded in a digital form so it can be directly displayed on digital readouts, fed to a digital computer, or stored on digital tapes and other digital storage devices. Using standard digital techniques one can scan the data at a very high rate, if required. The system described by Brown (1974) made a complete high resolution scan (16 bits) of three sensors within 30 ms.

Apart from transmitting the data themselves, the transmission line can also be used to test the performance and even calibrate the probe *in situ*. The *in situ* probe described by Ben-Yaakov and Kaplan (1968b) included several self-check data channels. Apart from checking the battery voltage and the stability of the voltage regulator, the system was capable of performing a calibration of the analog-to-frequency converter. This was achieved by recording the output frequency of the converter when it was fed by two highly stable voltages. The entire calibration curve could then be drawn by passing a straight line between the two points, taking advantage of the good linearity of the analog-to-frequency converter (Ben-Yaakov, 1968). Another calibration that could be performed *in situ* was the determination of the input offset voltage of the electrometer, by shorting its input with a high insulation resistance reed relay. The frequency reading for the shorted input was later subtracted from the frequency reading of the pH electrode signal, thereby obtaining a drift-free measurement.

When an electrical cable is not available, *in situ* recording must be employed. Ben-Yaakov and Kaplan (1971b) described an *in situ* marine electrochemical probe which included a magnetic tape recorder. The *in situ* recorder was designed around an entertainment-type tape recorder using a novel, digital recording technique (Ben-Yaakov, 1968b). Based on the original design and experience gained in many cruises, the recording system was redesigned using a specially constructed two-channel digital tape deck and associated electronics. Details of the newer design are given in Section 3.5.

The amount of data collected by a multi-parameter *in situ* probe can be too large to handle manually. For example, the *in situ* probe described by Ben-Yaakov and Kaplan (1971b) and Ben-Yaakov (1970) recorded data at a rate of 1 channel per 5 s and had a capacity of 8 h of continuous operation. The total number of data points per tape reel therefore exceeded 11 000 data points or 22 000 numbers, as each datum was recorded twice to increase reliability by redundancy. It is obvious that this amount of data could not be handled manually, especially since extensive data reduction was required (Ben-Yaakov, 1970). The system described by Ben-Yaakov and Kaplan (1968b) and Ben-Yaakov (1970) used a two-step data reduction scheme. First the magnetic tape was read into a PDP-8/L minicomputer which was used to prepare an edited perforated paper tape. The paper tape was then read back into the minicomputer and the data were processed by a FOCAL program. The processed data were finally printed out in a tabular form by a teletype which was also used as a plotter to present some of the data in a graphical form. A more sophisticated data reduction scheme was employed in the newer instrumentation system described in Section 3.5.

An important feature of the system described by Ben-Yaakov and Kaplan (1970) is the possibility of looking at the raw data as soon as the deep sea probe was brought back on board ship. From a practical point of view this feature is essential considering the vulnerability of oceanographic instruments and the very high cost of shiptime operation. The option of looking at the data as soon as the probe is recovered enables the investigator to base his operational decisions on the quality of the data obtained. Ideally, one should have the tools to process the data on board ship. The 'poor man's' solution would be to print out part of the raw data on a slow digital printer as soon as the probe returns from its journey.

As more marine electrochemical sensors are developed and refined, the need for reliable *in situ* data acquisition systems will increase. A number of companies are at present involved in the field of data acquisition and one can safely assume that a large variety of reliable systems will be commercially available in the future. It is also probably safe to assume that one will be able to adapt some of these systems to *in situ* marine electrochemical studies.

3.5. A DESIGN EXAMPLE: UCLA DEEP SEA PROBE

The deep sea probe described in this section was developed during an oceanographic research programme conducted at UCLA in collaboration with Professor I. R. Kaplan. The engineering activity of the programme was initiated in 1967 and was aimed at the development of a high-pressure pH electrode and associated electronic circuitry for *in situ* oceanographic studies (Ben-Yaakov and Kaplan, 1968a–c). Subsequently an *in situ* recorder was incorporated in the system and it was used to study the carbonate system in the oceans (Ben-Yaakov and Kaplan, 1971a,b). The field experience gained with the instrumentation system was then used to design the Deep Sea Probe (DSP) and associated surface units described here. Actual construction of the new system was completed in late 1972. The probe was then used in a number of studies to a maximum depth of 6000 m. A modified version of the probe was constructed for the GEOSECS programme and used in a number of GEOSECS profiles.

The UCLA system is composed of two major subsystems, the DSP and a data reader (DR). The function of the probe is to sample the input signal from *in situ* sensors, to digitize the signals, to record the data on a magnetic tape cassette, and/or feed the digitized information to a telemetry line. The function of the DR is to read out the information recorded on the magnetic tape by the *in situ* probe. However, the DR can also be used to monitor the function of the probe. That is, when the DR is interconnected to the probe, it will read out and display the information recorded by the probe. Furthermore, the DR is capable of controlling, through the interconnecting lines, the function of the probe. This feature enables one to check out completely the operation of the DSP just before lowering it into the ocean.

The information read off the magnetic tape by the DR is displayed on a LED display and is also available through the pins of an output connector. The information retrieved is then fed to a PDP-8/L minicomputer or transferred to a multitrack computer-compatible magnetic tape for further data processing.

3.5.1. Deep Sea Probe

The DSP consists of sensors, signal conditioners, analog multiplexer and digitizer, and a digital cassette recorder. A general view of the DSP with the

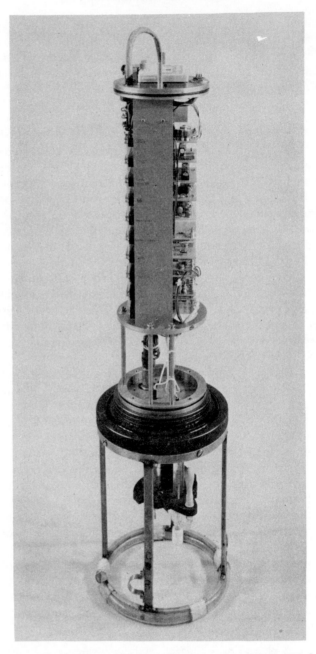

Figure 11. The UCLA Deep Sea Probe with pressure housing removed. The sensors are located at the bottom and the digital cassette deck is at the top

pressure housing removed is given in Figure 11. The sensors are located at the bottom and the digital magnetic cassette deck is at the top. As can be seen, the electronics are constructed on square printed circuit boards of about 10×10 cm.

The data acquisition system of the DSP was designed to record up to 16 channels of data on a standard cassette (Figure 12). The system accepts analog or digital inputs and records digital information on the magnetic tape cassette. Analog signals are digitized by an integrating digitizer.

By setting self-contained printed circuit switches, the data acquisition can be programmed to operate in one of three modes: (i) continuous operation; (ii) on/off mode; and (iii) a single scan mode. When in standby condition all subassemblies except the clock are turned off, thereby reducing the current drain to an extremely low value (50 μA).

Digitization is accomplished by an integrating bi-polar analogue-to-digital (A/D) converter. The relatively long integration period (approximately 0.5 s) ensures reliable conversion even of noisy signals. The digitized data are recorded serially in a true digital form on a standard (Philips) cassette. Two-channel recording provides high reliability and independence of tape speed. A total of 16 bits are recorded for each channel including synchronization and sign bits. A separate synchronization signal is recorded at the beginning of each scan. Tape movement is controlled by a single gearhead motor driving the take-up reel of the cassette. No capstan or pressure rollers are used.

In the original design, the DSP incorporated the following sensors: a thermistor, a pressure transducer, a reference electrode, and three pH electrodes. A single differential high-impedance amplifier (Ben-Yaakov and Ruth, 1974), was used to service the three glass electrodes. Connection of the electrodes to the amplifier were made via high insulation type dry reed relays. A fourth relay was used for checking the input offset voltage of the amplifier.

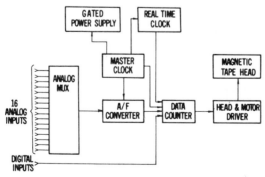

Figure 12. Block diagram of the data acquisition portion of the UCLA Deep Sea Probe

In most of the studies the DSP was used as an autonomous unit and lowered by a regular hydrographic cable. In a number of studies the DSP was used to conduct *in situ* experiments for periods of more than 10 h. During all this time data were continuously recorded on the cassette *in situ*. To facilitate operation with an electrical cable and thus permit examination of the collected data in real time, an interface unit was developed. The interface unit uses an optical coupler to avoid the problem of earth loops and possible polarization of the reference electrode (see Section 3.3).

3.5.2. Data reduction

Two methods have been used to process the data of the DSP: (a) direct processing by a minicomputer and (b) translation to an IBM compatible digital magnetic tape and reduction by the computer facility at UCLA (Figure 13). The first method was found to be satisfactory for moderate data loads. It has the advantage of in-house operation and is the least expensive, neglecting the original expense of acquiring the minicomputer. Unfortunately, data reduction by a minicomputer is a slow process and is not suitable for handling large masses of data. The bottleneck of the operation is the teletype (ASR 33) used as an output device. This problem was overcome by developing a hardware–software package that facilitates processing by a large computer installation.

A procedure was developed to transfer the data from the cassette to a 7-track IBM compatible format. The transfer of the data from the cassette to the 7-track tape was carried out by the minicomputer under the control of a program written in SYMBOLIC language. The program reads the data from the DR, codes it to IBM format, and transmits it to the 7-track recorder. The program also arranges the data in files and initiates record and file gap formats. Once transferred to the 7-track recorder the data were

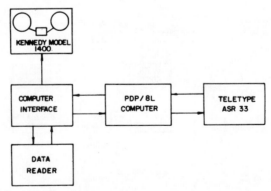

Figure 13. Block diagram of the data reduction system of the UCLA Deep Sea Probe

available in a format that could be handled by a large computer installation. Final data reduction was achieved by a FORTRAN IV program.

REFERENCES

Atwood, D. K., M. J. Kinard, M. J. Barcelona, and E. G. Johnson (1977). Comparison of polarographic electrode and Winkler titration determination of dissolved oxygen in oceanographic samples. *Deep Sea Res.*, **24**, 311–313.

Bates, R. G. (1964). *Determination of pH, Theory and Practice.* Wiley, New York.

Ben-Yaakov, S. (1968a). Analog to frequency converter is simple and accurate. *Electron. Des.*, **16**, July, 96–97.

Ben-Yaakov, S. (1968b). Data recording system for deep sea logging. *Proc. ITC*, **4**, 149–160.

Ben-Yaakov, S. (1970). An oceanographic instrumentation system for *in situ* measurements. *PhD Thesis*, University of California, Los Angeles, 343 pp.

Ben-Yaakov, S., and I. R. Kaplan (1968a). High pressure pH sensor for oceanographic applications. *Rev. Sci. Instrum.*, **39**, 1133–1138.

Ben-Yaakov, S., and I. R. Kaplan (1968b). A versatile probe for *in situ* oceanographic measurements. *J. Ocean Technol.*, **2**, 25–29.

Ben-Yaakov, S., and I. R. Kaplan (1968c). pH temperature profiles in ocean and lakes using *in situ* probe. *Limnol. Oceanogr.*, **13**, 688–693.

Ben-Yaakov, S., and I. R. Kaplan (1971a). Deep sea *in situ* calcium carbonate saturometry. *J. Geophys. Res.*, **76**, 722–731.

Ben-Yaakov, S., and I. R. Kaplan (1971b). An oceanographic instrumentation system for *in situ* applications. *Mar. Technol. Soc. J.*, **5**, 41–46.

Ben-Yaakov, S., and I. R. Kaplan (1973). Design and application of a deep sea pH sensor. In *Marine Chemistry* (Eds. T. B. Berkowitz, R. A. Horne, M. Banus, P. L. Howard, M. J. Pryor, and G. G. Whitnack). Electrochemical Society, Princeton, N.J., pp. 98–108.

Ben-Yaakov, S., and E. Ruth (1974). An improved in situ pH sensor for oceanographic and limnological applications. *Limnol. Oceanogr.*, **19**, 144–151.

Briggs, R., and M. Viney (1964). The design and performance of temperature compensated electrodes for oxygen measurements. *J. Sci. Instrum.*, **41**, 78–83.

Brown, N. L. (1968a). An *in situ* salinometer for use in the deep ocean. In *Marine Science Instrumentation*, Vol. 4 (Ed. F. Alt). Plenum Press, New York, pp. 563–578.

Brown, N. L. (1968b). The paraloc—a precise telemetry subcarrier oscillator. In *Marine Science Instrumentation*, Vol. 4 (Ed. F. Alt). Plenum Press, New York, pp. 106–112.

Brown, N. L. (1974). A precision CTD microprofiler. *Proc. IEEE Conf. Ocean 1974, Halifax, Nova Scotia*, Vol. 2, pp. 270–278.

Brown, N. L., and B. V. Hamon (1961). An inductive salinometer. *Deep Sea Res.*, **8**, 65–75.

Burr Brown, (1971). *Operational Amplifiers*, McGraw-Hill, New York.

Carpenter, J. H. (1965). The Chesapeake Bay Institute Technique for the Winkler dissolved oxygen method. *Limnol. Oceanogr.*, **10**, 141–143.

Clark, L. C., Jr., R. Wolf, D. Granger, and Z. Taylor (1953). Continuous recording of blood oxygen tension by polarography. *J. Appl. Physiol.*, **6**, 189–193.

Conti, U., and P. Wilde (1972). Diver operated *in situ* electrochemical measurements. *Mar. Technol. Soc. J.*, **6**, 17–23.

Fenwal, (1976). *Thermistor manual EMC-6*, Fenwal Electronics, Framingham, Massachusetts.

Hamon, B. V. (1956). A portable temperature–chlorinity bridge for estuarine investigations and sea water analysis. *J. Sci. Instrum.*, **33**, 329–333.

Hamon, B. V., and N. L. Brown (1958). A temperature–chlorinity–depth recorder for use in sea. *J. Sci. Instrum.*, **35**, 452–458.

Hoare, J. P. (1968). *The Electrochemistry of Oxygen*, Interscience, New York.

Jeter, H. W., E. Føyn, M. King, and L. I. Gordon (1972). The expendable bathyoxymeter. *Limnol. Oceanogr.*, **17**, 288–292.

Lambert, R. B., Sr., D. R. Kester, M. E. Q. Pilson, and K. E. Kenyon (1973). *In situ* dissolved oxygen measurements in the north and west Atlantic ocean. *J. Geophys. Res.*, **78**, 1479–1483.

Libovitz, A. (1966). Permeability of polymers to gases vapors and liquids. *Mod. Plast.*, **43**, 139–213.

Mancy, K. H., D. H. Okun, and C. N. Reilley (1962). A galvanic cell oxygen analyzer. *J. Electroanal. Chem.*, **4**, 65–92.

Pijanowski, B. S. (1975). Dissolved oxygen sensors—theory of operation, testing and calibration techniques. *Symposium on Chemistry and Physics of Aqueous Gas Solutions, Toronto, Canada, May 1975*, pp. 373–388.

Sawyer, D. T., and J. L. Roberts (1974). *Experimental Electrochemistry for Chemists*. Wiley, New York.

Van Landigham, J. W., and M. W. Greene (1971). An *in situ* molecular oxygen profiler. *Mar. Technol. Soc. J.*, **5**, 11–23.

Warner, T. B. (1972). Ion selective electrodes—properties and use in seawater. *Mar. Technol. Soc. J.*, **6**, 29–33.

Weiss, R. F. (1970). The solubility of nitrogen, oxygen and argon in water and seawater. *Deep Sea Res.*, **17**, 721–735.

Whitfield, M. (1971). A compact potentiometric sensor of novel design. *In situ* determination of pH, pS^{2-}, and *Eh*. *Limnol. Oceanogr.*, **16**, 829–837.

Whitfield, M. (1975). The electroanalytical chemistry of seawater. In *Chemical Oceanography*, 2nd edition, Vol. 4 (Eds. J. P. Riley and G. Skirrow). Academic Press, London, pp. 1–154.

Wilde, P., and P. W. Rodgers (1970). Electrochemical meter for activity measurements in natural environments. *Rev. Sci. Instrum.*, **41**, 356–359.

Marine Electrochemistry
Edited by M. Whitfield and D. Jagner
© 1981 John Wiley & Sons Ltd.

DANIEL JAGNER
Department of Analytical and Marine Chemistry,
University of Göteburg,
S-41296 Göteborg, Sweden

4

Computer Control and Processing of Electrochemical Measurements

4.1. INTRODUCTION

The concept of computer control of chemical measurements is less than twenty years old. At the beginning of its development, the cost of computer power made the computerization of instrumentation justifiable only for very expensive and complex equipment, such as nuclear magnetic resonance and mass spectrometers. The drastic decrease in the cost of computer hardware, by more than an order of magnitude in the last ten years, has resulted in the computerization of almost every kind of chemical instrumentation (Perrin, 1977), including such minor equipment as the pH meter (e.g. Orion Research Inc., Cambridge, Mass., USA) and the pipette (e.g. Hamilton Co., Reno, Nev., USA).

Some areas of electrochemistry, such as voltammetry, are particularly suitable for computerization, owing to the dynamic character of the measurements. In voltammetry the computer can be used to generate potential ramps and pulses of almost any shape and to measure, very rapidly, the instantaneous current response. Computerization has thus facilitated new approaches to electroanalytical measurements. In other areas of electrochemistry, such as potentiometry and coulometry, which are based on steady-state measurements, the use of computers is restricted to the automation of manual operations, such as the initiation of measurement, the registration of a steady-state signal, data reduction, and the presentation of the final results. Up to now, computers have had surprisingly little influence on the design of electrochemical instrumentation. Even in the area of voltammetry, computer applications have hitherto been directed towards the automation of already known techniques, such as pulse polarography and differential pulse anodic stripping voltammetry. Very few electrochemical instruments have been proposed in which the computer is an absolute necessity for the technique used. Presumably the greatest impact of computers in marine electrochemistry has yet to come.

4.2. COMPUTER LANGUAGES. REAL-TIME DATA ACQUISITION

All digital computers work with binary information. When computers were first introduced, the programmer had to translate every instruction and datum into binary code, a procedure known as machine code programming. The development of intermediate level computer languages, such as ASSEMBLER, and of high-level computer languages, such as ALGOL and FORTRAN, has greatly simplified computer programming. The introduction of interpretative languages (such as BASIC) for use with minicomputers enables formal programming errors to be corrected during the development of the program and has made computer programming even easier. When using high-level languages the computer programmer does, however, lose full control of computer operation since the high-level languages are word-orientated, i.e. the binary pattern of one computer word cannot be altered. This is of very little significance when the computer is used for solving arithmetical problems, but it is a considerable drawback when it is to be used for controlling the operation of electrochemical instruments. For example, before signals from an electrochemical cell can be interpreted by the computer they must be converted to a digital form by an analog-to-digital (A/D) converter (see also Chapter 3). The information coming from this type of converter, which is incorporated in all computerized instrumentation, is either in binary or in binary coded decimal (BCD) form and must be correctly interpreted by the controlling computer program. Furthermore, the computer is also used to activate various instrumental functions. This is normally done by activating different relays, each relay being connected to one bit of a computer word. The instructions for such operations must be given in machine code which can address individual bits of the computer word. Machine code programming is so time consuming that the advantages which can be obtained by computerization of an electrochemical measurement do not always compensate for the effort needed to construct the controlling computer program. No doubt this was one of the major reasons for the delay in introduction of computers in many electrochemical laboratories.

In the early 1970s, microcomputers became available commercially. These computers differ from the earlier minicomputers in that the central processing unit (CPU) is incorporated in a single chip, and they exploit different kinds of semiconductor memories, rather than the much more expensive core memories. The inexpensive microcomputer systems now available must, however, be programmed in machine code or with the aid of expensive programming systems, such as the Tektronix 8002 microcomputer development system. Indeed, the use of microcomputers in different electrochemical laboratories is hindered today to a very much greater extent by the time and effort needed to program the computer than by the cost of purchasing it. At present only electrochemical laboratories which are well equipped in terms of computer knowledge and electronic workshop facilities can take full advantage of the cheap microcomputer systems.

In the near future it is likely that almost all commercial electrochemical instruments will be controlled by microcomputer systems. A disadvantage is that the operator of such instrumentation may not always be fully aware of the digital manipulations that the computer performs on the primary electrochemical signals. The reason for this is that the control programs are likely to be incorporated in programmable read-only memories (PROMs), the contents of which are difficult to extract and even more difficult to interpret. This may well be a future source of error in electrochemical research and applications.

The real time data acquisition rate denotes the maximum number of measurements, at constant time intervals, which the computer system is capable of making per second. This rate is the single most important factor to be considered when a computer is to be incorporated into an electrochemical instrument. The demand on the real time capacity is very different in different types of electrochemical application. In potentiometric measurements, for example, it is normally acceptable if the computer system can monitor the potentiometric sensor at regular intervals of once or twice per second. In rapid scanning anodic stripping voltammetry the demand on real time capacity is considerably higher. At a linear potential ramp rate of $20 \, V \, s^{-1}$, for example, the whole stripping event is over within 50 ms. In order to obtain a satisfactory representation of the stripping current at different times (i.e. at different potentials) the computer system must be able to make at least 100 measurements of current at regular time intervals. Thus, a real time capacity of 2 kHz is required. During the 500 μs between successive measurements the computer should be able to perform a number of tasks, such as making the measurement, starting a real time clock, storing the datum in a suitable location, checking for possible memory overflow, checking whether the criteria for stopping the experiment are fulfilled, and reading the real time clock to check if a new measurement should be made. Thus, even the modest real time rate of 2 kHz puts great demands on the computer system and on the computer program operating the system.

The real time data acquisition rate of a computer system is intimately related to the computer language used to operate the system. In a high-level language such as BASIC, each instruction generates a great number of machine code operations, most of these being unnecessary for the particular task the computer has to fulfill at a particular moment. For example, if a minicomputer is programmed in BASIC to add all integers between 1 and 1000, this will take a very much longer time than if the same task is performed by the same computer using a optimum machine coded program. It is not possible, however, to state exactly by how much the computations are speeded up, since this differs from one BASIC interpreter to another, and also, to a minor extent, from one computer to another. ASSEMBLER language programming is an intermediate between programming in BASIC and in machine code, both with respect to program efficiency and time required to construct the program. Table 1 compares, on a relative scale, the merits of three different language levels and gives very approximate estimates of the maximum real time capacity associated with the different languages and the time required to construct the computer program.

As is apparent from Table 1, the short time required for programming in BASIC is counterbalanced by the limited real time data acquisition rate achievable in this language. Many computerized electrochemical instruments are programmed at two language levels, e.g. ASSEMBLER programming is used for instrument control and real time data acquisition and BASIC programming is used for data reduction. In this way the merits of both languages can be exploited. Of particular relevance in this connection is the possibility of linking ASSEMBLER subroutines, so-called drivers or handlers, into a main BASIC program (Anfält and Jagner, 1971). Until very

Table 1. Comparison between different programming languages

Computer language	Programming time, relative scale	Real time data acquisi- tion rate, kHz	Possible electro- chemical techniques
BASIC	1	0.1	Potentiometry, potentiometric titrations, coulometry, polarography
ASSEMBLER	50	20	(Differential) anodic stripping voltammetry
Machine Code	200	100	Potentiometric stripping analysis

recently the possibility of linking ASSEMBLER or machine code sub-routines into a main BASIC program had been limited to minicomputers. The introduction of PROM resident BASIC interpreters, which can be incorporated into a microcomputer system has, however, extended this to very small computer systems. Electrochemical applications in which AS-SEMBLER or machine code subroutines are linked into a main BASIC program were described by Anfält and Strandberg (1978) and by Rigdon *et al.* (1979).

An alternative method by which the simultaneous demand for rapid data acquisition and high-level language for data reduction can be met is by constructing a hierarchical system. In such a system the microprocessor is controlled by a host computer to which the microprocessor transfers the data after their acquisition for reduction and presentation. Hierarchical systems can be designed in different ways, as described in detail by Dessy (1977, 1978).

4.3. THE HARDWARE CONFIGURATION

The four main tasks that a computer system has to fulfill in connection with electrochemical instrumentation are

(i) initiation of the instrument and control of experimental performance, e.g. addition of titrant, generation of potential ramps, or charging of new samples;
(ii) measurement of sensor signals;
(iii) data reduction;
(iv) presentation of experimental results on a suitable medium.

Initiation of the instrument is achieved, for example, by means of rapid mercury-wetted relays. It is also possible to program the computer to generate potential ramps and potential pulses (e.g. Kryger *et al.*, 1975; Bos, 1978). Linear potential ramps or potential pulses of the same shape and frequency, such as square or triangular waves, are more suitably generated by separate electronic circuitry. In this way the computer power is concentrated on the tasks it is best suited for, namely the measurement of sensor signals and the processing of primary data. Furthermore, modern electronic circuitry generates uniform potential pulses much more accurately and precisely than is possible with modern digital computers. If the sensor response from one potential pulse influences the shape and size of the next potential pulse it is, however, necessary to allow the computer to control pulse generation. No applications of this rather advanced way of using the computer in electrochemistry, in which the computer optimizes experimental parameters while the experiment is in progress, have yet been reported. The main difficulty in such applications is that they simultaneously demand both a high real time data acquisition rate and considerable data reduction. This is not easily achieved by modern computers.

The hardware needed to register the potential of the sensor electrode (A/D converters or digital voltmeters) has been discussed in Chapter 3. Digital voltmeters are normally not able to make more than 500 A/D conversions per second and are thus suitable only for low data flow rates (*cf.* Table 1). Modern rapid A/D converters are able to make up to one million conversions per second. Another important aspect of A/D converters in computerized electrochemistry is the resolution of the converter. An 8-bit converter has a resolution of 1 in 256 and if this converter is to be used, for example, for potential measurements in the range ± 1.5 V, the resolution will be less than 10 mV, which is normally not sufficient. In the author's opinion, the extra cost associated with the incorporation of 10-bit, or preferably 12-bit, converters in computerized electrochemical instrumentation is usually well justified. A complication arises, however, when an 8-bit microprocessor is used, since such a processor is not able to read a 10- or 12-bit converter in a single step. This means that the data acquisition rate is decreased by at least a factor of two. Future 16-bit microprocessors will, no doubt, solve this problem.

Data reduction programs for computerized electrochemical measurements are usually simple arithmetically. They are usually composed of one or more of the following features:

  (i) addition of several signals in order to obtain improved signal-to-noise ratio by elimination of random noise;
 (ii) subtraction of one sensor signal from another in order to extract relevant information, e.g. subtraction of the sensor signal at pulse bottom from the sensor signal at pulse peak or, more often, subtraction of a measured background from a combined signal–background curve;
(iii) computation of peak areas or heights.

Only in a very few applications has more advanced data treatment been used. This has often been associated with the use of pattern recognition in the treatment of electrochemical data (Thomas *et al.*, 1976; Bos and Jasink, 1978).

The display unit is of utmost importance in a computerized electrochemical set-up. For example, the experimental results can either be displayed digitally on a teletype printer, or graphically on a plotter or a strip-chart recorder. Graphical displays are usually preferable since only a quick glance at a graph is necessary to see whether the experiment has been successful or not. A simple strip-chart recorder is, however, sufficient in most computerized electrochemical instruments. Even if the measuring event occurs so rapidly that a normal strip-chart recorder is not able to follow the signals, the computer can store these signals and display them at a slower rate to the strip-chart recorder. In fact, the incorporation of a microcomputer in an electrochemical instrument can decrease the total cost of the system by reducing the need for rapid display units, such as oscilloscopes. In addition, incorporation of a microcomputer increases the flexibility and capacity of the system considerably.

A dynamic graphical display for use in computerized electroanalysis has been described by Skov *et al.* (1976). This system, which was based on the use of six dynamic random access memories in combination with a normal television monitor, makes it possible to follow a computerized electrochemical experiment while it is in progress. The cost and effort of incorporating such a system into a computerized electrochemical system is, however, justified only in a research laboratory.

4.4. COMPUTERS SUITABLE FOR MARINE ELECTROCHEMICAL APPLICATIONS

Computers used in connection with marine electrochemical applications on board ship have to meet a number of special requirements, apart from being inexpensive. Such computer systems should be small, robust, and reliable. The computer must also be insensitive to voltage variations in the power supply and have low power requirements. All of these requirements are met by modern microcomputers, whereas minicomputers are often too expensive and too bulky. As pointed out above, microcomputers are, however, often difficult to program and, moreover, the high real time data acquisition rate may not be essential for the particular application under consideration. The recent introduction of 'personal computers' or 'hobby computers' such as the PET Commodore, shown in Figure 1, overcome to a great extent the initial programming difficulties facing the electrochemist inexperienced in programming. These computers provide not only a microprocessor and a high-level language interpreter at a very moderate cost but also a graphical display (*cf.* Figure 1). Even more important is the fact that most electrochemical instruments can be interfaced to the computer system (e.g. by means of GP-IB 'bus terminals'). Furthermore, the computer can be equipped with mass storage memories, such as magnetic tape or floppy discs. The computer capacity and system flexibility of these 'hobby computers' are equivalent to, if not better than, the advanced minicomputer systems of only ten years ago, while their cost is less than 10% of that of a minicomputer system. Since these computers have been introduced only recently, very few electrochemical applications have yet been reported.

Since 'hobby computers' are programmed with high-level languages they are less well adapted for rapid data acquisition. In applications where this is necessary there are several suitable microcomputer systems on the market, e.g. the INTEL SDK-85 system shown in Figure 2. Such a computer system

Figure 1. PET Commodore computer system

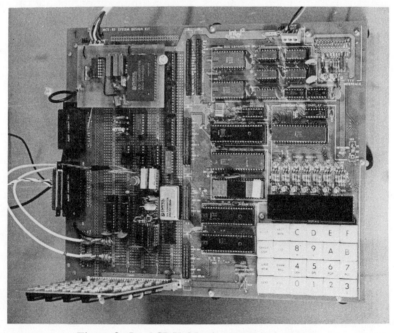

Figure 2. Intel SDK-85 microcomputer system

contains most of the peripherals needed for computerized electrochemistry such as an 8-bit processor, read-only memories (ROM), random-access memories (RAM), programmable read-only memories (PROM), teletype interface, a simple keyboard, and a 6-character alpha-numeric display. Furthermore, digital voltmeters or other A/D converters can be easily interfaced with the system. The total system cost is of the same order of magnitude as a high-quality glass electrode. The new 16-bit computer systems, e.g. INTEL SDK-86, are even better suited for electrochemistry. However, at their present state of development, such systems are difficult to program. Furthermore, very little use can be made of the computer system unless it is combined with a more powerful display unit. This, together with A/D converters and power supplies, makes the total cost for computerization of an electrochemical instrument several times greater than the cost of the computer system itself.

4.5. POTENTIOMETRIC TITRATIONS

Potentiometric titrations provided some of the earliest examples of computer-controlled electrochemical instrumentation. The reasons for this are that titration is a frequently used analytical technique and that the data flow is relatively slow. The earliest titrators were so-called off-line instruments, where the computer was only used to process the titration data subsequent to analysis. Hardware logic units were used to control the experiment and to present the primary data on a suitable medium, such as punched cards or magnetic tape (Jagner, 1970; Johansson and Pehrsson, 1970; Jagner, 1974). A fully computerized titrator was described by Anfält and Jagner (1971). The main parts of this titrator, which are shown in Figure 3, consist of an HP 2114 minicomputer with about an 8000-word 16-bit core memory, and a digital voltmeter capable of measuring eight different sensors in random access. An 8-bit relay register makes it possible to operate eight different syringe burettes, the volume of each titrant increment being controlled by a 50-Hz computer clock. The computer system of Anfält and Jagner (1971) was the first to demonstrate the suitability of BASIC as a controlling and processing language in electrochemical instrumentation. ASSEMBLER subroutines, controlling the digital voltmeter and the syringe burettes, were incorporated into the BASIC interpreter in such a way that BASIC programming could be used for the control of these peripherals. The use of BASIC limited the real time data acquisition rate to approximately 10 Hz. Owing to the ease with which the system could be programmed it was possible to incorporate evaluation methods such as the Gran (1952) linearization method (Chapter 7) into the computer program which processes the titration data.

The main purpose of the computer system of Anfält and Jagner (1971) was

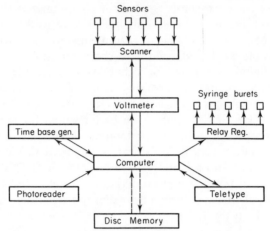

Figure 3. Main parts of the computerized titrator described by Anfält and Jagner (1971). Reproduced by permission of Elsevier Scientific Publishing Company

to automate potentiometric titration procedures for the determination of the major sea water components (*cf.* Chapter 7). The computer system was used for on-board analysis on several expeditions (Almgren *et al.*, 1977) and it was concluded that computerization both simplified the titration procedure and increased the analytical capacity for on-board analysis. Accuracy and precision were also enhanced but to a smaller extent than was expected. The main reason for this is that sample handling and temperature control are fundamental parameters for increasing precision and these are not improved by computerization. Computer systems can be reliable, even when used on-board ship. More than ten years after its construction, the computer system of Anfält and Jagner (1971) still works without malfunction. Owing to its bulkiness this computer system cannot compete, however, with modern computerized titrators.

Microcomputers have been incorporated in most modern commercial titrators and it is therefore seldom necessary for individual laboratories to construct their own titrators. The data processing programs in commercial titrators are, however, PROM resident and in practice it is impossible for the user of the system to alter the programming of the titrator. This is often a drawback in marine electroanalytical applications where very specialized titration procedures and evaluation methods are often desired. Frazer *et al.* (1975) have described a sophisticated computer system for potentiometric titrations using different ion-selective electrodes. The system is operated in the high-level language FOCAL and the primary titration data, or derived functions such as Gran plots (Gran, 1952), are displayed graphically on an oscilloscope screen. By means of a graphic pen the operator of the titrator can communicate with the computer system in an interactive mode and decide, for example, which experimental points are to be used for evaluation of the equivalence point. If a titration point is affected by random errors this is also easily seen on the graphical display. Frazer *et al.* (1977) have demonstrated the use of this computer system for the optimization of potentiometric titrations with respect to accuracy, precision and the time needed for performing the titration. The results obtained illustrate the importance of a rapid graphical display in connection with computerized electrochemistry (*cf.* Skov *et al.*, 1976). Rigdon *et al.* (1979) recently described a microcomputerized titrator built around an INTEL 8080 processor using an oscilloscope as graphical display. The titrator was programmed in BASIC II, which is a version of BASIC with a real time data acquisition capability. A ROM resident arithmetic processing unit was used for floating point arithmetic operations. Without this unit it would have been possible for the computer system to work only with integers.

Yamaguchi and Kusuyama (1979) have described a microcomputerized titrator built around the 16-bit processor LSI-11. The experimental results were displayed in digital form on a teletype writer. Very few details are given with respect to programming language and real time data acquisition capacity. The use of a teletype instead of a graphical display does, however, reduce the suitability for marine electroanalytical applications.

Gaarenstroom *et al.* (1978) have shown how a microcomputer, hierarchically controlled by a minicomputer, can be used for the potentiometric monitoring of sodium and potassium when these ions are simultaneously present in the sample in such high concentrations that they each interfere with the measurement of the other (*cf.* Chapter 7). The approach taken was to allow the microcomputer to generate, by means of controlling peristaltic pumps, a reference solution in which the sodium and potassium electrode potentials were the same as those in the sample. The measurements of Gaarenstroom *et al.* (1978) were performed in a flowing system and it seems likely that a similar approach could be used for minitoring sodium and potassium in saline waters.

All computerized titrators hitherto suggested have used the Gran (1952) linearization method for evaluation of the titration data. Horvai *et al.* (1978) have suggested a different method, based on a three-parameter curve fitting procedure, for use in standard addition titrations. The method is less general, however, than the Gran method and more complicated to program.

4.6. ANODIC STRIPPING VOLTAMMETRY AND POLAROGRAPHY

Great progress has been made in the field of anodic stripping voltammetry instrumentation during recent years (*cf.* Chapters 2 and 10). This progress is

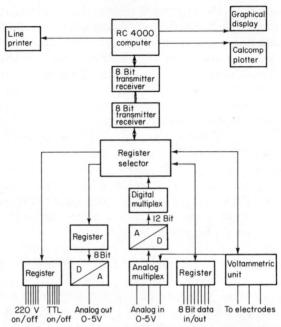

Figure 4. System for computerized electroanalysis. Kryger *et al.* (1975). Reproduced by permission of Elsevier Scientific Publishing Company

due to the use of advanced electronic circuitry, however, rather than to the use of computers. A versatile computer-operated system capable of making all types of voltammetric measurements has been described by Kryger *et al.* (1975). The system, which was built around a large computer, is shown in Figure 4 and details of the voltammetric and potentiostatic units are shown in Figure 5.

The main purpose of the computer system was to test different instrumental approaches to the electroanalysis of trace constituents in sea water. Once a suitable approach had been developed, this was then to be incorporated into a microcomputer system. Attempts were made to improve the detection limit and accuracy in differential anodic stripping voltammetry by digitally controlling the experiment. It was concluded, however, that computerization of differential anodic stripping voltammetry did not improve the results.

Kryger and Jagner (1975) suggested a multiple scanning technique for improving the detection limit of anodic stripping voltametry. In this technique, a rapid linear potential ramp was applied to the mercury film working electrode after pre-electrolysis and the current response was measured and stored in the computer core memory. Immediately after the scan was stopped the pre-electrolysis potential was again applied to the working electrode. In this way the analyte ions did not have time to migrate away from the working electrode and a major fraction was reduced into the mercury film. After a short delay of about one second, the reduced ions were re-oxidized during a new scan and the computer added the current response to that of the previous scan. The accumulated current after 10–30 consecutive scans was taken as the anodic stripping signal. This is illustrated in Figure 6, where the current *vs.* potential curves for 30 consecutive scans are shown (Kryger and Jagner, 1975). Each potential scan contributes a capacitance background, however, and multiple stripping therefore does not increase the signal-to-background ratio. In the work of Kryger and Jagner (1975), the capacitance background was measured separately after all of the amalgamated metals had been re-oxidized and given sufficient time to diffuse away from the working electrode surface. The background was scanned, and the current *vs.* potential accumulated in core memory.

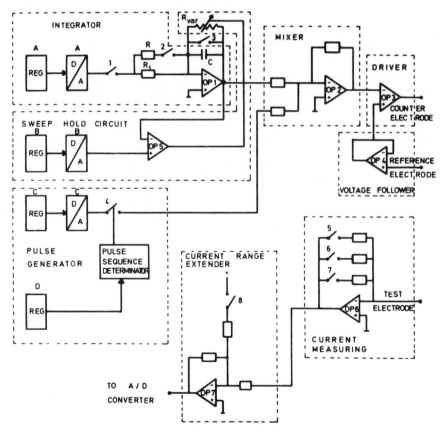

Figure 5. The voltammetric unit in the computerized electroanalytical system shown in Figure 4. Kryger *et al.* (1975)

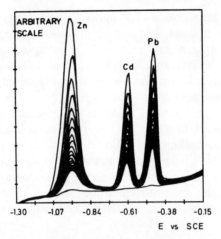

Figure 6. Multiple scanning anodic stripping voltammetry. Current response from thirty consecutive stripping cycles and one background cycle. Kryger and Jagner (1975). Reproduced by permission of Elsevier Scientific Publishing Company

Finally, the computer subtracted the background scan from the accumulated signal + background scans and displayed the result on a graphical screen or plotter. This is illustrated in Figure 7, where curve 1 is the accumulated stripping current from 40 potential scans and curve 2 the accumulated background current from 40 background scans. Curve 3 in

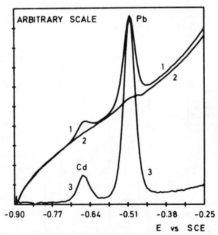

Figure 7. Multiple scanning anodic stripping voltammetry. Curve 1 represents the accumulated signals from forty analytical-background scans, curve 2 represents the accumulated signals from forty background scans, and curve 3 represents the difference between curves 1 and 2. Kryger and Jagner (1975). Reproduced by permission of Elsevier Scientific Publishing Company

Figure 7 is the difference between curves 1 and 2 and comprises the analytical signal. The multiple scanning technique was applied to the determination of zinc, cadmium, lead, and copper in sea water by Jagner and Kryger (1975).

Brown and Kowalski (1979) used a background correction procedure for anodic stripping voltammetry very similar to that of Kryger and Jagner (1975). Since their minicomputer-operated electrochemical system had a real time data acquisition capacity of approximately 15 kHz, they were able to obtain excellent results with only one analytical + background and one background scan. Brown and Kowalski (1979) claimed detection limits of the order of 10 ng dm^{-3} for Cd(II) and Pb(II).

Turner *et al.* (1979) used a minicomputer system for an extensive investigation of the optimum voltammetric stripping technique in computerized electroanalysis. They concluded that square-wave and differential pulse stripping were the optimum techniques with respect to sensitivity and time needed for the analysis. Turner *et al.* (1979) used a background subtraction procedure similar to that of Brown and Kowalski (1979).

Bos and Jasink (1978) used the linear learning machine method to determine cadmium, thallium, and lead in the 10^{-6}–10^{-8} M range. They were able to determine these elements with a relative precision of 10% even in cases where the presence of the elements could not be detected visually on the voltammograms. The learning machine method is difficult to program, however, and does not seem to be suitable for the analysis of samples with varying composition.

Bos (1976) used a PDP-11 minicomputer system for controlling direct current polarography experiments. Both experimental control and data processing were carried out in ASSEMBLER, and Bos concluded that ASSEMBLER programming was so tedious that the programming efforts were greater than the advantages gained by computerization. Bos (1978) has used a similar computer system in connection with Kalousek polarography.

Bond and Grabaric (1979) used a quadratic least-squares fit of the form

$$I = A(E - C)^2 + B \qquad (1)$$

to compensate for the background in differential pulse polarography where the background cannot be measured separately as in anodic stripping voltammetry. In equation 1, I denotes the current, E the potential applied to the dropping mercury electrode, and A, B, and C are constants, the values of which vary from one experiment to the next. Experimental data were collected by a PAR 174 Polarographic Analyzer interfaced to a PDP-11/10 minicomputer. The data collection and processing were performed in BASIC. According to the authors, the general form of equation 1 makes it possible to compensate not only for capacitance background but also for the presence of trace oxygen impurities.

4.7. POTENTIOMETRIC STRIPPING ANALYSIS

Potentiometric stripping analysis (see also Chapter 7) is based on the pre-concentration of trace metal analytes by means of potentiostatic reduction at a mercury film electrode as in anodic stripping voltammetry, i.e.

$$M(n) \xrightarrow{ne^-} M(Hg) \tag{2}$$

The method of re-oxidation of the amalgamated metals differs, however, from that of anodic stripping voltammetry. In potentiometric stripping analysis the potentiostatic circuitry is disconnected and the metals re-oxidized chemically according to

$$M(Hg) \xrightarrow{\text{oxidant}} M(n) \tag{3}$$

The flow of oxidant to the electrode surface is controlled by a constant rate of stirring at the electrode. The electrode potential *vs.* time curve comprises the analytical signal, as is illustrated for a Baltic sea water in Figure 1 of Chapter 7.

The dominant oxidizing agent in sea water is dissolved oxygen and if this is not removed, for example by bubbling with argon, the amalgamated metals will be re-oxidized according to

$$M(Hg) \xrightarrow{O_2(aq)} M(n) \tag{4}$$

where either water or hydrogen peroxide is formed, depending on the acidity of the sample and the redox potential of the amalgamated metal oxidized (Pelletier *et al.*, 1971; Buffle *et al.*, 1973). Since the concentration of dissolved oxygen in sea water is very high, the oxidation according to equation 4 will take place so rapidly that it cannot be registered on a normal strip-chart recorder. In order to overcome this problem Anfält and Strandberg (1978) constructed a fully computerized potentiometric stripping analyzer. The system was built around an INTEL SBC 80/10 single-board microcomputer and programmed in BASIC. The real time data acquisition rate was not high enough, however, to enable accurate measurements of potentiometric stripping curves in non-deoxygenated samples to be made (Jagner, 1979).

Granéli *et al.* (1980) used an INTEL SDK-85 (*cf.* Figure 2) computer system for potentiometric stripping analysis. The system had a real time capacity of 32 kHz. In the analysis the computer system first registers and stores in memory the potentiometric stripping curve after a suitable preelectrolysis time, e.g. 4 min. It then registers and stores in memory a background stripping curve obtained after only a few seconds of pre-electrolysis. The computer then subtracts the background scan from the

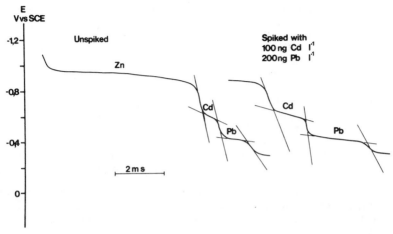

Figure 8. Computerized potentiometric stripping analysis in an acidified non-deoxygenated sea water sample. Four minutes of pre-electrolysis. Granéli *et al.* (1980)

analytical + background scan and displays, at a slower rate, the net result on a normal strip-chart recorder. The procedure is illustrated in Figure 8, where a non-deoxygenated acidified sea water sample, containing approximately $100 \, \text{ng dm}^{-3}$ of Cd(II) and $180 \, \text{ng dm}^{-3}$ of Pb(II) has been pre-electrolysed for 4 min prior to recording the potentiometric stripping curve. The right-hand curve in Figure 8 shows the potentiometric stripping curve recorded under the same experimental conditions after the standard addition of $100 \, \text{ng dm}^{-3}$ of Cd(II) and $200 \, \text{ng dm}^{-3}$ of Pb(II). It is apparent that computerized potentiometric stripping analysis can be used for low trace metal concentrations. Moreover, the total analysis time is reduced since de-oxygenation can be omitted. Since it does not require prior de-oxygenation, computerized potentiometric stripping analysis ought to be well suited to *in situ* measurements in sea water.

REFERENCES

Almgren, T., D. Dyrssen, and M. Strandberg (1977). Computerized high-precision titrations of some major constituents of sea water on board the R.V. Dmitry Mendeleev. *Deep-Sea Res.*, **24**, 345–364.

Anfält, T., and D. Jagner (1971). A computer processed semi-automatic titrator for high precision analysis. *Anal. Chim. Acta*, **57**, 177–183.

Anfält, T., and M. Strandberg (1978). A micro-computer system for potentiometric stripping analysis. *Anal. Chim. Acta*, **103**, 379–388.

Bond, A. M., and B. S. Grabaric (1979). Correction for background current in differential pulse, alternating current, and related polarographic techniques in the determination of low concentrations with computerised instrumentation. *Anal. Chem.*, **51**, 337–341.

Bos, M. (1976). The development of a fully computerized system for sampled d.c. polarography with standard interfacing. *Anal. Chim. Acta*, **81**, 21–30.

Bos, M., and G. Jasink (1978). The learning machine in quantitative chemical analysis. Part I. Anodic stripping voltammetry of cadmium, lead and thallium. *Anal. Chim. Acta*, **103**, 151–165.

Bos, H. (1978). Computerised Kalousek polarography. *Anal. Chim. Acta*, **103**, 367–378.

Brown, S. D., and B. R. Kowalski (1979). Minicomputer controlled, background-subtracted anodic stripping voltammetry. Evaluation of parameters and performance. *Anal. Chim. Acta*, **107**, 13–27.

Buffle, J., M. Pelletier, and D. Monnier (1973). Chronopotentiometric study of the oxidation of some diluted amalgams in the presence of traces of oxygen dissolved in aqueous solution. Part II. Study of influence of various parameters on the reaction rate. *J. Electroanal. Chem.*, **43**, 185–195.

Dessy, R. E. (1977). Computer net-working: a rational approach to lab automation. *Anal. Chem.*, **49**, 1100A–1106A.

Dessy, R. E. (1978). Small laboratory computer networks. *Anal. Chim. Acta*, **103**, 459–468.

Frazer, J. W., A. M. Kray, W. Selig, and R. Lim (1975). Interactive-experimentation employing ion-selective electrodes. *Anal. Chem.*, **47**, 869–875.

Frazer, J. W., W. Selig, and L. P. Rigdon (1977). Equivalence point determinations for automated potentiometric titrations with ion-selective electrodes. *Anal. Chem.*, **49**, 1250–1255.

Gaarenstroom, P. D., J. C. English, S. P. Perone, and J. W. Bixler (1978). Computer controlled interference correction for ion-selective electrode measurements in a flowing system. *Anal. Chem.*, **50**, 811–817.

Gran, G. (1952). Determination of the equivalence point in potentiometric titrations. Part II. *Analyst*, **77**, 661–671.

Graneli, A., D. Jagner, and M. Josefsson (1980). Microcomputer system for potentiometric stripping analysis. *Anal. Chem.*, **52**, 2220–2223.

Horvai, G., L. Domokas, and E. Pungor (1978). Novel computer evaluation of multiple standard addition with ion-selective electrodes. *Fresenius' Z. Anal. Chem.*, **292**, 132–134.

Jagner, D. (1970). A semi-automatic titrator for precision analysis. *Anal. Chim. Acta*, **50**, 15–22.

Jagner, D. (1974). Computers in titrimetry. *Microchem. J.*, **19**, 406–415.

Jagner, D., and L. Kryger (1975). Computerized electroanalysis. Part III. Multiple scanning anodic stripping voltametry and its application to sea water. *Anal. Chim. Acta*, **80**, 255–266.

Jagner, D. (1979). Potentiometric stripping analysis in non-deaerated samples. *Anal. Chem.*, **51**, 342–345.

Johansson, A., and L. Pehrsson (1970). Automatic titration by means of stepwise addition of equal volumes of titrant. *Analyst*, **95**, 652–656.

Kryger, L., D. Jagner, and H. J. Skov (1975). Computerized electroanalysis. Part I. Instrumentation and programming. *Anal. Chim. Acta*, **78**, 241–249.

Kryger, L., and D. Jagner (1975). Computerized electroanalysis. Part II. Multiple scanning and background subtraction. A new technique for stripping analysis. *Anal. Chim. Acta*, **78**, 251–260.

Pelletier, M., J. Buffle, and D. Monnier (1971). Étude chronopotentiométrique de l'oxydation de quelques amalgames dilués en présence de traces d'oxygène. *Chimia*, **25**, 61–62.

Perrin, D. D. (1977). Recent applications of digital computers in analytical chemistry, *Talanta*, **24,** 339–345.

Rigdon, L. P., C. L. Pomernacki, D. J. Balaban, and J. W. Frazer (1979). Automated potentiometric analysis with selective electrodes. *Anal. Chim. Acta*, **112,** 397–405.

Skov, H. J., L. Kryger, and D. Jagner (1976). Dynamical graphical display in computerised analytical procedures. *Anal. Chem.*, **48,** 933–937.

Thomas, Q. V., L. Kryger, and S. P. Perone (1976). computer-assisted optimization of anodic stripping voltammetry. *Anal. Chem.*, **48,** 761–766.

Turner, J. A., U. Eisner, and R. A. Osteryoung (1977). Pulsed voltammetric stripping at the thin-film mercury electrode. *Anal. Chim. Acta*, **90,** 25–34.

Yamaguchi, S., and T. Kusuyama (1979). Microcomputer-aided high-speed potentiometric titration system by linear titration plots. *Fresenius' Z. Anal. Chem.*, **295,** 256–259.

Part B   Applications

Marine Electrochemistry
Edited by M. Whitfield and D. Jagner
© 1981 John Wiley & Sons Ltd.

T. R. S. WILSON
Institute of Oceanographic Sciences,
Wormley, Godalming, Surrey, UK

5

Conductometry

GLOSSARY OF SYMBOLS

A	Cross-sectional area (equation 4)
c	General concentration term (equation 1)
C_d	Diffusion layer capacitance (Figure 1)
C_i	Inter-electrode capacitance (Figure 1)
k_c	Experimental constant (equation 2)
K	Conductance (equation 4)
l	Path length (equation 4)
R_c	Cell resistance (Figure 7)
R_s	Standard resistance (Figure 7)
R_t	Conductivity ratio (relative to 35‰ salinity) at temperature t (Table 2)
R_w	Resistive elements in equivalent circuit (Figure 1)
V_c	Comparator output voltage (Figure 7)
V_{ref}	Voltage of reference source (Figure 7)
Z_F	Faradaic component to electrode impedance (Figure 1)
z_i	Ion charge (equation 6)
κ	Measured conductivity (equation 1)
Λ	Equivalent molar conductivity (equation 1)
Λ_0	Limiting equivalent molar conductivity (equation 2)
Λ_0^+	Limiting equivalent cationic conductivity (equation 3)
Λ_0^-	Limiting equivalent anionic conductivity (equation 7)
Λ_i	Equivalent ionic conductivity in sea water (equation 7)
ρ_t	Relative density at temperature t (equation 10)
σ_t	Reduced density parameter at temperature t (equation 10)

5.1. INTRODUCTION

Sea water is a complex mixture of dissolved salts. Its conductivity is largely due to these solute molecules and, since the composition of 'sea salt' is relatively constant (Section 5.2), the measured conductivity is related closely to the 'sea salt' concentration. For this reason the measurement of conductivity is a major preoccupation of oceanographers, and this is a quantity measured more frequently than any other except temperature.

Similarly, because conductivity is a function of the total ionic concentrations, it is practically impossible to use the technique to investigate the concentration or behaviour of any single ion. For this reason, the measurement of total salt concentration is the only major application of conductometry in oceanography. The exceptions to this are a few analytical procedures which utilize, for instance, conductometric monitoring in titrimetric determinations. These are minor applications compared with the conductometric determination of total dissolved salts by direct measurements on sea water, and this chapter is devoted to a discussion of this single extremely important application of the technique.

The systematic study of the conductivity of electrolyte solutions owes much to the work of Kohlrausch, who developed the basic a.c. bridge techniques and who recommended basic instrumental designs and procedures which are still of value today. These empirical measures were found necessary to reduce the effects of polarization.

The careful work of Kohlrausch and co-workers led to the definition of equivalent conductivity:

$$\Lambda = \frac{\kappa}{c} \qquad (1)$$

where κ is the measured conductivity and c is the concentration in appropriate units. Observing the behaviour of this quantity for strong and weak electrolytes, Kohlrausch found that his data could be represented by the equation

$$\Lambda = \Lambda_0 - k_c c^{1/2} \qquad (2)$$

where k_c is an experimental constant. Λ_0 therefore represents the equivalent conductivity at infinite dilution. Since the difference in Λ_0 for different pairs of salts having a common ion was always constant, Kohlrausch concluded that the contribution of each ion was independent of the presence of other ions, so that for a salt of two monovalent ions

$$\Lambda_0 = \Lambda_0^+ + \Lambda_0^- \qquad (3)$$

where Λ_0^+ and Λ_0^- are the equivalent ionic conductivities at infinite dilution (Appendix XIII). This conclusion influenced greatly the ideas of Arrhenius on the theory of electrolytic solutions. The conductance theories of Debye and Hückel and of Onsager (see, for example, Onsager, 1927; Chen and Onsager, 1977) are essentially expansions of the second term of equation 2, in which the various factors influencing ionic mobility and ionic association are identified and given quantitative expression. However, the complexity and ionic strength of sea water prevent the direct theoretical analysis of its conductance, in the present state of understanding, so that these developments have not directly affected the progress of sea water conductometry. The remainder of this section deals with the concepts which have influenced practical developments.

Consider a right cylindrical cell in which the end surfaces are electrodes. If the flow of electricity through the solution is ohmic, then the conductance (which is the reciprocal of the resistance) may be expressed by the relationship

$$K = \kappa A l^{-1} \qquad (4)$$

where A is the cross-sectional area of the cell and l is the path length between electrodes. From equations 1 and 4, we may write

$$K = \Lambda c A l^{-1} \qquad (5)$$

so that the resistance of the solution is a function of the dimensions of the cell, the equivalent conductivity and the concentration. Since, from equation 3, the equivalent conductivity is itself an additive function of the individual ionic species present, then we can express the conductance, at infinite dilution, of a more complex solution of several ions in our simple cell by

$$K = Al^{-1} \sum_i c_i \Lambda_{0i} z_i \qquad (6)$$

where c_i is the ionic concentration, Λ_{0i} its equivalent conductance at infinite dilution and z_i is its charge.

The values of Λ_{0i} for the major ions in sea water are not too dissimilar so that small variations in the relative proportions of ions at constant normality would not be expected to vary the measured conductance greatly. However, sea water is a moderately concentrated solution in which the effects of ionic association are not negligible. Therefore, although it is possible to write for sea water

$$K = Al^{-1} \sum_i c_i \Lambda_i z_i \qquad (7)$$

(where Λ_i is the equivalent ionic conductance in sea water), it is not yet possible to evaluate Λ_i in complex mixtures of ions at moderate total concentrations, although empirically determined values have been tabulated (Connors and Park, 1967; Connors and Weyl, 1968). Fortunately, the relative ionic composition of sea water varies very little throughout the world ocean. Consequently, conductance may be used to measure the total dissolved salt content or salinity of sea water. The concept of salinity depends on the assumption of constant relative composition of sea salt (Chapter 1 and Section 5.2) and the limitations of this assumption will be discussed later (Section 5.4).

It is implicit in the form of equation 4 that the flow of current in an electrolyte solution is ohmic. This was not accepted by early workers: their practical results did not appear to support it because of the phenomenon of polarization, which is the most important single influence on the design of conductance experiments. It arises as a result of processes which occur at the electrodes when a current is passed. If no reaction occurs, the electrodes behave as if they consisted of a resistance and a capacitance element in parallel with each other and connected in series with the resistance of the solution. The polarization capacitance is directly proportional to the electrode area but is invariant with current density. Physically this capacitance probably arises as the result of the formation of a Helmholtz double layer at the electrode–solution interface (Section 1.3.3). In addition, inter-electrode capacitance is important in some cell designs: this appears as a capacitance in parallel with the solution resistance.

The electrode polarization resistance has been ascribed to a number of physical causes, their relative importance being determined by experimental conditions. The most important appear to be ionic diffusion effects arising in the diffuse Gouy layer which lies between the double layer and the bulk solution (Section 1.3.3), and Faradaic effects which are observed when electrode reactions occur. The former give rise to a resistive effect which varies inversely with the square root of the exciting frequency (Warburg impedance); it is analogous to the concentration polarization effect which is observed in simple electromotive cells. The Faradaic effects are non-linear and difficult to model in terms of simple equivalent circuits (Graham, 1974). They depend on the electrode reactions which occur, and so are strongly influenced by electrolyte solution composition and the electrode material. In some reactions, such as the formation of hydrogen gas at a platinum electrode, the physical state of the electrode surface is also important in determining the rate at which the reaction progresses for a given impressed voltage.

The simple two-electrode conductance cell may thus be expressed by the complex equivalent circuit shown in Figure 1. Only effects internal to the cell are included in this circuit: external effects such as resistive or capacitive coupling between electrode leads may also be significant in practical circuits.

Consideration of Figure 1 indicates the basic design elements which are important if accurate conductance values are to be obtained from a two-electrode cell. The choice of a high a.c. frequency minimizes the impedance of the diffusion layer capacitances (C_d) relative to the resistive elements in parallel (R_w); this choice also increases the influence of the inter-electrode capacitance (C_i), which must therefore be minimized by careful design. The use of a low current density reduces the relative importance of the Faradaic component (Z_F), so that electrodes of large surface area and low excitation voltage are favourable. It is also helpful to design the cell so that the resistance of the solution under study is large relative to the polarization effects.

Not unexpectedly, these design constraints have in practice often proved irksome, particularly to designers of *in situ* devices. The development of the inductive sensor head (Section 5.7.2), which permits a conductance measurement without electrodes, is one response which has been important in the development of oceanographic instrumentation. These devices are free from

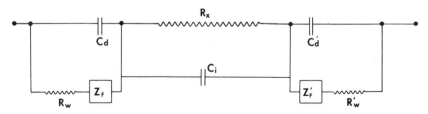

Figure 1. Equivalent-circuit representation of a two-electrode cell

polarization effects and exhibit high stability with time. They are subject, however, to proximity effects when the electric field around the head is disturbed or distorted by nearby objects.

Since the formation of a double layer and the initiation of Faradaic reactions are both results of the passage of current, it is possible to reduce the effect of polarization by the use of a four-electrode cell. In this cell the outer electrodes are connected to a known current source, while the voltage between two electrodes a known distance apart at some point in the current field is measured by a high-impedance technique. This design has been adopted in the most recent *in situ* microprofiling instruments and is described in Section 5.8.2.

5.2. THE SALINITY CONCEPT

The concept of salinity has pervaded oceanographic studies since systematic marine science began. Together with temperature, it is the basic parameter which must be measured, which is catalogued in data banks, and which is taken to characterize every sample which is drawn from the sea. It comes as something of a shock, therefore, to realise that this parameter is almost never measured directly (Section 5.5), has never been measured directly on any routine water samples with the accuracy considered acceptable in hydrographic studies (Section 5.3.2), and indeed probably never could be so measured (Section 5.4).

Why, then, is the concept so attractive? The major reason must be its superficial simplicity. The Second International Conference on the Exploration of the Sea resolved, in 1901, that 'by salinity is to be understood the total weight in grammes of solid matter dissolved in 1000 grammes of sea water' (see Cox, 1965; Wilson, 1975). The Conference itself recognized that this definition was not useful in practice because of the volatility of some of the halide salts formed on evaporation. They adopted a number of elaborations designed to arrive at a practical definition. In the years since, new methods have been developed to measure 'salinity' and the definition has been further elaborated. The apparent simplicity of the concept has long been recognized as something of an illusion. It has been proposed on more than one occasion that the concept is an unnecessary complication and should be discarded. It is, however, still used because it has proved to be a useful operational definition to workers in a wide range of disciplines, who find it convenient to think and behave as if salinity were an exact quantity, capable of exact definition. Because of this it is most important not to forget that salinity is actually a useful approximation, hedged about with assumptions. These are discussed in the following two sections.

5.3. HISTORICAL DEVELOPMENT OF THE
SALINITY CONCEPT

5.3.1. Early formulations

The term salinity was first used to describe the amount of sea salt per unit volume of sea water by Forchammer (1865). In the same monograph he also showed that the amounts of the major constituents were proportional to one another. If this were not true it would be necessary to analyse for each constituent separately in each sample; the extent to which the 'constancy of composition' holds for any sample is thus a close measure of the usefulness of the salinity concept in relation to that sample. For chloride, Forchammer determined the constant of proportionality to salinity to be 1.812 on average, although he appears to have utilized significantly different values when dealing with more limited areas of the world ocean (Wallace, 1974).

The Challenger Expedition of 1873–76 represents the start of oceanography as an organized science. The expedition chemist, Buchanan, collected a set of 77 water samples, representative of the world's oceans in geographical spread and in depth. These were analysed during the years 1878–81 by William Dittmar, who was at that time Professor of Chemistry at the University of Glasgow. His analyses were sufficiently complete, extensive, and reliable that they served as a basis for the concept of salinity for more than 80 years. His value for the ratio of chloride to total salts was 1.806, which is very close to the modern value (Section 5.3.4).

5.3.2. The 1901 definition

With the growth of interest in marine studies which followed the Challenger expedition, it soon became clear that a standard system for reporting the salt content of sea water was required. An International Commission under Professor Martin Knudsen was set up by the first International Conference for the Exploration of the Sea, at Stockholm in 1899, and its recommendations were adopted at the Second Conference at Christiana in 1901 (see Forch *et al.*, 1902). A considerable amount of very painstaking work was performed in these two years by the staff of the Commission. Although difficulties were experienced and their initial plans were somewhat modified, the quantity and quality of their results was such that they remained the means for relating chlorinity (see Sections 1.3.1 and 5.6.1), salinity, and density for the next 65 years. It was originally intended to measure these three parameters on a suite of 24 samples from the whole world ocean; in practice, however, the direct gravimetric determination of salinity presented difficulties, and for this parameter only ten analyses, on

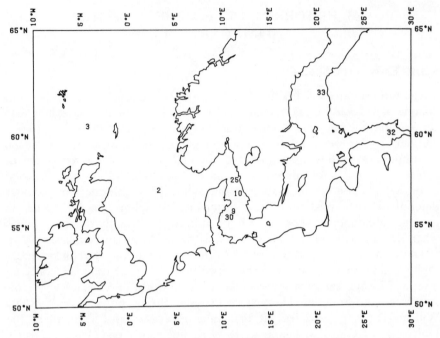

Figure 2. Positions from which surface samples were taken. These samples were
used to establish equation 8. Forch *et al.* (1902)

nine samples, were finally presented. Only two of these samples had
salinities close to that of open ocean water; apart from a single high-salinity
Red Sea sample, the rest formed a dilution series along the axis of the Baltic
(Figure 2).

The equation produced to represent the relation between the gravimetric
salinity of these samples and their chlorinity was

$$S = 1.805Cl + 0.030 \tag{8}$$

where S and Cl represent the salinity and chlorinity (Section 5.6.1), respectively, in parts per thousand by weight (‰). The presence of the constant
term was a consequence of the use of Baltic samples and resulted from the
dilution of sea water with river water containing dissolved solids other than
chloride. It was considered that the precision of the gravimetric method was
about 1 part in 2000 (or *ca.* 0.017‰ for sea water of 35‰). In his report
(Forch *et al.*, 1902), Knudsen indicated that further work using more precise
techniques would be necessary to define the relationship between chlorinity
and salinity with the accuracy needed for open ocean studies. This work was
never performed, probably because the tables produced (Knudsen, 1901)
made no reference to these reservations, which appear only in the report
(Wallace, 1974).

5.3.3. The 1940 redefinition

Since the 1901 decision effectively defined salinity in terms of chlorinity, the accuracy of determination of chlorinity became the factor limiting the precision with which salinity and specific gravity could be known. Unfortunately, the atomic weights of the elements involved in this determination were not accurately known in 1901. As these values were revised, so the calculated chlorinity for a given titre would change. This is obviously undesirable, as the atomic weight values used would need to be known before data from different sources could be compared. In 1937, therefore, chlorinity itself was redefined in terms of the mass of silver in grams needed just to precipitate the halogens in a stated mass (0.328 523 4 kg) of sea water (Jacobsen and Knudsen, 1940). The silver was of a stated high purity prepared by a known method. A sample, deposited with the Danish Hydrographic Laboratory at the time, is now held by the IAPSO Standard Sea Water Service. This is still the basic standard to which all salinity measurements are referred (but see Section 5.5.2).

5.3.4. The 1969 reformulation

The increasing use of conductometric methods, which have now completely replaced the chlorinity titration for routine high-precision oceanographic measurements, led to a careful examination of the concept and definition of salinity. An International Joint Panel was set up to consider these questions, and the accuracy of the data relating conductivity ratio to salinity (Section 5.5.1). The meetings of this panel were sponsored by UNESCO, which also published their reports (UNESCO, 1962, 1965, 1966). Because of the limited scope and accuracy of the earlier determinations, it was decided to re-determine all of the relevant parameters on a selection of sea waters collected from representative areas. A large number of carefully collected and stored sea water samples were analysed for chemical composition (Culkin and Cox, 1966; Morris and Riley, 1966; Riley and Tongudai, 1967), chlorinity, and conductivity ratio (Cox *et al.*, 1967). It was also hoped to determine the absolute conductivity of these samples, so that conductometric measurements could be directly related to the units of electric current and length, but the tragic early death of Dr. Roland Cox prevented this study from being concluded at that time (Section 5.9).

In the light of the results of this work, the International Joint Panel decided to alter the relationship between chlorinity and salinity hitherto used. The new relationship calculated was

$$S = 1.80655 Cl \qquad (9)$$

which in the oceanic salinity range gives results very close to the old relationship (Lyman, 1969). The panel further recommended that, in future, chlorinity determined by titration should be reported as such, while conductometric measurements should be used in conjunction with the International Oceanographic Tables to derive salinity. This aspect is discussed in greater detail in Section 5.5.

5.4. THE LIMITATIONS OF THE SALINITY CONCEPT.

The history of the salinity concept, discussed above, deals mainly with the practical difficulties of selecting a reproducible measurement technique which will give a result approximating to total dissolved salts. The fundamental limitations of the concept have only recently become the object of close study; this is a consequence of improvements in technology, particularly that of *in situ* conductivity measurement, and of the more demanding requirements of modern studies in marine physics.

The composition of a typical ocean surface water of 35‰ is given in Table 1, together with estimates of the accuracy with which each value is known

Table 1. The major ions of surface sea water, salinity 35‰

Ion*	g kg^{-1}	g kg^{-1}/S‰‡	Molarity	Accuracy, %§	Precision, %‖	Ref.¶
Cl$^-$	19.354	0.5529	0.5452	0.002	0.002	1
SO$_4^{2-}$	2.712	0.0631	0.0282	0.4	0.1	2, 3
Br$^-$	0.0673	0.00192	8.42×10^{-4}	2.0	0.2	4
F$^-$	0.0013	3.71×10^{-5}	6.84×10^{-5}	1.0	0.6	5
HCO$_3^-$†	0.1403	4.01×10^{-3}	2.3×10^{-3}	†	0.2	6, 7
Si(OH)$_4$†	0.00682	1.95×10^{-4}	7.1×10^{-5}	†	2.5	8
B(OH)$_3$	0.0253	7.25×10^{-5}	4.1×10^{-4}	5	0.5	9
Na$^+$	10.77	0.3078	0.4685	0.13	0.13	10, 11
Mg^{2+}	1.290	0.0368	0.0531	1.0	0.06	7, 10, 11, 12
Ca^{2+}	0.4121	0.0118	0.0103	0.75	0.1	10, 11, 13
K$^+$	0.399	0.0114	0.0102	0.5	0.12	10, 11, 14
Sr^{2+}	0.0079	2.27×10^{-4}	9.016×10^{-5}	1.0	0.2	15, 16

* The most important ionic form is quoted. Minor ionic forms, if any, are included in the abundance values given.
† Bicarbonate and silicate are not conservative: approximate values quoted.
‡ 35‰ salinity corresponds to 19.3739‰ chlorinity.
§ Accuracy with which composition of sea water is known: subjective assessment of published data. Includes effects of natural variation of composition if this is larger than analytical uncertainty.
‖ Precision is the quoted precision of the best available method (one standard deviation).
¶ 1, Reeburgh and Carpenter (1964); 2, Bather and Riley (1954); 3, Jagner (1970); 4, Morris and Riley (1966); 5, Anfält and Jagner (1971); 6, Edmond (1970); 7, Almgren *et al.* (1977); 8, Strickland and Parsons (1965); 9, Greenhalge and Riley (1962); 10, Culkin and Cox (1966); 11, Riley and Tongudai (1967); 12, Carpenter and Manella (1973); 13, Tsunogai *et al.* (1973); 14, Mangelsdorf and Wilson (1971); 15, Brass and Turekian (1972); 16, Brass and Turekian (1974).

from available data in the literature, and the quoted precision of the best available methods. It will be noted that in several instances the disagreement between different studies is larger than would be expected from the quoted values of precision. This reflects difficulties in the absolute calibration of the methods (Wilson, 1975); it is not appropriate to discuss these problems in detail here, but they should not be forgotten when any individual analytical study is being planned or evaluated.

The real variability of sea water composition is difficult to distinguish from apparent variations generated by analytical noise. Composition variations are small, but in many cases they are significant at the levels of accuracy required by modern hydrographic studies and made available by recent instrumental developments. Most of the variation in open ocean samples is caused by the dissolution of calcium carbonate and silica, and the diagenesis of organic carbon in deep ocean water (see Section 1.4.2). This water is formed by the sinking of surface waters in the North Atlantic and progresses over a period of 1000–2000 years to the deep North Pacific. Thus there is a systematic difference between Atlantic surface and Pacific deep waters. The contribution to the salinity of the Pacific deep waters by dissolution of particulate biogenic material has been considered by Brewer and Bradshaw (1975) and Millero *et al.* (1977). Calcium carbonate dissolution increases the conductometric salinity by about 0.001‰, and organic carbon oxidized to CO_2 contributes about 0.007‰. Hence, if conductometric salinity is to be used as a conservative tracer these corrections must be applied. Along its trajectory, the deep water increases in salinity by about 0.2‰ due to conservative mixing processes. The effects discussed here therefore amount to about 4% of the total signal.

Larger changes in the relative composition of sea water occur in more restricted seas. The classic example of this phenomenon occurs in the Baltic, where the influence of river outflow waters relatively rich in dissolved carbonate and calcium ions causes progressive deviation at lower salinities from the composition predicted on the basis of dilution of sea water with distilled water (Kremling 1970). The observed differences between the conductometric- and chlorinity-derived salinities can be explained in terms of this conservative mixing model (Millero and Kremling, 1976). A similar situation occurs in the Black Sea, where surface exchange is also restricted, with an excess of fresh water supplied by run-off. In the Red Sea there is also restricted exchange; here, however, there is little contribution from surface run-off so that the relative composition of Red Sea surface water is close to that of open ocean water, although the salinity is much higher. Some slight loss of calcium occurs, apparently because of the deposition of biogenic carbonate (Wilson, 1975).

Almgren *et al.* (1977) have published analyses for several stations in the Pacific which show variations of up to 2% in the magnesium to chlorinity ratio. As Almgren *et al.* point out, a variation of this magnitude implies a corresponding difference of 0.15–0.03‰ between the chlorinity-derived and the conductivity-derived salinity values for the same sample. This is larger than has been observed (Cox *et al.*, 1967). The observation is also at variance with those of Culkin and Cox (1966) and of Riley and Tongudai (1967) (see Table 1).

The difference chromatographic method, which compares a sample sea water with a standard directly (Mangelsdorf and Wilson, 1971) also failed to show significant variations in the magnesium content of sea water (the standard deviation of 880 samples from the Atlantic and Pacific oceans was ±0.09%) (Wilson, 1975; Mangelsdorf and Wilson, unpublished data). Of these samples, 345 were taken on a transect North to South which intersects the E–W transect of Almgren *et al.* at about 20°N 150°W. At this point the results of Almgren *et al.* indicate a variation of about 0.4% in the water column; the difference chromatographic data show a maximum (peak-to-peak) variation of 0.11% at this station. In view of the weight of evidence suggesting rather small variation in the Mg:Cl ratio, the results of Almgren *et al.* (1977) require confirmation. It is also true that the absolute value of this ratio is not as well characterized as would be wished (Table 1) (Wilson, 1975), so that further measurements of this parameter would be welcome.

In modern physical oceanography it is customary to quote salinity, measured conductometrically, to 0.001‰. However, the results cited in this section indicate that, in an absolute sense, this procedure may mask errors of up to ten times this value simply because variations in sea water composition are conventionally ignored. This obtains even if a perfect conductivity instrument introducing no error is used; such instruments are unfortunately not yet available (Section 5.8.4). It is true that the error is greatest in comparing results for water masses geographically distant from each other. Even so, oceanographically significant discrepancies have arisen from this cause (Poisson *et al.*, 1978), and such instances may be expected to become more frequent as instrumentation is improved.

Other factors which tend in practice to limit the accuracy and range of applicability of salinity data are discussed in Section 5.6, on methods of salinity measurement, and Sections 5.7 and 5.8, on instrumentation.

5.5. THE RELATIONSHIP OF SALINITY TO OTHER VARIABLES

5.5.1. Conductivity

Following the discussions of an International Joint Panel (UNESCO, 1962, 1965, 1966), conductivity measurement was adopted in 1969 (Wooster *et al.*, (1969) as the recommended method of comparing a sample salinity with that of a standard, replacing the chlorinity method which had been in use up to that time. Since absolute conductivity (in terms of basic units of length and time) is very difficult to measure (Section 5.9), a ratio technique was retained in which the conductivity of the sample is compared with that of IAPSO Standard Sea Water. The procedures for preparing and standardizing this water were not altered in 1969 (but see Section 5.5.2), so that the *absolute* definition of salinity in terms of a high-precision chlorinity titration against pure silver remained unchanged.

Consideration was given to the possibility of reporting a conductivity ratio directly, rather than a salinity derived from it, but this was rejected on grounds of continuity and convenience. A study of the relationship between conductivity ratio at 15°C and salinity (the latter derived from accurate chlorinity measurement) was undertaken (Cox *et al.*, 1967), and from these results a polynomial (Table 2) was calculated to enable this relationship to be applied to any conductivity ratio measured at this temperature.

Since many salinometers operate at temperatures other than 15°C, a second polynomial to allow the ratio at 15°C (R_{15}) to be related to that at any temperature ($t°C$) between 10 and 30°C (Table 2) was also calculated. The International Oceanographic Tables (UNESCO, 1966) include, *inter alia*, tabulated versions of these data for use when a computer is not available or appropriate.

Although the Panel report (Wooster *et al.*, 1969) and the International Oceanographic Tables refer to these expressions as a new 'definition of

Table 2. Procedure for the derivation of salinity from laboratory measurement of the conductivity ratio (R_t) of a sea water sample measured at temperature $t°C$

Definitions: $R_t = \dfrac{\kappa_{(S,t,0)}}{\kappa_{(35,t,0)}}$

$R_{15} = \dfrac{\kappa_{(S,15,0)}}{\kappa_{(35,15,0)}}$

$\Delta_{15}(t) = R_{15} - R_t$

where κ is the conductivity of the sample under the stated conditions of salinity, temperature (°C) and pressure (decibars) above atmospheric pressure.

Empirical relationships:

$$\Delta_{15}(t) = 10^{-5} R_t (R_t - 1)(t - 15)[96.7 - 72.0\, R_t + 37.3\, R_t^2 - (0.63 + 0.21\, R_t^2)(t - 15)]$$

$$S = -0.08996 + 28.2972\, R_{15} + 12.80832\, R_{15}^2 - 10.67869\, R_{15}^3$$
$$+ 5.98624\, R_{15}^4 - 1.32311\, R_{15}^5$$

Empirical relationships valid in the range 30‰ $< S <$ 40‰, 10°C $< t°C <$ 30°C and at a pressure of 1 atmosphere (10 1325 Pa)

---

salinity' it should be noted that they merely express sample conductivity in terms of that of a notional 35‰ water (Table 2). Since this cannot itself be defined except by reference to its chlorinity, it is more accurate to refer to the empirical polynomial (Table 2) as a reformulation, in terms of conductivity ratios, of the Jacobsen and Knudsen (1940) definition of chlorinity.

This international agreement, together with the universal availability of IAPSO Standard Sea Water, has provided a common basis which allows bench salinometer measurements made by different workers at different times to be compared with confidence. The improvements in *in situ* instrumentation which have been made over the past decade have, however, opened up another important area of study for which the situation is as yet considerably less satisfactory. By their nature, *in situ* devices frequently operate below the temperature range of the data of Cox *et al.* (1967). They also make measurements at high hydrostatic pressure. The available data on the relationship between conductivity ratio and pressure and temperature are sparse: in contrast, a wealth of routines have been developed for the reduction of conductivity, temperature and depth (CTD) data to salinity. Lewis and Perkin (1978) list twelve such routines in addition to the relationships mentioned above (Table 2). All use the data of Bradshaw and Schleicher (1965) for the effect of pressure. All but one utilize the data of Brown and Allentoft (1966), with or without that of Cox *et al.* (1967), to establish the conductivity–temperature relationship. Most are employed only by their originating institution. Walker and Chapman (1973), Walker (1976), and Lewis and Perkin (1978) have investigated the uniformity of these equations. For a salinity of about 30‰ a spread of 0.02‰ was found in the calculated results from the same input data. Since the attainable accuracy of modern *in situ* instruments has been shown by comparison with samples taken close to the sensor package to be at least ±0.003‰ (Fofonoff *et al.*, 1974), this is obviously an unsatisfactory state of affairs, moderated slightly by the universal practice of taking samples for bench analysis at several points on each profile.

Obviously a standard data reduction process should be agreed on internationally. Unfortunately, the data-base mentioned above is not adequate, and recent attempts to augment it have revealed minor but significant inconsistencies. Dauphinee and Klein (1977) have measured the temperature dependence of the conductivity ratio of IAPSO Standard Sea Water, and have found values different by 0.009‰ from the only comparable data set at 0°C, although the difference is small in the normal abyssal temperature range around 2°C. It is hoped that improved data will be available from several workers in the near future (Lewis and Perkin, 1978), and that it will then prove possible to achieve agreement on the preferred system for reduction of *in situ* data.

Because of the almost universal use of conductivity measurements to obtain salinity values, it is expected that the calibration of IAPSO Standard Sea Water will shortly be changed from a chlorinity-based system to one based on conductivity measurement (see Sections 5.5.3, 5.9, and 5.10.4).

5.5.2. Density

Density and density-related parameters are derived quantities basic to studies of the physics of ocean circulation. Although recent instrumental developments have improved and simplified the direct determination of density in the laboratory (Kratky *et al.*, 1969; Kremling, 1972), no method of making direct *in situ* measurement of density has yet proved sufficiently practical for wide adoption. *In situ* conductivity measurement has by contrast reached high levels of precision and accuracy, so that the degree to which the relationship between conductivity and density is known becomes a question of considerable practical importance.

Knudsen and co-workers (Knudsen, 1901; Forch *et al.*, 1902), who performed valuable early work on this subject, defined the quantity σ_t, which is conventionally used to express the density of sea water:

$$\sigma_t = (\rho_t - 1)1000 \qquad (10)$$

where ρ_t is the density of the sample at the temperature ($t°C$) of measurement related to that of water at 3.98°C, the temperature of maximum density. It is therefore, in fact, a specific gravity value and no absolute density values were used by Knudsen. The absolute density of pure water at 3.98°C is about 30 ppm below unity, the exact value depending on isotopic composition (Girard and Menaché, 1971). It is thus rather confusing to refer to σ_t as 'density', a usage unfortunately frequent in the oceanographic literature. Measurements of σ_t have been made by several groups of workers in response to the improved data requirements mentioned above. Cox *et al.*, (1970) reported that less scatter was found in the relationship between conductivity and density calculated from their results than in the relationship between chlorinity and density. Since they measured 50 samples of surface water collected from most of the important oceans and seas, it is probable that this finding is a consequence of the fact that both conductivity and density reflect total dissolved salt better than chlorinity, which is a measurement only of halogen ion content. Small variations in the relative composition of sea salt between samples, which would be expected for this sample set, would thus produce the observed increased noise level in the chlorinity to conductivity relation.

Cox *et al.* (1970) concluded that the results of Forch *et al.* (1902) were only about 0.006 in σ_t (6 ppm in specific gravity) below the values found by the new work. Since the isotopic composition of the standard distilled water used by Forch *et al.* is unknown, it is possible that this difference could be accounted for by a difference between this distilled water and that used as a standard by Cox *et al.* which was produced by a defined procedure and was of known isotopic composition (Menaché, 1971). Kremling (1972), Millero and Lepple (1973), Fofonoff and Bryden (1975), Millero *et al.* (1977), and Poisson *et al.* (1978) have conducted independent measurements on sea water samples from various sources. Their results agree with those of Cox *et al.* (1970) in showing that the tables of Forch *et al.* (1902) are accurate to about 1 part in 10^5.

At this level of accuracy effects due to variations in the ionic composition of sea water become significant. These arise in two ways: the addition or removal of material (i) affects the density directly and (ii) may also affect the conductometric salinity in the manner discussed above (Section 5.5.1). Hence the density calculated from the measured salinity may deviate from the true value. The direct effect on density may be calculated from a knowledge of the partial molar volume in sea water of the species which undergo variation (Brewer and Bradshaw, 1975; Millero *et al.*, 1976). Partial molar volume data for sea water were given by Duedall (1972) and Bradshaw (1973). As in the salinity error calculation, the changes which contribute the largest variation in true density are those due to the dissolution of calcium carbonate and silica and of CO_2 derived from the diagenesis of organic material. The total difference between the true density and that derived from conductometric salinity by the application of the standard salinity–density relationships is the sum of the conductometric error and the real change in density, and in the North Pacific deep water amounts to an increase of 11 ppm in true density compared with the conductometrically derived value. Brewer and Bradshaw note that this error may have been responsible for an apparent discontinuity in isopycnal (constant density) surfaces between the depths reported for the Atlantic (Lynn and Reid, 1968) and the Pacific (Reid and Lynn, 1971). At its largest this apparent discontinuity is of the order of 500 m and is difficult to explain except by consideration of changes in the conductivity to density relationship as the deep water passes along its trajectory between the two oceans.

5.5.3. The future of the salinity concept

The increasing use of the conductometric method and the improvements in its reproducibility have called into question the use of chlorinity as the basis for the definition of salinity and indeed the significance of the salinity concept itself. It seems probable that the latter will survive, since it is useful to a wide range of workers, many of whom are prepared to sacrifice some rigour and accuracy in the interests of conceptual simplicity (UNESCO, 1965). However, it is probable that for the highest accuracy, and especially for *in situ* measurements, the density and related parameters will increasingly be calculated directly from conductivity ratio measurements. It has been decided (Lewis and Perkin, 1978) that the use of a chlorinity measurement to define the composition of IAPSO Standard Sea Water will be superseded by the use of conductivity. The absolute conductivity standard needed to implement this procedure will be a solution of potassium chloride, measured under defined conditions. The concentration of this solution will be chosen so as to give a *de facto* continuity between the present standard and all future batches. Field procedures will thereby be unaffected by the change, the only changes in procedure being made at the Wormley Laboratory where IAPSO Standard Sea Water is now produced (Section 5.10.3). The advantages hoped for are an operationally simpler standardization procedure, and the removal of slight batch to batch variation in the conductivity of the standard sea water which occurs because of slight compositional variations at the same nominal chlorinity value (Millero *et al.*, 1977; Poisson *et al.*, 1976; see Figure 3). In addition, a unified series of equations has been produced, which permit the determination of salinity from conductivity ratio measurement over a wide range of salinity, temperature, and pressure (Table 3). These changes are currently under discussion by the International Joint Panel on Oceanographic Tables and Standards (Grasshoff *et al.*, 1978). If changes of the nature outlined are agreed, they

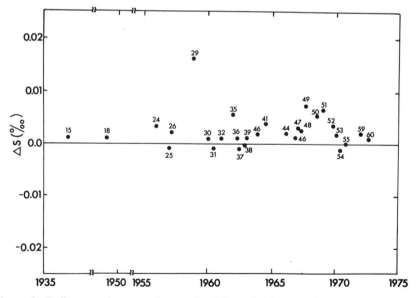

Figure 3. Differences between the conductivity salinities and the chlorinity salinities of Standard Sea Water produced at various dates, relative to batch P64. Millero *et al.* (1977). Reproduced by permission of Pergamon Press Ltd.

Table 3. Proposed redefinition of salinity in terms of conductivity ratio (R) temperature (°C), and pressure (decibars). Working equations for use with *in situ* and laboratory instruments (for full discussion see Perkin and Lewis 1980). In the oceanographic range $(S = 30\text{–}40\text{‰}, \ t = 0\text{–}30°C, \ P = 0\text{–}10^4 \text{ dbar})$ the overall standard deviation of this fit to the experimental data is estimated at 0.0015%

$$R = \frac{\kappa_{(S,t,P)}}{\kappa_{(35,15,0)}}$$

$$R_t = \frac{R}{R_p r_t}$$

where

$$R_p = 1 + P(A_1 + A_2 P + A_3 P^2)(1 + B_1 t + B_2 t^2 + B_3 R + B_4 Rt)^{-1}$$

$$r_t = c_0 + c_1 t + c_2 t^2 + c_3 t^3 + c_4 t^4$$

$$S = \sum_{j=0}^{5} a_j R_t^{j/2} + \frac{(t-15)}{1+k(t-15)} \sum_{j=0}^{5} b_j R_t^{j/2}$$

and

$A_1 = 2.070 \times 10^5$	$a_0 = 0.0080$
$A_2 = -6.370 \times 10^{-10}$	$a_1 = -0.1692$
$A_3 = 3.989 \times 10^{-15}$	$a_2 = 25.3851$
$B_1 = 3.426 \times 10^{-2}$	$a_3 = 14.0941$
$B_2 = 4.464 \times 10^{-4}$	$a_4 = -7.0261$
$B_3 = 4.215 \times 10^{-1}$	$a_5 = 2.7081$
$B_4 = -3.107 \times 10^{-3}$	$b_0 = 0.0005$
$c_0 = 6.766097 \times 10^{-1}$	$b_1 = -0.0056$
$c_1 = 2.00564 \times 10^{-2}$	$b_2 = -0.0066$
$c_2 = 1.104259 \times 10^{-4}$	$b_3 = -0.0375$
$c_3 = -6.9698 \times 10^{-7}$	$b_4 = 0.0636$
$c_4 = 1.0031 \times 10^{-9}$	$b_5 = -0.0144$
	$k = 0.0162$

are expected to be promulgated by the end of 1980, by which time it is intended that all batches of standard sea water will be labelled with the appropriate R_{15} value (Table 2), as well as chlorinity. A very comprehensive account of the investigations of the Joint Panel on Oceanographic Tables and Standards, together with their conclusions, will be found in the *IEEE Journal of Oceanic Engineering* (Vol. 5, No. 1, January 1980) which contains ten papers on the work of the Panel.

5.6. METHODS OF SALINITY MEASUREMENT

Any method used for the measurement of open ocean salinity must satisfy the following basic requirements. It should be accurate; an accuracy of 1 part in 10^4 in the field is adequate. It should not be affected by poor environmental conditions or operator fatigue. It should be mechanically reliable and rapid in use. For many studies it must be capable of *in situ* operation. Instruments based on the measurement of conductivity have proved most capable of meeting these stringent requirements. In this section, alternatives which have been used are summarized and the practical aspects of conductivity measurement are briefly discussed.

5.6.1. Chlorinity

Chlorinity measurement is carried out by the titration of the halide contained in the sample by addition of silver nitrate solution. A considerable number of methods based on this principle have been used (Wilson, 1975). At its most accurate, in the form used to standardize IAPSO Standard Sea Water, the method is capable of a precision of about 1 part in 10^5. However, in field conditions weighing is not practical, and using volume techniques a skilled worker may expect to return a precision no better than 5 parts in 10^4 ($\pm 0.01Cl‰$). The method is therefore not generally used for oceanic studies, being mainly of interest for historical reasons.

5.6.2. Optical methods

The refractive index of sea water is more closely related to the total dissolved solids and hence to density than any of the other parameters used to estimate salinity (Mangelsdorf, 1967). In theory this would recommend its use, but practical difficulties have limited the accuracy attainable in the field to about 3 parts in 10^4. Russian workers have made extensive use of shipboard refractive index measurement at this level of accuracy (Vel'mozhanya, 1960). No *in situ* device has yet been produced.

5.6.3. Other approaches

The laboratory method for the direct measurement of specific gravity mentioned above in Section 5.5.2 has been applied to the routine measurement of sea water samples (Kremling, 1972), where it has found particular usefulness in areas of restricted circulation where constancy of composition does not hold closely. The method is based on measurement of the frequency of vibration of a tube filled with sample and could in principle be applied to *in situ* measurement (Kuenzler, 1968). However, practical problems connected with the effect of temperature on the system have so far prevented the adoption of this approach.

The membrane salinometer, which depends on the measurement of the voltage developed across an ion-exchange membrane separating the sample from a standard sea water, has proved useful for estuarine measurements and for the measurement of interstitial waters of sediments (Sanders *et al.*, 1965; Mangelsdorf, 1967; Wilson, 1971). Gieskes (1967) has discussed the theory of these devices, which show greatest accuracy at low salinities and are thus not in general useful for open ocean measurements.

The remainder of this chapter will be devoted to conductometric salinometers for laboratory and *in situ* measurements.

5.7. THE DESIGN OF LABORATORY SALINOMETERS

In solution, ionic charge transport is the chief mechanism for conduction; in metallic conductors the mobile charged moieties are electrons. In order to connect a test element of sea water into an instrumental circuit it is therefore necessary to provide a means, the electrodes, by which charge may pass from one carrier type to the other in a quantitative and reproducible way. Ideally these electrodes would form a chemically reversible cell, completely 'transparent' to the passage of the insignificantly small measuring electric current (i.e. the two electrode reactions would be the same, but progress in opposite directions) (Lingane, 1958). In practice, a finite current must pass and in consequence polarization may occur.

Effects associated with electrode processes may dominate and control the current flow through the cell unless reduced by careful design (Section 5.1).

The second practical consideration which dominates the design of all conductive instruments is the effect of temperature. In contrast to electronic conduction, ionic conduction increases with increase of temperature by about 2.5% $°C^{-1}$ for sea water. Conductivity is almost directly proportional to total dissolved sea salt. Hence, for deep ocean studies at least, the temperature must be measured or controlled to better than $\pm 0.002°C$ or compensation for the effects of temperature variations greater than this must be introduced.

5.7.1. Temperature control

The requirement for accuracy in temperature control is in practice very difficult to satisfy. Until recently, designers had eased the problem by adopting a differential technique. In such a design a reference element is provided which has a temperature coefficient of conductivity matched as closely as possible to the sample. The sample and reference are then compared directly by a bridge circuit. Provided that care is taken to ensure that the sample and reference are at the same temperature at any instant during the measurement, the requirements for accuracy of temperature control are eased by about two orders of magnitude by this approach.

The reference element most usually adopted is a duplicate cell, identical with the sample cell, filled with a reference sea water. For oceanic studies this may be IAPSO Standard Sea Water. Unfortunately, the temperature coefficient of conductivity varies with salinity, so that for samples of considerably higher or lower salinity than the reference, temperature error again becomes significant. It may be minimized by the use of a more accurate thermostat, or by adopting a reference sea water closer in salinity to the samples under investigation. Operational routines used with this type of compensation system are mentioned in Section 5.10, where the design of particular instruments is discussed.

Although earlier thermostat salinometers ran at higher temperatures, most thermostat instruments have adopted 15°C as the bath operating temperature. This requires the use of a refrigeration unit which increases the bulk and cost of the instrument, but has the advantage that gas bubbles are less likely to form in the cell, since the solubility of gases is greater at low temperatures. The presence of such bubbles can render accurate measurement impossible by causing unpredictable changes in the effective geometry of the cell, so that it is worth going to considerable trouble to prevent their occurrence.

A new design of salinometer, manufactured by the Guildline company, has appeared. This instrument is noteworthy in that it relies for its accuracy on a very precise water-bath held at a known temperature chosen to be just above ambient. The bath stability is better than 0.001°C per day with a setting accuracy of 0.01°C, and a continuous flow of sample is used to minimize the possibility of bubble release (Section 5.10).

Attempts to reduce the size and cost of bench salinometers have resulted in non-thermostatted salinometers in which the reference element consists of a temperature-measuring device, such as a thermistor or platinum resistance thermometer, incorporated in a resistance network designed to match the temperature coefficient of conductivity of sea water. Considerable ingenuity has been exercised to produce circuits which mimic the behaviour of sea water well enough to permit the use of non-thermostatted cells and still give acceptable accuracy. Salinometers of this design, using inductive measuring cells, are probably the most commonly used at present. Their rapidity, simplicity, and reliability, together with the facts they can be carried in one hand and take up little bench space, account for their popularity. However, because the temperature compensation circuit cannot accurately simulate the behaviour of sea water of extreme salinities they are not suitable for use outside the normal oceanic salinity range if the utmost accuracy is required (Grasshoff and Hermann, 1976).

5.7.2. Cell design

Early thermostat salinometers required a cell resistance greater than 1000 Ω for reasons connected with the design of precise resistance bridges.

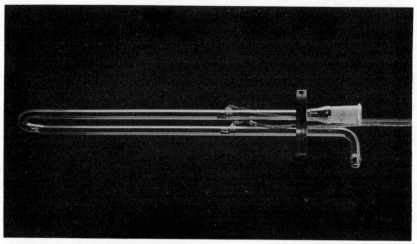

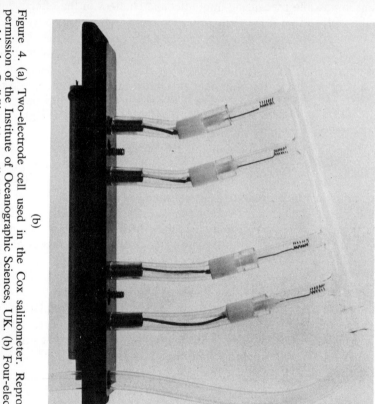

(b)

Figure 4. (a) Two-electrode cell used in the Cox salinometer. Reproduced by permission of the Institute of Oceanographic Sciences, UK. (b) Four-electrode cell used in the Guildline 'Autosal' salinometer. Reproduced by permission of Guildline Instruments, Ltd., Smiths Falls, Ontario, Canada

Consequently, the cells adopted had a considerable path length and small internal diameter (Figure 4a). For obvious reasons cells of this type are difficult to clean, are particularly susceptible to bubble formation, are not easy to manufacture to a uniform specification, and take some time to rinse, fill, and empty. Use of more sophisticated circuitry has permitted the design of cells which are free from some of these objections. In particular, the cell design used in the Guidline Co. instrument is much more compact (Figure 4b). It is a four-electrode design, in which the square-wave excitation current passes between the outer pair of electrodes, while the signal voltage between the inner pair is sensed by an amplifier of high input impedance so that very little current passes through these electrodes. This minimizes errors due to polarization. The current path is defined by the geometry of the main body of the cell, small variations in the side arms having less influence; since the main body can be cleaned with an ordinary bottle-brush without risk of damage to the electrodes, the problem of cell fouling by organic material from the sample sea water is much reduced.

Both of these cells utilize metal electrodes in contact with the sea water sample. An alternative system is the inductive sensor method, a schematic drawing of which is shown in Figure 5.

A transmitter and receiver coil are arranged side by side on a common axis inside a toroidal sensing head. An a.c. excitation field is generated by the transmitter coil; the degree of coupling between the coils, which is a function of the conductivity of the surrounding medium, determines the signal voltage sensed by the receiver coil. These cells are much more resistant to fouling than designs with exposed electrodes, and of course they can have few polarization problems. They are also mechanically strong and

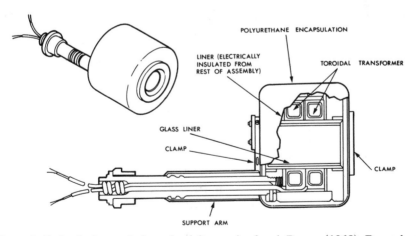

Figure 5. Inductively coupled conductivity sensing head. Brown (1968). Reproduced by permission of Compass Publications, Inc

stable. Because there is a practical lower limit to the size of the toroid, the cells require a rather larger sample than normal to rinse and fill them (about 150 cm³). This does not constitute a problem in normal applications.

5.7.3. Electronic design

In the variety of circuit designs employed, possibly the only common factor is the use of some form of a.c. system to avoid polarization. Early designs for laboratory work such as the instrument Wenner *et al.* (1930) developed for the International Ice Patrol, used circuits based on the simple resistance bridge. As this instrument was developed (Paquette, 1959) the basic circuit was elaborated in order to increase its stability and accuracy, and to decrease the effects of stray capacitance and earth leakage. Cox (1965) gives a useful account of this development.

Succeeding devices abandoned the resistance bridge concept in favour of the transformer bridge system, since these devices are more robust, more stable with time, and do not suffer from earth-leakage problems. (Cox, 1958, 1965). Later versions of the thermostat salinometer, and all commercial versions of the inductive bench salinometer (Brown and Hamon, 1961), have used this system, which is shown schematically in Figure 6. The voltage applied to the sample cell is varied by means of switched taps on the secondary of transformer T_1 until a null is obtained on the centre-zero galvanometer G. Since the voltage is proportional to the number of turns, the position of the tapped turn is a measure of the conductance of the sample relative to the standard. In practice, a more complex transformer system is used, in which successive decades are represented by a series of

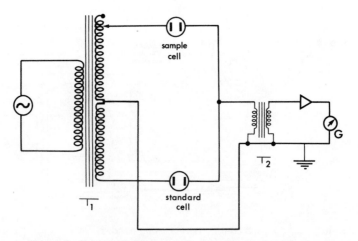

Figure 6. Schematic diagram of a transformer bridge circuit

inter-tapped transformers, each transformer energized by a tapped connection to the previous one corresponding to one tenth of the previous decade. To balance the bridge the operator selects the appropriate setting for each decade on a bank of ten-position rotary switches each connected to its own transformer tappings. When a null is obtained the appropriate conductivity ratio can be read directly from the settings of these switches. The advantage of this system over the simple arrangement shown in Figure 6 is that the latter would require a large transformer with several thousand taps to obtain the resolution needed for oceanic salinity measurement; the former can be built up in modular fashion, and gives the degree of resolution required with only five inter-tapped transformers.

The most recent type of bench salinometer to be widely adopted, the Guidline Autosal, due to Dauphinee (1968), has broken away from the basic bridge configuration. Use of the four-electrode cell permits the current passing between the two outer electrodes to be varied so as to produce a constant voltage between the inner voltage-sensing elements. A schematic diagram is shown in Figure 7. A square-wave voltage at 250 Hz passes through the cell R_c and a standard resistance R_s. Comparator A compares the voltage V_{ref} from a precision source with the voltage between the sensing electrodes V_c and adjusts the current source so as to make V_c equal to V_{ref}. The voltage V_s across R_s is now a linear analogue of the cell

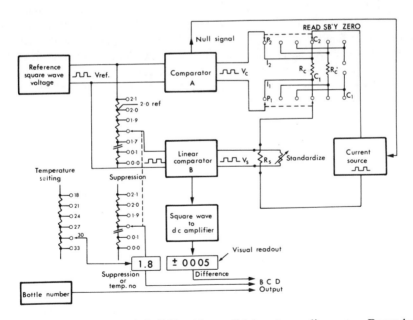

Figure 7. Block diagram of Guildline 'Autosal' laboratory salinometer. Reproduced by permission of Guildline Instruments, Ltd., and Compass Publications, Inc

conductivity. It is compared with the standard voltage V_{ref} by comparator B and the difference is displayed on a digital panel meter. By previous calibration with IAPSO Standard Sea Water the current parameters are so chosen that this difference reading is equal to $2R_t$, where R_t is the ratio of the sample conductivity to that of the IAPSO standard. Since the thermostat temperature (t) is accurately controlled and known, the salinity may be calculated from this reading (Table 2).

5.8. THE DESIGN OF *in situ* SALINOMETERS

5.8.1. Temperature Measurement

Because of the very high temperature coefficient of conductivity, the accuracy of temperature measurement must also be high. For reasons of response time the IAPSO Standard Sea Water reference cell system used to compensate for temperature effects in some bench devices cannot be applied to *in situ* measurements and hence the temperature must be measured to $\pm 0.002°C$ or better. The stability requirement which this imposes means that the only device which has been used successfully in this application is the resistance thermometer, usually with a platinum element. Much effort has been devoted to improving the response time of this sensor. The salinity measured at any instant is strongly dependent on the instantaneous temperature and conductivity readings so that these two sensors must have matched time constants if spurious transient signals are to be avoided. The conductivity response time can easily be made very short so that, in practice, it is the response time of the temperature sensor which controls the time resolution of the instrument. The precision micro-profiler developed by Brown (1974) uses a hybrid system in which the response time of the basic platinum thermometer is improved by adding to its output the differentiated output of a fast-response thermistor bridge. In this way a response time of 30 ms is achieved without sacrifice of long-term accuracy.

5.8.2. Cell design

Until recently, only inductive sensors (Figure 5) were used for the *in situ* measurement of conductivity, because of their robust construction, stability, and relative insensitivity to fouling (Brown, 1968). However, these cells are not easily scaled down to the small sizes demanded for the most detailed fine structure studies (Brown, 1974) and are subject to proximity effects. More recent designs have used four-electrode microcells of the types shown in Figures 8 and 9. In both of these designs attempts have been made to site the potential-sensing electrodes in regions of relatively low potential gradient, so that small shifts of electrode position (apparent or real) caused by

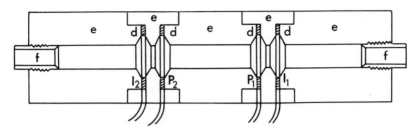

Figure 8. Four-electrode cell used in the *in situ* conductivity measuring instrument of Dauphinee (1972). Reproduced by permission of Instrument Society of America. Current electrodes I_1 and I_2 and potential electrodes P_1 and P_2 are stainless steel rings (d), held between sections of Pyrex tube (e). Cell resistance can be adjusted by movement of the threaded-end pieces (f)

shock or fouling have little effect on the voltage sensed. In the most recent version of the Guildline Co. instrument the electrodes are recessed even further into glass side arms in a manner similar to the cell of the Guildline Autosal bench salinometer (Figure 4b). These cells are pre-calibrated and can be exchanged in the field without opening the instrument pressure case.

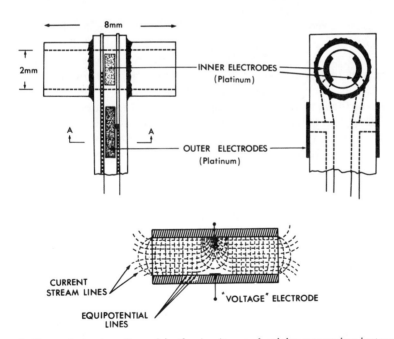

Figure 9. Four-electrode cell used in the *in situ* conductivity measuring instrument of Brown (1974). Reproduced by permission of the Institute of Electrical and Electronics Engineers, Inc

5.8.3. Electronic design

Because of the hostile environment and the high precision required, the measurement of salinity by *in situ* devices represents a considerable challenge to the electronic designer. The first such device to achieve wide acceptance was that due to Brown (1968). This design uses a hard-wired analogue system to compute the salinity from the measured conductivity and temperature before transmitting salinity, temperature and depth (pressure) information up the single conductor power supply and suspension cable. The data transmission system uses audio signals, a separate band being reserved for each parameter. Because of the limited resolution of this telemetry system the alternative approach, in which conductivity, temperature and pressure are telemetered and salinity is calculated on shipboard, was not considered practical at that time. A schematic diagram of the salinity bridge is shown in Figure 10. It can be seen that it is complex, requiring three temperature sensors and two pressure sensors. Because of mismatch between the temperature and conductivity sensor time constants, spurious salinity spikes are generated when strong thermal gradients are present. This defect of the circuit is difficult to rectify after the data have been collected, because numerical methods of correction depend on a very accurate knowledge of the various system time constants.

The output from the salinity bridge is made to control a phase shift oscillator which generates a frequency analogue of the salinity signal. This is transmitted to the ship by way of the power supply cable. The overall accuracy of the measurement, signal conditioning, and transmission system is quoted as equivalent to ±0.02‰. This accuracy is adequate for many studies, particularly in coastal waters, but greater accuracy is required for most deep-sea applications. Improved designs are now available which appear to be capable of routinely equalling the accuracy of shipboard measurement. These developments have been made possible by the introduction of digital techniques. The precision C.T.D. microprofiler (Brown, 1975) was the first of these devices to be developed. The three sensors and their interface circuits are excited by a 10-kHz sine wave and are so designed that their zero to full scale output range is 0–500 mV. Each is connected in turn to a precision digitizer and the parallel output from this device is converted to a serial format for transmission to the surface by frequency-shift-keyed telemetry. Because salinity and other derived parameters are often required in order to decide cruise strategy, the conductivity, temperature, and depth information are usually converted on-line to the derived parameters required by a shipboard computer.

5.8.4. Calibration

It is difficult and expensive to expose an instrument in the laboratory to the low temperatures and high pressures encountered in the field. Also, it is

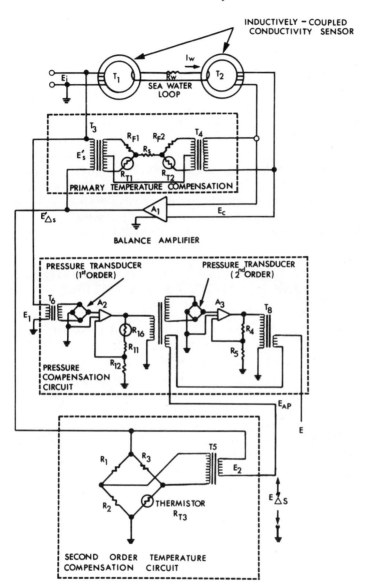

Figure 10. Schematic diagram of the *in situ* conductivity to salinity conversion circuit used in the STD instrument of Brown (1968). Reproduced by permission of Compass Publications, Inc

desirable to monitor the performance of an instrument closely during its use at sea, so that calibration shifts can be detected. For these reasons, the calibration procedure universally adopted is to take samples in conventional water bottles attached to the instrument package and to process these samples on board ship as a check on the performance of the instrument. In order to compensate for temperature effects the *in situ* temperature reading is checked by means of reversing thermometers attached to each bottle. This has the drawback that the thermometers require several minutes to come to equilibrium with their surroundings at each sampling depth, so that the rapidity and ease of measurement theoretically offered by the *in situ* device are in practice not fully realized because of the delays and complications of calibration. The present generation of instruments, however, require this quality control if their full potential is to be realized. The sampling is usually accomplished by means of a rosette multi-sampler, which allows one of a number of bottles to be triggered individually by means of an electrical signal from the surface. These devices add to the bulk and cost of the overside instrument assembly. There is no doubt that the most useful improvement in *in situ* technology at the present would be to reduce the frequency of calibrations needed, without sacrifice of data quality.

5.9. DETERMINATION OF THE ABSOLUTE CONDUCTIVITY OF SEA WATER

The use of conductivity to characterize sea waters in terms of their salinity has, within the past two decades, become almost universal, displacing all other methods. These measurements are only relative, however: the apparatus is usually calibrated with IAPSO Standard Sea Water, and the chlorinity given on the label is used to calculate its salinity by means of the standard relationship (equation 9). Because the chlorinity measurement is made by high-precision titration, the procedure can introduce error if the chlorinity to conductivity relationship assumed is not exactly that of the batch of standard sea water used. It appears that this error can be significant in practice, and consequently it would be desirable to characterize the standard sea water in terms of conductivity rather than chlorinity (Poisson *et al.*, 1978). This can be done in two ways: either the cell itself is calibrated by measurement of a potassium chloride solution of known concentration, or the cell can be constructed in such a way that the desired conductivity can be calculated directly from measurements of length and current flow made on the cell. In the former case the potassium chloride is being used as a transfer standard, the true absolute measurement having being made on a different potassium chloride solution at some time in the past. The accuracy is dependent on the accuracy with which the solution can be prepared and it is not easy to achieve the level of 1 part in 10^5 required. In the latter case a true absolute measurement is being made: this is preferable theoretically but is only practically helpful if it can be made routinely on the batches of IAPSO Standard Sea Water as they are prepared.

A development programme to achieve this end has been underway at the Wormley laboratory of the Institute of Oceanographic Sciences for some

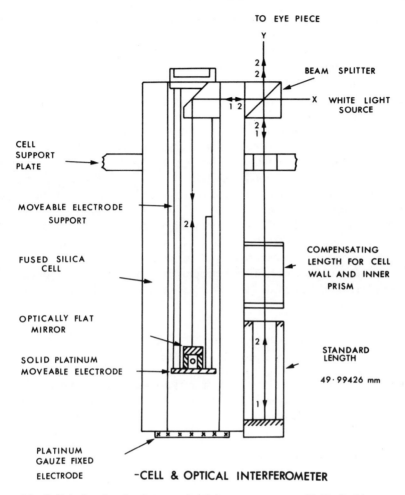

Figure 11. Cell design for absolute conductivity measurement. N. D. Smith, personal communication (1978). Reproduced by permission of the Institute of Oceanographic Sciences, UK

years. The cell adopted has a two-electrode arrangement so constructed that one electrode can be moved relative to the other (Figure 11). The cell is made from a fused silica tube of rectangular cross-section, with electrodes which fit the tube closely. The displacement of the moveable electrode can be determined to an accuracy of 0.05 μm by means of a Michelson interferometer. If two successive measurements are made with different electrode positions the difference in resistance depends only on the cell cross-sectional area, the distance moved, and the conductivity, provided that the electrodes are not moved so close as to cause overlap of the non-linear potential gradient regions close to each electrode. Since an accuracy of 1 part in 10^5 is sought, the temperature of the cell, which is held constant to within $\pm 10^{-4}\,°C$, must be known to an accuracy of at least $4 \times 10^{-4}\,°C$. A precision platinum resistance thermometer is used, calibrated against two triple point cells of water (0.0100°C) and one of phenoxybenzene (26.8685°C). The accuracy of this system appears to be about $\pm 10^{-4}\,°C$, which is more than adequate for the purpose.

The precision of conductivity measurements on replicate samples of IAPSO Standard Sea Water with this equipment is 2 parts in 10^6 and the overall accuracy is estimated to be 5 parts in 10^6, so that the design objectives have been achieved. The largest uncertainty in accuracy is contributed by the measurement of the cell cross-sectional area. It is hoped that this can be improved by a factor of five, so that the overall accuracy would be close to the precision. Measurements have been made on various batches of IAPSO Standard Sea Water at 15°C and also on potassium chloride solutions of various concentrations at various temperatures between 5 and 25°C, as part of an international programme to relate the conductivity of IAPSO Standard Sea Water to that of potassium chloride as accurately as possible.

5.10. PRACTICAL ASPECTS OF CONDUCTOMETRIC MEASUREMENTS

Particular aspects of the design of various instruments have been dealt with in preceding sections. This section deals with the development of instrumentation and with the practical aspects of using this equipment.

5.10.1. Laboratory instruments

The first successful sea-going salinometer was developed by the US National Bureau of Standards in conjunction with the US Coastguard for the work of the International Ice Patrol in the years 1922–1930 (Wenner *et al.*, 1930). As finally developed this instrument compared the resistances of two cells held at the same temperature (30 or 40°C) in a bath regulated to ±0.01°C. A resistance bridge was used, with telephone detection of the null point, the circuit being excited by an electromechanical device at a frequency of about 1000 Hz. The accuracy of the instrument was about ±0.02‰, comparable to routine chlorinity titration. The main advantage over the latter technique was that conductivity measurements were more easily made in bad sea conditions.

The basic design of this instrument was retained by Bradshaw and Schleicher (1956) and by Paquette (1958), who produced designs which took advantage of the improvement in electronic technology which had occurred in the interim. In particular, since the sensitivity of null detection was much improved, the cell current could be reduced below the level at which self-heating and polarization errors occurred. Other improvements introduced by these workers were the use of thermostat bath running at 15°C and the provision of multiple sample cells matched electrically by external trimming components and switched in turn into the bridge circuit. In this way the number of samples per hour is much improved, as it is not necessary to wait for each sample individually to come to thermal equilibrium. The whole batch of six or eight samples is equilibrated simultaneously.

The instrument described by Cox (1958) is based closely on those mentioned above, the major difference being the use of a ratio transformer bridge instead of the resistance design. Eight cells are provided, thermostatted at 15 ± 0.01°C. One of these is used as the reference cell and is filled with IAPSO Standard Sea Water. The others are filled with sample sea waters and, after allowing a short time for thermal equilibration, are switched in turn into the circuit and measured. Because the cells are externally trimmed to be electrically identical no corrections are needed, and the measured ratios may be used directly to determine the salinity from tables. It is possible to process about 180 samples per day on machines of this type, the accuracy corresponding to ± 0.003‰ in salinity at 35‰ (i.e. about 1 part in 10^4). The instrument is unaffected by ship movement and requires little training to use. This was a distinct improvement relative to the titration methods previously used, by which a skilled analyst might expect to process 50 samples a day to 0.01‰ accuracy, if he was not sea-sick.

The drawbacks of instruments of this type are chiefly complexity, cost, and size. These difficulties were overcome by the design of Brown and Hamon (1961), which utilized an inductive cell assembly, transformer ratio bridge, and a thermistor circuit designed to compensate for the variation in conductivity with temperature. This dispensed with the need for any kind of thermostat bath, and reduced the weight by an order of magnitude compared with the earlier designs. Because the circuit cannot cope with variations of more than about ± 3°C the samples should be stored close to the instrument for several hours, provided that ambient temperature changes are not large. The cell is filled with IAPSO Standard Sea Water and its temperature measured by a built-in thermistor thermometer. This temperature is then used to set the compensation circuit to cover the temperature range appropriate and standardization settings are made which give the transformer ratio setting appropriate to the standard. The cell is drained and refilled with the first sample, rinsing if the salinity of the sample is not very close to that of the previous contents of the cell. The bridge is balanced using the ratio dials: the temperature compensation and standardization settings are not changed during a run. The temperature of a sample is measured periodically to check that the samples are within ± 3°C of the set temperature. A known substandard is measured every 15 samples to check for drift; if this is unacceptable the instrument is restandardized.

Within the limitations imposed by the temperature compensation system this instrument is as accurate as its predecessors. It is faster in use, up to 45 samples an hour being possible with an experienced operator. Instruments of this type are the most commonly used at present, because of their relative simplicity, reliability, and low cost. They are not well suited to the precision measurement of samples above or below the usual oceanic salinity range of 30–36‰, because the assumed temperature correction is calculated for 35‰ water. Grasshoff and Hermann (1976) have reported variations of ±0.02% at 38‰ and ±0.05‰ at 8‰ when different instruments at different laboratories are used to analyse the same samples.

Dauphinee and Klein (1977) described an instrument in which the need for temperature compensation or a reference cell is avoided by the use of a highly accurate thermostat bath. The accuracy of this bath is better than 0.01°C per day, and the stability is given as better than 0.001°C per day. Provided that the sample and the standard do not differ in temperature by more than this amount it is possible to dispense with the reference cell, and indeed all other devices for temperature compensation, and simply to measure the ratio of their conductivities at the known set temperature. The International Oceanographic Tables can then be used to derive a value for the salinity of the sample.

Both the circuit design and the cell design of this instrument depart considerably from those of previous instruments. The sample flows continually through the cell after coming to temperature equilibrium by passing through a stainless-steel heat exchanger (Figure 12). The operator selects a bath temperature conveniently above ambient and calibrates the system using IAPSO Standard Sea Water. The sample is then run through the cell and the first two digits of the conductivity ratio are selected manually so that the last four digits are steadily illuminated on the digital panel meter display. The display then reads a value which corresponds to $2R_t$, where R_t is the ratio of the sample conductivity to that of the standard at the thermostat temperature t. The speed of operation is similar to that of the inductive design previously described, but as the stability of the thermostat and electronics is very high, standardization is required less frequently. Because the heat exchanger cannot equilibrate samples which deviate by more than a few degrees from ambient, samples should be stored close to the instrument for several hours, just as for the inductive device.

5.10.2. *In situ* instruments

Although several workers, notably Esterson (1957) and Siedler (1963), described instruments for the *in situ* measurement of conductivity, the first

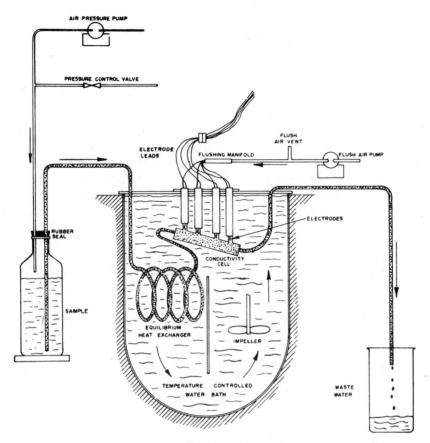

Figure 12. Flow diagram of the Guildline 'Autosal' laboratory salinometer. Reproduced by permission of Compass Publications, Inc., and Guildline Instruments, Ltd

device to achieve wide acceptance was that due to Brown (1968). This has been in routine use in many laboratories for the past 15 years, and has only been superseded recently by instruments of higher precision. The sensor and circuit design of this instrument have been described in Section 5.7.3. The instrument is capable of measurements down to 6000 m, being lowered on a 6 mm hydrographic wire with a single conductor core. The deck unit consists of power supply, telemetry decoder, and chart display, the salinity and temperature being displayed during the cast on an X_1, X_2, Y recorder. Although this form of display allows the performance of the instrument and the hydrographic structure to be closely monitored, the chart record is inconvenient and inaccurate as a data storage medium. Consequently, it has become common practice to digitize the data as they are obtained and to store this information on magnetic tape or by other suitable methods.

The accuracy of this instrument in use depends on the care which is devoted to calibration (Pingree, 1970). Typically, users report apparent accuracies about ±0.01‰ in salinity and ±0.01°C in temperature when the instrument reading is compared with samples taken at the same point. Their main use is therefore in surface waters and coastal areas where variability is high, rather than in pelagic waters. However, even in the deep ocean it is possible to study the effects of mixing processes which would not be observable by any previous technique (see, for example, Tait and Howe, 1968) because, although conventional sampling techniques are more accurate, they cannot provide continuous profiles with depth. For this reason, even though the need for calibration means that the *in situ* equipment is an addition rather than a substitute for the conventional shipboard salinity determination, the instrument has been accorded wide acceptance.

For deep ocean studies a higher accuracy is preferable, approaching the 0.001‰ level, which is the practical limit of the salinity concept. This need has been met by instruments using miniature multi-electrode conductivity cells and digital telemetry techniques (Dauphinee, 1972; Brown, 1975; Kroebel *et al.*, 1976). Such improvements in technology make it possible to obtain conductivity measurements with very high spatial and temporal resolution, and to telemeter these readings to the surface without loss of accuracy. Fofonoff *et al.* (1974) conducted an exhaustive evaluation of the prototype Brown instrument during the Mode series of cruises. From the results of laboratory checks and over 200 hydrographic stations occupied they concluded that the accuracy achieved was ±0.003‰ in salinity, ±0.0015°C in temperature, and ±1.5 decibars in pressure.

5.10.3. The preparation of IAPSO Standard Sea Water

IAPSO Standard Sea Water ('Copenhagen Water') has been the practical standard for oceanographic salinity determination since 1901. The calibration of present-day batches can be traced back, indirectly, to that date. The present method of standardization against silver was established in 1937 (Jacobsen and Kundsen, 1940). Preparation of the standard at the Danish Fisheries Laboratory, Charlottenlund, Denmark, was continued until 1975, when the operation was transferred to the Institute of Oceanographic Sciences in Britain (IAPSO Standard Sea Water Service, Institute of Oceanographic Sciences, Wormley, Godalming, Surrey GU8 5UB, England). The procedure for the preparation of the standard has been described by Hermann and Culkin, (1978).

The sea water used for the preparation of the standard is collected in the North East Atlantic as close as practicable to 60°N 20°W and transported to the Wormley Laboratory in 50-dm^3 polyethylene drums. At Wormley it is pumped through a 0.3-μm membrane filtration assembly into a 5-m^3

holding tank lined with PVC, and is circulated for several days through the filters and through a UV irradiation device to break down organic compounds. At the same time its temperature is raised to 26°C to reduce the dissolved gas content slightly. Doubly distilled water is cautiously added to adjust the salinity to a value close to 35‰. Finally, the water is sealed in ampoules made of resistant glass, which have previously been rinsed with distilled water and dried for 6 h at 120°C. One ampoule out of every 300 prepared is reserved for chlorinity determination. About 7000 ampoules are filled and sealed in one day, and this constitutes a batch. Current output is in excess of 20 000 ampoules per year.

The chlorinity is measured by a gravimetric/potentiometric method which superseded the classic Volhard method in 1969. Aliquots of the sea water and of a concentrated silver nitrate solution are weighed, using special weight-burettes designed by Hermann. The exact size of these aliquots is chosen so that there is a slight excess of sea water on mixing. This excess is then titrated against dilute silver nitrate solution, using the electrode system described by Hermann (1951) for end-point detection. A Primary Standard Sea Water, the chlorinity of which has been determined relative to the pure silver standard (Section 5.3.4), is used to calibrate the silver nitrate solutions. The standard deviation of the method is 3×10^{-4} in chlorinity; analyses carried out at Copenhagen and Wormley by this method have usually agreed to within 1×10^{-4} in chlorinity (Hermann and Culkin, 1972).

For reasons mentioned in Section 5.5.3 it is probable that in the near future batches of IAPSO Standard Sea Water will be produced with a conductivity ratio standardization, as well as one based on chlorinity. It is likely that the chlorinity determination outlined above will eventually be replaced by a conductivity comparison method based on the use of a potassium chloride solution of known concentration as a working standard (Lewis and Perkin, 1978; Poisson *et al.*, 1978).

5.11. ACKNOWLEDGEMENTS

I thank the many colleagues who have helped me by discussion of various aspects of this paper. In particular the assistance of Dr. F. Culkin, on matters pertaining to the Standard Sea Water Service, and Mr. N. Smith, on absolute conductivity measurement, is acknowledged with gratitude.

REFERENCES

Almgren, T., D. Dyrssen, and M. Strandberg (1977). Computerized high-precision titrations of some major constituents of sea water on board the R.V. "Dmitry Mendeleev". *Deep-Sea Res.*, **24**, 345–364.

Anfält, T., and D. Jagner (1971). A standard addition titration method for the potentiometric determination of fluoride in sea water. *Anal. Chim. Acta*, **53**, 13–22.

Bather, J. M., and J. P. Riley (1954). The chemistry of the Irish Sea. Part 1—The sulphate chlorinity ratio. *J. Cons. Perm. Int. Explor. Mer.*, **20**, 145–152

Brass, G. W., and K. K. Turekian (1972). Strontium distributions in sea water profiles from the Geosecs I (Pacific) and Geosecs II (Atlantic) test stations. *Earth Planet. Sci. Lett.*, **16**, 117–121.

Brass, G. W., and K. K. Turekian (1974). Strontium distribution in Geosecs oceanic profiles. *Earth Planet. Sci. Lett.*, **23**, 141–148.

Bradshaw, A. (1973). The effect of carbon dioxide on the specific volume of sea water. *Limnol. Oceanogr.*, **18**, 99–105.

Bradshaw, A., and K. E. Schleicher (1965). The effect of pressure on the electrical conductance of seawater. *Deep-Sea* Res., **12**, 151–162.

Brewer, P. G., and A. Bradshaw (1975). The effect of the non-ideal composition of sea water on salinity and density. *J. Mar. Res.*, **33**, 157–175.

Brown, N. L. (1968). An *in situ* salinometer for use in the deep ocean. In *Marine Sciences Instrumentation*, Vol. 4 (Ed. F. Alt). Plenum Press, New York, pp. 563–578.

Brown, N. L. (1974). A precision CTD microprofiler. In *Ocean '74 (IEEE International Conference on Engineering in the Ocean Environment)*, Vol. 2. *IEEE*, New York, pp. 270–278.

Brown, N. L., and B. Allentoft (1966). Salinity, conductivity and temperature relationships of seawater over the range of 0–50 p.p.t. *Bissett-Berman Corporation Report*, No. MJO 2003, p. 60.

Brown, N. L., and B. V. Hamon (1961). An inductive salinometer. *Deep-Sea Res.*, **8**, 65–75.

Carpenter, J. H., and M. E. Manella (1973). Magnesium to chlorinity ratios in seawater. *J. Geophys. Res.*, **78**, 3621–3626.

Chen, M.-S. and L. Onsager (1977). The generalized conductance equation. *J. Phys. Chem.*, **81**, 2017–2021.

Connors, D. N., and K. Park (1967). The partial equivalent conductance of electrolytes in seawater: a revision. *Deep-Sea Res.*, **14**, 481–484.

Connors, D. N., and P. K. Weyl (1968). The partial equivalent conductances of salts in seawater and the density/conductance relationship. *Limnol. Oceanogr.*, **13**, 39–50.

Cox, R. A. (1968). *The Thermostat Salinity Meter*. N.I.O. Internal Report No. C.2, March 1958.

Cox, R. A. (1965). The physical properties of sea water. In *Chemical Oceanography*, Vol. 1 (Eds. J. P. Riley and G. Skirrow). Academic Press, New York, pp. 73–120.

Cox, R. A., F. Culkin, and J. P. Riley (1967). The electrical conductivity/chlorinity relationship in natural sea water. *Deep-Sea Res.*, **14**, 203–220.

Cox, R. A., M. J. McCartney, and F. Culkin (1970). The specific gravity/salinity/temperature relationship in natural sea water. *Deep-Sea Res.*, **17**, 679–689.

Culkin, F. (1965). The major constituents of sea water. In *Chemical Oceanography*, Vol. 1 (Eds. J. P. Riley and G. Skirrow). Academic Press, New York, pp. 121–161.

Culkin, F., and R. A. Cox (1966). Sodium, potassium, magnesium, calcium and strontium in sea water. *Deep-Sea Res.*, **13**, 789–804.

Culkin, F., and F. E. Hermann (1978). The preparation and chlorinity calibration of standard seawater. *Deep-Sea Res.*, **25**, 1265–1270.

Dauphinee, T. M. (1968). *In situ* conductivity measurements using low-frequency square-wave A.C. In *Marine Sciences Instrumentation*, Vol. 4. (Ed. F. Alt). Plenum Press, New York, pp. 555–562.

Dauphinee, T. M. (1972). Equipment for rapid temperature–conductivity–depth surveys. In *Oceanology International '72: Conference Papers.* BPS Exhibitions Ltd., London, pp. 53–57.

Dauphinee, T. M., and H. P. Klein (1975). A new automated laboratory salinometer. *Sea Technol.* **16**, 23–25.

Dauphinee, T. M., and H. P. Klein (1977). The effect of temperature on the electrical conductivity of seawater. *Deep-Sea Res.*, **24**, 891–902.

Dittmar, W. (1884). Report on researches into the composition of ocean water, collected by H.M.S. "Challenger". *Report on the Scientific Results of the Voyage of H.M.S. "Challenger"*, (Physics and Chemistry), **1**, 1–251, H.M.S.O., London.

Duedall, I. W. (1972). The partial molal volume of calcium carbonate in sea water. *Geochim. Cosmochim. Acta,* **36**, 729–734.

Edmond, J. M. (1970). High precision determination of titration alkalinity and total carbon dioxide content of sea water by potentiometric titration. *Deep-Sea Res.*, **17**, 737–750.

Esterson, G. L. (1957). The induction conductivity indicator: a new method for conductivity measurement at sea. *Chesapeake Bay Inst. Johns Hopkins Univ. Tech. Rep.*, 14 (Ref 57–3), 163.

Fofonoff, N. P., and H. Bryden (1975). Specific gravity and density of sea water at atmospheric pressure. *J. Mar. Res.*, **33**, Suppl., 69–82.

Fofonoff, N. P., S. P. Hayes, and R. C. Millard (1974). W.H.O.I./Brown CTD microprofiler: methods of calibration and data handling. *Woods Hole Oceanogr. Inst. Tech. Rep.*, No. 74–89, p. 64.

Forch, C., M. Knudsen, and S. P. L. Sørensen (1902). Berichteuber die konstantenbestminugen zur Aufsteung den hydrographischen Tabellen. *Kongelige Danske Videnskabernes Selskabs Skrifter*, 6 Raekke, Naturvidensk og Mathem. Afd. XII, No. 1, p. 151.

Forchammer, G. (1865). On the composition of sea water in different parts of the ocean. *Phil. Trans. R. Soc. London*, **155**, 203–262.

Gieskes, J. M. T. M. (1967). Der Membran-Salzfühler als geeignetes Gerat sur Registrierung der Schichtung im Meere. *Kieler Meeresforsch.*, **23**, 75–79.

Girard, G., and M. Ménaché (1971). Variation de la masse volumique de l'eau en fonction de sa composition isotopique. *Metrologia*, **7**, 83–87.

Grasshoff, K., and F. Hermann (1976). Seventh report of the joint panel on oceanographic tables and standards, Grenoble, France, 2–5 September 1975, sponsored by UNESCO, ICES, SCOR, IAPSO. *UNESCO Tech. Pap. Mar. Sci.*, No. 24, p. 61.

Grasshoff, K., F. Culkin, L. P. Fofonoff, W. Kroebel, E. L. Lewis, O. Mamayev, M. Ménaché, F. Millero, A. Poisson, and C. K. Ross (1978). Eighth report of the Joint Panel on Oceanographic Tables and Standards, held Woods Hole, U.S.A., 23–25 May 1977. *UNESCO Tech. Pep. Mar. Sci.*, No. 28, p. 35.

Graham, D. C. (1947). The electrical double layer and the theory of electrocapillarity. *Chem. Rev.* **41**, 441–501.

Greenhalgh, R., and J. P. Riley (1963). Occurrence of abnormally high fluoride concentrations at depth in the oceans. *Nature*, **197**, 371–372.

Hermann, F. E. (1951). High accuracy potentiometric determination of chlorinity of sea water. *J. Cons. Perm. Int. Explor. Mer.* **17**, 223–230.

Hermann, F. E., and F. Culkin (1972). A check of the analysis of standard sea water. *International Council for the Exploration of the Sea, Contribution to Statutory Meeting*, C. M. 1972/C : 33, 3 pp.

Jacobsen, J. P., and M. Knudsen (1940). Urnormal 1937 or primary standard seawater 1937. *Pub. Sci. Ass. Oceanogra. Phys.*, No. 7, 38 pp.

Jagner, D. (1970). The determination of sulphate in sea water by means of photometric titration with hydrochloric acid in dimethyl sulphoxide. *Anal. Chim. Acta*, **52**, 483–490.

Knudsen, M. (1901). *Hydrographische Tabellen*. Buchdruckerei Bianco Luno, Copenhagen, p. 63.

Kratky, O., H. Leopold, and H. Stabinger (1969). Dichtemessungen an Flüssigkeiten und Gasen auf 10^{-6} g/cm^3 Praparatvolumen. *Z. Angew. Phys.*, **27**, 273–277.

Kremling, K. (1971). New method for measuring density of seawater. *Nature*, **229**, 109–110.

Kremling, K. (1972). Comparison of specific gravity in natural sea water from hydrographical tables and measurements by a new density instrument. *Deep-Sea Res.*, **19**, 377–383.

Kroebel, W., P. Diehl, L. Ginzkey, K.-H. Mahrt, J. Rathley, R. Siara, and Th. Schultz (1976). Die Kieler Multisonde der Jahre 1975/76, ihre Sensoren, Parameter mit Ergebnissen von Datenaufnahmen und Perspektiven fur ihre Auswertung. 1034–1046. In *Interocean '76*, Band 2. Dusseldorfer Meesegesellschaft mbH, Dusseldorf Paper 10 76–402.

Kuenzler, H. W. (1968). An experimental *in situ* densitometer. *Mass. Inst. Technol., Dep. Meterol., Fluid Dynam. Lab., Rep.* No. 68–3, p. 56.

Lewis, E. L., and R. G. Perkin (1978). Salinity: its definition and calculation. *J. Geophys. Res.*, **81**, 466–478.

Lingane, J. J. (1958). *Electroanalytical Chemistry*, 2nd edition. Interscience, New York, p. 669.

Lyman, J. (1969). Redefinition of salinity and chlorinity. *Limnol. Oceanogr.*, **14**, 928–929.

Lynn, R. J., and J. L. Reid (1968). Characteristics and circulation of deep and abyssal waters. *Deep-Sea Res.*, **15**, 577–598.

Mangelsdorf, P. C. (1967). Salinity measurements in estuaries. In *Estuaries* (Ed. G. H. Lauff). American Association for the Advancement of Science, Washington, D.C., pp. 71–79.

Mangelsdorf, P. C., and T. R. S. Wilson (1971). Difference chromatography of sea water. *J. Phys. Chem.*, **75**, 1418–1425.

Ménaché, M. (1971). Vérification, par analyse isotopique, de la validité de la méthode de Cox, McCartney et Culkin tendant à l'obtention d'un étalon de masse volumique. *Deep-Sea Res.*, **18**, 449–456.

Millero, F. J., and K. Kremling (1976). The densities of Baltic Sea waters. *Deep-Sea Res.*, **23**, 1129–1138.

Millero, F. J., and F. K. Lepple (1973). The density and expansibility of artificial seawater solutions from 0 to 40°C and 0 to 21% chlorinity. *Mar. Chem.*, **1**, 89–104.

Millero, F. J., P. Chetirkin, and F. Culkin (1977). The relative conductivity and density of standard seawaters. *Deep-Sea Res.*, **24**, 315–321.

Morris, A. W., and J. P. Riley (1964). The direct gravimetric determination of the salinity of sea water. *Deep-Sea Res.*, **11**, 899–904.

Morris, A. W., and J. P. Riley (1966). The bromide/chlorinity and sulphate/chlorinity ratio in sea water. *Deep-Sea Res.*, **13**, 699–705.

Onsager, L. (1927). Zur Theorie der Elektrolyte. II. *Phys. Z.*, **28**, 277–298.

Paquette, R. G. (1958). A modification of a Wenner–Smith–Soule salinity bridge for the determination of the salinity of seawater. *Univ. Wash. Dep. Oceanogr. Tech. Rep.*, No. 71 (Ref. 58–14).

Perkin, R. G., and E. L. Lewis (1980). The practical salinity scale 1978: fitting the data. *IEEE J. Oceanogr. Eng.*, **5**, 9–16.

Pingree, R. D. (1970). *In situ* measurements of salinity, conductivity and temperature. *Deep-Sea Res.*, **17**, 603–610.

Poisson, A., T. M. Dauphinee, C. K. Ross, and F. Culkin (1978). The reliability of standard seawater as an electrical conductivity standard. *Oceanol. Acta*, **1**, 425–433.

Reeburgh, W. S. (1965). Measurements of the electrical conductivity of seawater. *J. Mar. Res.*, **23**, 187–199.

Reeburgh, W. S., and J. H. Carpenter (1964). Determination of chlorinity using a differential potentiometric end point. *Limnol. Oceanogr.*, **9**, 589–591.

Reid, J. L., and R. J. Lynn (1971). On the influence of the Norwegian–Greenland and Weddell seas upon the bottom waters of the Indian and Pacific oceans. *Deep-Sea Res.*, **18**, 1063–1088.

Riley, J. P., and M. Tongudai (1967). The major cation/chlorinity ratios in sea water. *Chem. Geol.*, **2**, 263–269.

Rusby, J. S. M. (1967). Measurements of the refractive index of sea water relative to Copenhagen Standard Sea Water. *Deep-Sea Res.*, **14**, 427–439.

Sanders, H. L., P. C. Mangelsdorf, and G. R. Hampson (1965). Salinity and faunal distribution in the Pocasset River, Massachusetts. *Limnol. Oceanogr.*, **10**, Suppl, R216–R229.

Schleicher, K. E., and A. Bradshaw (1956). A conductivity bridge for the measurement of salinity of sea water. *J. Conseil*, **22**, 9–20.

Siedler, G. (1963). On the *in situ* measurement of temperature. *Deep-Sea Res.*, **10**, 269–277.

Strickland, J. D. H., and T. R. Parsons (1968). A practical handbook of seawater analysis. *Bull. Fish. Res. Bd Can.*, No. 167, p. 311.

Tait. R. I., and M. R. Howe (1968). Some observations of thermo-haline stratification in the deep ocean. *Deep-Sea Res.*, **15**, 275–280.

Tsunogai, S., H. Yamahata, S. Kudo, and O. Saito (1973). Calcium in the Pacific Ocean. *Deep-Sea Res.*, **20**, 717–726.

UNESCO (1962). First report of Joint Panel on Oceanographic Tables and Standards. *Tech. Pap. Mar. Sci.*, **1**, 29.

UNESCO (1965). Second report of Joint Panel on Oceanographic Tables and Standards. *Tech. Pap. Mar. Sci.*, **4**, 29.

UNESCO (1966). *International Oceanographic Tables*. National Institute of Oceanography of Great Britain and UNESCO, Paris p. 118.

Vel'mozhnaya, Yu. A. (1960). Interferometric determination of the salinity of sea water. *Tr. Morsk. Gidrofiz. Inst. SSSR*, **22**, 26–32.

Walker, E. R. (1973). A comparison of salinity equations. In *Proceedings Second S/T/D Conference and Workshop, January 24–26, 1973*. Plessey Environmental Systems, San Diego, Calif., pp. 1–17.

Walker, E. R., and K. D. Chapman (1973). Salinity–conductivity formulae compared. *Can. Dep. Environ. Mar. Sci. Direct., Pacific Mar. Sci. Rep.*, No. 73–5, p. 52.

Wallace, W. J. (1974). *The Development of the Chlorinity/Salinity Concept in Oceanography*. Elsevier, Amsterdam, p. 227.

Wenner, F., E. H. Smith, and F. M. Soule (1930). Apparatus for the determination aboardship of the salinity of sea water by the electrical conductivity method. *J. Res. Nat. Bur. Stand.*, **5**, 711–730.

Wilson, T. R. S. (1971). A portable flow-cell membrane salinometer. *Limnol. Oceanogr.*, **16**, 581–586.

Wilson, T. R. S. (1975). Salinity and the major elements of sea water. In *Chemical Oceanography*, Vol. 1, 2nd edition (Eds. J. P. Riley and G. Skirrow). Academic Press, London, pp. 365–413.

Wooster, W. S., A. J. Lee, and G. Dietrich (1969). Redefinition of salinity. *Deep-Sea Res.*, **16**, 321–322.

Marine Electrochemistry
Edited by M. Whitfield and D. Jagner
© 1981 John Wiley & Sons Ltd.

CHARLES H. CULBERSON

College of Marine Studies,
University of Delaware,
Newark, Delaware 19711,
USA

6

Direct Potentiometry

GLOSSARY OF SYMBOLS

a_M	Single ion activity of M (equation 2)
a'_H	Apparent hydrogen ion activity (equation 14)
A_T	Total (titration) alkalinity (equation 59)
c_M	Concentration of M in mol/kg sea water (M_w) (equation 21)
E_M, $E_M^{\ominus}$	Potential and standard potential of electrode M (equation 2)
E_j	Liquid junction potential (equation 5)
ΔE_j	Residual liquid junction potential (equation 12)
f_{MX}	Mean ionic activity coefficient of MX on the molar scale (equation 28)

f_M	Single ion activity coefficient of M on the molar scale (equation 3)
f'_M	Conventional (or apparent) single ion activity coefficient of M on the molar scale (equation 32)
g	$(RT/F) \ln 10$ (equation 10)
I_{tot}	Formal (total) ionic strength (equation 40)
$K^{\ominus}_{MX}$	Thermodynamic equilibrium constant at infinite dilution (equation 62)
K_i	Selectivity coefficient (equation 4)
m_x	Molality of x (equation 25)
m	Molal concentration $(mol\,kg^{-1})$ scale
M	Molar concentration $(mol\,dm^{-3})$ scale
M_W	Mol $(kg\ sea\ water)^{-1}$ scale (molinity)
pH	$-\log a_H$
pH(NBS), pH$'$	pH on NBS dilute buffer scale (equation 13)
pH(SWS)	pH on total hydrogen ion pH scale (equation 20)
pm_H	pH on free hydrogen ion pH scale (equation 29)
tot	Subscript for total concentration
[M]	Concentration of M in $mol\,dm^{-3}$ (M) (equation 17)
β_{MX}	Association constant of MX (equation 21)
λ	Conversion factor between apparent and thermodynamic hydrogen ion activity (equation 15)

6.1. THEORETICAL INTRODUCTION

This chapter covers the applications of direct potentiometry to marine chemistry. The use of ion-selective electrodes in sea water and sea water-like media is emphasized, and readers interested in more general applications of ion-selective electrodes are referred to the reviews by Ives and Janz (1961), Eisenman (1967), Durst (1969a), Whitfield (1971), Moody and Thomas (1971), Bates (1973), Covington (1974), Lakshminarayanaiah (1976), Koryta (1977), and Buck (1978).

The measurement of pH is the most common application of direct potentiometry to sea water, and much of the chapter is devoted to a discussion of pH measurements in sea water. In particular, the relative merits of high ionic strength sea water buffers and dilute buffers are considered, and it is shown that available field measurements do not support the claim that pH measurements based on sea water buffers are inherently more precise than those using dilute buffers. Nonetheless, there are significant advantages to be gained from the introduction of thermodynamically and stoichiometrically clearly defined concentration scales for hydrogen ions in sea water, particularly when data from different ionic media are being compared or when reaction rates and reaction mechanisms are being studied. These alternative scales will be discussed in some detail.

6.1.1. The cell potential

The potential of an electrode reversible to the general half-cell reaction

$$M \rightleftharpoons M^{n+} + ne^- \tag{1}$$

is given by the equation (Ives and Janz, 1961b)

$$E_M = E_M^{\ominus} + (RT/nF) \ln a_M \tag{2}$$

where $E_M^{\ominus}$ is the standard potential of the electrode and a_M is the activity of M^{n+} in solution.

$$a_M = [M^{n+}]f_M \tag{3}$$

Many ion-selective electrodes, such as the calcium liquid ion exchanger, respond to more than one ion and their potentials are better represented by the equation (Moody and Thomas, 1971)

$$E_M = E_M^{\ominus} + (RT/nF) \ln \left(a_M + \sum_j K_j a_j^{n/z} \right) \tag{4}$$

K_j is the selectivity coefficient, which defines the selectivity of the electrode for ion M^{n+} in the presence of ion J^{z+}. It must be emphasized that the selectivity coefficients in equation 4 are generally not constant but can vary greatly with solution composition and ionic strength (Whitfield and Leyendekkers, 1970).

The potential of an individual half-cell cannot be determined directly, but must be measured relative to some reference electrode. For the most common use of a reference electrode responsive to chloride—the silver/silver chloride or calomel electrodes—the reference electrode potential is

$$E_{ref} = E_{ref}^{\ominus} - (RT/F) \ln a_{Cl} - E_j \tag{5}$$

E_j, the liquid junction potential, is present only in cells with a liquid junction. The activity of the reference ion, chloride in this case, is constant in cells with a liquid junction; it may not be constant in cells without a liquid junction, in which the concentration of the reference ion may vary (Miller and Kester, 1976).

The measured potential of a cell is the difference between the potentials of the ion-selective and reference electrodes:

$$E = (E_M^{\ominus} - E_{ref}^{\ominus}) + (RT/nF) \ln a_M + (RT/nF) \ln a_{Cl}^n + E_j \tag{6}$$

The activity coefficients in equation 3 can be based on the infinite dilution activity scale (Bates, 1973) or on the constant ionic medium activity scale (Dyrssen and Hansson, 1973; Khoo et al., 1977a). Activity coefficients on the infinite dilution scale approach unity at infinite dilution in the pure solvent, while those on the constant ionic medium scale approach unity at infinite dilution in the pure ionic medium. As a first approximation the activity coefficients of free ions are independent of solution composition at constant ionic strength (Garrels, 1967), and in high ionic strength media such as sea water the activity coefficients in equation 3 can be incorporated into the standard potential for measurements at constant ionic strength. This allows changes in the cell potential (equation 6) to be interpreted directly in terms of ionic concentrations rather than activities.

In addition to the two activity scales, there are three concentration scales (molar, molal, and mol (kg solution)$^{-1}$) in common use in marine chemistry. Equations for converting concentrations and activities from one scale to another are given by MacIntyre (1976) (see Section 1.3.1).

6.1.2. pH scales for sea water

At present, three pH scales are used to report sea water pH measurements: the National Bureau of Standards (NBS) pH scale (Bates, 1973), and the constant ionic medium pH scales proposed by Hansson (1973a) and by Bates and Macaskill (1975). In addition to these three scales, the Sørenson pH scale (Bates, 1973) was often used for oceanic pH measurements prior to 1960.

The NBS pH scale

This is the most widely used pH scale in geochemistry and is based on the hydrogen ion activity on the infinite dilution activity scale. The NBS scale is defined by several buffers (Bates, 1973), the hydrogen ion activity of which was calculated from measurements with hydrogen and silver/silver chloride electrodes, using the Debye–Hückel theory to estimate the activity coefficient of chloride ion in the buffers. The use of non-thermodynamic assumptions to calculate the activity coefficient of chloride ion means that strictly the NBS scale is a conventional pH scale. Bates (1973) estimates that the NBS scale is accurate to ±0.01 pH unit. The five primary NBS buffers and their pH values at 25°C are listed in Table 1.

Of particular importance to geochemists is the fact that NBS 4.008, 6.865, 7.413, and 9.180 buffers are internally consistent for pH measurements using a free diffusion liquid junction with a concentrated KCl salt bridge.

Table 1. Properties of NBS primary standard buffers at 25°C. Additional buffers covering the pH range 1.68–12.47 are given by Bates (1973)

Buffer substance	Concentration (m)	pH
Potassium hydrogen phthalate	0.05	4.008
KH_2PO_4	0.025	6.865
$+Na_2HPO_4$	0.025	
KH_2PO_4	0.008695	7.413
$+Na_2HPO_4$	0.03043	
$Na_2B_4O_7 \cdot 10H_2O$(borax)	0.01	9.180

Paabo and Bates (Bates, 1973) standardized the cell

$$\text{Pt} \mid \text{H}_2(g) \mid \text{NBS buffer} \mid \text{KCl (satd.)} \mid \text{Hg}_2\text{Cl}_2\,(s) \mid \text{Hg (l)} \qquad (7)$$

with the equimolal phosphate buffer (pH 6.865) and then measured the pH of the other three buffers relative to pH 6.865 buffer. The difference between the measured and assigned pH values at 25°C (measured − assigned) were +0.004, +0.001, and +0.004 pH for the pH 4.008, 7.413, and 9.180 buffers. The small differences between the measured and assigned pH values mean that the liquid junction potential developed between KCl and each of the buffers is relatively constant and does not vary from buffer to buffer.

In sea water the interpretation of pH measurements in terms of hydrogen ion activity is complicated by the presence of a residual liquid junction potential when the reference electrode is transferred from buffer to sea water. The cell potentials in the NBS buffer and in sea water are as follows:

Buffer:

$$E_b = E^* + g \log (a_H)_b + E_{jb} \qquad (8)$$

Sea water:

$$E_x = E^* + g \log (a_H)_x + E_{jx} \qquad (9)$$

where

$$g = (RT/F) \ln 10 \qquad (10)$$

Subtracting equation 8 from equation 9, substituting pH for $-\log a_H$, and rearranging the result yields an equation for the pH of sea water in terms of the buffer pH, the measured potentials, and the difference in liquid junction potentials between sea water and buffer:

$$pH_x = pH_b + (E_b - E_x)/g - (E_{jb} - E_{jx})/g \qquad (11)$$

Of the quantities on the right-hand side of equation 11, pH_b is known and E_b and E_x can be measured. However, the residual liquid junction potential,

$$\Delta E_j = (E_{jb} - E_{jx}) \qquad (12)$$

cannot be measured. The unknown value of the residual liquid junction potential makes the interpretation of pH measurements in sea water impossible in terms of hydrogen ion activity.

The quantity pH_x' that is actually calculated when pH measurements in sea water are standardized with NBS buffers can be derived by rearranging equation 11:

$$pH_x' = pH_x + (E_{jb} - E_{jx})/g = pH_b + (E_b - E_x)/g \qquad (13)$$

In terms of hydrogen ion activity

$$a_H' = a_H 10^{\Delta E_j/g} \qquad (14)$$

Thus the apparent hydrogen ion activity, a'_H, calculated from pH measurements in sea water differs from the true activity, a_H, by an approximately constant amount equal to

$$\lambda = 10^{\Delta E_j/g} \tag{15}$$

Based on three independent methods, Hawley and Pytkowicz (1973) showed that the value of λ in sea water is about 1.13 relative to NBS pH 4.008 and 7.413 buffers. Similar calculations, using sea water values for the mean ionic activity coefficient of HCl (Khoo *et al.*, 1977a; $f_{HCl} = 0.730$), the single ion activity coefficient of chloride (Robinson and Bates, 1978; $f_{Cl^-} = 0.655$), and the conventional free hydrogen ion activity coefficient on the NBS scale (Bates and Macaskill, 1975; $f_{H^+} = 0.998$), yield a value $\lambda = 1.23$ at ionic strength 0.72 and 25°C. The difference between the two values for λ is primarily due to differences in the assumed values for f_{Cl^-}.

The difference between the apparent hydrogen ion activity, a'_H, and the true activity is not important in practice provided that the residual liquid junction potential remains constant. Johnson *et al.* (1977) emphasize that it is the reproducibility of pH measurements, rather than their accuracy relative to the hydrogen ion activity, that is important in practice for equilibrium studies. Consider the protolysis of boric acid in sea water:

$$B(OH)_3 + H_2O \rightarrow H^+ + B(OH)_4^- \tag{16}$$

$$[B(OH)_3]_{tot} = [B(OH)_3] + [B(OH)_4^-] \tag{17}$$

$$K'_a = a'_H \frac{[B(OH)_4^-]}{[B(OH)_3]} = \lambda a_H \frac{[B(OH)_4^-]}{[B(OH)_3]} = \lambda K_a \tag{18}$$

and assume that we wish to determine the borate ion concentration from measurements of pH and total boron. Rearrangement of equations 17 and 18 yields (Skirrow, 1975)

$$\begin{aligned}
[B(OH)_4^-] &= K'_a[B(OH)_3]_{tot}/(a'_H + K'_a) \\
&= \lambda K_a[B(OH)_3]_{tot}/(\lambda a_H + \lambda K_a) \\
&= K_a[B(OH)_3]_{tot}/(a_H + K_a)
\end{aligned} \tag{19}$$

The derivation of equation 19 shows that, within the reproducibility of liquid junction potentials, the value of λ has no effect on the calculated borate concentration provided that both a'_H and K'_a contain the same value of λ.

The use of the NBS pH scale in sea water has been criticized for two reasons. The first is that the pH scale is not thermodynamically well defined, owing to the indeterminate value of λ in equation 15. This is a theoretical, not a practical problem, since the difference between a'_H and a_H cancels when used with equilibrium constants determined on the same pH scale.

The second criticism is that low ionic strength NBS buffers can cause large, unreproducible changes in the residual liquid junction potential upon transferring electrodes from buffer to sea water. An unreproducible value of the liquid junction potential will cause the factor λ to vary, and thus introduce errors in any calculated parameters.

In an attempt to eliminate the effects of liquid junction potentials, Hansson (1973a) and Bates and Macaskill (1975) proposed two new pH scales for sea water based on the concept of sea water as a constant ionic medium.

The total hydrogen ion pH scale

To minimize the above objections to the NBS scale, Hansson (1973a) developed a pH scale for sea water based on the total hydrogen ion concentration. This pH scale has two potential advantages over the NBS scale. First, the total hydrogen ion concentration is thermodynamically well defined, whereas the conventional hydrogen ion activity is not. Secondly, the use of buffers prepared in synthetic sea water should eliminate unreproducible changes in the residual liquid junction potential between buffer and sea water, since the two solutions are nearly identical in composition.

Hansson (1973a) measured the total hydrogen ion concentration (Table 2) of an equimolar tris [tris(hydroxymethyl)aminomethane, $(HOCH_2)_3CNH_2$] buffer prepared in synthetic sea water (Table 3). The pH(SWS) of the buffer was determined by titration with HCl to a final pH of 3. The standard potential of the glass/reference electrode pair was calculated from the titration points past the equivalence point, where the total hydrogen ion concentration was known, and the initial pH(SWS) was then calculated from the potential before addition of HCl and from the standard potential. The effects of temperature and salinity on the pH(SWS) of Hansson's (1973a)

Table 2. Values of pH(SWS) in Hansson's (1973a) equimolar tris buffer as a function of temperature and salinity, calculated from equation 20 (Almgren *et al.*, 1975a). The values at 0 and 30°C are extrapolated

	t, °C						
S, ‰	0	5	10	15	20	25	30
10	8.844	8.673	8.508	8.348	8.194	8.045	7.901
15	8.857	8.684	8.518	8.357	8.201	8.051	7.905
20	8.870	8.696	8.527	8.365	8.208	8.057	7.910
25	8.883	8.707	8.537	8.374	8.215	8.063	7.915
30	8.896	8.718	8.547	8.382	8.223	8.068	7.919
35	8.908	8.730	8.557	8.391	8.230	8.074	7.924
40	8.921	8.741	8.567	8.399	8.237	8.080	7.929

Table 3. Composition (mM$_W$) of the tris buffer (35‰) used to establish the total hydrogen ion pH scale (Hansson, 1972). Salinities of 10, 20, 25, 30, and 40‰ were prepared by multiplying the given concentrations by 2/7, 4/7, 5/7, 6/7, or 8/7. The tris concentrations were maintained at 0.005 M$_W$ by adjusting the sodium concentration. The composition of natural sea water is given for comparison. The properties of tris are reviewed by Riddick (1961)

Solution species	Natural sea water	Tris buffer
Na^+	468.16	463
K^+	10.21	10
Mg^{2+}	53.24	54
$Ca^{2+} + Sr^{2+}$	10.39	10
$Cl^- + Br^- + F^- + HCO_3^-$	549.16	550
SO_4^{2-}	28.24	28
$B(OH)_3$	0.42	0
Tris		5
Tris–H^+		5

tris buffer are given by the following equation (Almgren *et al.*, 1975a):

$$pH(SWS) = -\log (c_{H^+} + c_{HSO_4^-})$$
$$= (2559.7 + 4.5S)/T - 0.5523 - 0.01391S \qquad (20)$$

which fits the experimental data with a standard deviation of 0.0035 pH unit.

Hansson's (1973a) pH scale is based on the total hydrogen ion concentration in synthetic sea water containing sulphate. Owing to the formation of HSO_4^-, the total hydrogen ion concentration differs from the concentration of free uncomplexed hydrogen ion. The association constant for HSO_4^- in synthetic sea water at 25°C and 35‰ (Khoo *et al.*, 1977a) is

$$\beta_{HSO_4^-} = \frac{c_{HSO_4^-}}{c_{H^+} c_{SO_4^{2-}}} = 12.37 \text{ M}_W^{-1} \qquad (21)$$

The total hydrogen ion pH scale retains its simplicity only so long as the ratio $c_{H^+,tot}/c_{H^+}$ remains constant, in which case the glass electrode potential is a function of $c_{H^+,tot}$. This is the case for acid–base titrations at constant temperature, pressure, and salinity. When Hansson's (1973a) synthetic sea water was titrated with HCl, part of the added hydrogen ions reacted with sulphate to form HSO_4^- and part entered the solution as free hydrogen ions:

$$c_{H^+,tot} = c_{H^+} + c_{HSO_4^-} = c_{H^+}(1 + \beta_{HSO_4^-} c_{SO_4^{2-}}) \qquad (22)$$

At pH values greater than 3, the amount of HSO_4^- formed is negligible compared with $c_{SO_4^{2-},tot}$ so that the stoichiometric sulphate concentration can be used in equation 22. Hansson's (1973a) synthetic sea water has $c_{SO_4^{2-},tot} =$

0.028 M_W, so that

$$c_{H^+,tot} = c_{H^+} + c_{HSO_4^-} = 1.346 c_{H^+} \qquad (23)$$

Thus the total hydrogen ion concentration is 35% greater than the concentration of free hydrogen ion at 25°C and 35‰. The ratio $c_{H^+,tot}/c_{H^+}$ varies with temperature and salinity, since $\beta_{HSO_4^-}$ is a function of temperature and salinity, and since $c_{SO_4^{2-},tot}$ is a function of salinity. The effect of temperature and salinity on the ratio of total to free hydrogen ion is shown in Table 4. As can be seen, the ratio is strongly temperature dependent and at 35‰ salinity it increases from 1.18 at 0°C to 1.45 at 35°C. The effect of temperature on pK_2 for carbonic acid as measured by Hansson (1973b) will, in fact, include the effect of temperature on the dissociation of bicarbonate ion *plus* the effect of temperature on the dissociation of HSO_4^-, a fact which complicates the interpretation of enthalpy and volume changes on the pH(SWS) scale for those not familiar with the concept of an ionic medium scale.

Consider the following cell:

$$Ag \mid AgCl \left| \begin{matrix} 0.0101 \text{ m HCl} \\ 0.4961 \text{ m NaCl} \\ 0.0263 \text{ m MgCl}_2 \\ 0.0101 \text{ m CaCl}_2 \\ 0.0304 \text{ m MgSO}_4 \end{matrix} \right| \text{Glass} \left| \begin{matrix} 0.0101 \text{ m HCl} \\ 0.4954 \text{ m NaCl} \end{matrix} \right| AgCl \mid Ag \qquad (24)$$

If we assume that the effect of pressure on f_{HCl} is the same in both solutions (a good assumption), then the change in potential of the above cell with

Table 4. Effect of temperature on the ratio of total to free hydrogen ion in Hansson's tris buffer calculated from equation 22. The association constant of HSO_4^- was taken from Khoo *et al.* (1977a). An asterisk indicates values calculated with extrapolated values of $\beta_{HSO_4^-}$

				t, °C				
S, ‰	0*	5	10	15	20	25	30	35
10*	1.082	1.092	1.105	1.119	1.136	1.155	1.178	1.205
15*	1.109	1.123	1.140	1.159	1.181	1.207	1.237	1.273
20	1.132	1.149	1.169	1.192	1.219	1.250	1.287	1.330
25	1.151	1.170	1.193	1.220	1.250	1.286	1.329	1.378
30	1.167	1.189	1.214	1.243	1.278	1.318	1.364	1.419
35	1.181	1.205	1.232	1.264	1.301	1.344	1.395	1.454
40	1.194	1.219	1.248	1.282	1.321	1.368	1.422	1.485
45	1.204	1.231	1.262	1.297	1.339	1.388	1.445	1.512

pressure is

$$(E_1 - E_p) = g \log \left\{ \frac{m_{H^+,1}}{m_{H^+,p}} \right\}_{\text{sulphate soln.}}$$

$$= g \log \left[\frac{(1 + m_{SO_4^{2-}} \beta_{HSO_4^-})_1}{(1 + m_{SO_4^{2-}} \beta_{HSO_4^-})_p} \right] \tag{25}$$

where E_1 is the potential of cell 24 at 1 atm and E_p the potential at p atm.

Distèche and Distèche (1967) measured the effect of pressure on the potential of this cell at 22°C. The total hydrogen ion concentration in both solutions is independent of pressure, and yet the potential of the cell decreases 3.5 mV when the pressure is increased from 1 to 1000 atm. The change in potential is linear with pressure and is due to the dissociation of hydrogen sulfate. In fact, the effect of pressure on $\beta_{HSO_4^-}$ at 22°C calculated from Distèche and Distèche's (1967) measurements and from the value of $\beta_{HSO_4^-}$ at 1 atm (Khoo *et al.* 1977a) is

$$\log (\beta_{HSO_4^-})_p = \log (\beta_{HSO_4^-})_1 - 3.631 \times 10^{-4}(p - 1) \tag{26}$$

where p is expressed in atmospheres.

Equations 22 and 26 were used to calculate the effect of pressure on the ratio of total to free hydrogen ion in Hansson's (1973a) synthetic sea water (Table 5). At 22°C and 35‰ salinity, the ratio $c_{H^+,\text{tot}}/c_{H^+}$ decreases from 1.32 at atmospheric pressure to 1.14 at 1000 atm. However, electrode measurements at high pressure (Culberson and Pytkowicz, 1968; Kester and Pytkowicz, 1970; Distèche, 1974) measure the ratio of the free ion concentration at pressure to its concentration at atmospheric pressure (equation 25). Since $\beta_{HSO_4^-}$ varies with pressure, high-pressure pH measurements on the total hydrogen ion pH scale will include the effect of pressure on the dissociation of HSO_4^- in the reference compartment of the high pressure cell. This will complicate the interpretation of high-pressure pH measurements in sea water (Distèche, 1974).

The above discussion indicates that for a proper application of the pH(SWS) scale it is necessary to know how changes in pressure and temperature influence the dissociation for hydrogen sulphate.

Table 5. The effect of pressure on the ratio of total to free hydrogen ion in Hansson's (1973a) synthetic sea water at 22°C and 35‰ salinity. Conditions: $c_{SO_4^{2-}} = 0.02902$ m; $\beta_{HSO_4^-} = 10.94$ m at 1 atm; pressure effect on $\beta_{HSO_4^-}$ from equation 26

Pressure (atm)*	1	200	400	600	800	1000
$c_{H^+,\text{tot}}/c_{H^+}$	1.317	1.269	1.227	1.192	1.163	1.138

*1 atm = 101 325 N m^{-2}.

The free hydrogen ion pH scale

In an attempt to eliminate the theoretical and practical objections to the NBS and total hydrogen ion pH scales, Bates and Macaskill (1975) proposed a pH scale for sea water based on the concentration of free uncomplexed hydrogen ions. This pH scale was further developed by Khoo *et al.* (1977a), who determined the standard potential of the hydrogen electrode, silver/silver chloride electrode pair in synthetic sea water, and by Ramette *et al.* (1977), who used Khoo *et al.*'s data to calibrate an equimolal sea water tris buffer in terms of the free hydrogen ion concentration.

Although the free hydrogen ion concentration is a simple concept, its determination in sea water is complicated by the formation of HSO_4^-. In contrast to total hydrogen ion concentration, the free hydrogen ion concentration cannot be determined from a simple HCl titration. To eliminate the effect of HSO_4^- on their measurements, Khoo *et al.* (1977a) standardized their electrodes in a chloride solution whose effective ionic strength and major cation concentrations approximated those of natural sea water (Table 6). They determined the potential of the cell

$$Pt \mid H_2(g, 1 \text{ atm}) \mid HCl\,(0.01 \text{ m}) \text{ in sulphate}-\text{free sea water} \mid AgCl \mid Ag \tag{27}$$

where

$$E = (E^\ominus - 2g \log f_{HCl}) - g \log m_{Cl} - g \log m_H \tag{28}$$

$$E = E^* - g \log m_{Cl} + g p m_H \tag{29}$$

in sulphate-free synthetic sea water (Table 6). The similarity in composition and ionic strength of the chloride solution and natural sea water means that f_{HCl} and E^* should be nearly identical in the two solutions. Khoo *et al.*

Table 6. Composition (mm_x) of synthetic sea waters used to standardize the free hydrogen ion pH scale at 35‰ salinity

Solution species	Natural sea water	Sulphate-free sea water	Equimolal tris buffer
Na^+	485.23	454.44	445.16
K^+	10.58	10.58	10.58
Mg^{2+}	55.18	55.18	55.18
$Ca^{2+} + Sr^{2+}$	10.77	10.77	10.77
H^+		10.00	
$Cl^- + Br^- + F^- + HCO_3^-$	569.18	606.92	569.12
SO_4^{2-}	29.27		29.26
$B(OH)_3$	0.43		
Tris			40.00
$Tris-H^+$			40.00

Table 7. Values of pm_H in 0.04 m equimolal tris buffer as a function of temperature and salinity (Ramette *et al.*, 1977). The values at 0°C are extrapolated

S, ‰	t, °C								
	0	5	10	15	20	25	30	35	40
30	8.966	8.798	8.635	8.479	8.329	8.185	8.047	7.914	7.788
35	8.983	8.815	8.652	8.496	8.345	8.201	8.062	7.930	7.803
40	9.001	8.832	8.669	8.512	8.361	8.216	8.078	7.945	7.818

(1977a) showed that differences in E^* between sea water with and without sulphate should be less than 0.2 mV.

Ramette *et al.* (1977) used the cell

$$\text{Pt} \mid \text{H}_2(\text{g, 1 atm}) \mid \text{Tris (m)} + \text{Tris–HCl (m) in sea water} \mid \text{AgCl} \mid \text{Ag} \tag{30}$$

in conjunction with values of E^* (Khoo *et al.*, 1977a) to calibrate an equimolal sea water tris buffer on the free hydrogen ion pH scale. They chose a 0.04 m equimolal tris buffer as their standard (Table 6), the pm_H of which (Table 7) is given by equation 31 between 30 and 40‰ salinity and 5–40°C.

$$pm_H = a + bS \tag{31}$$

where

$$a = 8.8621 - 0.033904t + 1.207 \times 10^{-4}t^2$$
$$b = 3.4617 \times 10^{-3} - 1.207 \times 10^{-5}t$$

and t is the temperature in °C. Equation 31 fits the experimental values of pm_H with a standard deviation of 0.0012 pm_H unit.

The compositions of the sea water tris buffers developed by Hansson (1973a) and Ramette *et al.* (1977) are very similar (compare Tables 3 and 6). They differ slightly in major ion concentrations, and in their tris concentrations (Hansson, 0.005 m; Ramette *et al.*, 0.04 m). The measurements of Ramette *et al.* (1977) show that the pm_H in the 0.04 m tris buffer is only 0.003 unit higher than in the 0.005 m tris buffer.

pH scales in simple salt solutions

Physico-chemical studies of sea water often require the use of chloride, perchlorate, or nitrate media to avoid the effects of unwanted solubility or ion-association reactions (Byrne, 1974; Byrne and Kester, 1974; Pytkowicz and Hawley, 1974; Atlas *et al.*, 1976; Dyrssen and Hansson, 1973). In these studies it is assumed that activity coefficients of free ions are functions of ionic strength only, and are independent of solution composition at constant ionic strength. The total and free hydrogen ion pH scales are identical in chloride, perchlorate, and nitrate media, and the only pH scales that have been used are the free hydrogen ion and the NBS pH scale. These two scales are related by the conventional (or 'apparent') single ion activity coefficient of hydrogen ion in each medium:

$$f'_H = a'_H/m_H \qquad (32)$$

The value of f'_H includes the effect of the residual liquid junction potential (see Section 6.3.6). Values of f'_H in chloride and nitrate media measured with a glass/saturated calomel electrode pair are listed in Table 8. The insolubility of $KClO_4$ prohibits the use of saturated calomel electrodes in $NaClO_4$, and acid–base measurements in $NaClO_4$ have used $NaNO_3$ (Byrne, 1974) or NH_4Cl (McBryde, 1969) salt bridges. Values of f'_H measured in perchlorate media with $NaNO_3$ and NH_4Cl salt bridges are also listed in Table 8 for comparison with measurements using saturated KCl.

Table 8 shows two features: (1) values of f'_H in KCl and KNO_3, measured with a saturated calomel electrode, are much smaller than values in NaCl, $NaNO_3$, $MgCl_2$, or $CaCl_2$; and (2) values of f'_H measured with a 0.68 m $NaNO_3$ salt bridge are much larger than values measured with concentrated

Table 8. Conventional single ion activity coefficients of hydrogen ion in simple salt solutions at ionic strength 0.7 M. The salt bridge is saturated KCl unless otherwise noted. ΔE_j is the residual liquid junction potential, E_j (pH 4 buffer/salt bridge) $-E_j$ (test soln./salt bridge), calculated from the Henderson equation (Bates, 1973) at 25°C

Salt	Molality	t, °C	f'_H	ΔE_j	Reference
NaCl	0.68	25	1.02	+3.2	Byrne (1974)
	0.69	20	1.01		Atlas *et al.* (1976)
KCl	0.68	25	0.81	+1.7	McBryde (1969)
	0.69	20	0.86		Atlas *et al.* (1976)
$MgCl_2$	0.23	20	0.99	+3.8	Atlas *et al.* (1976)
$CaCl_2$	0.23	20	0.98	+3.6	Atlas *et al.* (1976)
$Na/Mg/CaCl_2$		25	0.99		Culberson *et al.* (1970), $I = 0.68$ m
$NaNO_3$	0.68	25	1.05	+2.9	Byrne (1974)
KNO_3	0.68	25	0.81	+1.4	McBryde (1969)
NaCl	0.68	25	1.71	+15.3	Byrne (1974), 0.68 m $NaNO_3$ salt bridge
$NaNO_3$	0.68	25	1.55	+14.3	Byrne (1974), 0.68 m $NaNO_3$ salt bridge
$NaClO_4$	0.68	25	1.55	+13.4	Byrne (1974), 0.68 m $NaNO_3$ salt bridge
	0.68	25	0.96	+2.8	McBryde (1969), 3.5 m NH_4Cl salt bridge

KCl or NH_4Cl salt bridges. Both of these effects are related to changes in the residual liquid junction potential between the constant ionic strength media and the dilute buffers used to calculate a'_H. Due to the low mobilities of Na^+, Mg^{2+}, and Ca^{2+} relative to K^+ (see Section 1.3.2), the residual liquid junction potential between 0.68 m KCl or KNO_3 and NBS pH 4.008 buffer is about 1.8 mV less than that between NaCl, $NaNO_3$, $MgCl_2$, or $CaCl_2$ and the pH 4.008 buffer.

Likewise, the large values of f'_H measured with a 0.68 m $NaNO_3$ salt bridge are due to the liquid junction potential developed at the 0.68 m $NaNO_3$/dilute buffer interface. The value of f'_H in $NaClO_4$ drops back to its normal value when concentrated NH_4Cl is used as the salt bridge.

The assumption that activity coefficients are independent of solution composition at constant ionic strength is clearly not correct for f'_H due to liquid junction effects. Byrne (1974) emphasized that the free hydrogen ion concentration, rather than a'_H, is the variable that should be measured in studies where pH measurements between single salt solutions are to be compared.

Mathematical relationships between pH scales

In view of the fact that three different pH scales are in use for sea water pH measurements it is important to have conversion factors so that pH measurements on one scale can be converted to a second scale.

The difference in pH between the free and total hydrogen ion pH scales can be calculated from equation 22:

$$pm_H - pH(SWS) = \log (1 + c_{SO_4^{2-}} \beta_{HSO_4^-}) \tag{33}$$

where $\beta_{HSO_4^-}$ is given by (Khoo et al., 1977a)

$$\log \beta_{HSO_4^-} = 647.59/T - 6.3451 + 0.019085 T - 0.5208(I_{tot})^{1/2} \tag{34}$$

The pH difference between the two scales ranges from 0.06 at 5°C and 20‰ to 0.18 pH unit at 35°C and 45‰ salinity.

The NBS and free hydrogen ion pH scales are related through the apparent single ion activity coefficient of free hydrogen ion in sea water.

$$a'_H = f'_H m_H \tag{35}$$

$$pm_H - pH(NBS) = \log f'_H \tag{36}$$

Unfortunately, f'_H is known for only two salinities at 25°C (Table 9). It is a lucky coincidence that the two pH scales are almost identical at 35‰ and 25°C, where

$$pm_H - pH(NBS) = 0.003 \tag{37}$$

The NBS and the total hydrogen ion pH scales are related through the

Table 9. Measured values of the conventional single ion activity coefficient of free hydrogen ion in sea water at 25°C. Molal concentrations. The author measured the pH(NBS) of the 0.04 m sea water tris buffer (Ramette et al, 1977) using the pH assembly described in Section 6.3.1. At 25°C and 35‰ salinity, the measured value is
$$pH(NBS) = 8.198 \pm 0.003 \ (2\sigma)$$

Salinity %	f'_H	Reference
35.00	$1.007 \pm 0.007 \ (2\sigma)$	Culberson (1977, unpublished data)
34.31	$1.000 \pm 0.007 \ (2\sigma)$	Bates and Macaskill (1975), sea water with sulphate
34.6	$1.006 \pm 0.033 \ (2\sigma)$	Culberson et al. (1970), sulphate- -free sea water.
26.7	$0.967 \pm 0.006 \ (2\sigma)$	

conventional single ion activity coefficient of total hydrogen ion in sea water:

$$a'_H = f'_{H,tot}[H^+]_{tot} \tag{38}$$

$$pH(NBS) - pH(SWS) = -\log f'_{H,tot} \tag{39}$$

Measured values of $f'_{H,tot}$ in natural and synthetic sea water are compared in Table 10. The value of $f'_{H,tot}$ depends on the concentrations of sulphate and fluoride, since both of these ions react with hydrogen ion in the pH range 3–4 over which $f'_{H,tot}$ is determined (Culberson et al., 1970). The sulphate and fluoride concentrations of Hansson's (1973a) synthetic sea water (Table 3) differ from those of natural sea water and theoretical calculations (Table 10) show that values of $f'_{H,tot}$ in Hansson's synthetic sea water are about 1.8% larger than in natural sea water.

At 35‰ salinity, the measured values of $f'_{H,tot}$ in Table 10 have a range of 5.4%. Most of this variation is probably due to differences in the residual liquid junction potentials of the reference electrodes used in the several studies.

Strictly, the value of $f'_{H,tot}$ in equation (39) should be measured in Hansson's synthetic sea water. Unfortunately, the only measurements that cover the entire oceanic temperature and salinity range (Mehrbach et al., 1973; Culberson and Pytkowicz, 1973) were made in natural sea water. However, values of $f'_{H,tot}$ in the presence and absence of fluoride differ by less than 0.01 pH units (Table 10). Considering the discrepancies between the measured values of $f'_{H,tot}$ and the effect of fluoride, it is suggested that the values measured by Culberson and Pytkowicz (1973) be used to compare the NBS and total hydrogen ion pH scales until better data become available. To facilitate comparison of the two pH scales, the measured values of $f'_{H,tot}$ (Culberson and Pytkowicz, 1973) were fitted to the following equation, which fits the 20 experimental points with a relative standard

Table 10. Measured values of the conventional single ion activity coefficient of total hydrogen ion in sea water at 25°C. Molal concentrations. Values marked with an asterisk are extrapolated. The calculated values were calculated from the association constants of HSO_4^- ($\beta = 11.92$ M; Khoo *et al.*, 1977a) and HF ($\beta = 399$ M; Culberson *et al.*, 1970), and the conventional single ion activity coefficient of free hydrogen ion in seawater ($f'_H = 1.007$; Table 9)

Salinity ‰	$f'_{H,tot}$	Reference
35	0.669	Hansson (1973a), synthetic sea water without
20	0.715	fluoride
35	0.703	Culberson and Pytkowicz (1973), synthetic
30	0.697	sea water with fluoride
25	0.695	
20	0.696	
35	0.695*	Culberson *et al.* (1970), natural sea water
30	0.700	
25	0.705	
20	0.710*	
35	0.667	Mehrbach *et al.* (1973), natural sea water
30	0.662	
25	0.658	
20	0.662	
35	0.737	Calculated, sea water with fluoride
	0.749	Calculated, sea water without fluoride
	0.750	Calculated, Hansson's (1973a) synthetic
		sea water

deviation of 0.9% in $f'_{H,tot}$,

$$\log f'_{H,tot} = \frac{-A(I_{tot})^{1/2}}{[1 + d(I_{tot})^{1/2}]} + (B_0 + B_1 t)(I_{tot}) - \log(1 + m_{SO_4^{2-}}\beta_{HSO_4^-}) \quad (40)$$

where concentrations are molal and A = Debye–Huckel slope (Bates, 1973), $d = 2.16$, $B_0 = 0.1681$, $B_1 = 6.26 \times 10^{-4}$, and t is the temperature in °C. $\beta_{HSO_4^-}$ was calculated from equation 34. Bearing in mind the systematic differences between the measured values of $f'_{H,tot}$, the accuracy of equation 40 is no better than ±0.03 pH unit when it is used to convert pH(NBS) to pH(SWS).

pH scales: pros and cons

At present three pH scales are in use for sea water pH measurements. Considering the small number of marine chemists and the large expanse of ocean, this is two too many. The advantages and disadvantages of the three scales are compared in Table 11. The major problem with the NBS pH scale for equilibrium measurements in sea water is the unreproducibility of the residual liquid junction potential, not its lack of theoretical meaning. The

Table 11. Comparison of advantages and disadvantages of the NBS, free, and total hydrogen ion pH scales

Scale	Advantage	Disadvantage
NBS	Buffers are easily prepared, certified buffers available from a central laboratory, standard pH scale for geochemistry and biochemistry	Measured pH has no theoretical significance, residual liquid junction potential between dilute buffer and sea water depends on type of liquid junction
free H^+ concentration	Conceptually the best pH scale, minimizes residual liquid junction potential for salinities near 35‰	Sea water buffers are difficult to prepare, certified buffers not available from a central laboratory, no better than NBS scale for low and high salinity sea water, pH scale is unique to oceanography
total H^+ concentration	Total H^+ is a well defined quantity, minimizes residual liquid junction potential for salinities near 35‰	Sea water buffers are difficult to prepare, certified buffers not available from a central laboratory, no better than NBS scale for low and high salinity sea water, inclusion of HSO_4^- complicates interpretation of pressure and temperature effects on pH and pK values, pH scale is unique to oceanography

free and total hydrogen ion pH scales minimize the junction potential by using sea water buffers whose salinity matches that of open ocean sea water (30–40‰). However, hydrogen ion concentration scales have no practical advantages for pH measurements in low (estuaries) or high (brines) salinity environments in which many of the most important reactions controlling the composition of the oceans occur. Adoption of one of the hydrogen ion concentration scales will still require the use of the NBS pH scale for measurements in estuaries and brines, and will mean that oceanographers use a pH scale different from that used by geochemists and biochemists.

The major experimental advantage claimed for the hydrogen ion concentration scales is that they minimize changes in the residual liquid junction potential, and thus enhance the precision of pH measurements in sea water. This is true only if the composition of sea water and buffer are identical. It is not true in estuaries and brines whose composition is not known *a priori* and in which it is impossible to match the composition of standards and unknowns. The comparison of field data in Section 6.3.1 (Figures 3, 5, and 6) shows that, at present, there is no basis for the claim that pH measurements using sea water buffers are inherently more precise than those using dilute NBS buffers.

The author believes that it would be more profitable to design a reproducible liquid junction than to adopt the free or total hydrogen ion pH scales. A well designed, simple, and reproducible liquid junction that was applicable to pH measurements in other fluids would have more commercial appeal than would sea water buffers with their limited market.

6.1.3. The method of standard additions

The use of direct potentiometry to determine ionic concentrations in natural waters requires careful matching of standards and unknowns to eliminate spurious effects due to changes in ionic strength and complexation. Since the composition of many natural waters is not known *a priori*, it is frequently impossible to prepare standards with the same bulk composition as the unknown. In this situation, the method of standard additions may be used. Consider the potential of the cell

$$ISE(X) \mid sea\ water \mid KCl\,(saturated) \mid AgCl \mid Ag \qquad (41)$$

where ISE(X) is an ion-selective electrode reversible to ion X. The potential of this cell is

$$E_1 = E^0 + E_j + (g/n) \log ([X]_1 f_X) \qquad (42)$$

If a small volume of a concentrated solution of X is added to the sea water sample, the new potential is

$$E_2 = E^0 + E_j + (g/n) \log ([X]_2 f_X) \qquad (43)$$

and the new concentration of X is

$$[X]_2 = ([X]_1 v_1 + [X]_s v_s)/(v_1 + v_s) \qquad (44)$$

where v_1 is the initial volume, v_s is the volume of the spike, and $[X]_s$ is the concentration of the spike. Substituting equation 44 into equation 43 and subtracting equation 42 from equation 43 yields the following equation for $[X]_1$, the initial concentration of X (Brand and Rechnitz, 1970; Durst, 1969b):

$$[X]_1 = [X]_s [v_s (v_1 + v_s)^{-1}][10^{n(E_2 - E_1)/g} - v_1 (v_1 - v_s)^{-1}]^{-1} \qquad (45)$$

If the volume of the spike, v_s, is negligible compared with the initial volume, v_1, equation 45 reduces to

$$[X]_1 = [X]_s v_s v_1^{-1} [10^{n(E_2 - E_1)/g} - 1]^{-1} \qquad (46)$$

Standard addition techniques measure total concentrations, and certain experimental conditions must be met if equations 45 and 46 are to be valid (Warner, 1973; Durst, 1969b). The first is that changes in total ionic strength and sample volume upon addition of the spike must be small so that the values of the total single ion activity coefficients and the liquid junction potential remain constant. The second is that the electrode must respond only to the ion of interest, and not to other ions in solution. The third is that the slope of the electrode response must be constant and known. The fourth is that the degrees of complexation of X must be identical in the initial and spiked solutions; changes in the fraction of X that is complexed will affect the total activity coefficient of X and invalidate equations 45 and 46.

6.2. BEHAVIOUR OF ION-SELECTIVE ELECTRODES IN SEA WATER

6.2.1. Reference electrodes

The use of ion-selective electrodes to measure the activity or concentration of single ions in sea water requires a reference electrode whose potential is a known function of solution composition. The most common reference electrodes are calomel or Ag | AgCl electrodes which are coupled to the test solution through a salt bridge, usually containing concentrated KCl. However, in certain circumstances it is possible to use reference electrodes without liquid junctions. The preparation and use of reference electrodes are discussed in detail by Ives and Janz (1961a), Covington (1969), and Bates (1973).

Reference electrodes without liquid junctions

The most suitable reference electrodes without liquid junction are those which respond to one of the major ions and which are not subject to interference from other ions in sea water. The chloride, sodium, and perhaps fluoride electrodes best fit these criteria.

Ag | AgCl electrodes without a liquid junction are frequently used as reference electrodes for pH measurements in synthetic sea water (Distèche and Distèche, 1967; Culberson and Pytkowicz, 1968; Khoo *et al.*, 1977b; Ramette *et al.*, 1977). However, Ag | AgCl electrodes are subject to bromide interference (Pinching and Bates, 1946; Riley, 1965), which precludes their use for thermodynamic measurements in natural sea water which contains 0.87 mmolal bromide. Culberson (1968) studied the effect of bromide on the potential of electrolytically prepared Ag | AgCl electrodes and found that the potentials of Ag | AgCl electrodes did not return to their initial potential when bromide was removed from the test solution.

Miller and Kester (1976) used a solid-state chloride electrode, as part of a F^-/Cl^- electrode pair, to study fluoride ion pairing in NaF–NaCl solutions at ionic strength 0.7. Short-range interactions in electrolytes are primarily plus–minus interactions, and the specific interaction model (Whitfield, 1974a) predicts that the activity coefficients of F^- and Cl^- should be independent of composition in NaF–NaCl mixtures at constant ionic strength, since the minus–minus interactions between F^- and Cl^- are long-range Debye–Hückel type interactions. The potential of the F^-/Cl^- electrode pair depends on the ratio of fluoride to chloride activities. If the activity coefficients of F^- and Cl^- are constant, the activity coefficient term can be absorbed into the standard potential, and the potential of the F^-/Cl^- cell can be interpreted in terms of F^- and Cl^- concentrations.

Mixtures of the types, (a) MX–MY and (b) NX–MX are commonly used to study ion association equilibria, and the specific interaction model (Whitfield, 1974a) suggests that ion association measurements using the X/Y and N/M electrode pairs will help minimize errors due to variation of the activity coefficient at constant ionic strength.

Wilde and Rodgers (1970) demonstrated the possibility of using a sodium ion glass electrode as the reference electrode for pH measurements in sea water. The best use of a sodium ion reference electrode would be to study mixtures of type (b) in conjunction with a second cation-responsive electrode. However, a sodium ion reference electrode might also be useful for measurements in sediments and interstitial waters which can produce anomalous liquid junction potentials (Zhelesnova, 1962, 1964).

Any use of reference electrodes without a liquid junction must consider ion association reactions that might affect the potential of the reference electrode. This is especially important for mixtures of types (a) and (b) in which the extent of ion pairing may vary with the composition of the mixture. Fluoride ion associates with many cations (Miller and Kester, 1976) and the fluoride electrode is not a suitable reference electrode in these mixtures. However, it might be a suitable reference electrode for natural sea water, in which the $Ag|AgCl$ electrode cannot be used due to bromide interference.

Reference electrodes with liquid junctions

Calomel or $Ag|AgCl$ electrodes with a 3.5 M or saturated KCl filling solution are the most commonly used reference electrodes in oceanography. The potentials of calomel and $Ag|AgCl$ electrodes are well defined and reproducible (Ives and Janz, 1961a), and experimental problems encountered with these electrodes are usually associated with the liquid junction and not with the electrode itself.

The discussion of pH scales for sea water (Section 6.1.1) showed that uncertainty in the value of the residual liquid junction potential, ΔE_j, is the major problem in interpreting pH measurements in concentrated electrolyte solutions. As Johnson *et al.* (1977) have emphasized, it is the reproducibility of ΔE_j, not its absolute magnitude, which is the most important consideration for pH measurements involving concentrated KCl salt bridges. The same argument holds for many other pIon measurements. However, systematic errors between many pH (Section 6.3.1) and pK measurements (Buch, 1951; Lyman, 1956; Mehrbach *et al.*, 1973) suggest that the reproducibility of the residual liquid junction potential is often poor. These systematic errors suggest that ΔE_j depends on the construction and/or history of the liquid junction used in each investigation.

In an attempt to determine the magnitude of possible variations in ΔE_j, the author (Johnson *et al.*, 1977) compared residual liquid junction potentials of several commercial reference electrodes (Table 12).

Eight reference electrodes, filled with 3.5 M KCl, were mounted in a cell maintained at $25.00 \pm 0.05°C$, and their potentials were measured to within

Table 12. Differences in residual liquid junction potentials of commercial reference electrodes in various solutions. The tabulated values are the potential of the electrode pair in the test solution minus their potential in 3.5 M KCl. Most values are the result of two or three measurements with a scatter of ± 0.02 mV. The reference electrodes used were: 1 and 2, Sargent/Jena S-30080-15 (ceramic frit junction); 3 and 4, Beckman 39170 (asbestos-fibre junction); 5, Sargent/Jena S-30084-15 (ground-glass sleeve junction); 6, Sargent/Jena S-30080-10 (platinum annulus junction); 7, Sargent/Jena S-30072-15 (platinum annulus junction); 8, Leeds and Northrup 117208 (glass annulus junction). From Johnson *et al.* (1977), with permission of Pergamon Press, Ltd

| Test solution | Electrode pair | | | | | | |
	1–2	1–3	1–4	1–5	1–6	1–7	1–8
NBS pH 4.008 buffer	−0.07	+0.11	+0.15	+0.06			
NBS pH 7.413 buffer	−0.19	−0.37	−0.38	−0.47	+8.5	+1.99	+0.64
Artificial river water (pH 8.2)	−0.16	−0.33	−0.33	−0.43			
Sea water (34.8‰) (pH 8.3)	−0.02	−0.25	−0.36	−0.31	+1.56	+0.50	+0.19
Acidified sea water (pH 3.0)	0.00	−0.31	−0.38	−0.28			
0.668 M KCl	−0.01	−0.01	0.00	−0.03	+0.99	+0.58	+0.13
0.668 M NaCl	−0.02	−0.17	−0.25	−0.24			
0.668 M HCl	−0.02	−0.19	−0.78	−0.54			
0.67 M NaOH	−0.53	−0.33	+0.06	−0.04			
0.223 M MgCl$_2$	−0.01	−0.93	−1.22	−1.00	+0.57	+0.44	+0.12
0.223 M CaCl$_2$	−0.01	−0.74	−0.98	−0.82			
0.309 M Na$_2$SO$_4$	−0.04	+0.35	+0.43	+0.22	+2.11	+0.77	+0.01
0.474 M MgSO$_4$	−0.02	+0.04	0.00	−0.04			

±0.01 mV. The cell was designed so that it could be emptied and filled without removing it or the electrodes from the water-bath. Prior to use, all solutions were filtered through a 0.45-μm filter to remove particulate matter which might have clogged the junctions, and were partially degassed with an aspirator to prevent air bubbles from forming on the junctions. All solutions were brought to 25°C before the cell was filled.

At the beginning of each experiment, the cell was filled with 3.5 M KCl and the bias potential of each electrode was measured relative to the potential of a calomel electrode that was used as the standard reference electrode. The cell was then filled with test solution and the potential of each electrode was re-measured *versus* the standard calomel electrode. At the end of the experiment, the cell was refilled with 3.5 M KCl and the bias potentials of the electrodes re-measured to check for electrode drift.

If the residual liquid junction potentials of each electrode are identical, the potential difference of each electrode pair will remain constant when 3.5 M KCl is replaced with test solution. However, the potential difference of an electrode pair will change if the residual liquid junction potentials of the two electrodes are different. The measurements of ΔE_j (Table 12) show the following features:

1. The residual liquid junction potentials of supposedly 'identical' ceramic or asbestos junctions can differ by as much as 0.3 mV, and the differences are greater in acidic or basic media.
2. Electrodes with platinum annulus junctions show large changes in ΔE_j and are unsuitable for precise measurements in which the electrode potentials are compared in solutions which differ in composition. However, platinum junctions work well in situations, such as Gran titrations, which only require relative pIon values. In addition, metal junctions should never be used in reducing solutions containing sulphide.
3. The ground-glass sleeve and asbestos-fibre junctions behave similarly, but their potentials differ by as much as 1 mV from the residual liquid junction potentials of ceramic frit junctions. This difference may be related to flow-rate, since the flow-rates of the ceramic frit junctions were less than those of the asbestos and glass sleeve junctions.
4. Replicate measurements of ΔE_j (not shown) showed that the reproducibility of individual ceramic, glass sleeve, and asbestos-fibre junctions was ±0.03 mV.
5. The electrodes show changes in ΔE_j when 3.5 M KCl is replaced with 0.67 M KCl. This is theoretically impossible since the liquid junction potential between two solutions of the same salt is independent of the structure of the junction (MacInnes, 1961). This observation indicates that the material used to contain the junction contributes to the junction potential.

6. The potentials of the electrode pairs often drifted with time when 3.5 M KCl was replaced with test solution. This drift was especially pronounced in pH 7.413 buffer, river water, $MgCl_2$, and $CaCl_2$, and the drift rates in these solutions were as large as $0.06 \, mV \, h^{-1}$. This observation suggests that the small drifts often seen in pH measurements may be due to the reference electrode and not to the glass electrode.

In a second experiment, not shown in Table 12, it was found that the glass sleeve and asbestos-fibre junctions responded immediately when the pressure differential across the liquid junction was increased by 9×10^{-3} atm ($912 \, N \, m^{-2}$), but that the potential of a ceramic frit junction took about 15 min to reach a constant potential after the pressure of the internal filling solution was increased. The maximum change in potential upon pressurizing the internal filling solution of the ceramic, asbestos-fibre, and glass sleeve junctions was less than $0.03 \, mV$ for a pressure increase equivalent to 10 mm of water.

It is commonly found that the potentials of reference electrodes with liquid junctions are sensitive to stirring. The exact value of the stirring potential depends on the geometry of the electrode cell and on the test solution, but the values listed below for a ceramic frit junction (3.5 M KCl) are typical of observed stirring potentials:

Solution	Stirring potential $[E(\text{stir}) - E(\text{still})]$, mV
NBS pH 4.01 buffer	-0.17
NBS pH 6.86 buffer	-0.38
NBS pH 9.18 buffer	-1.10
0.1 M HCl	0.00
Sea water	-0.01

For comparison, Zirino (1975) reported that the stirring potentials of a ceramic frit combination electrode in phosphate buffer (pH 7) and in sea water were $+0.3 \, mV$ and $-0.3 \, mV$.

Certain types of Ag | AgCl electrodes are sensitive to light (Janz, 1961; Milward, 1969; Moody *et al.*, 1969) and they should be kept out of direct sunlight and used under conditions of constant illumination. This includes the Ag | AgCl internal reference electrodes of glass electrodes.

Standardization of pH measurements on the free and total hydrogen ion pH scales (Section 6.1.2) requires the use of synthetic sea water buffers whose salinities match as closely as possible those of the unknown sea water samples.

In practice, pH measurements on the free and total hydrogen ion pH scales are made with the cell

Glass | sea water or saline tris buffer (e.g. Table 6) |

$$\text{| saturated KCl | Ag|AgCl or calomel} \quad (47)$$

and only one sea water buffer is used to standardize the pH cell. It is assumed that changes in the residual liquid junction potential between the 35‰ tris buffer and sea water are negligible. This assumption could be verified by standardizing cell 47 with 35‰ tris buffer, and then measuring pH(SWS) or pm_H of buffers with other salinities. Variations in ΔE_j would cause differences between the measured and assigned values of pH(SWS) and pm_H. Since the above experiment has never been performed, the effect of salinity on ΔE_j (Table 13) was calculated from the Henderson equation (Bates, 1973; see Section 1.3.2). The salinities of most open ocean waters lie between 30 and 40‰, and the theoretical calculations show that the variation in E_j over this salinity range is only ± 0.2 mV when a saturated KCl salt bridge is used. The variation in E_j is ± 0.8 mV when a 35‰ salinity sea water salt bridge is used. The theoretical calculations suggest that systematic errors introduced by using a fixed salinity buffer and a saturated KCl salt bridge are almost negligible for open ocean waters. However, the calculations are only approximate and need to be checked experimentally.

6.2.2. Hydrogen ion electrodes

Although the hydrogen electrode is the standard for pH measurements, it is not suitable for field measurements because volatile gases such as CO_2 are stripped from solution, thus changing the pH. The hydrogen electrode is suitable for laboratory studies where it has been used to measure the acidity

Table 13. Effect of salinity on the liquid junction potential between sea water and saturated KCl or sea water and 35‰ salinity sea water. Calculated from the Henderson equation (Bates, 1973) at 25°C

	Liquid junction potential (E_j), mV	
Salinity, ‰	Saturated KCl	Sea water, 35‰
10	+1.1	+7.0
15	+0.6	+4.8
20	+0.3	+3.2
25	0.0	+1.9
30	−0.2	+0.9
35	−0.5	0.0
40	−0.7	−0.7
45	−0.9	−1.4

constants of boric acid (Hansson, 1973b), ammonia (Khoo *et al.*, 1977b), and tris(hydroxymethyl)aminomethane (Ramette *et al.*, 1977), and to establish the free hydrogen ion pH scale (Khoo *et al.*, 1977a). The hydrogen electrode is useful for measurements at high pH, at which glass electrodes are subject to sodium ion error. The preparation and use of hydrogen electrodes is discussed in detail by Hills and Ives (1961) and Bates (1973).

Almost all pH measurements in sea water are made with hydrogen ion sensitive glass electrodes, the properties of which are discussed by Bates (1973). The 'standard' potentials of glass electrodes are not as reproducible as that of the hydrogen electrode, and in addition glass electrodes are subject to errors due to non-Nerstian response, sodium ion interference at high pH, and changes in the asymmetry potential of the glass membrane. Nevertheless, glass electrodes are capable of high precision when used carefully (King and Prue, 1961).

Several studies (Table 14) have compared glass and hydrogen electrodes using the cell

$$\text{Pt} \mid \text{H}_2(\text{g}) \mid \text{sea water or test solution} \mid \text{Glass} \qquad (48)$$

whose potential is given by

$$E = E_{\text{H}_2}^{\ominus} - E_{\text{glass}}^{\ominus} = \text{constant} \qquad (49)$$

The hydrogen ion activity cancels in equation 49, and the potential of the cell is constant as long as the glass electrode behaves reversibly. The results in Table 14 and other studies (Zielen, 1963; Beck *et al.*, 1968) show that most glass electrodes behave theoretically, except at high pH where the sodium in sea water interferes. The Jena and Corning electrodes in Table 14 behave theoretically in sea water at pH values less than 9. Although these two electrodes exhibit a Nernstian slope, their slope measured in NBS

Table 14. Comparison of glass and hydrogen electrodes at 25°C. Hansson (1972) found that the sodium ion error increased slightly at low temperatures. Zielen (1963), MacInnes (1961), and Beck *et al.* (1968) compared many types of glass electrodes with the hydrogen electrode. The 'NBS slope' is the electrode slope calculated from measurements in NBS pH 4.008 and 7.413 buffers

Glass electrode	Medium	$E°$, (mV)	pH range for constant $E°$	NBS slope, %	Sodium ion error	Reference
Jena type U	0.5 M NaCl	604.3 ±0.10	2.1–8.5		0.3 mV at pH 9	Hansson (1972)
Jena type U	Sea water	638.63 ±0.10	2.4–8.9	99.1		Mehrbach *et al.* (1973)
	0.5 M NaCl	638.58 ±0.13	2.1–9.0			
	NBS pH 7.41	638.6				
Corning triple purpose	0.7 M NaCl	708.93 ±0.07	2.2–9.3	99.4	0.15 mV at pH 9.8 0.40 mV at pH 10.4	Culberson and Pytkowicz (1973)

buffers is significantly less than theoretical. This apparent discrepancy is due to changes in the residual liquid junction potential at the ceramic frit and asbestos-fibre junctions commonly used in reference electrodes. Bates (1973) (Sections 6.1.2 and 6.3.1) has shown that the residual liquid junction potential of NBS buffers is essentially constant when a free diffusion liquid junction is used. Distèche and Distèche (1967) and Culberson and Pytkowicz (1968) have shown that the glass electrode slope is independent of pressure.

The data in Table 14 also show that the potential of the Jena electrode is not affected by the major ions in sea water, in agreement with the data of MacInnes (1961) for Corning 015 glass. This implies that the major ions in sea water have no effect on the asymmetry potentials of glass electrodes.

Pytkowicz *et al.* (1966) and Johnson *et al.* (1977) evaluated the performance of several commercial glass and reference electrodes, and concluded that the 95% confidence interval for pH measurements in sea water on the NBS pH scale is approximately ±0.011 pH unit. Note that this is a measure of instrumental precision; sampling errors (Section 6.3.1) during field measurements can be much larger.

Even with the most careful treatment, the potential of cells containing glass electrodes often drifts slowly with time ($<0.6\,mV\,h^{-1}$) after the cell is placed in a new solution. This drift occurs in cells with and without liquid junction, and is worst in poorly buffered solutions. Possible causes of observed drifts are (1) contamination of the liquid junction with the previous solution, (2) adsorption or desorption of hydrogen ions and weak acids and bases on the cell walls, and (3) changes in the asymmetry potential of the glass membrane. Factor (1) can be minimized by a proper choice of liquid junction, and it can be eliminated by using a free diffusion junction which can be renewed for each measurement (Section 6.2.2). Factor (2) is difficult to eliminate, but it might be minimized by changing the material used to construct the cell, for instance by replacing glass with PTFE, by speeding up the analysis to shorten the contact time between the test solution and the cell walls, or by using a flow system.

Many procedures recommend drying the glass electrode with tissue when changing solutions. A better technique is to rinse the electrodes with an aliquot of the solution to be measured, since touching the glass membrane can affect the asymmetry potential. In addition, the best results are obtained when the electrodes are maintained at constant temperature, since temperature changes can strain the glass membrane and affect the asymmetry potential.

6.2.3. Cation-selective electrodes

Both sodium amalgam and sodium glass electrodes have been used to measure sodium ion activity and concentration in sea water. The sodium amalgam electrode (Platford, 1965a; Platford and Dafoe, 1965) behaves well in sea water, but in view of the excellent results obtainable with the sodium glass electrode (Gieskes, 1966; Kester, 1970; Christenson, 1973) and the fact that the amalgam electrode must be used in the absence of oxygen, the amalgam electrode does not appear to possess any advantages over the glass electrode.

In sea water, sodium glass electrodes have been used over the temperature range 2–45°C (Kester, 1970; Platford, 1965b) and at pressures to 1000 atm (Kester and Pytkowicz, 1970; Whitfield, 1969). The electrodes are not subject to K^+ or H^+ interference at the concentrations of these ions in natural sea water (Whitfield, 1975). The potentials of sodium glass electrodes are very stable when the electrodes are carefully handled: the drift of one electrode was less than $0.06 \, \text{mV h}^{-1}$ 5–15 min after immersion in a new solution (Christenson, 1973), and Kester (1970) shows a case in which the potential of an electrode remained constant to $\pm 0.01 \, \text{mV}$ over a 4-h period. Christenson (1973) and Kester (1970) found that the precision of repeated measurements of the same solution with sodium glass electrodes was about $\pm 0.02 \, \text{mV}$, and Kester (1970) rejected measurements which differed by more than $\pm 0.05 \, \text{mV}$.

Two types of potassium-selective electrodes—glass and valinomycin— have been used in sea water. However, potassium glass electrodes are subject to severe sodium ion interference ($K_{Na} = 0.1$), so that special techniques are required for their use in sea water (Garrels, 1967). The activity coefficient measurements of Christenson and Gieskes (1971) and Christenson (1973) in KCl/MCl_2 mixtures (M = Mg, Ca, or Ba) show that potassium glass electrodes are capable of high precision in the absence of sodium ion. The behaviour of valinomycin-based potassium electrodes in sea water was investigated by Anfalt and Jagner (1973), Cattrall *et al.* (1974), Miller (1974), and Savenko (1977a). In general, the response of these electrodes was sub-Nernstian when used in solutions containing sodium. For instance, the response of one electrode decreased to 58.7 mV/decade after 1 h in 0.6 M NaCl, and to 19.5 mV/decade after 96 h storage in the same solution. The sub-Nernstian response and associated drifts in electrode potential limit the usefulness of valinomycin-based electrodes for precise thermodynamic measurements in sea water. Nevertheless, these electrodes are useful for measurement of the total potassium concentration of sea water. Using direct potentiometry, Cattrall *et al.* (1974) bracketed sea water between two standards and achieved a precision of $\pm 2\%$ (2σ) in the potassium concentration. Based on 97 individual titrations, Anfalt and Jagner (1973) achieved a precision of $\pm 0.3\%$ (1 relative standard deviation of the mean) with a computer-controlled standard addition technique. However, the relative standard deviation of a single titration was only 2.6%.

Commercially available liquid ion exchanger electrodes for calcium and divalent ions have been used to measure the concentrations and activities of calcium and magnesium in sea water by direct potentiometry and by potentiometric titrations. However, these electrodes are subject to interference from the other major cations in sea water, and special precautions are required for their use in mixed electrolytes at high ionic strength.

The behaviour of the Orion 92-20 calcium electrode in sea water-like media has been investigated by Thompson and Ross (1966), Kester and Pytkowicz (1969), Kester (1970), Bradford (1968), Whitfield and Leyendekkers (1970), and Leyendekkers and Whitfield (1971). In addition, Briggs and Lilley (1974) and Lilley and Briggs (1976) discuss the precautions necessary for high-precision measurements with the liquid ion exchanger calcium electrode. The Orion calcium electrode suffers serious interference from Na^+ and Mg^{2+} in sea water. The approximate effect of interfering ions on the potential of the calcium electrode is (Rechnitz and Lin, 1968)

$$E_{Ca} = E_{Ca}^{\ominus} + \frac{g}{2} \log \left(a_{Ca} + K_{Mg} a_{Mg} + K_{Na} a_{Na}^2 \right) \qquad (50)$$

The selectivity coefficients in equation 50 are not constant but vary with the composition of the medium, even at constant ionic strength. In two-component mixtures at ionic strength 0.72 M and 25°C, $K_{Na} = 0.03$ in NaCl (0.54 M)/$CaCl_2$ (0.06 M) (Whitfield and Leyendekkers, 1970) and $K_{Mg} = 0.07$ in $MgCl_2$ (0.20 M)/$CaCl_2$ (0.04 M); (Leyendekkers and Whitfield, 1971). These values of K_{Na} and K_{Mg} may not be representative of their values in sea water. If K_{Na} and K_{Mg} are constant and greater than zero, equation 50 predicts that the empirical slope of the calcium electrode in mixed electrolytes,

$$\frac{g}{2} = \Delta E / \Delta pCa < 29.6 \, mV/decade \qquad (51)$$

should be less than theoretical. This is true in NaCl/$CaCl_2$ solutions ($g/2 = 25 \, mV/decade$; Bradford, 1968), but measured slopes in sea water (36 mV/decade) are greater than theoretical (Thompson and Ross, 1966; Bradford, 1968). This observation suggests that equation 50 is not a good description of the behaviour of the calcium electrode in sea water. The potential of the calcium electrode is more stable in NaCl than in sea water (Thompson and Ross, 1966; Bradford, 1968), and in sea water the potential of this electrode can vary as much as 1 mV in 6 h (Kester, 1970).

Direct potentiometric measurements with the calcium electrode in sea water must use the potential comparison method (Thompson and Ross, 1966; Kester and Pytkowicz, 1969) in which the electrode potential and activities of interfering ions are matched in the standard and unknown solutions.

Several experimental variables may affect the potential of the calcium liquid ion exchanger electrode. Rechnitz and Lin (1968) found that the potential of the electrode increased 1 mV per millimetre of immersion between 10 and 60 mm; however, Briggs and Lilley (1974) were unable to detect any potential changes for immersions between 5 and 20 mm. Stirring affects the measured potential, and potentials in stirred and unstirred low ionic strength solutions differ by about 0.2 mV (Briggs and Lilley, 1974; Rechnitz and Lin, 1968). Kester (1970) was unable to obtain reproducible potentials with a calcium electrode at low temperatures in sea water, and Whitfield (1969) found that the Orion calcium electrode was unsuitable for use at high pressures.

The behaviour of the Orion 92-32 divalent ion electrode in sea water was investigated by Thompson (1966) and Kester and Pytkowicz (1968, 1970). This electrode is equally selective to calcium and magnesium, and can be used to measure Mg^{2+} activity or concentration in sea water if the potential comparison method is used. The divalent ion electrode suffers from sodium interference, and in $NaCl/MgCl_2$ mixtures at 25°C and ionic strength 0.7 M the potential of one divalent electrode was given by (Kester and Pytkowicz, 1968),

$$E_{Mg} = E_{Mg}^{\ominus} + 28.92 \log [a_{Mg} + 0.043(a_{Na})^2] \qquad (52)$$

In contrast to the calcium electrode, the divalent cation electrode can be used at low temperatures (2°C), but Kester and Pytkowicz (1970) were unable to use it at high pressure due to unreproducible asymmetry potentials.

Copper-, cadmium-, and lead-selective ion electrodes consist of a mixture of silver sulphide and the appropriate metal sulphide. They are electrodes of the third kind (Buck, 1971) and respond to the activity of silver ion:

$$E_M = E_M^{\ominus} + g \log a_{Ag} \qquad (53)$$

Provided that copper, cadmium, or lead is in excess, their concentration controls the silver ion activity through the solubility products of the two sulphides, and the potential of the electrode is proportional to the activity of ion M:

$$E_M = E_M^{\ominus} + (g/n) \log a_M \qquad (54)$$

Several studies have been made of the behaviour of the copper ion electrode in sea water (Jasinski *et al.*, 1974; Rice and Jasinski, 1976) and in high ionic strength solutions of NaCl, KNO_3 (Oglesby *et al.*, 1977), and $NaClO_4$ (Hittinger, 1975; Paulson, 1978). Measurements in KNO_3 and $NaClO_4$ show that copper electrodes exhibit nearly Nernstian response (29.6 mV/decade) with measured slopes ranging from 28.6 to 33.2 mV/decade. The standard potentials of copper electrodes show long-term drifts of up to 10 mV/month in nitrate and perchlorate media (Oglesby *et al.*, 1977; Paulson, 1978). However, Paulson (1978) points out that much of the drift in the standard potential is due to the fact that the measurements must be extrapolated roughly 5 orders of magnitude to calculate $E^{\ominus}$. A more realistic estimate of the variation with time of the electrode potential is obtained by calculating the potential of the copper electrode at a copper concentration midway between the highest and lowest standards (Paulson, 1978). For a copper concentration of 5×10^{-6} M (in 0.7 M $NaClO_4$), Paulson (1978) found that the measured electrode potential varied ± 3 mV over a 2-month period, whereas the calculated standard potential varied ± 8 mV over the same period.

In contrast to the measurements in KNO_3 and $NaClO_4$, the slopes of copper electrodes in sea water (Jasinski *et al.*, 1974; Rice and Jasinski, 1976) and in NaCl (Oglesby *et al.*, 1977) are greater than that for a two-electron process (29.6 mV/decade), and approach the value for a one-electron process (59.2 mV/decade). The effect of chloride on the measured slope has been attributed to the precipitation of AgCl on to the membrane (Ross, 1969) and to the reduction of Cu^{2+} to Cu^+ in the presence of chloride (Westall *et al.*, 1977).

Copper-selective electrodes are sensitive to light and to stirring (Jasinski *et al.*, 1974; Rice and Jasinski, 1976) and these parameters must be controlled for precise results. The CuS/Ag_2S membrane has a finite solubility and, at low copper concentrations, dissolution of the membrane can add significant amounts of copper to the test solution. Paulson (1978) found that the amount of copper derived from the membrane was roughly 3×10^{-7} M in 0.7 M $NaClO_4$ at pH 4. The amount of copper, introduced from the membrane decreases as the total copper concentration increases, and becomes negligible at copper concentrations greater than 2×10^{-6} M (Paulson, 1978). The copper electrode responds quickly at 25°C and at 15°C, but at 5°C one electrode required more than 10 h to reach a stable potential in 5×10^{-5} M copper in 0.7 M $NaClO_4$ (Hittinger, 1975).

Copper electrodes based on mixed sulphides are of limited use for direct potentiometry in sea water due to the combined effects of low copper concentrations ($[Cu^{2+}]_{tot} < 1 \times 10^{-8}$ M; Boyle *et al.*, 1977) and chloride interference. Procedures for the determination of copper in sea water using ion-selective electrodes (Jasinski *et al.*, 1974; Rice and Jasinski, 1976) rely heavily on frequent electrode calibrations by atomic-absorption spectroscopy.

Not much is known about the behaviour of the lead and cadmium electrodes in sea water. The lead electrode is subject to chloride interference (Mascini, 1973), but the cadmium electrode is unaffected by the presence of chloride in sea water (Sunda *et al.*, 1978).

6.2.4. Anion-selective electrodes

The silver | silver chloride electrode (Section 6.2.1) is the standard chloride-selective electrode against which other chloride electrodes must be judged. Unfortunately, $Ag \mid AgCl$ electrodes are subject to bromide interference (Section 6.2.1), which prevents their use for precise thermodynamic measurements in natural sea water unless a salt bridge is used (Hansson, 1973b). The $Ag \mid AgCl$ electrode can be used, however, to detect the end-point in potentiometric titrations for which only relative potential changes are needed (Reeburgh and Carpenter, 1964).

The mixed crystal $(AgX/Ag_2S; X = Cl^-$ or $Br^-)$ chloride- and bromide-selective electrodes have been used in sea water by Miller and Kester (1976) and Savenko (1977b), respectively. Both electrodes behave theoretically (response $59.16 \, mV/decade$) in the absence of interfering ions but, in sea water, chloride interferes with the bromide electrode at bromide concentrations less than 8 mmolal (Savenko, 1977b). Miller and Kester (1976) do not report any bromide interference for the chloride mixed crystal electrode, but possible interference at the bromide concentration (0.87 mmolal) in natural sea water deserves to be checked. Electrodes containing $AgCl$, $AgBr$, or Ag_2S are frequently light sensitive, and for precise measurements they should be protected from direct sunlight and used under conditions of constant illumination. Srna (1974) studied the behaviour of a chloride liquid ion exchanger electrode in sea water.

With the exception of the pH-sensitive glass electrode, the lanthanum fluoride electrode has been used more than any other ion-selective electrode in sea water. Hydroxide is the only ion that interferes with the response of the fluoride electrode and, at sea water fluoride concentrations, hydroxide interference is not significant until the pH exceeds 8.5 (Kosov et al., 1977). Warner (1969) found that the response of the fluoride electrode was exactly theoretical $(59.16 \, mV/pF)$ at fluoride concentrations between 10^{-6} and $10^{-1} \, M$ (and probably to $1 \, M$) in $1 \, M$ NaCl and in pure water. Stirring has no effect on the potential of the fluoride electrode (Kosov et al., 1977), except in dilute solutions (Warner, 1969).

The standard potential of the fluoride electrode is remarkably constant: observed variations in the standard potential range from less than $0.25 \, mV$ in 8 h to $\pm 3 \, mV$ over several months (Warner, 1969). Changes in the standard potential have no effect on the Nernstian slope (Kosov et al., 1977). The lanthanum fluoride membrane is slightly soluble, and the amount of fluoride dissolved from the membrane of one electrode was $2 \times 10^{-11} \, mol \, h^{-1}$ (Warner and Bressan, 1973). The response of the fluoride electrode 'appears' to become non-theoretical (slope $< 59.2 \, mV/pF$) at fluoride concentrations less than $1 \times 10^{-6} \, M$ (Warner, 1969; Rix et al., 1976); this deviation is not real and is probably due to fluoride contamination from the reagents or from the electrode itself (Rix et al., 1976), since Baumann (1971) found that the fluoride electrode behaves theoretically at concentrations as low as $1 \times 10^{-9} \, M$ in suitably buffered solutions.

The only sulphate electrode that has been used in sea water is the lead amalgam | lead sulphate electrode used by Platford and Dafoe (1965) to measure the mean ionic activity coefficient of sodium sulphate. This electrode was used at temperatures between 0 and 50°C by LaMer and Parks (1933) and Cowperthwaite and LaMer (1931), but Platford and Dafoe (1965) were unable to use it below 15°C, at which temperature it required one day to equilibrate. The $Pb(Hg)/PbSO_4$ electrode cannot be used in the presence of dissolved oxygen (Ives and Smith, 1961) or in natural sea water, since carbonate must be absent to prevent lead carbonate precipitation. The solubility of $PbSO_4$ in sea water has not been measured, but theoretical calculations suggest that it is 0.6 mM at pH 5 due to the formation of lead chloride complexes (Zirino and Yamamoto, 1972). The solubility would be greater at higher pH where lead hydroxide complexes are important. For future use of the $Pb(Hg)$ | $PbSO_4$ electrode in sea water, one should consider the effect of lead sulphate solubility on the sulphate concentration of the test solution.

Silver | silver sulphide electrodes have been used to measure both sulphide (Berner, 1963; Green and Schnitker, 1974; Mor et al., 1975; Savenko, 1977c) and silver (Jagner and Árén, 1970) in sea water and marine sediments. The slope of the Ag | Ag_2S electrode response is theoretical or nearly theoretical over at least 12 orders of magnitude ($8 < pS < 20$; Berner, 1963; Savenko, 1977c). The preparation of silver | silver sulphide electrodes was described by Berner (1963) and Savenko (1977c) (see also Ives, 1961), who used these electrodes in sea water at temperatures between 10 and 35°C, at salinities between 5 and 35‰, and at pH values between 5 and 8.5. Note that values of the first apparent ionization constant of H_2S in sea water determined with the Ag | Ag_2S electrode (Savenko, 1977c) are in good agreement with spectrophotometric measurements (Goldhaber and Kaplan, 1975).

6.2.5. Gas-sensing electrodes

Gas-sensing electrodes differ from conventional ion-selective electrodes in that they respond to the activity (partial pressure) of a neutral molecule (e.g. CO_2 or NH_3) rather than the activity of an ion. These electrodes consist of a gas-permeable membrane surrounding a pH glass electrode. The pH of the internal filling solution is proportional to the partial pressure of the volatile gas in solution. The behaviour of the ammonia electrode in sea water was studied by Srna et al. (1973), Gilbert and Clay (1973), and Merks (1975). The response of the electrode is nearly Nernstian (59 mV/decade) at ammonia concentrations greater than 1×10^{-6} M, and sea water has no deleterious effects on the electrode performance (Srna et al., 1973; Gilbert and Clay, 1973). The electrode response is sub-Nernstian at NH_3 concentrations less than 1×10^{-6} M, possibly due to the long equilibration times required at low concentrations. One ammonia electrode required 60 min to equilibrate at an NH_3 concentration of 9×10^{-8} M (Gilbert and Clay, 1973). The electrode can be calibrated in pure water or in sea water, but if pure water standards are used, the salt effect on the activity coefficient of NH_3 must be considered (Gilbert and Clay, 1973; Merks, 1975). For identical ammonia concentrations, Gilbert and Clay (1973) found that the NH_3 partial pressure was 6% higher in 0.7 M NaCl than in pure water. This agrees with the theoretical calculations of Whitfield (1974b), but not with the measurements of Merks (1975). The ammonia electrode is temperature sensitive (Gilbert and Clay, 1973) and measurements should be made at constant temperature.

6.3. MARINE APPLICATIONS OF DIRECT POTENTIOMETRY

6.3.1. pH measurement

Sample collection and storage

Samples for pH measurement at sea are normally collected in water bottles during a hydrographic cast, and the time the samples remain in the water bottles prior to being brought on deck may vary from a few minutes to over an hour for deep cast. Experience has shown that samples stored in Nansen bottles, even those nominally coated with PTFE or epoxy resin, are unsatisfactory for precise pH measurements (Ivanenkov, 1964; Park, 1968; Takahashi *et al.*, 1970). Ivanenkov (1964) found that the pH increased by as much as 0.10–0.15 pH unit after storage for 2 h in an uncoated Nansen bottle, while Bruyevich (1966) showed that pH measurements on deep samples from unlined Nansen bottles are about 0.1 pH unit too high. Park (1968) found an error of ±0.04 pH unit due to imperfections in the lining of PTFE-coated Nansen bottles during a 7000-m cast. A detailed comparison of pH measurements from different water bottles was made during the 1969 GEOSECS intercalibration cruise (Takahashi *et al.*, 1970) and it was found that pH measurements on samples from epoxy-coated Nansen bottles were consistently higher and more scattered than those from plastic water bottles (Figure 1).

The pH increase in Nansen bottles is undoubtedly due to corrosion according to the general equation (Park, 1968)

$$2M + O_2 + 4HCO_3^- \rightarrow 2M^{2+} + 4CO_3^{2-} + 2H_2O \qquad (55)$$

where $M = Cu$ or Zn. The pH increase is accompanied by an increase in alkalinity and a decrease in oxygen concentration (Ivenenkov, 1964; Park, 1968). During the 1969 GEOSECS intercalibration cruise, the author observed an average increase of 0.029 mequiv kg^{-1} in alkalinity samples from Nansen bottles when compared with samples from plastic water

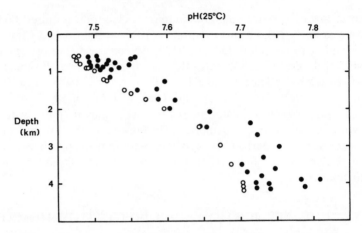

Figure 1. Effect of corrosion on pH(NBS) values at 25°C and atmospheric pressure measured during the 1969 GEOSECS intercalibration cruise (Takahashi *et al.*, 1970). The pH values from Nansen bottles (solid circles) are consistently higher than values from plastic water bottles (open circles). Reproduced by permission of the American Geophysical Union

bottles. Based on the above discussion, it must be concluded that Nansen or other metallic water bottles are unsuitable for precise pH and alkalinity measurements even when coated with PTFE or epoxy resin.

Samples for pH measurement should be drawn and analysed as soon as possible after the water bottles are brought on deck. Samples stored in plastic or glass containers are subject to degassing and loss of CO_2 as cold samples warm to room temperature (measured pH too high), to exchange of CO_2 with the atmosphere (measured pH too high for pCO_2 greater than 330 ppm), and to bacterial respiration (measured pH too low).

The effects of degassing and atmospheric CO_2 exchange are similar. Consider a sea water sample (35‰ salinity and 0°C) at equilibrium with atmospheric N_2, O_2, and Ar which is warmed to 25°C. At 0 and 25°C the combined N_2, O_2, and Ar solubilities (Weiss, 1970) are 22.0 and 13.4 cm^3 kg^{-1}, respectively.

If 1 kg of this initially air-saturated sea water is warmed to 25°C and allowed to degas, a gas volume of 8.6 cm^3 (25°C) will be created. Carbon dioxide from the sea water will diffuse into the newly created gas volume, and the pH of the sea water will increase. The error in the measured pH will be greatest for low pH and high pCO_2 samples. The lowest pH and highest pCO_2 observed in the open ocean occur in the oxygen minimum zone of the eastern tropical Pacific. Calculations (Table 15) show that equilibrating this sea water with the volume of degassed N_2, O_2, and Ar will cause the pH to increase by 0.004 pH unit. This is the maximum possible error in pH due to degassing because sea water from the O_2 minimum has the lowest pH

Table 15. Calculated pH changes due to degassing and to atmospheric exchange. The initial conditions were (degassing): 1 kg of 35‰ salinity sea water; equilibrated with atmospheric N_2, O_2, Ar at 0°C; warmed to 25°C and allowed to degas. Atmospheric exchange: 1 kg of 35‰ sea water at specified pH allowed to equilibrate with specified volume of a wet atmosphere ($pCO_2 = 330$ ppm). Alkalinity, 2.4 mequiv kg^{-1}, total boron $= 0.42$ mM_W; $K'_1 = 9.99 \times 10^{-7}$ M_W, $K_2 = 7.68 \times 10^{-10}$ M_W (Mehrbach *et al.*, 1973); $K'_B = 1.98 \times 10^{-9}$ M_W (Lyman, 1956); CO_2 solubility, $\alpha = 0.0284$ M_W (Weiss, 1974)

Initial conditions				
pH (25°C)	$c_{CO_2,tot}$, mM_W	pCO_2 ppm	Final pH (25°C)	Comments
7.4	2.427	3214	7.404	Degassing
7.8	2.283	1198	7.801	Degassing
8.2	2.085	411	8.200	Degassing
7.4	2.427	3214	7.403	1% air volume
			7.416	5% air volume
			7.430	10% air volume
			7.457	20% air volume

observed in the open ocean, is warmer than 0°C, and contains little dissolved oxygen.

Similar calculations for the case in which sea water from the oxygen minimum is equilibrated with various volumes of air are also shown in Table 15. The calculations show that pH errors due to degassing or air equilibration are small provided that the gas volume in contact with the sample is less than 1% of the sample volume. However, serious errors can result if the sample is allowed to exchange with larger volumes of air. Almgren *et al.* (1975a) report that 'CO$_2$ gas precipitation . . . can easily cause an increase of 0.03–0.04 pH unit without the appearance of visible bubbles'. The calculations in Table 15 suggest that the pH increases of 0.03–0.04 pH unit observed by Almgren *et al.* (1975a) are not due to sample degassing.

In general, samples for pH measurement should be drawn with the same precautions that are observed for drawing oxygen samples (see also Chapter 9), and they should be analysed as soon as possible—certainly within 1 h after being drawn.

Methods for pH measurement aboard ship

Techniques for pH measurements at sea have been discussed by Strickland and Parsons (1965), Park (1966), Takahashi *et al.* (1970), Zirino (1975), and Almgren *et al.* (1975a). In Park's (1966) procedure, pH samples are drawn in wide-mouthed 120-cm³ plastic bottles and rapidly warmed to

25°C in a water-bath. All air is excluded from the sample bottles. The pH of the samples is measured by inserting the glass and calomel electrodes directly into the 120-cm^3 bottles. The two electrodes, a thermometer, and a capillary tube for overflow are mounted in a rubber stopper. Inserting the stopper and electrodes into the sample bottle displaces sea water and prevents air from coming in contact with sample. This is the simplest of the above procedures, but it is capable of a precision of only ±0.02 pH unit (2σ) based on the reproducibility of replicate measurements (Park, 1966). Park's (1966) method is a modification of the procedure outlined by Strickland and Parsons (1965). In both methods the electrodes are calibrated with dilute NBS buffers.

Zirino (1975) used a micro combination electrode (5 mm diameter) mounted in a 25-cm^3 water-jacketed flask. Samples were drawn in 50-cm^3 glass syringes and injected into the jacketed flask through a rubber septum. The flask contained a PTFE-coated stirring bar and ports for overflow and discharge of the sea water sample. Measurements were made on stirred samples. The combination electrode was calibrated with dilute NBS buffers (pH 4.008, 7.413, 9.180) before each cast. After buffer calibration, the electrode was soaked in sea water for 15 min to allow the ceramic frit liquid junction to equilibrate with sea water. Zirino (1975) found that the ceramic frit junction was sensitive to stirring: potentials in pH 7 buffer were 0.3 mV more positive, while potentials in sea water were 0.3 mV more negative when stirring. The pH of stirred samples was 0.01 pH unit greater than that of unstirred sea water because of the stirring potentials in buffer and sea water. With this procedure, samples required 3–6 min for analysis, depending on the temperature of the injected sample. The precision of this technique was checked under field conditions over a 3-week period. The average deviation, calculated from 40 pairs of duplicates, was ±0.003 pH unit and 87% of the duplicates were within 0.004 pH unit of the individual mean.

Almgren *et al.* (1975a) developed a technique for shipboard pH measurements based on the sea water tris buffers and total hydrogen ion pH scale developed by Hansson (1973a). Water samples were drawn in 1-dm^3 glass bottles and a 20-cm^3 sample was then poured into a 25-cm^3 beaker. After temperature equilibration, the glass and calomel reference electrodes, which were mounted in a rubber stopper, were inserted into the 25-cm^3 beaker. The samples were not thermostated with a water-bath, but were placed in a metal temperature-equilibrating block which contained holes for two 25-cm^3 beakers (sample and buffer). The e.m.f. of the two solutions and the temperature of the block were measured, and pH(SWS) was calculated from the equation

$$\text{pH(SWS)} = \text{pH}_{\text{tris}} + (E_{\text{tris}} - E_{\text{SW}})/g \tag{56}$$

where pH_{tris} and g were evaluated at the temperature of the measurement.

One advantage of sea water buffers is that the liquid junction does not become contaminated with dilute NBS buffer and the junction potential reaches a steady state much more rapidly when going from buffer to sea water. For instance, Zirino (1975) found that reproducible potentials could not be obtained when going from sea water to NBS buffer until the sea salts had been rinsed from the ceramic frit junction of his electrode. A disadvantage of Almgren's method is that the samples are not measured at a uniform temperature. At station 652 (Almgren *et al.*, 1975a), the temperature of the deep samples was 3°C cooler than that of the surface samples, owing to differences in the initial temperatures of the samples. This problem could be eliminated by using a constant-temperature water-bath. Almgren *et al.* (1975a) do not give any information on the reproducibility of their procedure, so it is unclear if the use of sea water tris buffers actually improves the precision of pH measurements in sea water.

The author has developed a sea-going pH apparatus based on the Beckman (580621) micro blood pH assembly (Figure 2). The original version of this assembly (Takahashi *et al.*, 1970) contained a palladium annulus liquid junction, and many hours were required to reach a steady-state potential after dilute NBS buffer was replaced with sea water. In addition, the pH assembly was unsuitable for use in reducing environments containing sulphide, apparently because of redox reactions across the metal junction. To improve the behaviour of the cell, the palladium annulus junction was replaced by a free diffusion junction (Figure 2) which allowed the liquid junction to be flushed and reformed between buffer and sea water.

A new liquid junction is formed by flushing the capillary T-joint with fresh KCl (saturated or 3.5 M). A sample is then injected into the assembly, flushing the KCl, and forming a sharp junction about 3 mm beneath the T. A new junction is formed when changing from buffer to sea water, sea water to buffer, or from one buffer to another. However, it is not necessary to reform the junction for repeated measurements in sea water or buffer. Twenty successive sea water injections, without reforming the junction, changed the cell potential by only 0.1 mV.

The entire pH assembly is enclosed in a plastic water-jacket maintained at 25°C. Sea water samples are collected in 10-cm^3 plastic syringes immediately after each water bottle is brought on deck, and the syringes are warmed to 25°C in a water-bath and then injected into the assembly. The cell potential reaches a steady state within 30 s after the sea water injection and the pH is read 2–3 min later, and within 8–18 min after the sample was drawn. Each syringe is returned to the laboratory and placed in the water-bath before the next sample is drawn. Analysis commences after four samples have been drawn and the first syringe has had time to warm to 25°C. After this, the second and succeeding samples are injected into the cell each time a new sample is placed in the water-bath. The major

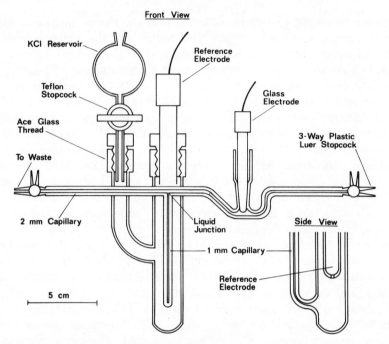

Figure 2. Electrode assembly for sea water pH measurements. The assembly is constructed from a Beckman (No. 580621) Micro Sensor Assembly in which the original liquid junction is replaced by a free diffusion junction. The KCl reservoir and reference electrode are mounted in threaded plastic to glass adapters (Ace Glass, Inc., Vineland, N.J., USA) for easy dismantling and cleaning. The three-way plastic Luer stopcock (Ace Glass) on the inlet mates directly with an identical stopcock on the sample syringes. The three-way stopcocks allow all air to be removed from the line before the sample is injected through the assembly. The assembly is permanently sealed in a plastic water-jacket for temperature control. The liquid junction is formed by opening the Teflon stopcock and allowing KCl to flow from the KCl reservoir, around the reference electrode, up the vertical 1 mm capillary, and out of the assembly through the three-way stopcock labelled waste. The Teflon stopcock is closed and a sample is injected through the assembly. The sample flushes excess KCl from the assembly, and forms a sharp junction 3 mm beneath the top of the vertical
1 mm capillary

constraint on the speed of analysis is the length of time it takes to draw each sample and return it to the laboratory. The assembly is calibrated before and after each cast with NBS pH 7.413 buffer, and the measured pH at 25°C is calculated from the equation

$$pH(NBS) = 7.413 + (E_b - E_{sw})/59.16 \qquad (57)$$

where E_b is the average buffer potential.

The behaviour of the pH assembly can be characterized by (1) the slope of the glass electrode response, (2) the stability of the standard potential, and (3) the response of the assembly when buffer is replaced with sea water.

The response of the assembly to pH changes depends on the slope of the glass electrode. Zielen (1963) compared Beckman general-purpose glass electrodes with a hydrogen electrode and showed that general-purpose electrodes exhibit the theoretical slope (59.157 mV/pH at 25°C). The empirical slopes of six separate pH assemblies containing free diffusion junctions, calculated by linear least squares from potential measurements in NBS pH 4.008, 7.413, and 9.180 buffers, ranged from 99.73 to 100.00% of the theoretical slope, with an average of 99.89% of the theoretical slope. For comparison, the slope of the original pH assembly which contained a palladium annulus junction ranged from 98.8 to 99.1% of the Nernstian slope.

The standard potential of the pH assembly is relatively constant when used ashore, but it often drifts slowly when used at sea. At sea, changes in the standard potential are often associated with ship motion and with large changes in laboratory temperature; the buffer potential sometimes shifts 0.5–1.0 mV during rough weather. In addition, the pH assembly is often subject to large temperature variations and rough handling during transport to the ship, and the potential of the assembly in the buffer solution often drifts 0.5 mV/day for the first few days after it is brought to 25°C aboard ship. The standard potential of one assembly showed a range of only 0.3 mV over a 2-month period in a shore-based laboratory. The behaviour of three assemblies at sea is summarized in Table 16. The difference in potential between successive buffer standardizations is a good measure of the uncertainty in the pH due to the use of NBS buffers to calibrate the assembly. In general, buffer calibrations before and after a cast differ by 0.1–0.2 mV (Table 16), and contribute an uncertainty of ±0.002 pH unit to the measured pH.

The response of the pH assembly when dilute buffer was replaced with sea water was determined by pH measurements in the sequence NBS buffer, sea water, NBS buffer. The assembly was equilibrated with dilute buffer or with sea water, and three successive 10-cm^3 portions of test solution were then injected into the assembly. The measured potential became constant very quickly for the sequence sea water, NBS buffer, and the potential of the second and third buffer injections agreed to ±0.1 mV when measured 3–5 min after injection. A similar rapid equilibration was observed for the sequence pH 9.180 buffer, sea water. However, longer equilibration times were required for the sequences pH 4.008 buffer, sea water and pH 7.413 buffer, sea water. The pH assembly must be conditioned in sea water at least half an hour after replacing pH 4.008 or 7.413 buffers with sea water. The drifts observed for the sequences pH 4.008 buffer, sea water and pH 7.413 buffer, sea water did not occur with sea water [pH(NBS) 8.2] buffered with

Table 16. Behaviour of the free diffusion pH assembly at sea. The pH measurements during Yaloc-69 and GEOSECS (1969) were made with the original assembly which contained a palladium annulus junction. They are included for comparison with the free diffusion junction

				Cruise	
	Yaloc-69	GEOSECS (1969)	Yaquina 7304C	Pacific GEOSECS (leg 6)	Knorr-71
Length of cruise, days	110	7	5	27	28
Average drift between buffer calibrations, mV			0.16	0.12	0.22
Maximum drift between buffer calibrations, mV			0.9	0.65	0.6
Average time between buffer calibrations, hours			11.2	8.4	3.2
Long term drift, mV/day			0.18	0.10	0.17
Precision (pH units) from duplicates (2σ)	0.007	0.005	0.001	0.002	0.005

borax and boric acid. This suggests that the drifts observed with natural sea water were due to its low buffer capacity and not to interactions of the major sea water ions with the glass electrode. The drifts when going from pH 4.008 or 7.413 buffer to sea water are not due to liquid junction effects since the free diffusion junction is renewed when buffer is replaced with sea water. The most likely explanation for these drifts is adsorption/desorption reactions involving hydrogen ion on the walls of the glass capillaries which comprise the pH cell.

Reproducibility of shipboard pH measurements

The vertical profiles in Figure 3, which were made with the free diffusion pH assembly, show the range of pH(NBS) normally encountered in the oceans. The correlation of pH and dissolved oxygen (Figure 4) is obvious and serves as a check on the quality of the pH data. Variations in the pH of deep vertical profiles that are uncorrelated with oxygen or with water mass structure probably indicate errors in the pH measurements.

The best test of any pH technique is the reproducibility of pH measurements at sea. Three possible tests of reproducibility are (1) the precision of

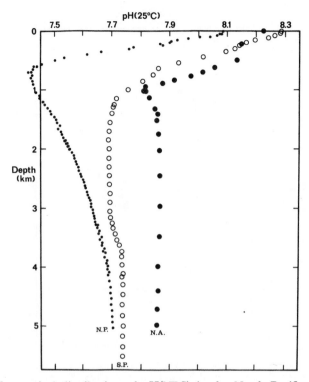

Figure 3. The vertical distribution of pH(NBS) in the North Pacific (NP), South Pacific (SP), and North Atlantic (NA) Oceans measured with the free diffusion micro pH assembly. The pH values are the measured values at 25°C and atmospheric pressure. The station positions are: NP, 38°21′N, 133°38′W, cruise Y7304C, Oregon State University; SP, 23°59′S, 174°26′W, GEOSECS station 269; NA, 36°49′N, 64°24′W, stations 11 and 13, cruise Knorr 71, Woods Hole Oceanographic Institution

replicate measurements, (2) the pH difference between overlapping bottles on successive casts, and (3) the test of 'oceanographic consistency'—the smoothness and lack of systematic differences between deep vertical profiles in the same region. The precision of replicate measurements is a measure of analytical and sampling errors after the samples arrive on deck, while factors (2) and (3) also include systematic errors between different casts and stations.

The precision of the various pH methods are compared in Table 17, from which it can be seen that pH measurements on the NBS pH scale are reproducible to ±0.005 pH unit. At present there are no published data on the precision of pH measurements using the total hydrogen ion scale (Hansson, 1973a). It should be noted that the precision of a particular method may vary from cruise to cruise and, in the author's experience, will

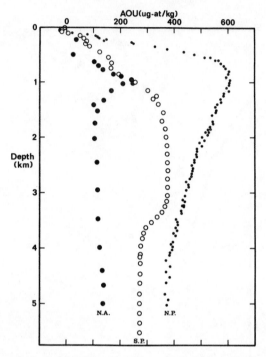

Figure 4. The vertical distribution of apparent oxygen utilization (AOU) in the North Pacific (NP), South Pacific (SP), and North Atlantic (NA) Oceans. The AOU is a measure of oxygen consumption, and each point is the difference between the oxygen solubility at that depth (Weiss, 1970; wet atmosphere) and the measured oxygen concentration. Station positions are the same as in Figure 3. The inverse correlation between pH(NBS) and AOU is due to the concurrent liberation of molecular CO_2 and consumption of oxygen during the decomposition of organic matter

probably be worse for measurements made by someone other than the originator of the method.

The pH of deep water varies only slowly with depth and position (Figure 3) and Figures 5 and 6 show pH values, at depths greater than 1500 m in the northeastern Pacific, measured by the procedures outlined above. Station positions are shown in Figure 7.

Measurements by Zirino (1975) in the northeastern tropical Pacific are plotted in Figure 5 for comparison with data from Yaloc-69 made with the

Table 17. Reproducibility of pH measurements at sea

Reference	Park (1966)	Smith (1973)	Zirino (1975)	Almgren *et al.* (1975a)	Culberson (Table 16)
Precision (2σ)	0.02	0.007	0.005	No data	0.002–0.005

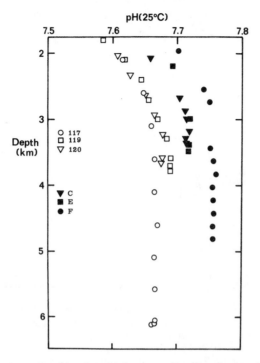

Figure 5. pH measurements in the northeastern Pacific. Comparison of pH(NBS) measurements by Zirino (1975; 1977, personal communication) (solid symbols) with those of the author (open symbols) during Yaloc-69 (Wyatt *et al.*, 1970). pH measurements during Yaloc-69 were made with the original micro pH assembly, which contained a palladium annulus liquid junction. The pH values are the measured values at 25°C and atmospheric pressure. The pH values measured with the original palladium annulus junction (Yaloc-69 and GEOSECS-1969, Figures 5 and 6) are internally consistent with those measured with the free diffusion junction (Y7304C, Figure 3). Station positions are shown in Figure 7

original pH assembly. The precision of Zirino's (1975) measurements along any one vertical profile is excellent. However, there is an offset of 0.035 pH unit between the deep water values at stations C/E and F, which is not reflected in the concentration of dissolved oxygen nor in the water mass structure at these stations. Zirino's (1975) values at station F are 0.09 pH unit greater than values at station 117 measured during Yaloc-69. Both of these stations were taken in the Mid-America Trench, and pH values below the sill depth (3500 m) should be identical at both stations.

Measurements by Almgren *et al.* (1975a,b, 1977) in the northeastern tropical Pacific are plotted in Figure 6 for comparison with pH(NBS) measurements made during Yaloc-69 and GEOSECS (1969) using the original pH assembly. Almgren *et al.*'s (1975a,b, 1977) data are based on

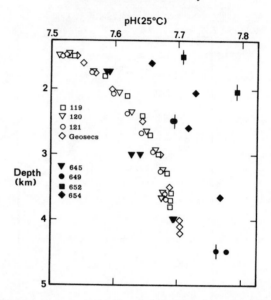

Figure 6. pH measurements in the northeastern Pacific. Comparison of pH(NBS) values calculated from the data of Almgren *et al.* (1975a,b; 1977) (solid symbols) with those measured by the author (open symbols) during Yaloc-69 (Wyatt *et al.*, 1970) and GEOSECS-1969 (Takahashi *et al.*, 1970) using the original micro pH assembly, which contained a palladium annulus liquid junction. For the data of Almgren *et al.*, a vertical line represents a directly measured pH value; all other values were calculated from their measurements of alkalinity and total inorganic carbon (see text for details). Station positions are shown in Figure 7

the total hydrogen ion pH scale developed by Hansson (1973a); their values of pH(SWS) were converted to pH(NBS) using the relationship

$$pH(NBS) = pH(SWS) + 0.124 \tag{58}$$

which was experimentally measured by the author (Table 9). Almgren *et al.* (1975a,b, 1977) report two types of pH measurements: (1) directly measured values (Almgren *et al.*, 1975a; stations 469 and 652) at room temperature (22–23°C); and (2) values at the *in situ* temperature (Almgren *et al.*, 1975b) calculated from their alkalinity and total carbon dioxide measurements. The author has corrected their data to 25°C using the procedure and equations given in their 1977 paper. Almgren *et al.* (1975a) have only four directly measured pH values at depths greater than 1500 m. To increase the number of points available for comparison, their calculated pH values are also shown in Figure 6. At station 649, the average difference between 20 measured and calculated values was −0.004 pH(SWS) units, with a range of −0.015 to +0.020 pH unit (Almgren *et al.*, 1977).

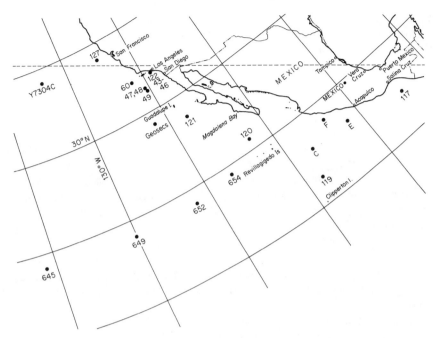

Figure 7. Station locations for the oceanographic stations in Figures 1–11

Comparison of the data in Figure 6 shows that Almgren *et al.*'s (1975a,b, 1977) measurements are higher and more scattered than those made during Yaloc-69 and GEOSECS (1969). There are several possible reasons for the discrepancies between the two sets of data: (1) their measurements are from stations to the west of the stations occupied by the author; (2) the conversion factor between the two pH scales (0.124 pH unit at 25°C) depends on the residual liquid junction potential which depends on the type of liquid junction used to measure pH; and (3) analytical and sampling errors.

The oxygen concentration in deep water increases slowly toward the west in the North Pacific (Reid, 1965). However, the increase in dissolved oxygen (O_2) between the coast of Central America and Hawaii is 5 μ mol kg^{-1} or less at depths greater than 2000 m, and Almgren *et al.*'s (1975a,b, 1977) pH values should be within 0.01 pH unit of those measured by the author.

An error in the conversion factor between the two pH scales would cause a constant offset between the two sets of data, but would not cause the scatter evident in Figure 6. The scatter in Almgren *et al.*'s (1975a,b, 1977) data suggests that their method is subject to analytical and/or sampling errors of at least ±0.05 pH unit.

The data in Figures 5 and 6 illustrate the problems involved in pH measurements at sea. In particular, the data in Figures 3, 5, and 6 do not support the contention (Hansson, 1973a; Almgren *et al.*, 1975a) that pH measurements using high ionic strength buffers are inherently more precise than those using NBS buffers.

Suggestions for reporting sea water pH measurements

For convenience, oceanic pH measurements are usually made aboard ship at room temperature and at 1 atm ($101\,325\ N\ m^{-2}$); they are then corrected to the *in situ* temperature and pressure with *in situ* values of the dissociation constants of carbonic and boric acids. At present, there is no standard method for reporting pH data. Some studies (Culberson and Pytkowicz, 1970) report the measured values at 25°C and 1 atm, others (Almgren *et al.*, 1975b) report pH values corrected to the *in situ* temperature at 1 atm, while still others (Dietrich *et al.*, 1966) report values corrected to the *in situ* temperature and pressure. The values of the dissociation constants, and thus the pH corrections, are subject to change as better data become available, and the best procedure for reporting field data is to report the measured data at room temperature and 1 atm, and to let each user apply his own pH corrections.

In any case the pH scale that was used, the temperature of the measurements, and the source of any temperature and pressure corrections to the data should be specified.

6.3.2. Alkalinity—the single acid addition technique

The total alkalinity (A_T) of sea water is equal to the equivalent concentration of anions of weak acids which are not charge balanced by hydrogen ions (Gieskes, 1974):

$$A_T = [HCO_3^-] + 2[CO_3^{2-}] + [B(OH)_4^-] + [Si(OH)_3O^-] + 2[HPO_4^{2-}]$$
$$+ 3[PO_4^{3-}] + [NH_3] + [OH^-] - [H^+] \tag{59}$$

Carbonic acid is the most important weak acid in sea water, and the total alkalinity is experimentally defined as the amount of strong acid required to titrate $1\ dm^3$ of sea water to the bicarbonate end-point (Gieskes, 1974). Acids with dissociation constants greater than that of carbonic acid (HSO_4^- and HF) are not included in the definition of alkalinity, since they are not appreciably protonated at the bicarbonate end-point.

The total alkalinity is commonly determined by titration of sea water with strong acid using an H^+-sensitive glass electrode to detect the end-point (Edmond, 1970; Culberson *et al.*, 1970; Almgren *et al.*, 1977). Colorimetric indicators are sometimes used for end-point detection (Graneli and Anfält, 1977). Two procedures involving glass electrodes are in general use. The first (Culberson *et al.*, 1970; Anderson and Robinson, 1946) is essentially a single point titration, while the second (Edmond, 1970; Almgren *et al.*, 1977) involves a Gran (1952) extrapolation of the points on the titration curve that fall after the bicarbonate end-point. In this section, we will discuss the single acid addition technique. Gran alkalinity titrations are discussed in Chapter 7.

In the single acid addition method (Culberson *et al.*, 1970), v_a ($\approx 30 \, cm^3$) of 0.01 M HCl are added to v_{sw} ($\approx 100 \, cm^3$) of sea water, air is bubbled through the solution to remove carbon dioxide, and the pH(NBS) of the mixture is measured at 25°C. The total alkalinity is calculated from the equation

$$A_T = 1000v_a \cdot 0.01/v_{sw} - 1000(v_{sw} + v_a)a'_H/v_{sw}f'_{H,tot} \qquad (60)$$

i.e.

$$A_T = [H^+]_{added} - [H^+]_{remaining}$$

where $a'_H = 10^{-pH(NBS)}$ and $f'_{H,tot}$ = conventional total hydrogen ion activity coefficient (see Section 6.1.1).

The glass/reference electrode pair is standardized in NBS pH 4.008 buffer before and after each set of samples, or about once every hour. The major advantages of the single addition technique are speed and simplicity. About nine samples an hour can be analysed, and the only equipment needed is two pipettes, a cheap aquarium pump to purge CO_2, electrodes, and a pH meter.

CO_2 is removed from the acidified sample because the protolysis of carbonic acid affects the pH of the acidified sample through the equilibrium

$$CO_2 + H_2O \rightleftharpoons H^+ + HCO_3^-$$

The measured pH is too low in the presence of dissolved carbon dioxide, and the alkalinity calculated from equation 60 is too small. Table 18 shows the error in the calculated alkalinity due to the presence of CO_2, and also the effect of errors in the measured pH and in $f'_{H,tot}$ on the total alkalinity. It is not necessary to remove all CO_2 from solution; removing only 90% of the initial CO_2 will cause errors in alkalinity of less than 0.05% (Table 18).

The accuracy and precision of the single addition technique has been investigated by several workers (Table 19). The results in this table show that the precision and accuracy of the single addition method rival those of the Gran titration (Edmond, 1970; Chapter 7). The major constraint on the precision of this method is the reproducibility of the pH measurements, and for the most precise results the electrodes should be thermostatted and a pH/mV meter sensitive to ±0.1 mV must be used.

The accuracy of the method also depends on the accuracy of the acid molarity and on the value of $f'_{H,tot}$. The dilute HCl used in the single addition technique is usually prepared from HCl which has been pre-standardized by the manufacturer. Some brands of pre-standardized HCl are

Table 18. Effect of CO_2 and analytical errors on the alkalinity calculated from equation 60. Initial conditions: 25°C; 100 cm^3 of 35‰ sea water ($c_{CO_2,tot} = 2.3$ mM) plus 30 cm^3 of 0.01 M HCl; $f'_H = 0.741$; $K_1 = 0.9 \times 10^{-6}$ M

Property	Values				
pH in absence of CO_2	3.9	3.7	3.5	3.3	3.1
Total alkalinity (mequiv l^{-1})	2.779	2.650	2.445	2.121	1.606
Error in A_T due to presence of CO_2, %	−0.56	−0.38	−0.27	−0.19	−0.16
Error in A_T due to 0.2 mv pH increase, %	+0.06	+0.10	+0.18	+0.32	+0.67
Error in A_T due to 5% decrease in f'_H, %	−0.40	−0.66	−1.15	−2.12	−4.53

not accurate (Koroleff, 1965, 1966; Grasshoff, 1966) and only brands guaranteed accurate to 0.1% should be used to determine alkalinity.

The value of $f'_{H,tot}$ in equation 60 was determined by Culberson et al. (1970). Other determinations of $f'_{H,tot}$ are discussed in Section 6.1.1 (see also Almgren and Fonselius, 1976). The original values for $f'_{H,tot}$ (Culberson et al., 1970) were calculated using the empirical slope of the glass/calomel electrode pair (99.6% of theoretical) determined from measurements in NBS pH 4.008 and 7.413 buffers. Recent work (Zielen, 1963; Hansson, 1973b; Mehrbach et al., 1973; Culberson and Pytkowicz, 1973) has shown that glass electrodes exhibit the Nernstian slope, and that apparent deviations from the Nernstian slope are due to changes in the residual liquid junction potential between buffers. Therefore, the original activity coefficients (Culberson et al., 1970) have been recalculated using the theoretical slope. The new values (Table 20) have been recalculated relative to both NBS pH 7.413 and 4.008 buffers.

Table 19. Precision and accuracy of the single acid addition alkalinity method. The precision is ±2 relative standard deviations. The accuracy of the technique was determined by measuring the alkalinity of sea water samples with known alkalinities. The accuracy is the percentage difference between the measured and the known alkalinities, 100(measured − known)/known

Precision, %	Number of replicates	Accuracy, %	Number of replicates	Reference
0.3	9, 5, 3, 3	+0.24	4	Culberson et al. (1970)
0.3	133 triplicates	—	—	Smith (1973)
0.2	5	−0.30	5	Johnson et al. (1977)

Table 20. Original (Culberson *et al.*, 1970) and recalculated values of f'_H (molar concentrations) at 25°C. The author recommends that the values based on NBS pH 4.008 buffer be used to calculate alkalinity from equation 60

Conditions	Recalculated values		Original data
Buffer pH	4.008	7.413	Both buffers,
Slope	59.16	59.16	empirical slope
Salinity before dilution, ‰:			
29.78	0.737	0.713	0.747
34.55	0.734	0.710	0.741
40.87	0.730	0.706	0.738

The author recommends that electrodes used in the single addition method be standardized in NBS pH 4.008 buffer, and that activity coefficients based on the theoretical slope and pH 4.008 buffer be used in equation 60. The average value, $f'_{H,tot} = 0.734$ can be used for initial salinities between 29 and 41‰. The pH of the acidified solutions should be calculated from the equation

$$pH(NBS) = 4.008 - (E_{sw} - E_{4.0})/59.16 \qquad (61)$$

6.3.3. Standard addition techniques

Ion-selective electrodes and the method of standard additions have been used to measure the concentrations of sodium (Garrels, 1967), fluoride (Warner, 1971, 1973; Anfalt and Jagner, 1971; Rix *et al.*, 1976), ammonia (Srna *et al.*, 1973; Gilbert and Clay, 1973; Merks, 1975), and calcium (Bradford, 1968) in sea water. Garrels (1967) obtained a precision (2σ) of ±4% in the measured sodium concentration by adding small spikes (0.01–0.04 M) of NaCl to sea water and measuring the change in potential of a sodium glass electrode. The potential changes upon addition of the spikes were small (0.50–2.08 mV) and an electrometer sensitive to ±0.01 mV was required for the measurements.

Warner (1971, 1973) measured fluoride in sea water with a standard addition technique and the lanthanum fluoride electrode. The electrode potential was measured in a mixture of 25 cm³ of sea water plus 5 cm³ of a total ionic strength adjustment buffer, and enough NaF solution (1.7×10^{-6} mol F^-) to double the initial fluoride concentration was then added to the mixture. The initial fluoride concentration was calculated from equation 45 and the measured potential difference (*ca.* 18 mV). The relative standard deviation of this method is 0.8% (Warner, 1973) and it is not subject to systematic error. The precision of the standard addition technique for fluoride is equal to that for direct potentiometric and colorimetric fluoride analysis (Warner *et al.*, 1975). Anfält and Jagner (1971) eliminated the total ionic strength adjustment buffer from their procedure, and instead acidified the sea water to pH 6.6 with dilute hydrochloric acid.

Standard addition techniques for determining ammonia with the ammonia electrode were developed by Srna *et al.* (1973) and Gilbert and Clay (1973). In each of these procedures, NaOH is added to the sea water sample (pH >11) to convert NH_4^+ to NH_3, the potential of the ammonia electrode is measured in the basic solution, and a spike of ammonia equal to the initial concentration is added to the solution. The electrode potential is re-measured and the initial ammonia concentration is calculated from equation 45. Each analysis requires about 5 min. The results in Table 21 (Srna *et al.*, 1973) illustrate the precision of the method for various ammonia concentrations in sea water.

Magnesium hydroxide and calcium carbonate precipitate at the high pH required in the standard addition method, and the electrode membrane must be cleaned periodically with dilute hydrochloric acid to remove precipitated material.

The above procedures use electrodes which are specific for the ion of interest and suffer no interference from other ions in sea water. In contrast, Bradford (1968) tested the calcium liquid ion exchanger, which is subject to sodium and magnesium interference, for possible use with a standard addition technique in sea water. The calcium electrode did not behave satisfactorily and in sea water the calculated calcium concentrations were from 5% (at 30‰ salinity) to 30% (5‰ salinity) too low. In contrast, the calculated calcium concentrations in magnesium free synthetic sea water were consistently 25% too large for salinities between 4 and 27‰.

Table 21. Precision and accuracy of a standard addition method for ammonia in sea water (32‰ salinity) using the ammonia gas electrode (Srna *et al.*, 1973)

Average concentration, M	Average concentration found by standard addition, M	Relative standard deviation, %	Number of replicates
1.00×10^{-6}	0.90×10^{-6}	6	8
1.00×10^{-5}	0.97×10^{-5}	7	8
1.00×10^{-4}	1.03×10^{-4}	2	8
1.00×10^{-3}	1.03×10^{-3}	2	8
1.00×10^{-2}	1.02×10^{-2}	1	8

6.3.4. Saturometer techniques

Saturometers consist of an ion-selective electrode surrounded by a sparingly soluble salt, one of whose ions is in equilibrium with the ion to which the electrode responds. The two most common examples are the carbonate (Weyl, 1961; Ingle *et al.*, 1973) and brucite (Pytkowicz and Gates, 1968) saturometers.

The carbonate saturometer consists of a glass/reference electrode pair in which calcite (Ingle *et al.*, 1973) or aragonite (Hawley and Pytkowicz, 1969) crystals are packed around the glass electrode. The cell is filled with sea water of known pH and alkalinity, and the pH of the interstitial sea water is monitored until a constant value is obtained which indicates that the sea water surrounding the glass electrode has reached equilibrium with calcium carbonate. The solubility of the carbonate phase can be calculated from the difference between the initial and final pH. The advantages of the carbonate saturometer over other solubility techniques are (1) rapid equilibration due to the large surface to volume ratio, and (2) the fact that the saturometer is easily adapted for measurements at high pressure, both in the laboratory (Ingle, 1975; Hawley and Pytkowicz, 1969) and *in situ* (Ben-Yaakov *et al.*, 1974).

The brucite saturometer consists of a glass electrode surrounded by solid $Mg(OH)_2$ (Pytkowicz and Gates, 1968; Gates, 1969). This saturometer is actually an electrode of the third kind which responds to magnesium through the following equilibria:

$$Mg(OH)_2(s) \rightleftharpoons Mg^{2+} + 2OH^- \qquad K_S^\ominus = a_{Mg}a_{OH}^2 \qquad (62)$$

$$H_2O \rightleftharpoons H^+ + OH^- \qquad K_W^\ominus = a_H a_{OH}/a_{H_2O} \qquad (63)$$

Solving equations 62 and 63 for a_H and substituting the result into the half-cell equation for the glass electrode yields an equation for the potential of the brucite saturometer in terms of the magnesium activity:

$$E = E^\ominus + (g/2) \log (K_W^{\ominus 2} a_{H_2O}^2 / K_S^\ominus) + (g/2) \log a_{Mg} \qquad (64)$$

or

$$E = \text{constant} + (g/2) \log a_{Mg} \qquad (65)$$

Pytkowicz and Gates (1968) used this saturometer to determine the amount of magnesium sulphate ion pairing in synthetic sea water. Certain precautions are necessary when using the brucite saturometer. First, sea water in equilibrium with $Mg(OH)_2$ is basic [pH(NBS) = 9.33 at 25°C; Pytkowicz and Gates, 1968] and carbonate must be excluded from the test solutions, since carbonate can be incorporated into the crystal lattice of brucite. Secondly, it is difficult to prepare well crystallized brucite and recent solubility product measurements range from $pK_S^\ominus = 10.88$ (McGee and

Hostetler, 1973) to 11.15 (Hostetler, 1963). Natural samples of brucite are well crystallized, but they often contain carbonate impurities, especially on their surface. Their carbonate content should be checked before use; one natural sample analysed by the author contained 2 mol-% of carbonate, even after the surface had been rinsed with dilute hydrochloric acid to remove surface contamination. A sample of reagent-grade $Mg(OH)_2$ contained 0.4 mol-% of $CaCO_3$.

Although it has never been used, a fluorite saturometer consisting of a fluoride electrode surrounded by solid CaF_2 appears feasible:

$$CaF_2 \rightleftharpoons Ca^{2+} + 2F^- \qquad K_S^\ominus = a_{Ca} a_F^2 \qquad (66)$$

An equation for the potential of the fluorite saturometer can be derived from equation 66 and the half-cell equation for the fluoride electrode:

$$E = E^\ominus - (g/2) \log K_S^\ominus + (g/2) \log a_{Ca}$$
$$= \text{constant} + (g/2) \log a_{Ca} \qquad (67)$$

Thus the fluorite saturometer responds to the calcium ion activity. It might also be possible to use MgF_2 as the solid phase, and design a saturometer responsive to magnesium. However, MgF_2 is more soluble then CaF_2, and calcium would have to be excluded in any experiments involving solid MgF_2.

6.3.5. Mean ionic activity coefficients

Activity coefficient measurements in sea water are of two types: (1) measurement of thermodynamically valid mean ionic activity coefficients using cells without a liquid junction, and (2) measurement of conventional single ion activity coefficients using non-thermodynamic assumptions and cells with liquid junctions.

The potentiometric measurement of mean ionic activity coefficients in sea water is limited to electrodes which are free of interference from other ions in sea water, *viz.* H^+, Na^+, Cl^-, Br^-, F^-, and SO_4^{2-}. At present, NaCl, Na_2SO_4, and HCl are the only components whose activity coefficients have been measured in sea water by potentiometric methods. Platford (1965a) and Gieskes (1966) measured the mean ionic activity coefficient of NaCl in synthetic sea water using the cell

Na^+ glass or amalgam electrode | synthetic sea water or NaCl | AgCl | Ag
$$(68)$$

for which

$$E = E^\ominus + g \log (f_{NaCl}^2 [Na^+][Cl^-]) \qquad (69)$$

The standard potential, $E^\ominus$, was determined in NaCl solutions of known activity, and f_{NaCl} in sea water was calculated from the measured potential in

Table 22. Mean ionic activity coefficients in sea water (25°C and 35‰ salinity) measured by direct potentiometry. The activity coefficient of Na_2SO_4 (Platford and Dafoe, 1965) has been recalculated using activity coefficients of pure Na_2SO_4 from Pitzer and Mayorga (1973). The temperature and salinity dependence of the activity coefficients is given in the original papers

Salt	f_{MX}	Comments	Reference
NaCl	0.672	Na^+(amalgam); Ag $\vert$ AgCl	Platford (1965a)
	0.668	Na^+ glass; Ag $\vert$ AgCl	Gieskes (1966)
Na_2SO_4	0.385	Na^+(amalgam); Pb(amalgam) $\vert$ lead sulphate	Platford and Dafoe (1965)
HCl	0.627	H_2(gas); Ag $\vert$ AgCl	Khoo *et al.* (1977a)

sea water (Table 22). Their procedure is typical of all activity coefficient measurements that have been made in sea water: the cell is standardized in pure solutions of the components—NaCl, Na_2SO_4, or HCl—of known activity, transferred to sea water, and the mean ionic activity coefficient in sea water calculated from the potential measured in sea water and the concentrations of the relevant ions.

Platford (1965a) and Platford and Dafoe (1965) used the potential comparison method to standardize their cells. In this technique, two standards whose activities bracket that of the salt in sea water are used to calibrate the cell. The advantage of this method is that it avoids long extrapolations, and minimizes errors due to electrode slopes which differ from the theoretical value.

6.3.6. Conventional single ion activity coefficients

Conventional single ion activity coefficients are measured with the cell

$$ISE(X) \mid \text{sea water or standard} \mid KCl \text{ (saturated or 3.5 M)} \mid AgCl \mid Ag$$
$$(70)$$

for which

$$E = \text{constant} + E_j + (g/n) \log (f_x m_x) \tag{71}$$

and

$$f_{x,sw} = f_{x,std}(m_{x,std}/m_{x,sw}) 10^{(\Delta E - \Delta E_j)n/g} \tag{72}$$

The calculation of single ion activity coefficients from equation 72 requires non-thermodynamic assumptions about the values of $f_{x,std}$ and/or ΔE_j.

Conventional single ion activity coefficients for ions in sea water are listed in Table 23. To provide a self-consistent set of data, these values have been recalculated from literature data using the mean salt method (Garrels, 1967) to estimate single ion activity coefficients in the standards.

Table 23. Conventional single ion activity coefficients in sea water at 25°C and 35‰ salinity. Activity coefficients of pure solutions were taken from Robinson and Stokes (1968). Single ion activity coefficients in the standard solutions were calculated from the mean salt method unless otherwise noted. The letters PC indicate the potential comparison method; N indicates that the theoretical slope (RT/nF) was used in the calculations. An asterisk indicates the best value

Ion	f_i	Standard	Comments	Reference
H^+	0.703	NBS buffer		Culberson and Pytkowicz (1973)
	0.631*	0.01 M HCl	From f_{HCl} and f_{Cl} given below	Khoo *et al.* (1977a)
Na^+	0.72	0.48 m NaCl	PC	Platford (1965b)
	0.71	0.10 m NaCl		Garrels (1967)
	0.723	0.50 m NaCl	PC, without E_j correction	Savenko (1977a)
	0.734*	0.50 m NaCl	PC, corrected for E_j	
K^+	0.64	Unknown	34.3‰ salinity	Garrels and Thompson (1962)
Mg^{2+}	0.256*	Synthetic sea water	PC, 34.8‰ salinity	Kester (1970)
	0.257	without sulphate	Differential brucite solubility	Pytkowicz and Gates (1968)
Ca^{2+}	0.228*	Synthetic sea water without sulphate	PC, 34.8‰ salinity	Kester (1970)
	0.204	NBS buffer	Aragonite solubility, 34.5‰	Berner (1976)
	0.175		Gypsum solubility and $f_{SO_4^{2-}}$ given below	Culberson *et al.* (1978)
Cl^-	0.623*	0.67 m NaCl	N, 34.8‰	Kester (1970)
Br^-	0.645*	0.005–0.1 m KBr	PC, corrected for E_j	Savenko (1977b)
	0.725	0.005–0.1 m KBr	PC, without E_j correction	
	0.737	0.005 m KBr	N, without E_j correction	
OH^-	0.218	NBS buffer	From $f_{H^+} = 0.703$	Culberson and Pytkowicz (1973)
	0.243*		From $f_{H^+} = 0.631$	
SO_4^{2-}	0.106	Na_2SO_4	From $f_{Na_2SO_4}$ and f_{Na} given above	Platford and Dafoe (1965)

Several methods have been used to calculate single ion activity coefficients. In the simplest method (f_{Na^+}; Garrels, 1967), cell 70 is standardized in a dilute single salt solution in which $f_{x,std}$ can be accurately calculated from the Debye–Hückel theory or the mean salt method. This procedure eliminates uncertainty in the value of $f_{x,std}$ at the cost of introducing an unknown value of the residual liquid junction potential, which is assumed to be zero. In the potential comparison method (f_{Na^+}; Platford, 1965b), the cell is standardized in two solutions whose activities bracket that of sea water and the activity in sea water is interpolated from the measured potential in sea water and the potentials in the two standards. In this procedure ΔE (equation 72) is zero and errors due to non-Nernstian behaviour are minimized. However, the ionic strength and composition of the standards may not match sea water and the residual liquid junction potential (ΔE_j in equation 72) may not be zero.

In a third method (f_{Cl^-}; Kester, 1970), the cell is standardized in a solution whose ionic strength and composition match sea water. This method eliminates most of the residual liquid junction potential but it involves estimating single ion activity coefficients at high ionic strength, a procedure which is subject to error.

Savenko (1977b,d) measured the single ion activity coefficients of Na^+ and Br^- in sea water by the potential comparison method, and he used the Henderson equation (Bates, 1973) to eliminate the effect of the residual liquid junction potential on his calculations. In Table 23, Savenko's (1977b,d) results are used to compare values of f_i calculated by the various procedures described above. The results for f_{Br^-}, which was standardized in dilute KBr solutions, show that activity coefficients calculated by ignoring the residual liquid junction potential are 12% higher than values calculated using the Henderson equation (Section 1.3.2) for liquid junction potentials (Bates, 1973); an increase in agreement with the results of Hawley and Pytkowicz (1973) (Section 6.1.2) who found a 13% increase in f_{H^+} due to the residual liquid junction potential between dilute buffers and sea water. The values of f_{Na^+} do not show such an increase because the NaCl standards are close to the ionic strength of sea water.

The magnesium and calcium activity coefficients calculated from the data of Kester (1970) deserve special consideration because the calcium and magnesium electrodes are subject to interference from the other major cations in sea water (Section 6.2.3). To minimize interferences, the electrodes were standardized in sodium/magnesium/calcium chloride solutions in which the activities of the interfering ions were identical with their values in sea water; the activity of magnesium or calcium in sea water was then determined by the potential comparison method.

Also shown in Table 23 are values of $f_{Mg^{2+}}$ and $f_{Ca^{2+}}$ calculated from the solubilities of brucite [$Mg(OH)_2$; Pytkowicz and Gates, 1968], aragonite ($CaCO_3$; Berner, 1976), and gypsum ($CaSO_4 \cdot 2H_2O$; Culberson *et al.*, 1978) in sea water.

The major uncertainty in the values of $f_{Mg^{2+}}$ and $f_{Ca^{2+}}$ calculated from the solubilities of brucite and aragonite is due to errors in the values of $K_S^{\ominus}$ and pH(NBS). The values of $f_{Ca^{2+}}$ calculated from the results of Berner (1965) ($f_{Ca^{2+}} = 0.144$; $K_S^{\ominus}$ from Berner, 1976) differ from those in Table 23 by 41% because the equilibrium pH value reported by Berner (1965) is 0.07 pH unit higher than the value in Berner (1976) for the same experimental conditions. The activity coefficient calculated from the solubility of aragonite depends on the square of the apparent hydrogen ion activity, and should be a factor of $(1.13)^2 = 1.28$ larger than values calculated by the mean salt method due to the residual liquid junction potential.

6.3.7. Major ion association

The chemical model for sea water developed by Garrels and Thompson (1962) stimulated interest in the speciation of ions in sea water. Direct potentiometry and potentiometric titrations are ideal methods for determining speciation at high ionic strengths, since with proper experimental design ion-selective electrodes can be used to measure free ionic concentrations. The speciation of both major (Kester, 1970; Elgquist and Wedborg, 1978) and minor (Paulson, 1978; Sunda et al., 1978) constituents of sea water have been determined by potentiometric measurements. However, experimental procedures for determining the speciation of major and minor constituents are often different, because significant changes must be made in the bulk chemical composition of the medium to study speciation of the major ions.

The association constants of the major cations (Na^+, Mg^{2+}, Ca^{2+}) with the major anions [Cl^-, SO_4^{2-}, HCO_3^-, $B(OH)_4^-$, F^-] were measured by Pytkowicz and Kester (1969), Kester and Pytkowicz (1968, 1969, 1970), Elgquist (1970), Dyrssen and Hansson (1973), Byrne and Kester (1974), Pytkowicz and Hawley (1974), Miller and Kester (1976), and Elgquist and Wedborg (1978) using ion-selective electrodes. In addition to these measurements, Thompson and Ross (1966), Thompson (1966), Pytkowicz and Gates (1968), and Miller and Kester (1976) measured the free concentrations of Mg^{2+}, Ca^{2+}, and F^- in sea water using ion-selective electrodes.

The experimental procedures of the above studies are basically similar. The electrodes are calibrated in a chloride or perchlorate solution in which it is assumed that no association occurs, and this solution is then titrated with a second solution containing the ligand of interest. The ionic strength is held constant during the titration to minimize changes in free ion activity coefficients due to changes in solution composition. The use of chloride media to standardize the electrodes (Kester and Pytkowicz, 1968, 1969, 1970; Pytkowicz and Hawley, 1974) assumes that the concentrations of chloride ion pairs such as NaCl and $MgCl^+$ are negligible. However, several later papers (Elgquist and Wedborg, 1975, 1978; Johnson and Pytkowicz, 1978) present evidence that chloride ion pairs with the major cations exist in sea water. The importance of these chloride ion pairs must be resolved before meaningful calculations of major ion speciation in sea water can be made.

Although all studies agree that the ionic strength must be held constant during speciation studies, opinion differs as to whether the formal (Elgquist and Wedborg, 1978) or the effective (Kester and Pytkowicz, 1968, 1969, 1970) ionic strength should be kept constant. The difference between the effective and formal ionic strengths can be substantial. For instance, the formal ionic strength of a pure $MgSO_4$ solution with an effective ionic strength of 0.67 M is 1.84 M ($\beta_{MgSO_4} = 10.5$ M; Kester, 1970), while the formal ionic strength of a pure Na_2SO_4 solution with an effective ionic strength of 0.67 M is 1.01 M ($\beta_{NaSO_4^-} = 2.02$ M^{-1}; Kester, 1970). The effective ionic strength is calculated assuming that ionic interactions involving dipolar ion pairs such as $NaSO_4^-$ are the same as those for simple univalent ions such as Cl^- and Na^+, and that ionic interactions involving uncharged dipolar molecules such as $MgSO_4$ with other ions are negligible. These assumptions may not be realistic since the average distance (11 Å; 1.1 nm; Robinson and Stokes, 1968) between positive and negative ions for a 1–1 electrolyte at ionic strength 0.7 M is not much greater than the distance between magnesium and sulphate ions in a hydrated $MgSO_4$ ion pair.

At present, our knowledge of the association constants for the major ions is both incomplete and contradictory. $NaSO_4^-$ is the only ion pair for which recent values of the association constant agree ($\beta = 2.0\,M^{-1}$, Pytkowicz and Kester, 1968; $\beta = 1.8\,M^{-1}$, Elgquist and Wedborg, 1978), and for which the effects of temperature, ionic strength, and pressure on the association constant are known (Kester and Pytkowicz, 1970). The presence of $MgCl^+$ ion pairs has a significant effect on calculated values of β_{MgSO_4}; recent values range from $6\,M^{-1}$ (Elgquist and Wedborg, 1978) to $10\,M^{-1}$ (Kester and Pytkowicz, 1968) depending on whether or not chloride ion pairs are included in the calculations. Magnesium sulphate association in sea water has been studied as a function of temperature and ionic strength (Kester and Pytkowicz, 1970), but its pressure dependence at high ionic strength has not been measured. Kester and Pytkowicz (1969) measured β_{CaSO_4} at 25°C and effective ionic strength 0.67 M. These are the only potentiometric measurements of β_{CaSO_4} in sea water, although Elgquist and Wedborg (1975) have calculated β_{CaSO_4} from solubility measurements.

Literature data for the association constants of the major cations with chloride and sulphate at 25°C, 1 atm, and ionic strength 0.7 M are given in Table 24. The lack of agreement for magnesium and calcium complexes in this table emphasizes the need for more accurate information on sulphate and chloride complexation in sea water.

Table 24. Association constants for $NaSO_4^-$, $MgSO_4$, $CaSO_4$, $MgCl^+$, and $CaCl^+$ ion pairs in synthetic sea water at 25°C. The constants are defined in terms of free concentrations

Ion pair	β_{MX},M^{-1}	Ionic strength	References
$NaSO_4^-$	2.02	$I_e = 0.69$ m	Pytkowicz and Kester (1969)
	1.22	$I_t = 0.86$ M	Elgquist and Wedborg (1975)
	1.8	$I_t = 0.70$ M	Elgquist and Wedborg (1978)
$MgSO_4$	10.2	$I_e = 0.67$ m	Kester and Pytkowicz (1968)
	7	$I_t = 0.70$ M	Elgquist and Wedborg (1974)
	12.3	$I_t = 0.86$ M	Elgquist and Wedborg (1975)
	6.3	$I_t = 0.70$ M	Elgquist and Wedborg (1978)
$CaSO_4$	10.8	$I_e = 0.67$ m	Kester and Pytkowicz (1969)
	30.6	$I_t = 0.86$ M	Elgquist and Wedborg (1975)
$MgCl^+$	0.48	$I_t = 0.86$ M	Elgquist and Wedborg (1975)
	0.34	$I_t = 0.70$ M	Elgquist and Wedborg (1978)
$CaCl^+$	1.20	$I_t = 0.86$ M	Elgquist and Wedborg (1975)

Table 25. Activity coefficient of $NaClO_4$ in $HClO_4/NaClO_4$ and $LiClO_4/NaClO_4$ mixtures at ionic strength 0.7 m and 25°C. Calculated from the equations of Pitzer and Kim (1974)

$NaClO_4$, m	f_{NaClO_4}	
	$HClO_4/NaClO_4$	$LiClO_4/NaClO_4$
0.70	0.646	0.646
0.35	0.683	0.687
0.00	0.723	0.731

The experimental determination of association constants for the major ions involves two assumptions: (1) that activity coefficients of free ions are independent of solution composition at constant ionic strength, and (2) that alkali and alkaline earth metal chloride or perchlorate solutions are free of ion pairing. The first assumption is often incorrect. Consider the alkali metal perchlorates, which are generally assumed to be non-associated in aqueous solutions (Elgquist and Wedborg, 1978). There is not much data on activity coefficients in mixtures of perchlorates, but the activity coefficients of $NaClO_4$ in $HClO_4/NaClO_4$ and $LiClO_4/NaClO_4$ mixtures can be calculated from the mixed electrolyte equations of Pitzer and Kim (1974). The calculations at ionic strength 0.7 M (Table 25) show that f_{NaClO_4} increases by about 12% over the full range of composition. These data do not support the assumption that free ion activity coefficients are independent of solution composition at constant ionic strength.

6.3.8. Direct potentiometry at high pressure: laboratory measurements

Pressure increases about 1 atm for each 10 m of depth in the ocean, and the average pressure at the sea floor is 400 atm (40.5 MN m^{-2}). The pressure in deep trenches can exceed 1100 atm (111.5 MN m^{-2}) (Mantyla and Reid, 1978). Direct potentiometry is the most direct method for studying the effects of pressure on chemical reactions in sea water and much effort has been expended in adapting ion-selective electrodes for use at high pressures, both in the laboratory and *in situ*. The main limitation on the use of direct potentiometry at high pressures is the lack of suitable electrodes. At present only pH and sodium glass electrodes, the silver | silver chloride electrode, and the lanthanum fluoride electrode have proved useful for high-pressure measurements in sea water. The calcium (Whitfield, 1969) and divalent ion (Kester and Pytkowicz, 1970) electrodes are not sufficiently stable at high pressure.

Distèche (1959) demonstrated that pH glass electrodes behaved well at pressures up to 1000 atm, and he made the first direct measurement of the effect of pressure on the pH of sea water. Since that time, pH glass electrodes have been used to determine the effect of pressure on the dissociation constants of water (Hammann, 1963; Whitfield, 1972) and carbonic and boric acids (Distèche and Distèche, 1967; Distèche, 1974; Culberson and Pytkowicz, 1968), and on the solubility products of calcite and aragonite (Ingle, 1975; Hawley and Pytkowicz, 1969). The feasibility of using the lanthanum fluoride electrode at high pressure was demonstrated by Brewer and Spencer (1970).

Electrode measurements at high pressure in the laboratory are essentially differential measurements in which the difference in potential, $E_1 - E_p$, between the cell at pressure p and at 1 atm is measured. The effect of pressure on the potential of the cell

$$Ag \mid AgCl \mid reference\ solution\ (0.71\ \text{M NaCl} + 0.01\ \text{M HCl}) \mid$$

$$glass \mid sea\ water \mid AgCl \mid Ag \quad (73)$$

is given by (Culberson, 1968)

$$E_1 - E_p = g \log \left[\left(\frac{a_{H,p}}{a_{H,1}} \right)_{sw} \left(\frac{a_{H,1}}{a_{H,p}} \right)_{ref} \right]$$

$$+ g \log \left[\left(\frac{a_{Cl,p}}{a_{Cl,1}} \right)_{sw} \left(\frac{a_{Cl,1}}{a_{Cl,p}} \right)_{ref} \right] + \Delta E_{asym} \quad (74)$$

where ΔE_{asym} is the change in asymmetry potential of cell 73 with pressure. The ionic strengths (and thus activity coefficients) of the reference solution and sea water are nearly equal, and if we assume that chloride is unassociated in sea water, the chloride term in equation 74 is zero. Furthermore, the hydrogen ion molality is constant in the reference solution, and the effect of pressure on f_{H^+} in the two compartments should be identical. Thus equation 74 reduces to

$$E_1 - E_p = g \log \left(\frac{[H^+]_p}{[H^+]_1} \right)_{sw} + \Delta E_{asym} \quad (75)$$

Cell 73 is a concentration cell and the asymmetry potential is the sum of the bias potentials of the Ag | AgCl and glass electrodes when both electrode compartments are filled with reference solution.

Whitfield (1969, 1970) made a thorough study of the effect of pressure on the asymmetry potentials of glass electrodes and found that pressure had the least effect on electrodes with flat membranes. At 1000 atm, ΔE_{asym} was typically 0.1 mV for electrodes with flat membranes, while bulb-type glass electrodes may have values of ΔE_{asym} greater than 1 mV. The asymmetry potential of a particular cell often changes with repeated pressurization, apparently due to deterioration of the $Ag \mid AgCl$ electrodes (Distèche, 1959, 1962; Distèche and Distèche, 1965), which must be replaced once a week during intensive experimentation (Distèche, 1962). The most constant asymmetry potentials are obtained after the cell has been pressurized several times (Distèche and Distèche, 1965). Kester (1970) found that, at 1000 atm and 1.5°C, ΔE_{asym} was approximately 1.0 mV for a cell with sodium glass and $Ag \mid AgCl$ electrodes. Both Whitfield (1969) and Kester (1970) found that the asymmetry potentials of calcium and divalent cation liquid ion exchanger electrodes were too large and erratic for use at high pressure. For pH and sodium glass electrodes and for $Ag \mid AgCl$ electrodes, the slope of the electrode response is unaffected by pressure (Distèche, 1959; Culberson, 1968; Whitfield, 1969; Kester, 1970).

High-pressure cells undergo more or less adiabatic temperature changes during compression and decompression (Distèche, 1959; Whitfield, 1969) and approximately 30 min are required to restore thermal equilibrium after a pressure change. For details of cell design and experimental technique the reader is referred to Distèche (1959, 1962), Distèche and Distèche (1965), Culberson and Pytkowicz (1968), Culberson (1968), Whitfield (1969, 1970, 1972), Kester (1970), and Ingle (1975).

6.3.9. Direct potentiometry: *in situ*

In situ analysis of sea water constituents offers several potential advantages over traditional discrete water sampling techniques. *In situ* measurements allow continuous profiles to be obtained, and *in situ* measurements of temperature, salinity, oxygen (Lambert *et al.*, 1973) and pH (Ben-Yaakov *et al.*, 1974) show fine structure which cannot be resolved with discrete water sampling. Furthermore, direct *in situ* measurements may eliminate sampling errors, such as gas exchange and water bottle corrosion, which plague shipboard analysis of trace constituents.

Unfortunately, the lack of suitable electrodes limits the use of direct potentiometry for *in situ* measurements. At present, the only electrodes which are sufficiently specific for *in situ* analysis are the pH and sodium glass electrodes, the lanthanum fluoride electrode, and the silver | silver sulphide electrode. Sodium is present at high concentrations and is a conservative constituent of sea water; except for cases such as the Red Sea (Degens and Ross, 1969) and Ocra Basin (Shokes *et al.*, 1977) brines, its concentration can be determined more accurately by chlorinity or conductivity measurements (Chapter 5) than by *in situ* electrode measurements. Likewise, fluoride is conservative or nearly conservative (Warner *et al.*, 1975) in sea water, and *in situ* fluoride measurements seem to offer little advantage over discrete measurements. The situation is different, however, for hydrogen and sulphide ions. These species are non-conservative and their concentrations show large variations.

The most extensive series of *in situ* measurements are those of Ben-Yaakov (Ben-Yaakov *et al.*, 1974; Ben-Yaakov and Kaplan, 1968a, 1971a) in the northeastern Pacific using a pressure-compensated pH electrode (Ben-Yaakov and Kaplan, 1968b, 1973; Ben-Yaakov and Ruth, 1974) and a submersible instrument package (Ben-Yaakov and Kaplan, 1968c, 1971b). By attaching a carbonate saturometer (Section 6.3.4) to the *in situ* pH assembly, Ben-Yaakov was able to measure the saturation of the water column with respect to calcium carbonate (Ben-Yaakov *et al.*, 1974) in addition to measuring the *in situ* pH. In Figures 8 and 9 *in situ* pH profiles (Ben-Yaakov *et al.*, 1974) in the northeast Pacific are compared with pH values calculated from conventional pH measurements made aboard ship at 1 atm and 25°C. With the exceptions of stations 43 and 48 (Ben-Yaakov *et al.*, 1974), the agreement of the measured and calculated *in situ* pH values is excellent.

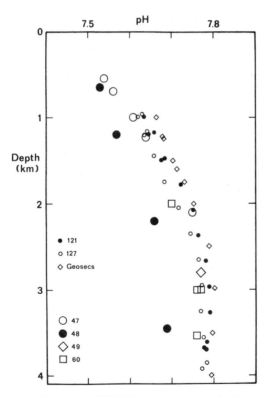

Figure 8. Comparison of *in situ* pH(NBS) measurements in the northeastern Pacific (Ben-Yaakov *et al.*, 1974; large symbols) with *in situ* pH values (small symbols) calculated from pH measurements at 25°C and atmospheric pressure (Wyatt *et al.*, 1970) made with the original micro pH assembly. Station positions are shown in Figure 7. Adapted from Ben-Yaakov *et al.* (1974); reproduced by permission of Pergamon Press Ltd

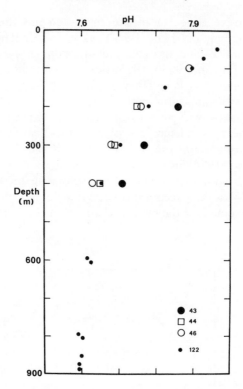

Figure 9. Comparison of *in situ* pH(NBS) measurements in the Santa Monica Basin (Ben-Yaakov *et al.*, 1974; large symbols) with *in situ* pH values (small symbols) calculated from pH measurements at 25°C and atmospheric pressure (Wyatt *et al.*, 1970) made with the original micro pH assembly. Station positions are shown in Figure 7

The major limitation on the accuracy of *in situ* potentiometric measurements is the calibration of the electrode assembly. Conventional pH measurements are made at constant temperature and pressure, and the electrodes are frequently standardized in buffer. Calibration of *in situ* measurements is complicated since the assembly must be calibrated as a function of temperature and pressure, a process which is inconvenient aboard ship. The procedure used by Ben-Yaakov and Ruth (1974) was to perform the pressure and temperature calibrations ashore, and then to perform a 'one point' calibration at atmospheric pressure and ambient temperature aboard ship. The slope of the pH *versus* temperature curve and the effect of pressure on the asymmetry potential were assumed to be constant with time. Any difference between the calibration ashore and the 'one point' calibration at sea was attributed to changes in the standard and/or asymmetry potentials of the glass and silver|silver chloride electrodes. The *in situ* pH was calculated from the equation

$$pH_x(t, p) = pH_b(t, 1) - g[E_x(t, p) - \Delta E_b^{\ominus} - \Delta E_{asym} - E_b(t, 1)] \tag{76}$$

where

$pH_x(t, p) = $ *in situ* pH on NBS pH scale;
$pH_b(t, 1) = $ buffer pH at *in situ* temperature and 1 atm;
$E_x(t, p) = $ electrode potential *in situ*;
$E_b(t, 1) = $ electrode potential in buffer at *in situ* temperature and 1 atm;
$\Delta E_b^{\ominus} = $ change in standard potential of cell between shipboard and shore-based calibrations;
$\Delta E_{asym} = $ change in asymmetry potential with pressure.

The buffer calibration ashore is made at 1°C intervals between 0 and 25°C, and the measured potential in buffer is fitted to a second-order polynomial:

$$E_b(t, 1) = E_b^{\ominus} + at + bt^2 \tag{77}$$

The constants a and b are assumed to be invariant with time, and the 'one point' calibration aboard ship is used to adjust the value of E_b^0 for changes in the standard and asymmetry potentials of the cell. The change in asymmetry potential with pressure ($\Delta E_{asym} = 1$–2 mV/670 atm) of Ben-Yaakov's *in situ* sensor is larger than the values reported by Distèche and Distèche (1965) and by Culberson (1968) for laboratory high-pressure pH cells ($\Delta E_{asym} = 0$–1 mV/1000 atm).

Based on a comparison of pH values for up and down traces, and for repeated casts in the same areas, Ben-Yaakov and Kaplan (1968a,b) and Ben-Yaakov *et al.* (1974) quote a precision of ±0.02 pH unit for their *in situ* measurements. However, the data in Figures 8 and 9 suggest that the reproducibility for repeated casts is of the order of ±0.04 pH unit. The more or less constant offset of pH values for casts 43 and 48 from the other stations suggests that these two stations suffer from calibration problems. Note that the lower pH values from station 48 are not reflected in lower values of percentage calcite saturation (Figure 10), as would be expected if these pH values were real.

The good agreement between the measured and calculated pH profiles in Figures 8 and 9 imply that the temperature and pressure coefficients of the apparent constants of carbonic and boric acids used to calculate the *in situ* pH are correct (Ben-Yaakov *et al.*, 1974).

In contrast to the *in situ* pH measurements, the measured and calculated profiles of calcite saturation in the northeastern Pacific show large discrepancies. The data plotted in Figures 10 and 11 show two significant features:

1. The depth at which the water column changes from super- to undersaturation is the same for the measured and calculated profiles.

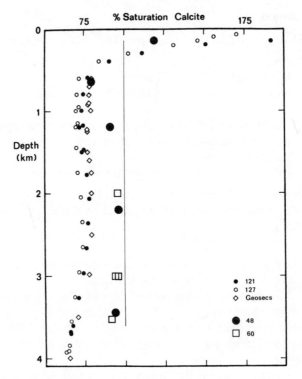

Figure 10. Calcite saturation in the northeastern Pacific. Comparison of *in situ* measurements (Ben-Yaakov *et al.*, 1974; large symbols) with values (small symbols) calculated from pH and alkalinity measurements aboard ship (Wyatt *et al.*, 1970). The solubility product of calcite was taken from Ingle (1975) and Ingle *et al.* (1973). Station positions are shown in Figure 7. Adapted from Ben-Yaakov *et al.* (1974); reproduced by permission of Pergamon Press Ltd

2. The directly measured values of percentage saturation (Ben-Yaakov *et al*, 1974) are less supersaturated near the surface and less undersaturated at depth than the calculated values.

The discrepancy between the measured and calculated values is not due to errors in the value of $[CO_3^{2-}]$ used to calculate the *in situ* ion product of calcium carbonate, since the measured and calculated *in situ* pH values are the same. If the discrepancy between the measured and calculated percentage saturations is caused by errors in the calculated values, it must be due to an error in the solubility product of calcite used in the calculations. In order to account for the differences in Figures 10 and 11, the solubility product used in the calculations would have to be too small in the region of supersaturation and too large in the undersaturated region. This possible

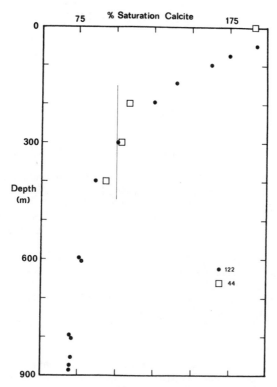

Figure 11. Calcite saturation in the Santa Monica Basin. Comparison of *in situ* measurements (Ben-Yaakov *et al.*, 1974; large symbols) with values (small symbols) calculated from pH and alkalinity measurements aboard ship (Wyatt *et al.*, 1970). The solubility product of calcite was taken from Ingle (1975) and Ingle *et al.* (1973). Station positions are shown in Figure 7

error in the solubility product is not simply due to uncertainties in the effect of pressure on the solubility of calcite (Ingle, 1975), since the discrepancy between the measured and calculated profiles is also present in surface waters where the effect of pressure is nil.

An alternative and equally likely explanation for the discrepancy is that insufficient time (15 min) was allowed for the *in situ* measurements to reach equilibrium with calcite. Insufficient equilibration time would cause supersaturated waters to appear less supersaturated, and undersaturated water to appear less undersaturated than they actually were.

In addition to the work of Ben-Yaakov and Kaplan, Wilde and co-workers (Conti *et al.*, 1971; Conti and Wilde, 1972, 1973; Conti, 1972) have developed *in situ* pH and sulphide electrodes for use in estuarine and coastal waters.

6.4. ACKNOWLEDGEMENTS

This work was supported by the Office of Naval Research (Contract N00014-77-C-0290) and the Oceanography Section of the National Science Foundation (OCE78-07488).

REFERENCES

Almgren, T., and S. H. Fonselius (1976). Determination of alkalinity and total carbonate. In *Methods of Sea Water Analysis* (Ed. K. Grasshoff). Verlag Chemie, Weinheim, pp. 97–115.

Almgren, T., D. Dyrssen, and M. Strandberg (1975a). Determination of pH on the moles per kg seawater scale (M_w). *Deep-Sea Res.*, **22**, 635–646.

Almgren, T., D. Dyrssen, and M. Strandberg (1975b). Data from the Swedish participation in the 9th cruise of Dmitry Mendeleev. In *Report on the Chemistry of Seawater XV*. Department of Analytical Chemistry, University of Goteborg, 33 pp.

Almgren, T., D. Dyrssen, and M. Strandberg (1977). Computerized high-precision titrations of some major constituents of seawater on board R.V. Dmitry Mendeleev. *Deep-Sea Res.*, **24**, 345–364.

Anderson, D. H., and R. J. Robinson (1946). Rapid electrometric determination of the alkalinity of sea water. *Ind. Eng. Chem.*, **18**, 767–769.

Anfält, T., and D. Jagner (1971). A standard addition method for the potentiometric determination of fluoride in sea water. *Anal. Chim. Acta*, **53**, 13–22.

Anfält, T., and D. Jagner (1973). The potentiometric titration of potassium in sea water with a valinomycin electrode. *Anal. Chim. Acta*, **66**, 152–155.

Atlas, E., C. Culberson, and R. M. Pytkowicz (1976). Phosphate association with Na^+, Ca^{++}, and Mg^{++} in seawater. *Mar. Chem.*, **4**, 243–254.

Bates, R. G. (1973). *Determination of pH, Theory and Practice*. Wiley, New York, 479 pp.

Bates, R. G., and J. B. Macaskill (1975). Acid–base measurements in sea water. In: *Analytical Methods in Oceanography* (Ed. R. P. Gibbs, Jr.). Advances in Chemistry Series, No. 147, American Chemical Society, Washington, D.C.

Baumann, E. W. (1971). Sensitivity of the fluoride-selective electrode below the micromolar range. *Anal. Chim. Acta*, **54**, 189–197.

Beck, W. H., A. E. Bottom, and A. K. Covington (1968). Errors of glass electrodes in certain standard buffer solutions at high discrimination. *Anal. Chem.*, **40**, 501–505.

Ben-Yaakov, S., and I. R. Kaplan (1968a). pH-temperature profiles in ocean and lakes using an *in situ* probe. *Limnol. Oceanogr.*, **13**, 688–693.

Ben-Yaakov, S. and I. R. Kaplan (1968b). High pressure pH sensor for oceanographic applications. *Rev. Sci. Instrum.*, **39**, 1133–1138.

Ben-Yaakov, S., and I. R. Kaplan (1968c). A versatile probe for *in situ* oceanographic measurements. *J. Ocean Technol.*, **2**, 25–29.

Ben-Yaakov, S., and I. R. Kaplan (1971a), Deep-sea *in situ* calcium carbonate saturometry. *J. Geophys. Res.*, **76**, 722–731.

Ben-Yaakov, S., and I. R. Kaplan (1971b). An oceanographic instrumentation system for *in situ* application. *Mar. Tech. Soc. J.*, **5**, 41–46.

Ben-Yaakov, S., and I. R. Kaplan (1973). Design and application of a deep sea pH sensor. In *Marine Electrochemistry* (Eds. J. B. Berkowitz, R. A. Horne, M. Banus,

P. L. Howard, M. J. Pryor and G. C. Witnack). Electrochemical Society, Princeton, N.J.

Ben-Yaakov, S., and E. Ruth (1974). An improved *in situ* pH sensor for oceanographic and limnological applications. *Limnol. Oceangr.*, **19**, 144–151.

Ben-Yaakov, S., E. Ruth, and I. R. Kaplan (1974). Calcium carbonate saturation in northeastern Pacific: *in situ* determination and geochemical implications. *Deep-Sea Res.*, **21**, 229–243.

Berner, R. A. (1963). Electrode studies of hydrogen sulfide in marine sediments. *Geochim. Cosmochim. Acta*, **27**, 563–575.

Berner, R. A. (1965). Activity coefficients of bicarbonate, carbonate and calcium ions in sea water. *Geochim. Cosmochim. Acta*, **29**, 947–965.

Berner, R. A. (1976). The solubility of calcite and aragonite in sea water at atmospheric pressure and 34.5‰ salinity. *Am. J. Sci.*, **276**, 713–730.

Boyle, E., F. R. Sclater, and J. M. Edmond (1977). The distribution of dissolved copper in the Pacific. *Earth Planet. Sci. Lett.*, **37**, 38–54.

Bradford, W. L. (1968). Calcium analysis in sea water by an ion sensitive electrode. *MS Thesis*, Oregon State University, Corvallis, Ore., 48 pp.

Brand, M. J. D., and G. A. Rechnitz (1970). Computer approach to ion-selective electrode potentiometry by standard addition methods. *Anal. Chem.*, **42**, 1172–1177.

Brewer, P. G., and D. W. Spencer (1970). An *in-situ* device for the measurement of fluoride in sea water. *Woods Hole Oceanogr. Inst. Tech. Rep.*, No. 70–21, Woods Hole, Mass.

Briggs, C. C., and T. H. Lilley (1974). A rigorous test of a calcium ion-exchange membrane electrode. *J. Chem. Thermodyn.*, **6**, 599–607.

Bruyevich, S. W., Ed. (1966). *The Pacific Ocean, Vol. 3, Chemistry of the Pacific Ocean*, translated by I. Evans. US Naval Oceanographic Office, Washington, D.C.

Buch, K. (1951). Das Kohlensaure Gleichgewichtssystem im Meerwasser. *Merentutkimuslaitoksen Julk. Havsforskningsinst. Skr.*, **151**, 1–18.

Buck, R. P. (1971). Potentiometry: pH measurements and ion selective electrodes. In *Techniques of Chemistry*, Vol. 1, Part 2A (Eds. A. Weissberger and B. W. Rossiter). Wiley–Interscience, New York.

Buck, R. P. (1978). Ion selective electrodes. *Anal. Chem.*, **50**, 17R–29R.

Byrne, R. H., Jr. (1974). Iron speciation and solubility in sea water *PhD Thesis*, University of Rhode Island, Kingston, R.I., 205 pp.

Byrne, R. H., Jr., and D. R. Kester (1974). Inorganic speciation of boron in sea water. *J. Mar. Res.*, **32**, 119–127.

Cattrall, R. W., S. Tribuzio, and H. Freiser (1974). Potassium ion responsive coated wire electrode based on valinomycin. *Anal. Chem.*, **46**, 2223–2224.

Christenson, P. G. (1973). Activity coefficients of HCl, NaCl, and KCl in several mixed electrolyte solutions at 25°C. *J. Chem. Eng. Data*, **18**, 286–288.

Christenson, P. G., and J. M. Gieskes (1971). Activity coefficients of KCl in several mixed electrolyte solutions at 25°C. *J. Chem. Eng. Data*, **16**, 398–400.

Conti, U. (1972). Study of an underwater environmental monitor. *PhD Thesis*, University of California, Berkeley, 184 pp.

Conti, U., P. Wilde, and T. L. Richards (1971). Towed vehicle for constant depth and bottom contouring operations. Paper No. OTC 1456, Offshore Technology Conference, Houston, Texas.

Conti, U., and P. Wilde (1972). Diver-operated *in situ* electrochemical measurements. *Mar. Tech. Soc. J.*, **6**, 17–23.

Conti, U., and P. Wilde (1973). Self-contained towable underwater system for environmental monitoring. In *Oceans 73. Proceedings of I.E.E.E. Conference on Engineering in Ocean Environment, Seattle*, pp 428–437.

Covington, A. K. (1969). Reference electrodes. In *Ion-selective Electrodes* (Ed. R. A. Durst). National Bureau of Standards Special Publication 314, US Government Printing Office, Washington, D.C.

Covington, A. K. (1974). Ion-selective electrodes. *CRC Crit. Rev. Anal. Chem.*, **3**, 355–406.

Cowperthwaite, I. A., and V. K. LaMer (1931). The electromotive force of the cell $Zn(s)/ZnSO_4(m)/PbSO_4(s)/Pb(s)$. An experimental determination of the temperature coefficient of the ion-size parameter in the theory of Debye and Hückel. *J. Am. Chem. Soc.*, **53**, 4333–4348.

Culberson, C. H. (1968). Pressure dependence of the apparent dissociation constants of carbonic and boric acids in seawater. *MS Thesis*, Oregon State University, Corvallis, Ore., 85 pp.

Culberson, C., and R. M. Pytkowicz (1968). Effect of pressure on carbonic acid, boric acid, and the pH in seawater. *Limnol. Oceanogr.*, **13**, 403–417.

Culberson, C., and R. M. Pytkowicz (1970). Oxygen–total carbon dioxide correlation in the eastern Pacific Ocean. *J. Ocean. Soc. Jap.*, **26**, 95–100.

Culberson, C., R. M. Pytkowicz, and J. E. Hawley (1970). Seawater alkalinity determination by the pH method. *J. Mar. Res.*, **28**, 15–21.

Culberson, C. H., and R. M. Pytkowicz (1973). Ionization of water in sea water. *Mar. Chem.* **1**, 309–316.

Culberson, C. H., G. Latham, and R. G. Bates (1978). Solubilities and activity coefficients of calcium and strontium sulfates in synthetic seawater at 0.5 and 25°C. *J. Phys. Chem.*, **82**, 2693–2699.

Degens, E., and D. Ross, Eds. (1969). *Hot Brines and Recent Heavy Metal Deposits in the Red Sea.* Springer-Verlag, New York.

Dietrich, G., W. Duing, K. Grasshoff, and P. H. Koske (1966). Physical and chemical data according to the observations of the R/V Meteor in the Indian Ocean 1964/65. *Meteor Forschungsergbnisse*, Reihe A, No. 2.

Distèche, A. (1959). pH measurements with a glass electrode withstanding 1500 kg/cm² hydrostatic pressure. *Rev. Sci. Instrum.*, **30**, 474–478.

Distèche, A. (1962). Electrochemical measurements at high pressures. *J. Electrochem. Soc.* **109**, 1084–1092.

Distèche, A. (1974). The effect of pressure on dissociation constants and its temperature dependency. In *The Sea*, Vol. 5, (Ed. E. D. Goldberg). Wiley–Interscience, New York.

Distèche, A., and S. Distèche (1965). The effects of pressure on pH and dissociation constants from measurements with buffered and unbuffered glass electrode cells. *J. Electrochem. Soc.*, **112**, 350–354.

Distèche, A., and S. Distèche (1967). The effect of pressure on the dissociation of carbonic acid from measurements with buffered glass electrode cells. *J. Electrochem. Soc.*, **114**, 330–340.

Durst, R. A., Ed. (1969a). *Ion-selective Electrodes*. National Bureau of Standards, Special Publication 314, US Government Printing Office, Washington, D.C.

Durst, R. A. (1969b). Analytical techniques and applications of ion-selective electrodes. In *Ion-selective Electrodes*, (Ed. R. A. Durst). National Bureau of Standards Special Publication 314, US Government Printing Office, Washington, D.C., pp. 375–414.

Dyrssen, D., and I. Hansson (1973). Ionic medium effects in seawater—a comparison of acidity constants of carbonic acid and boric acid in sodium chloride and synthetic seawater. *Mar. Chem.*, **1**, 137–149.

Edmond, J. M. (1970). High precision determination of titration alkalinity and total carbon dioxide content of sea water by potentiometric titration. *Deep-Sea Res.*, **17**, 737–750.

Eisenman, G., Ed. (1967). *Glass Electrodes for Hydrogen and Other Cations*. Marcel Dekker, New York, 582 pp.

Elgquist, B., and M. Wedborg (1970). Determination of the stability constants of MgF^+ and CaF^+ using a fluoride ion selective electrode. *J. Inorg. Nucl. Chem.*, **32**, 937–944.

Elgquist, B., and M. Wedborg (1974). Sulphate complexation in sea water. *Mar. Chem.*, **2**, 1–15.

Elgquist, B., and M. Wedborg (1975). Stability of ion pairs from gypsum solubility: degree of ion pair formation between the major constituents of sea water. *Mar. Chem.*, **3**, 215–225.

Elgquist, B., and M. Wedborg (1978). Stability constants of $NaSO_4^-$, $MgSO_4$, MgF^+, $MgCl^+$ ion pairs at the ionic strength of seawater by potentiometry. *Mar. Chem.*, **6**, 243–252.

Garrels, R. M. (1967). Ion-sensitive electrodes and individual ion activity coefficients. In *Glass Electrodes for Hydrogen and Other Cations*, (Ed. G. Eisenman). Marcel Dekker, New York, pp. 344–361.

Garrels, R. M., and M. E. Thompson (1962). A chemical model for sea water at 25°C and one atmosphere total pressure. *Am. J. Sci.*, **260**, 57–66.

Gates, R. F. (1969). Magnesium sulfate ion association in sea water. *MS Thesis*, Oregon State University, Corvallis Ore., 39 pp.

Gieskes, J. M. (1966). The activity coefficients of sodium chloride in mixed electrolyte solutions at 25°C. *Z. Phys. Chem.*, *N. F.*, **50**, 78–90.

Gieskes, J. M. (1974). The alkalinity–total carbon dioxide system in sea water. In *The Sea*, Vol. 5 (Ed. E. D. Goldberg). Wiley–Interscience, New York, pp. 123–151.

Gilbert, T. R., and A. M. Clay (1973). Determination of ammonia in aquaria and in sea water using the ammonia electrode. *Anal. Chem.*, **45**, 1757–1759.

Goldhaber, M. B., and I. R. Kaplan (1975). Apparent dissociation constants of hydrogen sulfide in chloride solutions. *Mar. Chem.*, **3**, 83–104.

Gran, G. (1952). Determination of the equivalence point in potentiometric titrations. Part II. *Analyst*, **77**, 661–671.

Graneli, A., and T. Anfalt (1977). A simple automatic phototitrator for the determination of total carbonate and total alkalinity of sea water. *Anal. Chim. Acta*, **91**, 175–180.

Grasshoff, K. (1966). On the chemistry of the Red Sea and the Inner Gulf of Aden according to observations of the R/V Metor during the International Indian Ocean Expedition 1964/65. *Meteor Forschungsergebinisse, Reihe A*, No. 6.

Green, E. J., and D. Schnitker (1974). The direct titration of water soluble sulfide in estuarine muds of Montsweag Bay, Maine. *Mar. Chem.*, **2**, 111–124.

Hamann, S. D. (1963). The ionization of water at high pressures *J. Phys. Chem.*, **67**, 2233–2235.

Hansson, I. (1972). An analytical approach to the carbonate system in sea water. *PhD Thesis*, University of Goteborg, Sweden.

Hansson, I. (1973a). A new set of pH-scales and standard buffers for sea water. *Deep-Sea Res.*, **20**, 479–491.

Hansson, I. (1973b). A new set of acidity constants for carbonic acid and boric acid in sea water. *Deep-Sea Res.*, **20**, 461–478.

Hawley, J., and R. M. Pytkowicz (1969). Solubility of calcium carbonate in sea water at high pressures and 2°C. *Geochim. Cosmochim. Acta*, **33**, 1557–1561.

Hawley, J. E., and R. M. Pytkowicz (1973). Interpretation of pH measurements in concentrated electrolyte solutions. *Mar. Chem.*, **1**, 245–250.

Hills, G. J., and D. J. G. Ives (1961). The hydrogen electrode. In *Reference Electrodes, Theory and Practice*. (Eds. D. J. G. Ives and G. J. Janz). Academic Press, New York, pp. 71–126.

Hittinger, R. C. (1975). Determination of cupric hydroxide ion pair stability constants at 0.7 ionic strength. *MS Thesis*, University of Rhode Island, Kingston, R.I., 95 pp.

Hostetler, P. B. (1963). The stability and surface energy of brucite in water at 25°C. *Am. J. Sci.*, **261**, 238–258.

Ingle, S. E. (1975). Solubility of calcite in the ocean. *Mar. Chem.*, **3**, 301–319.

Ingle, S. E., C. H. Culberson, J. E. Hawley, and R. M. Pytkowicz (1973). The solubility of calcite in sea water at atmospheric pressure and 35‰ salinity. *Mar. Chem.*, **1**, 295–307.

Ivanenkov, V. N. (1964). Influence of the inner surface of a Nansen bottle on changes in concentration of chemical elements in a water sample. *Tr. Inst. Okeanol. Akad. Nauk SSSR*, **64**, 80–84.

Ives, D. J. G. (1961). Oxide, oxygen, and sulfide electrodes. In *Reference Electrodes, Theory and Practice*. (Eds. D. J. G. Ives and G. J. Janz). Academic Press, New York, pp. 322–392.

Ives, D. J. G., and G. J. Janz, Eds. (1961a). *Reference Electrodes: Theory and Practice*. Academic Press, New York. 651 pp.

Ives, D. J. G., and G. J. Janz (1961b). General and theoretical introduction. In *Reference Electrodes, Theory and Practice* (Eds. D. J. G. Ives and G. J. Janz). Academic Press, New York, pp. 1–70

Ives, D. J. G., and F. R. Smith (1961). Electrodes reversible to sulfate ions. In *Reference Electrodes, Theory and Practice* (Eds. D. J. G. Ives and G. J. Janz). Academic Press, New York, pp. 393–410.

Jagner, D., and K. Aren (1970). A rapid semi-automatic method for the determination of the total halide concentration in sea water by means of potentiometric titration. *Anal. Chim. Acta*, **52**, 491–499.

Janz, G. J. (1961). Silver–silver halide electrodes. In *Reference Electrodes, Theory and Practice* (Eds. D. J. G. Ives and G. J. Janz). Academic Press, New York, pp. 179–230.

Jasinski, R., I. Trachtenberg, and D. Andrychuk (1974). Potentiometric measurement of copper in sea water with ion selective electrodes. *Anal. Chem.*, **46**, 364–369.

Johnson, K. S., R. Voll, C. A. Curtis, and R. M. Pytkowicz (1977). A critical examination of the NBS pH scale and the determination of titration alkalinity. *Deep-Sea Res.*, **24**, 915–926.

Johnson, K. S., and R. M. Pytkowicz (1978). Chloride ion pairs in seawater. *Trans. Am. Geophys. Un.*, **59**, 307.

Kester, D. R. (1970). Ion association of sodium, magnesium, and calcium with sulfate in aqueous solutions. *PhD Thesis*, Oregon State University, Corvallis, Ore., 116 pp.

Kester, D. R., and R. M. Pytkowicz (1968). Magnesium sulfate association at 25°C in synthetic seawater. *Limnol. Oceanogr.*, **13**, 670–674.

Kester, D. R., and R. M. Pytkowicz (1969). Sodium, magnesium, and calcium sulfate ion-pairs in seawater at 25°C. *Limnol. Oceanogr.*, **14**, 686–692.

Kester, D. R., and R. M. Pytkowicz (1970). Effect of temperature and pressure on sulfate ion association in sea water. *Geochim. Cosmochim. Acta*, **34**, 1039–1051.

Khoo, K. H., R. W. Ramette, C. H. Culberson, and R. G. Bates (1977a). Determination tion of hydrogen ion concentrations in seawater from 5 to 40°C: standard potentials at salinities from 20 to 45‰. *Anal. Chem.*, **49**, 29–34.

Khoo, K. H., C. H. Culberson, and R. G. Bates (1977b). Thermodynamics of the dissociation of ammonium ion in seawater from 5 to 40°C. *J. Solution Chem.*, **6**, 281–290.

King, E. J., and J. E. Prue (1961). Precise measurements with the glass electrode, of the ionization constants of benzoic, phenylacetic, and β-phenyl-o-propionic acids at 25°C. *J. Chem. Soc.*, (1961) 275–279.

Koroleff, F. (1965). The results of alkalinity intercalibration measurements in Copenhagen. *UNESCO Tech. Pap. Mar. Sci.*, No. 3, Report on the intercalibration measurements in Copenhagen, 9–13 June 1965.

Koroleff, F. (1966). Intercalibration of alkalinity, Copenhagen, September 1966. *UNESCO Tech. Pap. Mar. Sci.*, No. 9, Report on intercalibration measurements, Leningrad, 24–28 May 1966; Copenhagen, September, 1966.

Koryta, J. (1977). Theory and applications of ion-selective electrodes. Part II. *Anal. Chim. Acta*, **91**, 1–85.

Kosov, A. Y., P. D. Novikov, O. T. Krylov, M. P. Nesterova, and A. F. Vitvinova (1977). Potentiometric determination of fluorine in sea water. *Oceanology*, **16**, 467–469.

Lakshminarayanaiah, M. (1976). *Membrane Electrodes*. Academic Press, New York, 368 pp.

Lambert, R. B., Jr., D. R. Kester, M. E. Q. Pilson, and K. E. Kenyon (1973). *In situ* dissolved oxygen measurements in the North and West Atlantic Ocean. *J. Geophys. Res.*, **78**, 1479–1483.

LaMer, V. K., and W. G. Parks (1933). The partial and integral heats of dilution of cadmium sulfate solutions from electromotive force measurements. *J. Am. Chem. Soc.*, **55**, 4343–4355.

Leyendekkers, J. V., and M. Whitfield (1971). Liquid ion exchange electrodes in mixed electrolyte solutions. *Anal. Chem.*, **43**, 322–327.

Lilley, T. H., and C. C. Briggs (1976). Activity coefficients of calcium sulfate in water at 25°C. *Proc. R. Soc. London A*, **349**, 355–368.

Lyman, J. (1956). Buffer mechanism of sea water. *PhD Thesis*, University of California, Los Angeles, Calif., 196 pp.

McBryde, W. A. E. (1969). The pH meter as a hydrogen-ion concentration probe. *Analyst*, **94**, 337–346.

McGee, K. A., and P. B. Hostetler (1973). Stability constants for $MgOH^+$ and brucite below 100°C. *Trans. Am. Geophys. Un.*, **54**, 487.

MacInnes, D. A. (1961). *The Principles of Electrochemistry*. Dover Publications, New York, 478 pp.

MacIntyre, F. (1976). Concentration scales: a plea for physico-chemical data. *Mar. Chem.*, **4**, 205–224.

Mantyla, A. W., and J. L. Reid (1978). Measurements of water characteristics at depths greater than 10 km in the Marianas Trench. *Deep-Sea Res.*, **25**, 169–173.

Mascini, M. (1973). Titration of sulfate in mineral waters and sea water using the solid state lead electrode. *Analyst*, **98**, 325–328.

Mehrbach, C., C. H. Culberson, J. E. Hawley, and R. M. Pytkowicz (1973). Measurement of the apparent dissociation constants of carbonic acid in seawater at atmospheric pressure. *Limnol. Oceanogr.*, **18**, 897–907.

Merks, A. G. A. (1975). Determination of ammonia in sea water with an ion-selective electrode. *Neth. J. Sea Res.*, **9**, 371–375.

Miller, G. R. (1974). Fluoride in sea water: distribution in the North Atlantic Ocean anf formation of ion pairs with sodium. *PhD Thesis*, University of Rhode Island, Kingston, R.I., 117 pp.

Miller, G. R., and D. R. Kester (1976). Sodium fluoride ion-pairs in seawater. *Mar. Chem.*, **4**, 67–82.

Milward, A. F. (1969). Effects of light on glass pH electrodes. *Analyst*, **94**, 154–155.

Moody, G. J., R. B. Oke, and J. D. R. Thomas (1969). The influence of light on silver–silver chloride electrodes. *Analyst*, **94**, 803–804.

Moody, G. J., and J. D. R. Thomas (1971). *Selective Ion Sensitive Electrodes*. Merrow Publishing Co., Watford, Herts., England, 140 pp.

Mor, E., V. Scotto, G. Marcenaro, and G. Alabiso (1975). The use of membrane electrodes in the determination of sulfides in seawater. *Anal. Chim. Acta*, **75**, 159–167.

Oglesby, G. B., W. C. Duer, and F. J. Millero (1977). Effect of chloride ion and ionic strength on the response of a copper(II) ion-selective electrode. *Anal. Chem.*, **49**, 877–879.

Park, K. (1966). Surface pH of the northeastern Pacific ocean. *J. Oceanol. Soc. Korea*, **1**, 1–6.

Park, P. K. (1968). Alteration of alkalinity, pH, and salinity by metallic water samplers. *Deep-Sea Res.*, **15**, 721–722.

Paulson, A. J. (1978). Potentiometric studies of cupric hydroxide complexation. *MS Thesis*, University of Rhode Island, Kingston, R.I., 102 pp.

Pinching, G. D., and R. G. Bates (1946). Purification of sodium chloride and potassium chloride for use in electrochemical work, and the determination of bromide. *J. Res. Nat. Bur. Stand.*, **37**, 311–319.

Pitzer, K. S., and G. Mayorga (1973). Thermodynamics of electrolytes. II. Activity and osmotic coefficients for strong electrolytes with one or both ions univalent. *J. Phys. Chem.*, **77**, 2300–2308.

Pitzer, K. S., and J. J. Kim (1974). Thermodynamics of electrolytes. IV. Activity and osmotic coefficients for mixed electrolytes. *J. Am. Chem. Soc.*, **96**, 5701–5707.

Platford, R. F. (1965a). The activity coefficient of sodium chloride in sea water. *J. Mar. Res.*, **23**, 55–62.

Platford, R. F. (1965b). Activity coefficient of the sodium ion in sea water. *J. Fish. Res. Bd. Can.*, **22**, 885–889.

Platford, R. F., and T. Dafoe (1965). The activity coefficient of sodium sulfate in sea water. *J. Mar. Res.*, **23**, 63–68.

Pytkowicz, R. M., D. R. Kester, and B. C. Burgener (1966). Reproducibility of pH measurements in sea water. *Limnol. Oceanogr.*, **11**, 417–419.

Pytkowicz, R. M., and R. Gates (1968). Magnesium sulfate interactions in sea water from solubility measurements. *Science*, **161**, 690–691.

Pytkowicz, R. M., and D. R. Kester (1969). Harned's rule behavior of $NaCl–Na_2SO_4$ solutions explained by an ion association model. *Am. J. Sci.*, 217–229.

Pytkowicz, R. M., and J. E. Hawley (1974). Bicarbonate and carbonate ion-pairs and a model of seawater at 25°C. *Limnol. Oceanogr.*, **19**, 223–234.

Ramette, R. W., C. H. Culberson, and R. G. Bates (1977). Acid–base properties of tris(hydroxymethyl)aminomethane buffers in sea water from 5 to 40°C. *Anal. Chem.*, **49**, 867–870.

Rechnitz, G. A., and Z. F. Lin (1968). Potentiometric measurements with calcium-selective liquid–liquid membrane electrodes. *Anal. Chem.*, **40**, 696–699.

Reeburgh, W. S., and J. H. Carpenter (1964). Determination of chlorinity using a differential potentiometric end point. *Limnol. Oceanogr.*, **9**, 589–591.

Reid, J. L., Jr. (1965). *Intermediate Waters of the Pacific Ocean*. Johns Hopkins Press, Baltimore, 85 pp.

Rice, G. K., and R. J. Jasinski (1976). Monitoring dissolved copper in sea water by means of ion-selective electrodes. In *Accuracy in Trace Analysis: Sampling, Sample Handling, Analysis*, Vol. 2 (Ed. P. D. LaFluer). National Bureau of Standards Special Publication 422, US Government Printing Office, Washington, D.C., pp. 899–915.

Riddick, J. A. (1961). Amine buffers as acidimetric standards. *Ann. N. Y. Acad. Sci.*, **92**, 357–365.

Riley, J. P. (1965). Analytical chemistry of sea water. In *Chemical Oceanography*, Vol. 2 (Eds. J. P. Riley and G. Skirrow). Academic Press, London, pp. 295–424.

Rix, C. J., A. M. Bond, and J. D. Smith (1976). Direct determination of fluoride in sea water with a fluoride selective ion electrode by a method of standard additions. *Anal. Chem.*, **48**, 1236–1239.

Robinson, R. A., and R. H. Stokes (1968). *Electrolyte Solutions*. Butterworths, London, 571 pp.

Robinson, R. A., and R. G. Bates (1978). Ionic activity coefficients in aqueous mixtures of NaCl and $MgCl_2$. *Mar. Chem.*, **6**, 327–333.

Ross, J. W., Jr. (1969). Solid-state and liquid membrane ion-selective electrodes. In *Ion-selective Electrodes*. (Eds. R. A. Durst). National Bureau of Standards Special Publication 314, US Government Printing Office, Washington, D.C., pp. 57–88.

Savenko, V. S. (1977a). Potassium determinations in sea water with aid of membrane selective electrode. *Oceanology*, **17**, 1123–1127.

Savenko, V. S. (1977b). Utilization of bromide ion-selective electrodes in geochemical investigations. *Oceanology*, **16**, 473–476.

Savenko, V. S. (1977c). The dissociation of hydrogen sulfide in sea water. *Oceanology*, **16**, 347–350.

Savenko, V. S. (1977d). Use of sodium-selective electrodes in marine geochemical studies. *Oceanology*, **17**, 43–45.

Shokes, R. F., P. K. Trabant, B. J. Presley, and B. F. Reid (1977). Anoxic, hypersaline basin in the northern Gulf of Mexico. *Science*, **196**, 1443–1446.

Skirrow, G. (1975). The Dissolved Gases—Carbon Dioxide. In *Chemical Oceanography*, Vol. 2, 2nd ed. (Eds. J. P. Riley and G. Skirrow). Academic Press, London, pp. 1–192.

Smith, S. V. (1973). Carbon dioxide dynamics: a record of organic carbon production, respiration, and calcification in the Eniwetok reef flat community. *Limnol. Oceanogr.*, **18**, 106–120.

Srna, R. (1974). The use of ion specific electrodes for chemical monitoring of marine systems. Part II. A rapid method of detecting changes in the relative concentration of chloride and divalent ions in seawater. *Coll. Mar. Stud. Univ. Delaware Report.*, DEL-SG-11-74, 27 pp.

Srna, R. F., C. Epifanio, M. Hartman, G. Pruder, and A. Stubbs (1973). The use of ion specific electrodes for chemical monitoring of marine systems. I. The ammonia electrode as a sensitive water quality indicator probe for recirculating mariculture systems. *Coll. Mar. Stud. Univ. Delaware Rep.*, No. DEL-SG-14-73.

Strickland, J. D. H., and T. R. Parsons (1965). A manual of sea water analysis. *Fish. Res. Bd. Can. Bull.*, No. 125.

Sunda, W. G., D. W. Engel, and R. M. Thuotte (1978). Effect of chemical speciation in toxicity of cadmium to grass shrimp, *Palaemonetes pugio*: importance of free cadmium ion. *Environ. Sci. Technol.*, **12**, 409–413.

Takahashi, T., R. F. Weiss, C. H. Culberson, J. M. Edmond, D. E. Hammond, C. S. Wong, Y. H. Li, and A. E. Bainbridge (1970). A carbonate chemistry profile at the 1969 Geosecs intercalibration station in the eastern Pacific Ocean. *J. Geophys. Res.*, **75**, 7648–7666.

Thompson, M. E. (1966). Magnesium in sea water: an electrode measurement. *Science*, **153**, 866–867.

Thompson, M. E., and J. W. Ross (1966). Calcium in sea water by electrode measurement. *Science*, **154**, 1643–1644.

Warner, T. B. (1969). Lanthanum fluoride electrode response in water and in sodium chloride. *Anal. Chem.*, **41**, 527–529.

Warner, T. B. (1971). Normal fluoride content of sea water. *Deep-Sea Res.*, **18**, 1255–1263.

Warner, T. B. (1973). Fluoride analysis in sea water and in other complex natural waters using an ion-selective electrode—techniques, potentialities, limitations. In *Chemical Analysis of the Environment and Other Modern Techniques* (Eds. S. Ahuja, E. M. Cohen, T. J. Kneip, J. L. Lambert, and G. Zweig). Plenum Press, New York, pp. 229–240.

Warner, T. B., and D. J. Bressan (1973). Direct measurement of less than 1 part-per-billion fluoride in rain, fog, and aerosols with an ion-selective electrode. *Anal. Chim. Acta*, **63**, 165–173.

Warner, T. B., M. M. Jones, G. R. Miller, and D. R. Kester (1975). Fluoride in sea water: intercalibration study based on electrometric and spectrophotometric methods. *Anal. Chim. Acta*, **77**, 223–228.

Weiss, R. F. (1970). The solubility of nitrogen, oxygen, and argon in water and sea water. *Deep-Sea Res.*, **17**, 721–735.

Weiss, R. F. (1974). Carbon dioxide in water and seawater: the solubility of a non-ideal gas. *Mar. Chem.*, **2**, 203–215.

Westall, J. C., F. M. M. Morel, and D. N. Hume (1977). Measurement of copper in saline waters by ion selective electrode. *Trans. Am. Geophys. Un.*, **50**, 1156.

Weyl, P. K. (1961). The carbonate saturometer. *J. Geol.*, **69**, 32–44.

Whitfield, M. (1969). Multicell assemblies for studying ion-selective electrodes at high pressures. *J. Electrochem. Soc.*, **116**, 1042–1046.

Whitfield, M. (1970). The effect of membrane geometry on the performance of glass-electrode cells. *Electrochim. Acta*, **15**, 83–96.

Whitfield, M. (1971). *Ion Selective Electrodes for the Analysis of Natural Waters.* AMSA Handbook No. 2, Australian Marine Sciences Association, Sydney, 130 pp.

Whitfield, M. (1972). Self-ionization of water in dilute sodium chloride solutions from 5–35°C and 1–2000 bars. *J. Chem. Eng. Data*, **17**, 124–128.

Whitfield, M. (1974a). A comprehensive specific interaction model for sea water—calculation of the osmotic coefficient. *Deep-Sea Res.*, **21**, 57–67.

Whitfield, M. (1974b). The hydrolysis of ammonium ions in sea water—a theoretical study. *J. Mar. Biol. Ass. U.K.*, **54**, 565–580.

Whitfield, M. (1975). The electroanalytical chemistry of sea water. In *Chemical Oceanography*, Vol. 4, 2nd ed. (Eds. J. P. Riley and G. Skirrow). Academic Press, London, pp. 1–154.

Whitfield, M., and J. V. Leyendekkers (1970). Selectivity characteristics of a calcium selective ion exchange electrode in the system calcium(II)–sodium(I)–chloride(I)–water. *Anal. Chem.*, **42**, 444–448.

Wilde, P., and P. W. Rodgers (1970). Electrochemical meter for activity measurements in natural environments. *Rev. Sci. Instrum.*, **41**, 356–359.

Wyatt, B., W. Gilbert, L. Gordon, and D. Barstow (1970). *Hydrographic Data from*

Oregon Waters 1969. Data Report No. 42, Dept. Oceanography, Oregon State Univ., Corvallis, Ore., 155 pp.

Zhelesnova, A. A. (1962). On the suspension effect in connection with the pH determination of sea sediments. *Tr. Inst. Okeanol. Akad. Nauk SSSR*, **54,** 83–99.

Zhelesnova, A. A. (1964). The suspension effect in pH determination of sea sediments. *Tr. Inst. Okeanol. Akad. Nauk SSSR*, **67,** 135–140.

Zielen, A. J. (1963). The elimination of liquid junction potentials with the glass electrode. *J. Phys. Chem.*, **67,** 1474–1479.

Zirino, A. (1975). Measurement of the apparent pH of sea water with a combination microelectrode. *Limnol. Oceanogr.*, **20,** 654–657.

Zirino, A., and S. Yamamoto (1972). A pH dependent model for the chemical speciation of copper, zinc, cadmium, and lead in sea water. *Limnol. Oceanogr.*, **17,** 661–671.

Marine Electrochemistry
Edited by M. Whitfield and D. Jagner
© 1981 John Wiley & Sons Ltd.

DANIEL JAGNER
Department of Analytical and Marine Chemistry,
University of Göteborg,
S-41296 Göteborg, Sweden

7

Potentiometric Titrations

GLOSSARY OF SYMBOLS

A_c	Carbonate alkalinity (equation 58)
A_t	Total alkalinity (equation 33)
C_T	Total inorganic carbon content (equation 53)
F_1, F_2	Gran functions (equations 19 and 23)
f_M	Conventional single-ion activity coefficient of M on the molar scale (equation 3)
g	Nernstian slope factor ($RT \ln 10/F$, equation 2)
K	Stoichiometric stability constant (equation 5)
K_{X_i}	Selectivity coefficient (equation 24)

M_t	Molarity of titrant (equation 12)
v	Incremental volume (equation 12)
v_{eq}	Equivalence volume (equation 12)
v_0	Starting volume (equation 12)
[M]	Concentration of M in $mol\,l^{-1}$ (equation 3)
β_{MX}	Stepwise formation constant for MX (equation 61)

7.1. INTRODUCTION

With the exception of gravimetric analysis, titration is the oldest analytical procedure. Owing to its well established theoretical background, its inherent high accuracy and precision, and the ease with which it can be automated, the titration technique is still frequently used to solve many analytical problems. Furthermore, titration procedures are often used to determine thermodynamic constants in solution.

In its analytical application a titration is based on the monitoring of the free concentration of one or several species during the successive addition of B in the titration reaction

$$n\,A + m\,B \rightleftharpoons A_n B_m \tag{1}$$

where the integers n and m are known. A is the component to be determined and B is a reagent forming either a very insoluble compound with A or a soluble compound with a high stability constant. Moreover, the reagent B should preferably be selective towards A so that other components present in the sample do not interfere with the main titration reaction. The combination of high stability or insolubility and chemical selectivity is, however, seldom met with in practice.

In order to be able to determine the equivalence volume in reaction 1, either the increase in concentration of B or $A_n B_m$ or the decrease in concentration of A must be monitored during the titration. A great number of different analytical techniques, e.g. photometry, conductometry, and voltammetry, have been exploited for this purpose. Potentiometric sensors provide unique advantages as titration monitors in that they are normally capable of measuring variations in activity over a large range (5–10 decades). A drawback of potentiometric sensors however, is that they provide a logarithmic rather than a linear signal. This complicates the calculation of the equivalence volume. On the other hand, experimental data from the whole titration curve can be exploited in the evaluation, thus increasing precision and accuracy.

7.2. POTENTIOMETRIC TITRATION *versus* SINGLE-POINT MEASUREMENT

For the potentiometric titration of a sea water component to be feasible there must exist either a potentiometric sensor for this component or a

Table 1. Relative precision (%) in concentration measurement by means of single-point potentiometry

Reproducibility of potential, mV	Precision of concentration, %	
	Monovalent ion	Divalent ion
±0.5	±2	±7.5
±2	±4	±15

sensor selective for the reagent used during the titration, e.g. a silver-selective electrode in the titration of sea water chlorinity. If there exists a component-selective potentiometric sensor there are two ways in which the component can be determined, namely titration, and single-point measurement with a pre-calibrated electrode. There is, however, a fundamental difference between the results obtained by titration and those obtained by single-point measurement in that they measure total and free concentrations, respectively. Only if the free concentration of the particular component is equal to the total concentration are the two techniques likely to give the same result. Theoretically however, it is possible to determine the total concentration of a component by means of single-point measurement provided that the sensor electrode is calibrated, by plotting electrode e.m.f. *versus* known total concentration of the component, for a solution of exactly the same composition as that of the sample. Even though it is possible to prepare accurate synthetic sea water samples, titration procedures are normally to be preferred in connection with total concentration determinations owing to their superior precision. The precisions of potentiometric single-point measurements, assuming reproducibility of potential of ±0.5 and ±2 mV, are shown in Table 1.

7.3. STANDARD ADDITION TITRATIONS

Most potentiometric sensors are capable of measuring free concentrations down to 10^{-10} M. Owing mainly to electrode kinetics, they are, however, not able to do so unless the total concentration of the component is 10^{-6} M or higher. This limits the number of sea water components that can be determined by potentiometric titration methods. Thus, no sea water component with a total concentration less than 10^{-6} M can be determined by the addition of a titrant (see Chapter 1, Table 1).

Sea water components with total concentrations in the 10^{-6}–10^{-4} M range can, however, be determined by a standard addition procedure. In this procedure the titrant contains a known concentration of the component to be determined. The change in sensor electrode potential caused by the addition of titrant is measured. For an ion, A^{n+}, the potential of the electrode couple, E mV, is

$$E = E' + (g/n) \log a_{A^{n+}} + E_j \tag{2}$$

where $g = RT \ln 10/F$ ($= 59.16$ mV per decade at 25°C) and E_j is the sum of the liquid junction potentials. Since the ionic composition is not altered appreciably by the addition of titrant to a sea water sample, equation 2 can be written as

$$E = E_1 + (g/n) \log [A^{n+}] \tag{3}$$

where $E_1 = E' + E_j + (g/n) \log f_A$. From equation 3 it can be seen that, if the concentration of the monovalent ion A^+ is doubled at 25°C, the electrode couple potential will increase by approximately 18 mV since $g = 59.16$ mV. Provided that the dilution caused by the titrant can be neglected, a rough estimate of the total concentration of A^+ can thus be obtained by adding A^+ titrant until the electrode potential has increased by 18 mV. For a divalent ion, A^{2+}, the potential increase should be approximately 9 mV. If a more accurate result is required, several standard additions must be performed and the results evaluated by the Gran (1952) linearization method described below (Section 7.6.3).

It should be emphasized that the standard addition titration procedure normally yields the total concentration of the component under consideration, irrespective of whether the component is complexed by other sample components or not. Assuming that A^+ is complexed by the sample component L^{2-} according to

$$A^+ + L^{2-} \rightleftharpoons AL^- \tag{4}$$

then

$$\frac{[AL^-]}{[A^+]} = K[L^{2-}] \tag{5}$$

where K is the stoichiometric stability constant. Assuming, moreover, that the total concentration of L^{2-} is considerably greater than the total concentration of A^+, so that a titrant addition of A^+ does not affect the concentration of L^{2-} to any appreciable extent, either by reaction or by dilution, equation 3 can be written as

$$E = E_1 + (g/n) \log [A^+]_{tot} - (g/n) \log (1 + K[L^{2-}]) \tag{6}$$

where $[A^+]_{tot} = [A^+] + [AL^-]$ or, by analogy with equation 3,

$$E = E_2 + (g/n) \log [A^+]_{tot} \tag{7}$$

where $E_2 = E_1 - (g/n) \log (1 + K[L^{2-}])$.

The standard addition titration technique should thus be used with great care. It is suitable only in cases where the component is non-complexed or where one has a previous knowledge that the dominant complexing agent is present in at least a 50-fold excess of the component to be titrated. If the total concentrations of the component and the complexing agent are of the same order of magnitude, the standard addition titration will yield a result somewhere between the free and the total concentration of the component.

7.4. POTENTIOMETRIC TITRATION FOLLOWING PRE-CONCENTRATION

In order to determine components with total concentrations below 10^{-4} M by means of potentiometric titration, such components must be pre-concentrated prior to titration. This may be accomplished, for example, by ion exchange, solvent extraction or coprecipitation. Few such procedures have, however, found any application in the analysis of saline waters. The only potentiometric titration method involving pre-concentration is that based on potentiometric stripping analysis (Jagner and Årén, 1978). In this technique the trace metal analytes are pre-concentrated by potentiostatic reduction and amalgamation in a mercury film on a glassy carbon working electrode surface, according to

$$M(II) \xrightarrow{2e^-} M(Hg) \tag{8}$$

in a de-oxygenated sample containing 1–20 mg dm^{-3} of Hg(II) (*cf.* anodic stripping voltammetry, Chapter 10). After pre-concentration, the potentiostatic circuitry is disconnected and the reduced elements are titrated by the mercury (II) ions:

$$M(Hg) \xrightarrow{Hg(II)} M(II) + Hg \tag{9}$$

which are transported at constant rate to the electrode surface by means of controlled stirring of the sample. If the electrode potential is measured during the re-oxidation of the amalgamated metals, a potentiometric titration curve, E mV $vs.$ time, will be obtained. A potentiometric stripping curve, obtained after 45 min of potentiostatic pre-concentration at -1.25 V $vs.$ SCE in a Baltic sea water sample containing approximately 0.8 μg dm^{-3} of zinc (II); 0.2 μg dm^{-3} of cadmium (II) and 1.0 μg dm^{-3} of lead (II), is shown in Figure 1 (Jagner and Årén, 1979).

The basic instrumentation for potentiometric stripping analysis is simple in the sense that an instrument can be assembled from standard laboratory equipment (Jagner, 1978). Figure 2 shows a block diagram of the main parts of a simple instrument for potentiometric stripping analysis. A computerized potentiometric stripping analyser, which can be used for trace metal determinations in non-deoxygenated samples, has been described in Chapter 4.

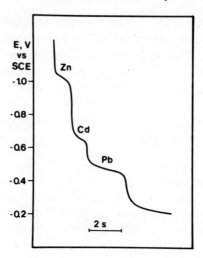

Figure 1. Potentiometric stripping curve obtained after 45 min of potentiostatic pre-concentration in an acidified Baltic sea water sample containing $0.8\ \mu g\,dm^{-3}$ of zinc (II), $0.2\ \mu g\,dm^{-3}$ of cadmium (II), and $1.0\ \mu g\,dm^{-3}$ of lead (II)

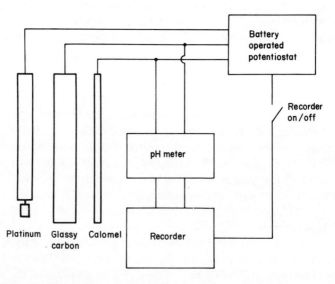

Figure 2. Block diagram of the main parts of a simple instrument for potentiometric stripping analysis

7.5. MAIN TYPES OF AUTOMATIC TITRATIONS

The basic equipment required for a potentiometric titration consists of a burette, a high-impedance input millivoltmeter and a suitable electrode couple. Since manual registration of a potentiometric titration curve is time consuming, titrators were the first analytical instrumentation to be automated by modern electronic circuitry. The first semi-automatic potentiometric titration was described more than 30 years ago (see discussion in Squirrell, 1964).

Automatic titrators have been constructed to give either continuous or incremental addition of titrant. Although the latter is more easily adapted to modern digital electronics, the first-mentioned type is still that most frequently used. Fully computerized titrators allowing the user to write his own evaluation program, as described in Chapter 4, are not yet available commercially.

7.5.1. Continuous addition of titrant

Automatic titrators operating according to the principle of continuous addition of titrant are less suitable in connection with marine applications, where the utmost accuracy and precision are normally required. The main reason for this is that the e.m.f. of the sensor electrode does not represent an equilibrium value, owing to time lags in titrant mixing, chemical reactions, and sensor response. For this reason the evaluation of the equivalence volume has to be performed according to empirical methods rather than according to titration theory. Furthermore, such titrators can hardly be used for the determination of equilibrium constants.

7.5.2. Stepwise addition of titrant

The introduction of digital electronic circuitory has made possible the construction of so-called 'digital titrators'. In these instruments titrant increments, of pre-programmable magnitude, are added stepwise to the titration vessel (Johansson and Pehrsson, 1970). After each addition the titrator waits for an equilibrium potential reading, which is then transferred to either a strip-chart recorder or a digital printout, e.g. a teletype. The criteria for equilibrium are, however, often chosen arbitrarily. In most cases it is assumed that an equilibrium value has been reached 10–20 s after a titrant increment. Even though this may not always be true, titrators based on the stepwise addition of titrant are required in almost all marine applications of potentiometric titrations. Only by stepwise addition of titrant is it possible to exploit linearization computer programs to evaluate the equivalence point (Section 7.6). Incremental addition of titrant is necessary in connection with multiple standard addition titrations in which the Gran (1952) linearization method is used for evaluation.

7.5.3. The inverse burette

In many titration applications buffer agents and masking substances have to be added to the sample prior to titration. These additions can be automated by the use of several syringe burettes. Alternatively, an inverse burette, as shown in Figure 3, can be used as titrator. In this titrator, sample, reagents, and titrant are sucked successively into the burette syringe through magnetic valves (Granéli and Anfält, 1976). Mixing is achieved by means of a PTFE-coated magnetic ball. In the titrator shown in Figure 3 five inlet or outlet channels have been incorporated. One channel is used for the introduction of the sample and one for the stepwise addition of titrant, leaving two channels for the addition of masking and buffering agents and one channel for the emptying of the syringe after titration.

In addition to being easy to automate and low in cost, the inverse burette offers the advantage of gas-tight operation. Furthermore, its size and shape

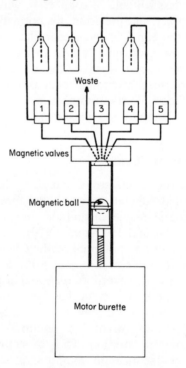

Figure 3. The inverse burette

make it very suitable for operation on board ship. Its main disadvantage is that it is not well suited for potentiometric applications since it is difficult to introduce the electrode couple into the syringe. For this reason the inverse burette has so far been used only for the photometric titration of sea water alkalinity (Granéli and Anfält, 1976).

7.6. METHODS OF EVALUATING POTENTIOMETRIC TITRATION CURVES

Titration theory is well developed and the errors inherent in different ways of evaluating potentiometric titration curves have been investigated both practically and theoretically (e.g. Meites and Meites, 1967; Anfält and Jagner, 1971b). The various methods used differ mainly in the number of experimental points which are exploited in the evaluation procedure. In titrations to a pre-chosen sensor electrode potential only one experimental point is used, while derivative methods exploit a number of experimental data points in the vicinity of the equivalence volume. Gran (1952) linearization methods often exploit all experimental data collected during the titration.

7.6.1. Titration to a pre-selected sensor electrode potential ('dead-stop' titrations)

Titration to a pre-selected electrode potential is the classical way of evaluating analytical potentiometric titration curves. The equivalence volume electrode potential is determined experimentally by titrating a standard sample of known composition. The procedure is illustrated in Figure 4, which shows the potentiometric titration of 24.90 cm^3 of a Baltic sea water of 5.3‰ salinity using 0.1001 M silver nitrate as titrant. A silver wire was used as sensor which, in combination with a suitable reference electrode, measured the electrode couple potential as

$$E = E_1 + g \log [Ag^+] \tag{10}$$

The main titration reaction before the equivalence point was

$$Ag^+ + X^- \rightleftharpoons AgX(s) \quad (X = Cl^-, Br^-) \tag{11}$$

As can be seen from Figure 4, the very low solubility of silver halides causes the potential of the sensor electrode to change very rapidly in the vicinity of the equivalence point (Dyrssen *et al.*, 1966). It would seem as if only an approximate knowledge of sensor potential at the equivalence point, i.e. ±10 mV, would be necessary to locate the equivalence point with satisfactory precision. In practice it is, however, very difficult and time consuming to register experimental points very close to the equivalence point. The sensor potential is not stable and extremely small increments of titrant must be

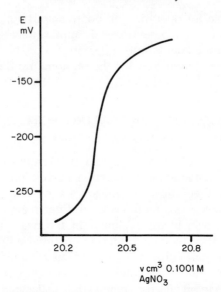

Figure 4. Potentiometric titration of 24.90 cm^3 of a 5.3‰ salinity Baltic sea water sample with silver nitrate, using a silver wire as sensor electrode

added. This evaluation method is thus suitable only if maximum precision is not sought.

7.6.2. Derivative methods

In derivative methods it is necessary to use stepwise addition of titrant. Normally the ratio $\Delta E/\Delta v$ is plotted against v cm^3 of titrant added and the maximum is located graphically. This is shown in Figure 5, where the derivative method has been used to evaluate the potentiometric data of Figure 4. The derivative technique is capable of locating the equivalence point with high precision. Furthermore, experimental data in regions remote from the equivalence point, where stable sensor potentials are easier to obtain, can be exploited.

7.6.3. The Gran linearization method

In order to obtain maximum precision in an analytical potentiometric titration, all experimental data should be exploited in the evaluation process. Since, however, there is always a logarithmic relationship between the concentration of the component measured and the sensor signal, use of all experimental data would normally have to involve an antilogarithmic procedure. Alternatively, a powerful minimization program such as LETAGROP

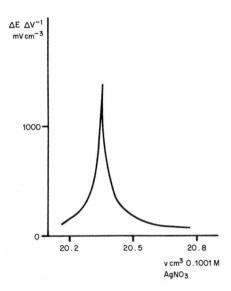

Figure 5. The derivative method used on the primary potential data shown in Figure 4

(Sillén and Warnqvist, 1969), operating on the primary E mV and v cm^3 of titrant data, could be used. Owing to the complexity of such programs and to the cost of operating them, they have hitherto only been used for the evaluation of stability constants from potentiometric data.

All titration curves can be linearized, although in some cases the linearization functions can be complicated (Anfält *et al.*, 1968). This is the principle of the Gran (1952) method, which is frequently exploited in marine applications. In its simplest form the derivation of Gran plots is based on the assumption that there exists a dominant main reaction in each part of a potentiometric titration curve. The first step in the derivation of the Gran plot is to identify this main reaction, e.g. for the data shown in Figure 4 the main reaction before the equivalence point is

$$Ag^+ + X^- \rightleftharpoons AgX(s) \qquad (X = Cl^-, Br^-) \qquad (11)$$

The second step is to set up the mass balance for the components involved in the titration reaction. With reference to the data in Figure 4, the mass balance before equivalence can be expressed as

$$(v_0 + v)[X^-] = (v_{eq} - v)M_t \qquad (12)$$

where v_{eq} is the equivalence volume, v_0 denotes the initial volume of the sample, M_t the molarity of the titrant and $[X^-]$ the halide concentration at this point in the titration. In other words, the total amount of untitrated halide, $(v_0 + v)[X^-]$ mmol, is equal to the total amount of halide,

$v_{eq}M_t$ mmol, minus the amount of halide which has already been titrated, vM_t mmol.

The third step is to formulate the equilibrium equation

$$[Ag^+][X^-] = K \tag{13}$$

where K is a solubility constant. Since Ag^+ is the ion to which the sensor electrode responds, the fourth step of the derivation is to replace $[X^-]$ in equation 12 with $[Ag^+]$ using equation 13. Thus

$$K(v_0 + v)[Ag^+]^{-1} = (v_{eq} - v)M_t \tag{14}$$

The final step in the derivation of the Gran plot is to replace the concentration of the ion measured by the experimental e.m.f. values, E mV. Since

$$E = E' + g \log a_{Ag^+}$$

or, assuming no change in the liquid junction potentials, reference electrode potential, or activity coefficient for silver(I),

$$E = E_1 + g \log [Ag^+] \tag{15}$$

the proportionality between the free silver ion concentration and the e.m.f. values may be expressed as

$$[Ag^+] \propto 10^{E/g} \tag{16}$$

or

$$[Ag^+] \propto 10^{(E-E_1)/g} \tag{17}$$

where the constant E_1 can be chosen arbitrarily. Consequently, by combining equations 14 and 17 we find that

$$(v_0 + v)10^{(E_1-E)/g} \propto (v_{eq} - v)M_t \tag{18}$$

Consequently, if the function F_1,

$$F_1 = (v_0 + v)10^{(E_1-E)/g} \tag{19}$$

is plotted against v cm^3 of titrant added, a straight line will be obtained which intersects the v axis at v_{eq} when F_1 is extrapolated to zero, i.e.

$$v \to v_{eq} \quad \text{as} \quad F_1 \to 0 \tag{20}$$

In this way the equivalence volume can be determined by linear extrapolation using all data for $v < v_{eq}$ and without any previous knowledge of the E_1 value of the measuring electrode couple. In Figure 6 the function F_1 has been applied to the experimental data of Figure 4.

The 'main reaction' after the equivalence point is the increase in the free silver ion concentration due to the addition of excess titrant. The mass balance equation is

$$(v_0 + v)[Ag^+] = (v_{eq} - v)M_t \tag{21}$$

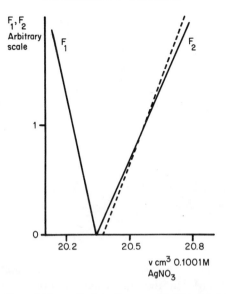

Figure 6. Gran plots F_1 and F_2 applied to the experimental data shown in Figure 4. The dotted line is the F_2 plot calculated for 17°C instead of the correct 22°C value

and by combining equations 17 and 21

$$(v_0 + v)10^{(E-E_1)/g} \propto (v_{eq} - v)M_t \tag{22}$$

Thus, if the function F_2,

$$F_2 = (v_0 + v)10^{(E-E_1)/g} \tag{23}$$

in which the constant E_1 can be assigned any numerical value, is plotted against v cm³ of titrant added, a straight line will be obtained which intersects the v-axis at $v = v_{eq}$ when F_2 is extrapolated to zero. This is illustrated in Figure 6.

The accuracy of the Gran evaluation method depends not only on the validity of the assumption that the reaction considered is dominant, but also on the perfect Nernstian behaviour of the sensor electrode. Even a slight deviation from a Nernstian response may result in serious systematic errors. This is illustrated by the dotted line in Figure 6, where a temperature of 17°C has been used in the function F_2 instead of the true sample temperature of 22°C, i.e. a g-value of 57.57 mV instead of the true value of 58.56 mV.

Very high precision can be obtained by the Gran extrapolation method. In laboratory work and in on-board analysis relative precisions of 0.02% and 0.05%, respectively, can be attained (Jagner and Årén, 1970). In on-board analysis, where samples cannot be weighed, the main limiting factor is the measurement of the amount of sample. Temperature changes, either in the titration vessel, causing an erroneous value of the Nernst factor, or in the titrant storage vessel, causing changes in titrant density and thus its molarity, can also decrease the precision considerably.

7.7. MARINE ANALYTICAL APPLICATIONS

7.7.1. Alkali metals

During the last decade a great number of new membrane electrodes sensitive to different alkali metal ions have been suggested. These electrodes measure the M^+ activity according to

$$E = E' + g \log \left(a_{M^+} + \sum_j K_{X_j} (a_{X_j^{n+}})^{1/n} \right) \qquad (24)$$

where X_j^{n+} is an interfering cation and K_{X_j} is the selectivity coefficient for this ion relative to M^+ in the particular membrane under consideration (Durst, 1969). Table 2 summarizes the membrane composition and the selectivity coefficients for some neutral carrier membranes and the selectivity coefficients for some glass membrane electrodes sensitive to alkali metal ions (Amman *et al.*, 1976).

The total lithium concentration in ocean sea water is approximately 0.02 mM (Riley, 1965). As can be seen from Table 2, it is not possible to measure lithium in sea water with a neutral carrier or a glass membrane, owing to interference from sodium and potassium ions.

At a sea water pH of 8 it would be possible to measure sodium ions, with a total concentration of approximately 0.5 M, potentiometrically. The e.m.f. contribution, E_K mV, due to the presence of potassium of total concentration 0.01 M can be estimated as

$$E_K = 59.16 \log (0.5 + 0.001 \times 0.01) - 59.16 \log 0.5 < 0.001$$

at 25°C, assuming that the free sodium concentration is close to the total concentration. In fact, the sodium concentration could decrease by almost two orders of magnitude before interference from potassium became appreciable. Unfortunately, however, there is no selective titration reagent for sodium. A standard addition titration of diluted sea water samples gives a relative standard deviation of only 1–3%, mainly due to changes in activity coefficients and in the reference liquid junction potential (Jagner, 1977).

Table 2 indicates that it would be possible to measure potassium potentiometrically in sea water. During a titration with, e.g. tetraphenylborate or valinomycin, the potassium concentration would, of course, decrease and the relative contributions from sodium, magnesium, and calcium ions to the

Table 2. Membrane electrodes sensitive to alkali metal cations

Ion	Membrane	Neutral carrier	Logarithm of selectivity coefficient	Logarithm of selectivity coefficient for glass membrane electrode
Li$^+$	Tris(2-ethylhexyl) phosphate in PVC	Li$^+$	$K_H = -0.1$ $K_{Na} = -1.4$ $K_K = -2.2$ $K_{Ca} = -3.3$ $K_{Mg} = -3.7$	$K_{Na} = -0.3$ $K_K = -4.1$
Na$^+$	Dibutyl sebacate in PVC	Na$^+$	$K_H = -0.3$ $K_K = -0.4$ $K_{Mg} = -3.2$ $K_{Ca} = -2.9$	$K_H = 3$ $K_K = -3$
K$^+$	Dinonyl phthalate	Valinomycin in PVC	$K_H = -5.0$ $K_{Na} = -5.6$ $K_{Mg} = -5.2$ $K_{Ca} = -4.7$	$K_H = 0.4$ $K_{Na} = -1.0$

membrane potential would increase. In the vicinity of the equivalence point these contributions to the membrane potential would dominate. Consequently, Anfält and Jagner (1973) attempted to determine potassium in sea water by means of a standard addition procedure using the Gran plot (*cf.* equation 23):

$$F_2 = (v_0 + v)10^{E/g}$$

to evaluate the potassium concentration. A Philips IS 560K potassium electrode with diphenyl ether as membrane solvent was used as sensor. The main difficulty in the standard addition titration of potassium is the change in the Nernst factor of the sensor electrode on operation in sea water owing to penetration of sodium and chloride ions into the membrane. After operation for 10 h in a sea water sample of 12‰ salinity at 22°C the Nernst factor had decreased from 58.7 to 54.2 mV. This effect necessitated frequent re-calibration of the electrode couple and limited the relative precision of the analysis to 2.6%.

The total rubidium and caesium concentrations in sea water are below 10^{-6} M and are not measurable by potentiometric methods.

7.7.2. Alkaline earth metals

The total beryllium and barium concentrations in sea water are below 10^{-6} M and these elements consequently cannot be measured by potentiometric methods.

Great efforts have been made during the last decade to develop a membrane electrode which is selective for magnesium ions. All membranes suggested hitherto, however, exploit liquid ion exchangers as the selective agent (Buck, 1978). This means that serious interferences are to be expected from ions which are closely related chemically to magnesium, such as calcium and zinc. In fact, most membranes suggested for magnesium are equally selective for calcium (Durst, 1969). For this reason these electrodes are often called 'water hardness electrodes'.

Magnesium can, however, be determined by potentiometric titration using hydroxide ions as the reagent and a hydrogen ion sensitive glass electrode as the sensor (Jagner, 1967). The initial step in the analysis is to add a known amount of hydrochloric acid in slight excess of the sample alkalinity and then boil off the carbon dioxide formed, i.e.

$$CO_3^{2-} + HCO_3^- \xrightarrow{H^+} CO_2(g) \tag{25}$$

Mannitol, H_2L, is then added to the sample in order to complex boric acid according to

$$B(OH)_3 + 2H_2L \rightleftharpoons BL_2^- + 3H_2O + H^+ \tag{26}$$

and the pH of the sample is adjusted to 7–8. In this pH region all boric acid has been titrated by the sodium hydroxide and no magnesium hydroxide has been formed. Finally, magnesium is titrated as

$$Mg^{2+} + 2OH^- \rightleftharpoons Mg(OH)_2(s) \tag{27}$$

In sea water this reaction is accompanied by the reaction

$$Mg^{2+} + 2F^- \rightleftharpoons MgF_2(s) \tag{28}$$

or

$$Mg^{2+} + F^- + OH^- \rightleftharpoons Mg(OH)F(s) \tag{29}$$

i.e. the result has to be corrected not only for sample alkalinity but also for the concentration of fluoride. Alkalinity has to be determined by a separate titration but it is sufficient to estimate the fluoride concentration from the sample salinity (see Section 1.3.1). The equivalence volume is determined by

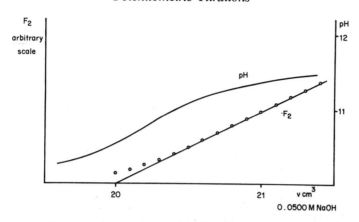

Figure 7. Potentiometric titration of a 100 cm³ sea water sample of 32.6‰ salinity with 0.500 M sodium hydroxide solution. The Gran plot $F_2 = (100+v)10^{(\text{pH}-P)}$ has been used to evaluate the equivalence point

Gran linearization of the experimental data after the equivalence point according to

$$F_2 = (v_0 + v)10^{-E/g}$$

or, if pH is measured,

$$F_2 = (v_0 + v)10^{\text{pH}-P} \tag{30}$$

where P is an arbitrary constant, i.e. no calibration of the glass electrode is necessary. It is, however, essential that the sample temperature is properly adjusted on the pH meter in order to obtain a correct value for the Nernst slope of the glass electrode.

Figure 7 shows the titration curve and the Gran plot for the titration of 100 cm³ of sea water of salinity 32.6‰ with 0.500 M sodium hydroxide.

Calcium is the element which has been most thoroughly investigated with respect to ion-selective membrane electrodes, mostly because of its clinical importance. Table 3 summarizes the characteristics of three major types of

Table 3. Characteristics of some calcium-selective membrane electrodes

Reference	Manufacturer	Solvent/membrane	Selective agent	K_H	K_{Na}	K_{Mg}
Ross (1967)	Orion	Di(n-octylphenyl)phosphate/ cellulose acetate	Liquid ion exchanger	10^5	10^{-3}	0.01
Růžička et al. (1973)	Radiometer	Di(n-octylphenyl)phosphate/ PVC	Liquid ion exchanger	10^4	6×10^{-6}	2×10^{-4}
Amman et al. (1976)	Philips	o-Nitrophenyl octyl ether/ PVC	Neutral carrier	3×10^{-5}	2×10^{-4}	3×10^{-5}

calcium electrodes with respect to membrane composition and selectivity, the selectivity coefficients being expressed as

$$E = E' + (g/2) \log \left(a_{Ca^{2+}} + \sum_j K_{X_j} (a_{X_j^{n+}})^{2/n} \right) \tag{31}$$

Table 3 also specifies manufacturers of these types of electrodes. The commercial product may, however, deviate considerably from that outlined in the original article.

Figure 8 shows the main parts of a calcium-selective membrane electrode as suggested by Ross (1967).

As can be seen from Table 3, the membrane electrodes can measure the calcium concentration of approximately 0.01 M in sea water without any serious interference from the two most likely sources, sodium and magnesium, with total concentrations of 0.5 and 0.05 M, respectively. In a titration situation, the calcium concentration decreases relative to the concentrations of the interfering ions and the interference becomes more serious the closer one approaches the equivalence point. When 90% of the titrant needed for equivalence has been added, the calcium concentration has been decreased by one order of magnitude, or even more if dilution is taken into account. At this calcium concentration the interference from magnesium becomes serious for the Růžička electrode, and very serious for the Ross electrode, while for the Amman neutral carrier electrode sodium is the most serious interferent. Even though the sodium interference can be diminished by dilution of the sea water sample prior to analysis (*cf.* equation 31), it must be concluded that, if calcium is to be titrated potentiometrically, only the first part of the titration, e.g. up to 90% of v_{eq}, can be used to evaluate

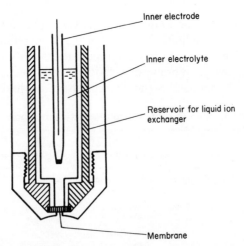

Figure 8. Main parts of a calcium-selective membrane electrode (Ross, 1967)

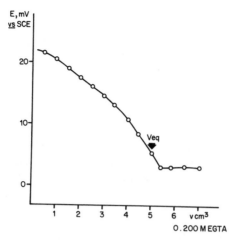

Figure 9. Potentiometric titration of 100 cm³ of Copenhagen Standard Sea Water with EGTA at pH 10 using an Orion membrane electrode as sensor

the equivalence volume. This makes Gran (1952) linearization the only possible method.

Figure 9 shows the titration curve obtained when 100 cm³ of Copenhagen Standard Sea Water are titrated with v cm³ of 0.200 M EGTA [ethylene glycol-2-(2-aminoethyl)tetraacetic acid] at pH 10. The equivalence point has been evaluated with the Gran plot F_1 (Dyrssen *et al.*, 1968):

$$F_1 = (100 + v)10^{2E/g}$$

(*cf*. equation 19). An Orion 92–20 calcium electrode was used as sensor (Ross, 1967). Since only a limited part of the titration curve could be exploited in the evaluation, the relative precision was not better than 2%. Later experiments with the Růžička electrode (Jagner, 1977) showed that this electrode was capable of a relative precision of 1%. This is, however, not as good as the photometric titration procedures.

7.7.3. Group 3B and 4B elements

Of the Group 3B and 4B elements only borate and carbonate species have concentrations above 10^{-6} M and are thus suitable for determination by means of potentiometric titration. Since these species have acid–base properties, titration with a strong acid or base, using a glass pH electrode, is feasible.

Since boric acid is very weak it cannot be titrated directly in sea water. Moreover, the carbonate species would interfere seriously. By the addition of polyhydroxy compounds with 1,2-*cis*-hydroxy groups, such as mannitol (H_2L), boric acid can, however, be made to act as a stronger acid according to the reaction

$$B(OH)_3 + 2H_2L \rightleftharpoons BL_2^- + 3H_2O + H^+ \tag{32}$$

In order to titrate boric acid it is necessary, however, first to remove all carbonate species by adding an excess of hydrochloric acid and boiling off the carbon dioxide thus formed. The excess of acid is then titrated with carbonate-free sodium hydroxide to a pH of approximately 6, and subsequently, mannitol is added to a total concentration of approximately 0.05 M. The boric acid–mannitol complex is then titrated with sodium hydroxide to an equivalence point at approximately pH 8 (*cf.* Gast and Thompson, 1958). The equivalence point can be evaluated by using the Gran plot (*cf.* equation 23):

$$F_2 = (v_0 + v)10^{E/g}$$

for the titration data after the equivalence point. No investigation using the Gran plot to evaluate the equivalence point has, however, hitherto been reported. Owing to the low boric acid concentration and to the difficulties in handling carbonate-free sodium hydroxide, a relative precision better than 1% cannot be expected, however. This is of the same order of magnitude as the spectrophotometric method of Hulthe *et al.* (1970).

The potentiometric titration of sea water carbonate species has been thoroughly investigated. The computer program HALTAFALL (Ingri *et al.*, 1967) has been used to simulate the potentiometric titration curve, v cm^3 of titrant *vs.* E mV, when a sea water sample is titrated with hydrochloric acid using a glass pH electrode as sensor (Hansson and Jagner, 1973). The aim of these calculations was to find the most accurate Gran plots for the evaluation of total alkalinity and total carbonate. Hansson and Jagner (1973) included the four main protolytic species, carbonate, borate, fluoride, and sulphate, in their calculations. In a more extensive model by Dyrssen *et al.* (1979), silicate and phosphate were also included. Table 4 shows the total concentrations of major sea water protolytic species used in these calculations, one set of total concentrations representing an ocean surface water and one a deep sea sample. Table 5 shows the stability constants for the protonation reactions of the major sea water protolytes at 35‰ salinity and 25°C. The total alkalinity, A_t, of the sample, i.e. the amount of H$^+$ (moles) per unit volume (or weight) of sea water needed to titrate the protolytic species to H$_2$CO$_3$, B(OH)$_3$, H$_2$PO$_4^-$, and Si(OH)$_3^-$, is defined as

$$A_t = 2[CO_3^{2-}] + [B(OH)_4^-] + [SiO_3^{2-}] + 2[PO_4^{3-}] - [H^+]_{tot,\,initial} \tag{33}$$

in the computer program. When the total concentrations of all components are introduced, together with the protonation constants, HALTAFALL calculates the equilibrium concentration of all species in the sample. This is

Table 4. Total concentrations of major sea water protolytic species in typical surface and deep waters. These total concentrations have been used in the computer calculations whose results are shown in Figures 10–12

Protolyte	Total concentration in surface water, mM_W	Total concentration in deep water, mM_W
SO_4^{2-}	28.23	28.23
CO_3^{2-}	2.00	2.40
$B(OH)_4^-$	0.412	0.412
F^-	0.073	0.073
SiO_3^{2-}	0.005	0.150
PO_4^{3-}	0.0001	0.003
$[H]_{tot, initial}$	2.117	2.868
Total alkalinity	2.300 mequiv kg^{-1}	2.500 mequiv kg^{-1}

illustrated for a surface water sample in Figure 10. At the v value equal to zero in Figure 10 the concentrations of the different protolytes in an untitrated surface water are shown. Figure 10 also shows the distribution of the different species when a 100 g sample of a typical surface water (*cf.* Table 4) is titrated with v ml of 0.7000 M hydrochloric acid. Figure 11 shows the changes in $-\log[H^+]$ *vs.* v ml of 0.7000 M hydrochloric as 100 g of surface water and 100 g of deep water are titrated. The calculated curves in Figure 11 correspond to the experimental titration curve which would be obtained in the titration of sea water, provided that the total concentrations are those of Table 4 and that the sensor glass electrode couple follows the

Table 5. Protonation constants for the major sea water protolytes

Protonation reaction	Stability constant (25°C and 35‰ salinity)	Reference
$OH^- + H^+ \rightleftharpoons H_2O$	$4.786 \times 10^{-14}\ M_W^{-2}$	Hansson (1973)
$SO_4^{2-} + H^+ \rightleftharpoons HSO_4^-$	$11.48\ M_W^{-1}$	Dyrssen and Hansson (1973)
$CO_3^{2-} + H^+ \rightleftharpoons HCO_3^-$	$1.197 \times 10^9\ M_W^{-1}$	Hansson (1973)
$HCO_3^- + H^+ \rightleftharpoons H_2CO_3$	$9.840 \times 10^5\ M_W^{-1}$	Almgren *et al.* (1975)
$B(OH)_4^- + H^+ \rightleftharpoons B(OH)_3$	$5.521 \times 10^8\ M_W^{-1}$	Hansson (1973).
$F^- + H^+ \rightleftharpoons HF$	$398\ M_W^{-1}$	Smith and Martell (1976)
$SiO_3^{2-} + H^+ \rightleftharpoons Si(OH)_3^- + OH^-$	1.995×10^5	Dyrssen (1975)
$PO_4^{3-} + H^+ \rightleftharpoons HPO_4^{2-}$	$1.202 \times 10^9\ M_W^{-1}$	Johansson and Wedborg (1979)
$HPO_4^{2-} + H^+ \rightleftharpoons H_2PO_4^-$	$1.199 \times 10^6\ M_W^{-1}$	Johansson and Wedborg (1979)
$H_2PO_4^- + H^+ \rightleftharpoons H_3PO_4$	$54.95\ M_W^{-1}$	Kester and Pytkowicz (1967)

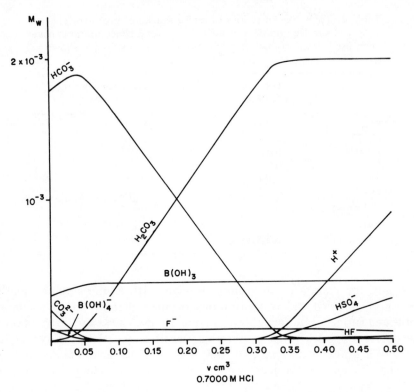

Figure 10. Potentiometric titration of 100 g of surface water (*cf.* Table 4) with v cm^3 of 0.7000 M hydrochloric acid

Nernst equation

$$E = E_1 + g \log [H^+] \qquad (34)$$

throughout the whole titration. Consequently, Figures 10 and 11 can be used to estimate the best way of evaluating the two equivalence points on the titration curve, one, v_1 cm^3, corresponding to titration to bicarbonate, and the other, v_2 cm^3, corresponding to titration to carbonic acid. Derivative methods, in which $\Delta \log [H^+]/\Delta v$ or $\Delta \log E/\Delta v$ is plotted *vs.* v cm^3 of titrant, are not very suitable since the inflexion points in Figure 11 are not well defined. For this reason, Dyrssen and Sillén (1967) suggested the use of Gran plots for the evaluation of the equivalence points. For experimental data for which $v > v_2$ cm^3, the plot F_2,

$$F_2 = (v_0 + v)10^{E/g}$$

(*cf.* equation 23) was suggested for the evaluation of v_2. For v values

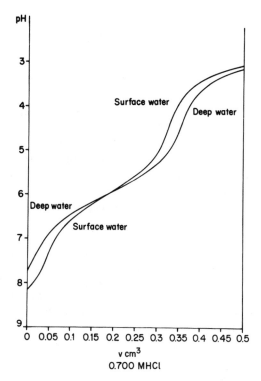

Figure 11. Potentiometric titration of 100 g of surface and deep sea water with v cm³ of 0.7000 M hydrochloric acid. The total concentrations are specified in Table 4

between v_1 and v_2, the Gran plot F_1,

$$F_1 = (v_2 - v)10^{E/g} \qquad (35)$$

was suggested, i.e. a prior knowledge of v_2 was necessary in order to use F_1. The derivation of F_1 was based on the assumption that the dominant reaction in the region between v_1 and v_2 is

$$HCO_3^- + H^+ \rightleftharpoons H_2CO_3 \qquad (36)$$

from which it follows that, in the interval $v_1 < v < v_2$,

$$[HCO_3^-] = M_t(v_2 - v)(v_0 + v) \qquad (37)$$

and

$$[H_2CO_3] = M_t(v - v_1)/(v_0 + v) \qquad (38)$$

where M_t is the concentration of the hydrochloric acid titrant and v_0 the initial sample volume. Since

$$[H_2CO_3] = \text{constant } [H^+][HCO_3^-] \qquad (39)$$

it follows that
$$(v_2 - v)[H^+] \propto v - v_1 \qquad (40)$$

and, by combining equations 34 and 40, the Gran plot F_2 is obtained. In the paper by Dyrssen and Sillén (1967), all concentrations were expressed as moles per kilogram of sample, M_W; this does not affect the derivation of the Gran plots.

As pointed out by Hansson and Jagner (1973), the Gran plots F_1 and F_2 are based on the assumption that no side-reactions occur in the volume intervals $v_1 < v < v_2$ and $v > v_2$ cm^3. From Figure 10 it can be seen, however, that this assumption is not absolutely true. In the interval $v > v_2$ the main reaction, the increase in the free concentration of hydrogen ions, is accompanied by the side-reactions

$$SO_4^{2-} + H^+ \rightleftharpoons HSO_4^- \qquad (41)$$

$$F^- + H^+ \rightleftharpoons HF \qquad (42)$$

$$HCO_3^- + H^+ \rightleftharpoons H_2CO_3 \qquad (43)$$

$$H_2PO_4^- + H^+ \rightleftharpoons H_3PO_4 \qquad (44)$$

where equation 43 corresponds to the titration of the bicarbonate which has not been titrated at $v = v_2$ and reaction 44 is of minor importance when compared with reactions 41–43. This is illustrated in Figure 12, where the

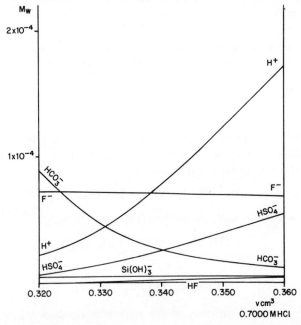

Figure 12. Potentiometric titration of 100 g of surface water with v cm^3 of 0.7000 M hydrochloric acid (*cf.* Table 4 and Figure 10)

concentration of the protolytic species of Figure 10 are shown in more detail in the neighbourhood of the second equivalence point, $v_2 = 0.3286$ cm^3. Consequently, a modified Gran plot F_2'

$$F_2' = (100 + v)([H^+] + [HSO_4^-] + [HF] + [H_3PO_4] - [HCO_3^-]) \qquad (45)$$

with

$$F_2' \propto (v - v_2) \qquad (46)$$

would locate v_2 with improved accuracy.

In the interval $v_1 < v < v_2$ (*cf.* Figure 10), the main reaction given in equation 36 is accompanied by the reactions

$$CO_3^{2-} + H^+ \rightleftharpoons HCO_3^- \qquad (47)$$

$$B(OH)_4^- + H^+ \rightleftharpoons B(OH)_3 \qquad (48)$$

$$OH^- + H^+ \rightleftharpoons H_2O \qquad (49)$$

$$HPO_4^{2-} + H^+ \rightleftharpoons H_2PO_4^- \qquad (50)$$

as well as the reactions given in equations 41–43. A modified Gran plot, F_1' would thus take the form

$$F_1' = (v_0 + v)([H^+] + [HSO_4^-] + [HF] + [H_2PO_4^-] + 2[H_3PO_4] + [H_2CO_3]$$
$$- [B(OH)_4^-] - [OH^-]) \qquad (51)$$

where the terms [HF] and [H$_3$PO$_4$] are of negligible importance. The modified plot F_1' would fulfill

$$F_1' \propto (v - v_1) \qquad (52)$$

and, if v_1 and v_2 are determined accurately, the total carbonate content, C_T mmol, of the sample can be calculated as

$$C_T = M_t(v_2 - v_1) \qquad (53)$$

and then expressed either in mol kg^{-1} or mol dm^{-3}.

In an experimental situation the species formed in the reactions given in equations 41–44 and 47–50 cannot be calculated however, unless one has a knowledge of E^0 (*cf.* equation 34) and approximate values of the total concentrations of all protolytic species and their protonation constants. The total concentrations for some of the major protolytic species, such as sulphate, borate, and fluoride, can be estimated from the known relation between their total concentrations and sample salinity (Section 1.3.1). The concentrations of phosphate and silicate are normally determined separately. The protonation constants are, moreover, known fairly accurately (*cf.* Table 5). An approximate initial value of E_1 (equation 34) can, moreover, be derived by applying F_2 to the experimental data, E mV *vs.* v cm^3, and computing E_1 from the approximate value of v_2 thus obtained, as

$$[H^+] = 10^{(E-E_1)/g} \tag{54}$$

where $[H^+]$ is obtained from

$$[H^+] + [HSO_4^-] = M_t(v_2 - v)/(v_0 + v) \tag{55}$$

so that

$$[H^+] = [M_t(v_2 - v)/(v_0 + v)]/(1 + 11.48 \times 0.028)$$

where 11.48 M_W is the stability constant for HSO_4^- and 0.028 M_W is the total sulphate concentration (*cf.* Tables 4 and 5). In the derivation of this equation it has been assumed that the free concentration of sulphate is very close to the total sulphate concentration. Figure 10 shows that the concentration of hydrogen sulphate can be neglected in comparison with the free sulphate concentration. Once an approximate value of v_2 has been calculated, an initial approximate value of v_1 can be obtained from F_1 and, consequently, an approximate value of $[CO_3^{2-}]_{tot}$ obtained from equation 53. An improved value of v_2 can thus be obtained from F_2' since $[HSO_4^-]$, $[HF]$, $[HCO_3^-]$, and $[H_3PO_4]$ can be calculated as

$$[HSO_4^-] = [SO_4]_{tot}/(1 + 1.148[H^+]) \tag{56}$$

$$[HCO_3^-] = [CO_3^{2-}]_{tot} \Big/ \left(1 + \frac{1.178 \times 10^{15}}{1.197 \times 10^9}[H^+]\right) \tag{57}$$

the protonation constants and total concentrations being those specified in Tables 4 and 5. The concentrations of all other protolytic species required in the calculation of F_2' and F_1' can be derived by analogy with equations 56 and 57. Consequently, as pointed out by Hansson and Jagner (1973), the modified Gran plots can be used to determine total alkalinity and total carbonate if an iterative calculation process is exploited. The initial step in the iteration is to calculate approximate values of v_1, v_2, and E from the F_1 and F_2 plots of Dyrssen and Sillén (1967), and then calculate F_2' and F_1', yielding improved values of v_1, v_2, E_1, and $[CO_3^{2-}]_{tot}$. The iterations, which must be performed by means of computer calculations, should be continued until the relative change in either v_2, v_1, or $[CO_3^{2-}]_{tot}$ is less than, say, 0.01% from one computation to the next.

The calculations by Hansson and Jagner (1973) show that it is theoretically possible to locate the equivalence points v_1 and v_2 with an accuracy better than 0.1% with the modified Gran plots, even if the uncertainty in the stability constants is of the order of ± 0.08 logarithmic unit and the uncertainty in sample temperature, (and therefore in the Nernst factor) is $\pm 2°C$. The main advantage of the modified Gran plots is, however, that all the experimental data, even those close to the equivalence points, can be used in

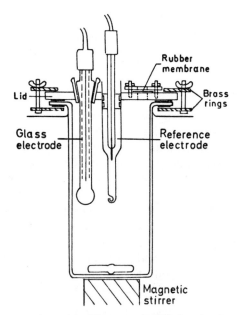

Figure 13. Titration vessel used by Hansson (1972) for the determination of acidity constants of carbonic acid in sea water. The burette tip and pressure equilibration piston are not shown

the evaluation. This means that the precision in the titration determination of total and carbonate alkalinity is increased considerably.

Almgren *et al.* (1977) were the first to apply the modified Gran plots to on-board analysis. They used a closed titration vessel, of approximate total volume $150 \, \text{cm}^3$, resembling that originally suggested by Hansson (1972) (Figure 13). On the basis of duplicate determinations, Almgren *et al.* (1977) concluded that total and carbonate alkalinities could be determined with relative precisions of 0.08 and 0.13%, respectively. In order to increase the precision, they used a thermistor to monitor the sample temperature at each titration point to ensure correct values of the Nernst factor. Furthermore, they used a fully computerized titrator which was able to carry out the titration and evaluate the equivalence points by the iterative procedure.

Almgren *et al.* (1977) calculated the carbonate alkalinity, A_c, from the total alkalinity, A_t (*cf.* equation 33), as

$$A_c = A_t - [B(OH)_4^-]_{tot} \tag{58}$$

where $[B(OH)_4^-]_{tot}$ denotes all the borate species, ion pairs included, in the sample prior to titration. A value of $[B(OH)_4^-]_{tot}$ was calculated from an estimated value of the total boron concentration in the sample, the pH value of the sample and a known value of the acidity constant of boric acid in sea water medium (Hansson, 1973). Once A_c had been determined, the concentrations of CO_3^{2-}, HCO_3^-, and $CO_2 + H_2CO_3$ were calculated from sample pH and known acidity constants for boric acid (Hansson, 1973).

7.7.4. Group 5B to 7B elements

Several attempts have been made to design a potentiometric sensor for nitrate ions. The most successful attempt hitherto is probably that of Hansen *et al.* (1977) using tetraoctylammonium bromide as ion exchanger in a PVC membrane and dibutyl phthalate as solvent. Chloride ions, however, interfere with such membranes, the selectivity coefficient being of the order of 5×10^{-3} (*cf.* equation 24). This rules out the possibility of using such electrodes in saline waters, even at very high nitrate concentrations.

In the ammonia probe, which is outlined schematically in Figure 14, a small volume of highly concentrated internal ammonium chloride electrolyte surrounds a glass pH electrode and is protected from direct contact with the sample by means of a gas-permeable membrane. In the internal electrolyte the equilibrium

$$NH_3 + H^+ \rightleftharpoons NH_4^+ \tag{59}$$

is affected by the sample concentration of ammonia, since the equilibrium condition

$$[NH_3]_{sample} = [NH_3]_{internal\,electrolyte} \tag{60}$$

is established rapidly. Thus, changes in sample ammonia concentration affect the pH of the internal electrolyte according to equation 59. In this way, ammonia concentrations down to 10^{-6} M can be monitored. This detection

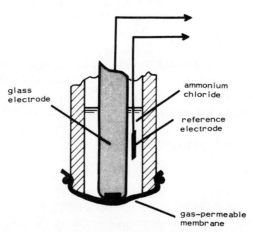

Figure 14. The ammonia probe

limit implies that only samples which are high in total ammonia, i.e. $NH_4^+ + NH_3$, can be handled and that a potentiometric titration must be performed by a standard addition procedure. Gilbert and Clay (1973) determined ammonia in coastal waters by means of a single-point standard addition procedure after buffering the sea water sample to $pH > 11$ by the addition of sodium hydroxide pellets. Their results indicate that the ammonia probe can be used for monitoring the potentiometric standard addition titration of total ammonia, provided that the sample concentration is above 2 μM. For lower sample concentrations the electrode response is slow and non-Nernstian. Moreover, the ammonium chloride inner electrolyte leaks slowly into the sample, giving rise to erroneous readings at low ammonia concentrations. By modifying the composition of the internal electrolyte, Garside *et al.* (1978) were able, however, to decrease the detection limit to approximately 0.2 μM in the potentiometric standard addition titration procedure.

Sulphate is one of the major ionic constituents of sea water and could thus be determined by direct titration. No selective potentiometric sensor for sulphate has yet been constructed, however, and the concentration increase of the reagent must therefore be followed. Mascini (1973) used a lead sulphide electrode to monitor the potentiometric titration of sulphate in sea water using lead nitrate as reagent. Interferences from chloride and hydrogen carbonate were eliminated by passing the sample through two ion-exchange columns, the first in silver form and the second in acid form. This makes the method elaborate and decreases its relative precision to approximately 3%. Levins (1971) suggested a type of barium-selective electrode based on the tetraphenylborate salt of a derivative of a poly(ethylene glycol) neutral carrier, using p-nitroethylbenzene as solvent in a PVC membrane. The Nernstian range of the electrode was $10^{-5} - 1$ M for Ba^{2+}. Since the electrode had selectivity constants of less than 10^{-4} for calcium and magnesium, Levins (1971) suggested that the electrode would be a suitable sensor for the potentiometric titration of sulphate. Ouzounian and Michard (1978) used a slightly modified version of Levin's electrode (Orion, Model 92–32) for the potentiometric titration of sulphate in sea water, using barium nitrate as reagent. Very few experimental data were reported. From the selectivity coefficients in sea water medium (9.4×10^{-4} for calcium and 1.8×10^{-4} for magnesium), it can be concluded that only potentiometric data after the equivalence point can be used in the evaluation. Moreover, a Gran (1952) extrapolation procedure would probably give erroneous results, owing to alkaline earth metal interference. Evaluation of the equivalence point by derivative methods would give the most accurate results and an approximate relative precision of ±2%.

The fluoride-selective lanthanum trifluoride membrane electrode (Frant and Ross, 1966) is, with the exception of the glass pH electrode, the most reliable membrane electrode. It has a dynamic range between 10^{-6} and $1\,\text{M}$ of free fluoride, the lower limit being set by the solubility of the membrane. The total fluoride concentration in sea water of 35‰ salinity is approximately $7 \times 10^{-5}\,\text{M}$, i.e. well within the dynamic range of the fluoride electrode. Owing mainly to lack of a suitable reagent, a potentiometric titration of fluoride in sea water must be performed as a standard addition titration (Anfält and Jagner, 1971a). The main fluoride species in sea water are F^-, MgF^+, and CaF^+, i.e.

$$[F^-]_{tot} = [F^-] + [MgF^+] + [CaF^+] = [F^-](1 + \beta_{MgF}[Mg^{2+}] + \beta_{CaF}[Ca^{2+}])$$
(61)

where, at 35‰ salinity and 25°C

$$1 + \beta_{MgF}[Mg^{2+}] + \beta_{CaF}[Ca^{2+}] = 2.07$$
(62)

according to Elgquist (1970). The Nernst equation for the fluoride electrode couple in sea water:

$$E = E_1 - g \log [F^-]$$

can consequently be rearranged as

$$E = E_{sal} - g \log [F^-]_{tot}$$
(63)

where

$$E_{sal} = E_1 + \log (1 + \beta_{MgF}[Mg^{2+}] + \beta_{CaF}[Ca^{2+}])$$
(64)

i.e. E_{sal} is dependent on the free concentrations of magnesium and calcium and, assuming a constant relationship between total magnesium and calcium and sample salinity, dependent on sample salinity.

During the standard addition titration, small fluoride aliquots, of the order of $2 \times 10^{-5}\,\text{M}$, are added. These aliquots do not change the free concentrations of magnesium and calcium to any appreciable extent, provided that dilution can be neglected. Equation 63 then implies that the Gran plot (*cf.* equation 23),

$$F_2 = (v_0 + v)10^{-E/g}$$

can be used to evaluate the total fluoride concentration of the sample. The negative intercept on the v-axis, $v_{eq}\,\text{cm}^3$, when F_2 is extrapolated to zero, gives the total concentration of fluoride in the sample as

$$[F^-]_{tot} = \frac{v_{eq}M_t}{v_0}$$
(65)

where M_t is the concentration of the sodium fluoride titrant and $v_0\,\text{cm}^3$ the initial sample volume.

Equation 63 indicates that the total concentration of fluoride in a sea water sample can be estimated from a single potential reading, provided that a value of E_{sal} is known for the relevant sample salinity and temperature (*cf.* Warner, 1969; Kosov *et al.*, 1976b). A value of E_{sal} must, however, be determined by means of a standard addition titration procedure and single-point reading is thus suitable only if a great number of samples of approximately the same salinity are to be analysed.

According to Anfält and Jagner (1971a), the relative precision in the standard addition titration of total fluoride is of the order of magnitude 1% in the salinity range 20–35‰. The same precision was also obtained by Rix *et al.* (1976) using this standard addition procedure.

Determination of sea water chlorinity by means of potentiometric titration using a silver wire as sensor electrode was first suggested by Hermann (1951). When silver ions are added to a sea water sample, the two reactions (*cf.* equation 11)

$$Ag^+ + Cl^- \rightleftharpoons AgCl(s)$$

$$Ag^+ + Br^- \rightleftharpoons AgBr(s)$$

will occur simultaneously (Jagner and Årén, 1970). The formation of silver iodide is of negligible importance. Owing to the low solubilities of silver halides, a well pronounced potential jump will occur at the equivalence point (*cf.* Figure 4). Consequently, the equivalence point can be located with high precision using derivative methods, as is shown in Figure 5. Hermann (1951) was able to reach a relative precision of the order of 0.01%. The use of the Gran plot F_2 (*cf.* equation 23) for the experimental data after the equivalence point, as shown in Figure 6, in fact yields a poorer relative precision (Jagner and Årén, 1970). This is mostly attributed to lack of temperature control during titration (*cf.* Figure 6) and to adsorption of soluble Ag^+ species on the solid silver halides. Increased precision can be obtained by monitoring the sample temperature, as shown by Almgren *et al.* (1977).

Direct potentiometric titration of bromide in sea water is not feasible owing to the lack of a suitable reagent and a sufficiently selective sensor electrode. All the successful potentiometric sensors for bromide have hitherto been based, directly or indirectly, on the silver/silver bromide system. This means that the selectivity of the bromide sensor couple is governed by the solubility product of the silver salt of the interfering anion as

$$E = E_1 + g \log \left([Br^-] + \frac{K_{s,AgBr}}{K_{s,AgX}} [X^-] \right) \tag{66}$$

where $K_{s,AgBr}$ is the solubility product of silver bromide. In sea water, in which the chloride concentration is approximately 500 times greater than the bromide concentration, a silver/silver halide potentiometric sensor will respond equally to chloride ions and to bromide ions, since the ratio $K_{s,AgCl}/K_{s,AgBr}$ is also approximately 500 (*cf.* equation 11). This rules out the possibility of determining bromide in sea water accurately by means of a potentiometric standard addition procedure. The only attempt to titrate bromide potentiometrically was described by Savenko (1976).

7.8. STUDY OF IONIC INTERACTIONS IN SEA WATER

The form of a potentiometric titration curve, E mV *vs.* v cm^3 of titrant added, is governed mainly by

  (i) the total concentration and the activity coefficient of the ion under study;

  (ii) the total concentration and activity coefficients of sample species forming complexes (or ion pairs) with the ion under study;

  (iii) the stability constants for all complexes (or ion pairs) which are formed between sample species

  (iv) the contribution to sensor electrode potential from ions other than that under study, according to equation 31; and

  (v) the liquid junction potentials.

In the analytical applications discussed above, the aim was to determine the total concentration of the particular component. For this reason the experimental conditions were chosen so that the ionic activity coefficients and liquid junction potentials did not alter appreciably during titration. Obviously, by potentiometrically titrating a sample containing known total concentrations of all components, the stability constants for complexes and/or variations in activity coefficients can be studied. Potentiometric titration is one of the most frequently used methods for the study of complex formation and its theory, and applications have been discussed in detail (see, for example, Rossotti and Rossotti, 1961; Rossotti, 1969). The relative importance of potentiometric titration as a technique for determining stability constants can be seen from Sillén and Martell (1964, 1971). In sea water media potentiometric titration has been used mainly to study weak interactions between the major ionic constituents and the formation of proton complexes of minor constituents.

7.8.1. Major ionic species

The major ionic species of sea water are Na$^+$, K$^+$, Ca^{2+}, Mg^{2+}, Cl$^-$, and SO$_4^{2-}$, which together contribute more than 99.5% to the total ionic strength, 0.714 M or 0.698 $_{MW}$, of a 35‰ salinity sea water. Interaction

Table 6. Some recent studies of the interaction between major ionic constituents of sea water by means of potentiometric titration methods at the ionic strength of 35‰ salinity sea water

System	Sensor electrode	Remark	Reference
$Na^+-Mg^{2+}-Ca^{2+}-SO_4^{2-}$	pH, Na^+ glass membrane	Single-point titration	Kester and Pytkowicz (1969)
$Na^+-Ca^{2+}-Cl^-$	Ca^{2+}, liquid ion-exchange membrane		Whitfield and Leyendekkers (1970)
$H^+-SO_4^{2-}$	pH, glass membrane	Single-point titration	Culberson *et al.* (1970)
$Mg^{2+}-H^+-OH^-$	pH, glass membrane, hydrogen electrode		Dyrssen and Hansson (1973)
$Na^+-SO_4^{2-}$	Na^+, glass membrane		Kosov *et al.* (1976a)
$Ca^{2+}-SO_4^{2-}$	Ca^{2+}, liquid ion-exchange membrane		Kosov (1976)
$H^+-Na^+-K^+-Ca^{2+}-Mg^{2+}-Cl^-$	pH, glass membrane, hydrogen electrode	Single-point titration	Johnson and Pytkowicz (1978)
$Na^+-Mg^{2+}-Cl^--SO_4^{2-}$	Na^+, glass membrane; F^-, LaF_3 membrane		Elgquist and Wedborg (1978)
$Ca^{2+}-SO_4^{2-}$	Ca^{2+}, liquid ion-exchange PVC membrane		Elgquist and Wedborg (1979)

between these ions have been studied experimentally by many techniques, and the results have been interpreted in terms of different physico-chemical models, including ion-pair formation (Wedborg, 1979). Whitfield (1975) gave a detailed discussion of the experimental techniques and the different physico-chemical models used. During the last decade, potentiometric titration techniques have played an increasingly important role in the study of ion–ion interactions in sea water, mainly owing to the development of new membrane electrodes (*cf.* Table 2). These membrane electrodes are, however, much more sensitive to interferences than the electrodes used in the classical investigations, e.g. the hydrogen electrode, the glass pH electrode and the silver/silver chloride electrode (Whitfield and Leyendekkers, 1970; Leyendekkers and Whitfield, 1971; Amman *et al.*, 1976). The reproducibility of the new membrane electrodes is, in most experimental situations, not better than ± 0.1 mV, compared with ± 0.01 mV for the classical electrodes. It is beyond the scope of this chapter to discuss the interpretation of these ion–ion interaction experiments (see Sections 6.3.5 to 6.3.7). Table 6 lists some of the recent work in which potentiometric titration has been employed in sea water media, or in media closely resembling sea water and at the ionic strength of sea water.

7.8.2. Minor and trace constituents

In contrast to the weak interactions between the major ionic species, most minor and trace constituents of sea water interact with other sea water constituents in a way which can be described in terms of complexation, i.e. with stability constants of the order of magnitude of 100 or more (*cf.* Table 5). Consequently, variations in ionic strength and in temperature will be of relatively less importance for these strong interactions. Experimental results obtained in media other than sea water can therefore, as a first approximation, be regarded as being relevant to sea water. This is particularly true for the strong complexes formed between some trace elements in sea water and the major anions in sea water. Sea water, with its large number of components, is much too complicated a medium for the determination of stability constants for these interactions by means of potentiometric titration, and one has to restrict oneself to systems that contain only one, or possibly two ligands. The combined effect of the many possible ligands in sea water can then be estimated from computer calculations using suitable computer programs (Ingri *et al.*, 1967; Perrin and Sayce, 1967; Fardy and Sylva, 1978).

A large number of stability constants have been determined by different techniques, including potentiometric titrations, and the results have been summarized by Sillén and Martell (1964, 1971). Critical selection of stability constants has been made by Smith and Martell (1976) and by an IUPAC commission on equilibrium data (Beck, 1977). These equilibrium constants have been exploited for computer modelling of the speciation of sea water components by, e.g. Dyrssen and Wedborg (1974), Kester (1975), and Batley and Florence (1976). The equilibrium constants used in these computer models have only rarely been determined in the sea water medium and the ionic strength used in the determination of the stability constants is often different from that of sea water. The main drawback of the computer calculations carried out hitherto is, however, that the programs do not include ion-exchange equilibria and adsorption/desorption effects between dissolved sea water species and solid particles present in sea water. Future applications of potentiometric titration will most likely focus on the study of the interactions between dissolved sea water species and solid phases present in the medium.

REFERENCES

Almgren, T., D. Dyrssen, and M. Strandberg (1975). Determination of pH on the moles per kg sea water scale (M_w). *Deep-Sea Res.*, **22**, 635–646.

Almgren, T., D. Dyrssen, and M. Strandberg (1977). Computerized high precision titrations of some major constituents of sea water. *Deep-Sea Res.*, **24**, 345–364.

Amman, D., R. Bissig, Z. Cimerman, U. Fiedler, M. Güggi, W. E. Morf, M. Oehme, H. Osswald, E. Pretsch, and W. Simon (1976). Synthetic neutral carriers for cations. In *Proceedings of the International Workshop on Ion and Enzyme Electrodes in Biology and Medicine* (Eds. M. Kessler, L. C. Clark Jr., D. W. Lübbers, I. A. Silver, and W. Simon), Urban and Schwarzenberg, Munich, Berlin, Vienna, pp. 22–37.

Anfält, T., D. Dyrssen, and D. Jagner (1968). Species formed in the potentiometric titration of fluoride with thorium or lanthanum nitrate and functions suitable for the evaluation of the equivalence point. *Anal. Chim. Acta*, **43**, 487–499.

Anfält, T., and D. Jagner (1971a). A standard addition titration method for the potentiometric determination of fluoride in sea water. *Anal. Chim. Acta*, **53**, 13–22.

Anfält, T., and D. Jagner (1971b). The precision and accuracy of some current methods for potentiometric end-point determination. *Anal. Chim. Acta*, **57**, 165–176.

Anfält, T., and D. Jagner (1973). The potentiometric titration of potassium in sea water with a valinomycin electrode. *Anal. Chim. Acta*, **66**, 152–155.

Batley, G. E., and T. M. Florence (1976). Determination of the chemical forms of dissolved cadmium, lead and copper in seawater. *Mar. Chem.*, **4**, 347–352.

Beck, M. T. (1977). Critical evaluation of equilibrium constants in solution. Stability constants of metal complexes. *Pure Appl. Chem.*, **49**, 129–135.

Buck, R. P. (1978). Ion selective electrodes. *Anal. Chem.*, **50**, 17R–29R.

Culberson, C., R. N. Pytkowicz, and J. E. Hawkey (1970). Sea water alkalinity by the pH method. *Mar. Res.*, **28**, 15–21.

Durst, D., Ed. (1969). *Ion-selective Electrodes*. National Bureau of Standards Special Publication 314, US Government Printing Office, Washington, D.C.

Dyrssen, D. (1975). Constituent interactions in sea water. *Proc. Anal. Div. Chem. Soc.*, **12**, 111–115.

Dyrssen, D., and I. Hansson (1973). Ionic medium effects in sea water—a comparison of acidity constants of carbonic acid and boric acid in sodium chloride and synthetic sea-water. *Mar. Chem.*, **8**, 137–149.

Dyrssen, D., D. Jagner, and H. Johansson (1968). On the potentiometric titration of calcium in sea water using a calcium selective membrane electrode. *Reports on the Chemistry of Sea Water*, V. Department of Analytical Chemistry, University of Göteborg.

Dyrssen, D., D. Jagner, and O. Johansson (1979) Computer generated data sets for the potentiometric titration of total alkalinity in a typical surface seawater sample and a typical bottom seawater sample. In *Workshop of Oceanic CO_2 Standardization* (Eds. G. Östlund and D. Dyrssen), La Jolla, California, 1980, mimeograph.

Dyrssen, D., D. Jagner, D. Svedung, F. Wengelin, and K. Årén (1966). Progress report on potentiometric titrations of alkalinity and chlorinity of sea water. *Report on the Chemistry of Sea Water*. II. Department of Analytical Chemistry, University of Göteborg.

Dyrssen, D., and L. G. Sillén (1967). Alkalinity and total carbonate in sea water. A plea for p–T-independent data. *Tellus*, **19**, 113–121.

Dyrssen, D., and M. Wedborg (1974). Equilibrium calculations of the speciation of elements in sea water. In *The Sea*, Vol. 5 (Ed. E. D. Goldberg). Wiley–Interscience, New York, pp. 181–195.

Elgquist, B. (1970). Determination of the stability constants of MgF^+ and CaF^+ using a fluoride ion sensitive electrode. *J. Inorg. Nucl. Chem.*, **32**, 937–944.

Elgquist, B., and M. Wedborg (1978). Stability constants of $NaSO_4^-$, $MgSO_4$, MgF^+, $MgCl^+$ ion pairs at the ionic strength of sea water by potentiometry. *Mar. Chem.*, **6**, 243–252.

Elgquist, B., and M. Wedborg (1979). Stability of the calcium sulphate ion pair at the ionic strength of seawater by potentiometry. *Mar. Chem.*, **7**, 273–280.

Fardy, J. J., and R. N. Sylva (1978). SIAS, a computer program for the generalised calculation of speciation in mixed metal-ligand aqueous systems. *Aust. Atomic Energy Comm. Rep.*, AAEC/E445.

Frant, M. S., and J. W. Ross (1966). Electrode for sensing fluoride ion activity in solution. *Science*, **154**, 1553–1554.

Garside, C., G. Hull, and S. Murray (1978). Determination of submicromolar concentrations of ammonia in natural waters by a standard addition method using a gas-sensing electrode. *Limnol. Oceanogr.*, **23**, 1073–1076.

Gast, J. A., and T. G. Thompson (1958). Determination of alkalinity and borate concentration of sea water. *Anal. Chem.*, **30**, 1549–1551.

Gilbert, T. R., and A. M. Clay (1973). Determination of ammonia in aquaria and in sea water using the ammonia electrode. *Anal. Chem.*, **45**, 1757–1759.

Gran, G. (1952). Determination of the equivalence point in potentiometric titrations, Part II. *Analyst*, **77**, 661–671.

Granéli, A., and T. Anfält (1976). A simple automatic phototitrator for the determination of total carbonate and total alkalinity of sea water. *Anal. Chim. Acta*, **91**, 175–180.

Hansen, E. H., A. K. Ghose, and J. Růžička (1977). Flow injection analysis of environmental samples for nitrate using an ion-selective electrode. *Analyst*, **102**, 705–713.

Hansson, I. (1972). An analytical approach to the carbonate system in sea water. *Thesis*, University of Göteborg.

Hansson, I. (1973). A new set of acidity constants for carbonic acid and boric acid in seawater. *Deep-Sea Res.*, **20**, 461–478.

Hansson, I., and D. Jagner (1973). Evaluation of the accuracy of Gran plots by means of computer calculations. Application to the potentiometric titration of the total alkalinity and carbonate content in sea water. *Anal. Chim. Acta*, **65**, 363–373.

Hermann, F. E. (1951). High accuracy potentiometric determination of chlorinity of sea water. *J. Cons. Perm. Int. Explor. Mer.*, **17**, 223–230.

Hulthe, P., L. Uppström, and G. Östling (1970). An automatic procedure for the determination of boron in sea water. *Anal. Chim. Acta*, **51**, 31–37.

Ingri, N., W. Kakotowicz, L. G. Sillén, and B. Warnqvist (1967). HALTAFALL, a general program for calculating the composition of equilibrium mixtures. *Talanta*, **14**, 1261–1286.

Jagner, D. (1967). A potentiometric titration of magnesium in sea water. *Report on the Chemistry of Sea Water, III*. Department of Analytical Chemistry, University of Göteborg.

Jagner, D. (1977). Unpublished results.

Jagner, D. (1978). Instrumental approach to potentiometric stripping analysis of some heavy metals. *Anal. Chem.*, **50**, 1924–1929.

Jagner, D., and K. Årén (1970). A rapid semi-automatic method for the determination of the total halide concentration in sea water by means of potentiometric titration. *Anal. Chim. Acta*, **52**, 491–499.

Jagner, D., and K. Årén (1978). Derivative potentiometric stripping analysis with a thin film of mercury on a glassy carbon electrode. *Anal. Chim. Acta*, **100**, 375–388.

Jagner, D., and K. Årén (1979). Potentiometric stripping analysis for zinc, cadmium, lead and copper in sea water. *Anal. Chim. Acta*, **107**, 29–35.

Johansson, A., and L. Pehrsson (1970). Automatic titration by stepwise addition of equal volumes of titrant. *Analyst*, **95**, 652–656.

Johansson, O., and M. Wedborg (1979). Stability constants of phosphoric acid in seawater of 5–40‰ salinity and temperature of 5–25°C. *Mar. Chem.*, **8**, 57–69.

Johnson, K. S., and R. M. Pytkowicz (1978). Ion association of Cl^- with H^+, Na^+, K^+, Ca^{2+} and Mg^{2+} in aqueous solutions at 25°C. *Am. J. Sci.*, **278**, 1428–1447.

Kester, D. R. (1975). Chemical speciation in seawater. In *The Nature of Seawater* (Ed. E. D. Goldberg). Report of the Dahlem Workshop, Dahlem Konferenzen, Berlin, pp. 17–43.

Kester, D. R., and R. M. Pytkowicz (1967). Determination of the apparent dissociation constants of phosphoric acid in seawater. *Limnol. Oceanogr.*, **12**, 243–252.

Kester, D., and R. M. Pytkowicz (1969). Sodium, magnesium and calcium sulphate ion pairs in sea water at 25°C. *Limnol. Oceanogr.*, **14**, 689–692.

Kosov, A. E. (1976). Potentiometric titration studies on the association between Ca^{2+} and SO_4^{2-} in synthetic sea water (in Russian). *Vinity* (Moscow), No. 4480.

Kosov, A. E., O. T. Krylov, and P. D. Novikov (1976a). Potentiometric titration studies of the association between Na^+ and SO_4^{2-} in synthetic sea water (in Russian). *Vinity* (Moscow), No. 4478.

Kosov, A. E., P. D. Novikov, O. T. Krylov, M. P. Nesterova, and A. F. Litvinova (1976b). A potentiometric method for the determination of F^- in sea water (in Russian). *Okeanologiya*, **16**, 5–15.

Levins, R. J. (1971). Barium ion-selective electrode based on a neutral carrier complex. *Anal. Chem.*, **43**, 1045–1047.

Leyendekkers, J. V., and M. Whitfield (1971). Liquid ion exchange electrodes in mixed electrolyte solutions. *Anal. Chem.*, **43**, 322–326.

Mascini, M. (1973). Titration of sulfate in mineral waters and sea water using the solid-state lead electrode. *Anal. Chim. Acta*, **98**, 325–328.

Meites, L., and T. Meites (1967). Theory of titration curves. *Anal. Chim. Acta*, **37**, 1–11.

Ouzounian, G., and G. Michard (1978). Dosage des sulfates dans les eaux naturelles a l'aide d'une electrode selective au baryum. *Anal. Chim. Acta*, **96**, 405–409.

Perrin, D. D., and I. G. Sayce (1967). Computer calculations of equilibrium concentrations in mixtures of metal ions and complexing species. *Talanta*, **14**, 833–841.

Riley, J. P. (1965). Analytical chemistry of sea water. In *Chemical Oceanography*, Vol. 2 (Eds. J. P. Riley and G. Skirrow). Academic Press, London, pp. 343–361.

Rix, C. J., A. M. Bond, and J. D. Smith (1976). Direct determination of fluoride in sea water with a fluoride selective ion electrode by a method of standard addition. *Anal. Chem.*, **48**, 1236–1239.

Ross, J. W. (1967). Calcium-selective electrode with liquid ion-exchanger. *Science*, **156**, 1378–1379.

Rossotti, H. (1969). *Chemical Applications of Potentiometry*. Van Nostrand, London.

Rossotti, J. C., and Rossotti, H. (1961). *The Determination of Stability Constants*. McGraw-Hill, New York.

Růžička, J., E. H. Hansen, and J. C. Tjell (1973). Selectrode—the universal ion-selective electrode, Part IV. *Anal. Chim. Acta*, **67**, 155–178.

Savenko, V. S. (1976). Use of ion-selective electrodes in geochemical determination studies using a bromide electrode (in Russian). *Okeanologiya*, **16**, 825–829.

Sillén, L. G., and A. E. Martell (1964, 1971). *Stability Constants of Metal-Ion Complexes*. Chemical Society Special Publications, Nos. 17 and 25 (supplement), Chemical Society, London.

Sillén, L. G., and B. Warnqvist (1969). High-speed computers as a supplement to graphical methods, VI. *Ark. Kemi*, **31**, 315–339.

Smith, R. M., and A. E. Martell (1976). *Critical Stability Constants*. Plenum Press, New York.

Squirrell, D. C. M. (1964). *Automated Methods in Volumetric Analysis*. Hilger and Watts, London.

Warner, T. B. (1969). Lanthanum fluoride electrode response in water and sodium chloride. *Anal. Chem.*, **41,** 527–529.

Wedborg, M. (1979). Studies on complexation in seawater. *Thesis*, University of Göteborg.

Whitfield, M. (1975). Sea water as an electrolyte solution. In *Chemical Oceanography*, Vol. 1 (Eds. J. P. Riley and G. Skirrow). Academic Press, London, pp. 43–171.

Whitfield, M., and J. V. Leyendekkers (1970). Selectivity characteristics of a calcium selective ion-exchange electrode in the system calcium(II)–sodium(I)–chloride(I). *Anal. Chem.*, **42,** 445–448.

8

Electrodeposition

E. E. BROOKS
Department of Chemistry,
Howard University,
Washington, D.C. 20059, USA

and H. B. MARK, JR.
Department of Chemistry,
University of Cincinnati,
Cincinnati, Ohio 45221, USA

GLOSSARY OF SYMBOLS

A	Surface area of working electrode (equation 3)
c_b	Bulk molar concentration of electroactive species (equation 5)
c_0	Initial molar concentration of electroactive species (equation 1)
c_t	Molar concentration of electroactive species at time t (equation 1)
D	Diffusion coefficient (equation 3)
F	Faraday (equation 5)
i_L	Limiting deposition current (equation 5)
k	Deposition rate constant (equation 1)
K^*	Stoichiometric equilibrium constant (equation 6)
M_W	Mol kg^{-1} sea water (Section 8.1)

n	Stoichiometric number of electrons transferred in cell reaction (equation 5)
t	Time (equation 1)
V	Volume of solution (equation 3)
δ	Diffusion layer thickness (equation 3)
ν	Kinematic viscosity (equation 4)
ω	Angular velocity (equation 4)

8.1. INTRODUCTION

In the past few years a substantial amount of research has been carried out on the chemical analysis of complex natural systems. The effect of man's impact on the environment and its delicate ecological balances has been recognized as a major cause for concern. One of the most difficult problems facing the investigators in such areas of research is the determination of the concentrations or the changes in concentration of trace metal ions (levels below the $mg\,dm^{-3}$ (parts per million) range in sea water systems (Mitchell, 1957; Robertson, 1968; Horne, 1969).

Of course, the major problem encountered in studying the effects of various concentration levels of trace metals on a system is one of analytical chemistry (Mark, 1973; Brewer and Spencer, 1970). Nothing can really be done unless true analytical descriptions of the system can be obtained. Therefore, the methods applied to the analysis of natural systems must be sensitive, selective, and applicable to the real problems of trace level determinations in these systems. Such methods are generally more complex and are subject to more interferences than those performed under laboratory development conditions (Hume, 1967), but the methods must first be developed in the laboratory. In spite of wide differences in approach to the problem of trace metal analysis, most investigators agree that the problem cannot be resolved unless better analytical techniques are developed and their quantitative and qualitative limits proved. Only then can we provide the necessary correlations with the ecological balances and other physical parameters in natural water systems that are needed to obtain a true analytical description of these systems.

Because of the difficulties involved in the analysis of these systems at the trace level, the methods developed should be viewed with some degree of scepticism. They must be carefully evaluated and intercalibrated before they can be accepted as reliable. A prime example of discrepancies that can occur between different methods was cited by Hume (1975) for an inter-laboratory comparison of lead in a carefully preserved sample of sea water. The results of several laboratories, reporting averages obtained by atomic-absorption spectrometry and anodic stripping voltammetry for lead, ranged from 2.4×10^{-10} to $6.3 \times 10^{-9}\,M_W$. The accepted value was $6.8 \times 10^{-11} \pm 1.4 \times 10^{-11}\,M_W$, obtained by isotope dilution mass spectrometry.

Another inter-laboratory study conducted by Rottschafer *et al.* (1971)

showed as much as 50% variation in the mercury content of homogenized fish tissue. Fifteen laboratories from industrial, government, and academic areas participated, and the methods used were atomic-absorption and X-ray fluorescence spectrometry, and destructive and non-destructive neutron activation analysis.

Brewer and Spencer (1970) of Woods Hole Oceanographic Institution conducted an involved inter-calibration study of 13 elements in sea water. Special care was taken to eliminate sampling, handling, and storage as error sources. The coefficient of variation ranged from 2.5% for strontium to 33.5% for nickel, with copper and lead values of 20.4% and 29.2%, respectively. An attempt was made to correlate the coefficient of variation with relative abundance of the element in the sea water; however, no conclusion of this sort could be drawn from the data.

8.2. PRE-CONCENTRATION AND *IN SITU* ELECTROCHEMICAL PRE-CONCENTRATION SAMPLING TECHNIQUES

8.2.1. Introduction

In trace analysis, the elements to be determined may be considered to exist in a matrix (Mizuike, 1965). If the matrix does not interfere with the analytical technique, and the species of interest is present at a sufficiently high concentration (as far as the instrumental response signal is concerned) the species may be determined by direct analysis. This is usually not the case for trace metal analysis, however. In addition to possible matrix interferences, the concentration of these metals or metal ions is normally too low for most direct analytical methods. Because of this, most environmental samples must be pre-concentrated before analysis.

The most common methods used for pre-concentration are solution evaporation, solvent extraction, precipitation, ion exchange, and electrochemical methods. These methods have been discussed in detail (Mark, 1970; Mizuike, 1965; Andelman and Caruso, 1971). Unfortunately, there are certain inherent sources of error in sample collection and handling in most of these methods. The methods that require chemical or physical manipulation, such as evaporation or extraction and precipitation methods, may be easily contaminated from the reagents and laboratory utensils used to pre-concentrate or handle the samples (Hume, 1967). Also, these methods may modify the physico-chemical characteristics of the species under investigation, which may lead to an incorrect interpretation of the data obtained (Schimpff, 1971). Electrochemical and ion-exchange methods are less susceptible to contamination or modification of the species of interest. The kinetics of ion-exchange methods, though, are rather slow.

Anodic stripping voltammetry (ASV), with detection limits down to 10^{-10} M, is one of the few analytical methods sensitive enough for direct or *in situ* analysis. Initially this technique would be thought of as being ideal for *in situ* analysis; however, inherent problems, such as poor resolution due to peak overlap (a result of a narrow analytical dissolution window), limit the method's usefulness as an *in situ* analytical technique for natural water systems where many ions may be present. These problems will be considered in detail in Chapter 10. Although ASV has drawbacks with respect to dissolution peak resolution in complex systems, the use of the electrochemically controlled potential step as an *in situ* pre-concentration sampling method is perfectly applicable (Mark, 1973).

The work presented in this chapter is based on the premise that the major problems of trace analysis in the environment are those of sampling, sample handling, and storage, and there currently exist sufficient instrumental analytical methods, such as atomic-absorption and X-ray fluorescence spectrometry and neutron activation analysis, once an appropriate sample can be presented for analysis (Mark, 1970; Mark *et al.*, 1973). Therefore, *in situ* pre-concentration sampling techniques in which the metal ions are electrochemically deposited on the surface of wax-impregnated graphite (WIG) electrodes from the sea water system should produce samples that are (1) representative of the exact environment from which they came, (2) inert to chemical and physical changes with time and handling, and (3) readily available for non-destructive analytical methods, or destructive methods if deemed necessary. Initially it might be asked, 'why not place the emphasis on developing a complete *in situ* analytical method in which the total analysis is done in the system?' Attempts have been made to do this (Schimpff, 1971; Whitfield 1975; Mancy *et al.*, 1962; Kester *et al.*, 1973; Whitfield, 1971). However, at present a complete system presents many problems, and is too expensive. A complete system may be considered a long-term solution to the problem of trace metal analysis where an *in situ* sampling system is an interim solution.

This chapter is concerned with electrodeposition techniques that are adaptable to *in situ* pre-concentration methods. Consequently, these electrodeposition methods must be conducted with controlled-current or controlled-potential electrolysis at solid electrode substrates. Some of the solid electrodes being studied today have been reviewed by Brooks and Mark (1975). Of these, pyrolytic graphite (PG), wax-impregnated graphite (WIG), and glassy carbon electrodes show the most promise.

8.2.2. Electrochemical pre-concentration methods

One of several methods that has the potential for adaptation to *in situ* pre-concentration was developed by Vassos *et al.* (1973) for study under laboratory conditions. The study involved investigating the technique of electrochemically depositing metals on a pyrolytic graphite electrode (PGE) substrate with subsequent analysis by X-ray fluorescence (XRF) spectrometry. This method is very similar to that introduced by Mark (1970) for neutron activation analysis of metal deposited films on these electrode substrates.

Experimentally the metals were deposited on the electrode using galvanostatic control. This allowed several cells to be operated in series. The yield of hard-to-deposit species was maximized using the galvanostatic technique, while the rate of hydrogen generation at the electrode was controlled. After the completion of the deposition process, the surface of the electrodes was sprayed with Krylon acrylic lacquer, for protection, and the discs were cleaved with a razor blade for storage or analysis. The method gave good results for Cu, Hg, Zn, Ni, and Co, and fair results for Cr, for samples ranging between 6.5×10^{-7} and 1.0×10^{-4} M total concentration.

Another method that may be adapted to *in situ* pre-concentration sampling was developed by Holcombe and Sacks (1973), Sacks and Holcombe (1974), and Thomas and Sacks (1978). The method involves electrochemically depositing trace metals on the surface of a thin silver wire. Then, via a high-voltage capacitive discharge excitation source, the wire is exploded in a specially designed chamber, and an emission spectrographic pattern is recorded for the deposited metals. The authors reported absolute detection limits for Cd, Ni, and Pb of 10, 10, and 15 ng, respectively. These reported values represent relative detection limits in the solution in the electrolysis cell of 8.9×10^{-9}, 3.4×10^{-8}, and 4.8×10^{-9} M for Cd, Ni, and Pb, respectively, based on the particular cell geometry, solution volume, and plating time.

Other electrochemical methods, such as anodic stripping voltammetry at mercury–graphite tubular electrodes (Lieberman and Zirino, 1974) and anodic stripping voltammetry with sample collection at tubular and ring-disc electrodes (Schieffer and Blaedel, 1977; Johnson and Allen, 1973; Laser and Ariel, 1974), may help pave the way towards the *in situ* electrochemical analysis of some metals. However, at present, experimental designs using these techniques require deoxygenation of the sample, which is not practical for *in situ* work. In addition to the problems previously mentioned, another daunting problem that must be resolved before these methods can become viable is the effect that the complex sea water matrix has on the collection response signals observed for the metals while operating under *in situ* conditions. Laboratory studies such as those conducted by Valenta *et al.* (1977) will undoubtedly bring about a better understanding of the effects of sea water with its complex biomatrix on solubilized metals.

8.2.3. *In situ* pre-concentration sampling methods

Preliminary studies of the controlled-potential electrodeposition techniques for *in situ* pre-concentration sampling [employing a pyrolytic graphite

(PGE) working electrode, an Ag|AgCl reference electrode, and a platinum auxiliary electrode] were carried out by Mark *et al.* (1973). The electrode surfaces were in contact with the water system and the studies were conducted under quiescent conditions (giving a diffusion-controlled mass transport process) to evaluate the effects of electrolysis time and potential variations, and electrode film handling. The results of these studies, based on the calculated efficiencies, indicated that a convection–diffusion (stirred) process was necessary.

Flow and closed volume cell approaches

In order to increase the deposition efficiencies, a flow-through electrolysis cell was developed. A wax-impregnated graphite electrode (WIGE) was used in the flow cell instead of the PGE as the working electrode. This allowed about a 50-fold increase in surface area at a fraction of the cost of using a PGE, with little or no sacrifice in deposition characteristics. An increase in deposition efficiency was noticed but there were other inherent problems that would restrict the use of a flow cell system for *in situ* sampling: (1) the efficiencies were still lower than expected, and (2) as the solution flowed between the working and auxiliary electrode, a current density gradient was created at the leading edge of the electrodes, which led to non-uniform leading edge plating on the electrode surface.

A flow cell system can be viewed as a stirred solution which should follow the Lingane equation for first order electrode kinetics (Lingane, 1948; Delahay, 1952; Reilley and Murray, 1963):

$$c_t = c_0 e^{-kt} \tag{1}$$

or

$$\ln \frac{c_t}{c_0} = -kt \tag{2}$$

where c_t is the molar concentration of the electroactive species at time t, c_0 is its initial concentration in bulk solution, and t is the deposition time (minutes). The rate constant k is defined by

$$k = DA/V\delta \tag{3}$$

where D is the diffusion coefficient of the electroactive species, A is the surface area of the working electrode, V is the volume of solution (the volume of the cell), and δ is the diffusion layer thickness. D is a function of the temperature, pressure, and salinity. The rate constant (k) would have to be known for each *in situ* deposition in order to relate the amount of metal deposited to the concentration of the metal ions present in the system.

One way around this problem is to use a closed volume cell and employ an exhaustive electrolysis technique, using a rotating electrode for rapid deposition.

In a closed cell of known volume as long as the deposition is allowed to go to completion, the mass of material deposited is a function of that volume. Of course, the time required for exhaustive electrolysis is primarily a function of δ, which is dependent on the angular velocity of the rotating electrode.

8.3. ROTATING AUXILIARY ELECTRODE CELL

8.3.1. Cell design

The cell was constructed from 2.5 inch (65 mm) diameter Plexiglas (polyacrylate) rod. It was designed so that it could be machined into four basic components which screw together and are isolated by O-rings (Figure 1). For laboratory studies, this arrangement allowed one part of the cell to be modified without having to machine an entirely new cell. The basic components are the working electrode holder, the working electrode holder-

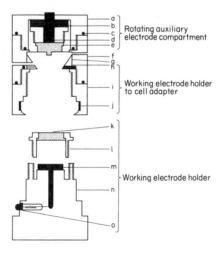

Figure 1. The electrolysis cell. (a) Auxiliary electrode holder; (b) coupling magnet; (c) O-rings (all solid circles); (d) epoxy-impregnated bearing and electrode contact; (e) rotating auxiliary electrode; (f) cell compartment; (g) wedge insert; (h) Ag|AgCl reference electrode; (i) working electrode adapter; (j) metal contact for working electrode; (k) working electrode; (l) working electrode holder; (m) working electrode contact post; (n) working electrode holder; (o) spring loaded contact. From Brooks and Mark (1977)

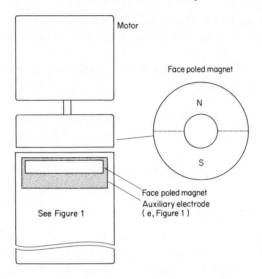

Figure 2. The magnetic coupling system. From Brooks and Mark (1977)

to-cell adapter, the cell wedge insert, the cell compartment, and the rotating auxiliary electrode compartment (Figure 1).

A tapered cell compartment was used instead of a cylindrical shape to help eliminate undesirable hydrodynamic turbulence, which will be discussed later.

To rotate the auxiliary electrode, a 12-V automotive type $\frac{1}{12}$ horsepower motor (the type of motor used to drive automotive air conditioning and heater fans) was used with a magnetically coupled system as shown in Figure 2. These motors, however, used in the analogue electronic mode, will not maintain a constant rotation speed or give a linear response to changes in the applied voltage owing to frictional effects caused by temperature changes while the motor is running. To remedy this problem, the motor was interfaced with a phase-locked loop (PLL).

8.3.2. Phase locked loop

A PLL is a frequency-controlling electronic instrument which compares a response signal from the controlled device to an external reference frequency and holds the two in phase. Therefore, if the reference frequency is tunable, the rotation speed of the controlled device will be variable. The PLL will assure that regardless of the load the motor will maintain a constant speed throughout its power range.

The PLL employed is a modified version of that described by Means

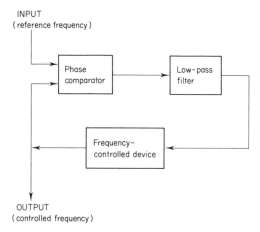

INPUT
(reference frequency)

Phase
comparator

Low-pass
filter

Frequency-
controlled device

OUTPUT
(controlled frequency)

Figure 3. Block diagram of Phase Locked Loop (PLL) for controlling the rotation speed of the auxiliary electrode

(1970). A block diagram is shown in Figure 3. There are essentially three elements:

1. The phase comparator, which produces an output signal proportional to the difference between the reference input frequency and the controlled output frequency from the device being driven;
2. The low pass filter, which is used to control the rate at which the controlled device can change frequencies and also to smooth the input signal to the frequency-controlled device;
3. The frequency-controlled device, which is the combination of the motor, the operational amplifier, and the power amplifier, together with the tachometer sensor which converts the rotation speed of the motor to a frequency.

The return of the signal from the frequency-controlled device to the phase comparator completes the feedback loop.

A reed switch, was used as a tachometer sensor to obtain the feedback signal from the rotating magnet attached to the shaft of the motor. The reed switch opens and closes as the rotating magnet changes from the north to the south pole. The pulses are obtained by allowing a current to flow through the reed switch when the contacts are closed, and this signal is sent to a pulse shaper.

A pulse shaper circuit was used on both the reed switch and the reference source before the signals were transmitted to the PLL. It consisted of a monostable device which ensured that the high portion of the two pulse signals were of the same pulse duration. Therefore, any change in frequency occurred by a change in the time duration of the low portion of the signal and hence the number pulses per unit time.

8.3.3. Hydrodynamic considerations

Rotating disc electrode systems have been studied by Levich (1962) and others (Milsaps and Pohlhausen, 1952; Riddiford, 1966; Opekar and Beran, 1976; Bruckenstein and Miller, 1977), and also have been described by Adams (1969) and Albery and Hitchman (1971). However, in these systems, the working electrode is allowed to rotate and the solution reservoir is considered to be 'infinitely large' with respect to the surface area of the electrode. This is not the case in this particular cell system. One can envisage the hydrodynamics of this system as an infinite number of planes rotating about the y'-axis as a function of the angular velocity (ω) (see Figure 4). In addition to the rotation about the y'-axis, the periphery of these planes (the cell boundaries) must be circulating in a steady-state manner in order to maintain, hydrodynamically, a conservation of mass about this axis.

As the solution is being dragged across the surface of the electrode, the hydrodynamic boundary layer created should be very similar to the Nernst diffusion layer model for stirred solutions. The thickness (δ) of the hydrodynamic boundary layer in the Levich model is approximated by

$$\delta \approx 3\left(\frac{\nu}{\omega}\right)^{1/2} \tag{4}$$

where ν is the kinematic viscosity of the solution. The transport rate of a species across the hydrodynamic boundary layer to the surface of the electrode is a function of the thickness of this layer (which is related to the angular velocity by equation 4) and the diffusion rate of the species. Once inside the boundary layer, the process is viewed as being a diffusion process

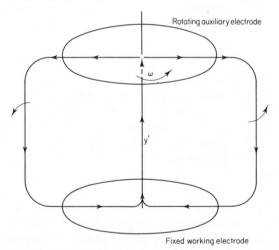

Figure 4. Hydrodynamic convection currents in deposition cell. y', rotation axis of auxiliary electrode

only. The act of getting the species to the boundary layer is primarily a convective or hydrodynamic process because the stirring action of the rotating disc will supersede any natural diffusion processes in the bulk of the solution. The two systems combine to give a hydrodynamic convective diffusion process which provides the 'mass transport' for the electrochemical reaction.

The rate of mass transport is determined by the angular velocity of the rotating electrode. This in turn dictates the current level measured for the electrodic process. By imposing the appropriate boundary conditions ($c = c_b$ as $y \to \infty$ and $c = 0$ at $y = 0$), Levich (1962) derived an equation that relates the hydrodynamic process to the electrochemical process so that

$$i_L = 0.62 \, n \, FAc_b \, D^{2/3} \nu^{-1/6} \omega^{1/2} \qquad (5)$$

where i_L is the limiting current, n is the stoichiometric number of electrons transferred, F is the Faraday, and c_b is the bulk concentration of the electroactive species. The other terms have been defined previously.

Although the cell designed for this work differs from the Levich rotating disc system because it has finite boundaries and the working electrode is stationary while the auxiliary electrode rotates, similarities are also observed that will be discussed later.

8.4. INVESTIGATION OF DIFFERENT AUXILIARY ELECTRODE SUBSTRATES

8.4.1. Cell characteristics

The normal operation of the electrodeposition cell was to draw approximately 30 cm^3 of the electrolyte solution to be analysed into an attached 50-cm^3 syringe through the three-way stopcock (D_f, Figure 5), and pump it into the cell displacing all the air. The rotating electrode drive motor was turned on and the solution was potentiostated for an appropriate time (at a potential of about -1.00 V *versus* the saturated calomel electrode, SCE). At the end of the electrodeposition the solution was emptied from the cell into a 50-cm^3 plastic storage bottle and was analysed by differential pulse anodic stripping voltammetry (DPASV), using a PAR 174A Polarographic Analyzer, to determine the amount of cation removed. The results of the sample run were compared with a stripping voltammogram of the stock to determine the percentage of the cation deposited on the surface of the working electrode.

The Cu(II) $\to$ Cu(0) system was chosen to evaluate the cell characteristics, since Vassos *et al.* (1965) had previously observed that copper forms reproducible deposits on graphite electrode substrates under stirred conditions. A wax-impregnated auxiliary electrode with a flat surface was initially

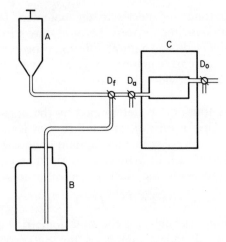

Figure 5. Electrolysis cell with filling syringe and reservoir. A, Syringe; B, solution reservoir; C, cell; D_f, cell filling stopcock; D_a, air inlet stopcock for emptying cell; D_o, cell outlet stopcock. From Brooks and Mark (1977)

used. However, based on angular velocity–time studies, it was found that the convective forces supplying the cations to the working electrode surface were not sufficient to obtain exhaustive electrolysis in a reasonable length of time (20–30 min). Therefore, it was replaced with a paddle design (see Figure 1). Typical results for a 3.1×10^{-4} M Cu(II) solution in a cylindrical cell are shown in Figure 6. Although this concentration is not at the trace

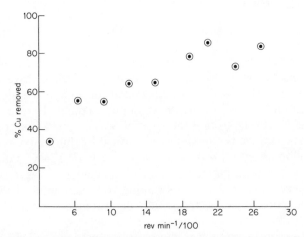

Figure 6. Percentage of copper removed from a 3.1×10^{-4} M copper solution as a function of auxiliary electrode rotation speed (revolutions per minute in a cylindrical cell). The values are for a 20-min deposition time with wax-impregnated graphite (WIG) auxiliary and working electrodes

level, it was used to observe the mass transport effects in the cell. For a 20–min deposition a maximum of about 85% of the Cu(II) is removed. The curve is not smooth and there is a series of breaks over the angular velocity range. It was postulated that this condition was caused by cell turbulence created from the abrupt boundaries of the cylindrically shaped cell compartment. Because the effects of turbulence interfered with the formation of a uniformly deposited film on the working electrode surface, a tapered cell insert was used in an attempt to reduce the degree of turbulent flow in the cell (see Figure 1). The reduction of cell turbulence was reflected in an improved uniformity of the deposited film.

It was also observed that the cell current initially increased to a maximum, remained there for a short time, and then decreased by a first-order decay. A plot of the limiting current (i_L) *versus* $\omega^{1/2}$ gave a linear response curve when the insert was used (see Figure 7b) and in contrast to the non-linear relationship observed when the insert was not present (Figure 7a). According to Levich's theory, a hydrodynamic condition of laminar flow exists in a cell when i_L is proportional to $\omega^{1/2}$, while ω raised to other powers describes

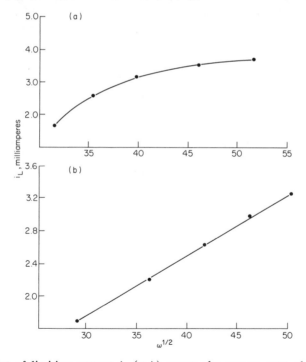

Figure 7. Plot of limiting current i_L (mA) *versus* the square root of the angular velocity $\omega^{1/2}$ (rev min^{-1}) for a cylindrical cell (a) and a tapered cell (b). Copper is deposited from a 3.1×10^{-4} M Cu(II) solution using WIG working and auxiliary electrodes

a state of turbulent flow. This criterion can be used as an approximate method for determining if a condition of laminar flow exists in the cell. It should be realized, however, that the boundary conditions that exist for the Levich system are different from those imposed on the system described here, and no hard and fast rules should be deduced from such plots.

8.4.2. Wax-impregnated graphite auxiliary electrode

Using the cell insert, the results of a series of deposition runs were obtained for a 1.57×10^{-4} M solution of Cu(II) at a constant angular velocity of 850 rev min^{-1} by varying the deposition time from 2.5 to 40.0 min (Table 1). From these results it is appeared that 20 min was an adequate time for the deposition, where a maximum of about 93% of the Cu(II) was removed from solution. Then a series of deposition experiments were run for a constant time of 20 min and varying the angular velocity from 850 to 2500 rev min^{-1}. As can be seen in Table 2, there is essentially no improvement in the percentage of Cu(II) removed from solution by increasing the angular velocity of the rotating electrode in this series of 20-min depositions. There did appear to be a slight hydrodynamic transition occurring at 1300 and 1750 rev min^{-1}, but the effects are much smaller than those observed earlier.

The important fact seen in these two tables is that after 20 minutes plating time the percentage of Cu(II) removed from solution, has become essentially independent of both the time and the angular velocity. This means that the maximum percentage of Cu(II) that can be removed from solution, since 100% removal is not achieved, is controlled by some intrinsic phenomenon of the electrodeposition process itself. It was proposed that this phenomenon could be an equilibrium process caused by the formation of soluble chlorine in the cell. For example, if Cu(II) is being reduced at the working

Table 1. Percentage of Cu(II) removed from 1.57×10^{-4} M solution at 850 rev min^{-1} using wax-impregnated graphite working and auxiliary electrodes (from Brooks and Mark, 1977)

Time	Cu(II) removed, %
2.5	16.07
5.0	41.07
10.0	85.27
15.0	88.39
20.0	92.75
30.0	92.53
40.0	92.86

Table 2. Percentage of Cu(II) removed from 1.57×10^{-4} M solution with a 20-min deposition time (cell construction as in Table 1) (from Brooks and Mark, 1977)

Rev min^{-1}	Cu(II) removed, %
850	93.39
1300	92.15
1750	91.24
2130	93.39
2500	93.80

electrode (the cathode) by a two-electron transfer process:

$$Cu^{2+} + 2e^- \rightarrow Cu^0$$

then, in order to maintain a conservation of charge in the cell, there must be a species oxidized at the auxiliary electrode or anode. Since the solution medium contains a rather high concentration (0.5 M) of Cl$^-$ ions, it is possible that chloride is being oxidized at the anode by the following process:

$$2Cl^- \rightarrow Cl_2 + 2e^-$$

This possibility is further supported by the work of Anderson and Tallman (1976), who found that soluble chlorine is generated at the surface of an epoxy-impregnated graphite electrode (EIGE) if the potential is sufficiently positive. The EIG electrode has electrochemical characteristics very similar to those of the WIG electrode.

The soluble chlorine in return re-oxidizes the Cu(0) when a certain concentration level is reached according to the reaction

$$Cl_2 + Cu^0 \rightleftharpoons 2Cl^- + Cu^{2+}$$

for which

$$K^* = \frac{[Cl^-]^2[Cu^{2+}]}{[Cl_2]} \tag{6}$$

The activity of Cu(0) is, of course, equal to unity.

To test the hypothesis, two 20-min depositions were run in solutions containing 1.57×10^{-5} M copper at 850 rev min^{-1} and they resulted in the removal of 38.8% and 36.3% of Cu(II), respectively, from solution. The concentration of Cu(II) remaining in solution was 9.63×10^{-6} and 1.00×10^{-5} M, respectively. For the 1.57×10^{-4} M solution with a 20-min deposition, 1.14×10^{-5} and 1.04×10^{-5} M (taken from Tables 1 and 2, respectively) of copper remained in solution. It is apparent, in comparing the values for

1.57×10^{-5} and 1.57×10^{-4} M solutions and considering the fact that the rate of deposition should be independent of concentration for a first-order process, that there is some type of equilibrium process taking place in the cell. If this equilibrium process is caused by the formation of chlorine at the auxiliary electrode, then it may be possible to shift the chlorine overpotential (the chlorine formation voltage with respect to the potential of the anode) to a more positive potential by changing the type of material used for the anode (Brooks and Mark, 1977).

8.4.3. Platinum and tantalum auxiliary electrodes

The function of the auxiliary electrode is to supply the current necessary to maintain the desired potential at the surface of the working electrode. During a cathodic deposition process, the potential of this electrode is positive with respect to the cathode, so the material used must be inert to oxidation while at the same time possess a relatively low resistivity. Commonly used materials are gold, platinum, and silver (in chloride-free solutions).

A platinum auxiliary electrode was constructed to give a paddle of the same height as the WIG auxiliary electrode previously used. A platinum wire was secured in the paddle crease or fold for electrical contact and the foil was glued to a pre-machined plastic disc with Varian Torr Seal resin. The excess platinum foil was trimmed away with scissors and filed to its final shape.

The same procedure as before was followed for a series of depositions from a 1.57×10^{-5} M Cu(II) solution, which were run at 850 rev min^{-1} with deposition times ranging from 2.5 to 40 min. The results of these depositions in terms of the percentage of Cu(II) removed from solution are shown in Figure 8. For a 20-min deposition time, approximately 82% of the Cu(II) was removed, which corresponded to about 2.80×10^{-6} M of Cu(II) left in solution. The experiment was repeated for a 1.57×10^{-4} M solution to compare the amount of copper removed with platinum with that with a WIG electrode (see Table 3). It can be seen from these results and Figure 8 that there was an improvement in the amount of Cu(II) removed from solution, especially from a 1.57×10^{-5} M solution. Although the use of platinum did appear to shift the equilibrium or the chlorine overpotential compared with that observed with the WIG electrode, it was not a sufficient shift to enable concentrations below 1.57×10^{-6} M solutions to be deposited. However, these results did suggest that it was possible to control the problem by using other metals as auxiliary electrodes.

Tantalum and titanium are less commonly used as anode materials in electrochemistry, but have been used commercially to generate chlorine gas from brines (Tomashov and Zalivalov, 1970; Klein, 1908). Both of these

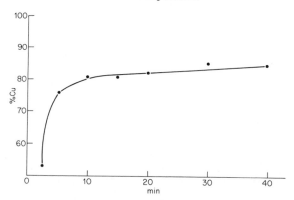

Figure 8. Plot of percentage of copper removed from solution *versus* deposition time for the electrolysis of a 1.57×10^{-5} M Cu(II) solution. A tapered cell was used with a platinum auxiliary electrode rotating at 850 rev min^{-1}. From Brooks and Mark (1977)

metals are reported to be inert in chloride media (Farbenfabriken Bayer, 1970). Titanium is a hard, rigid metal and it was thought that this metal would be difficult to fabricate into a paddle electrode. Therefore, a 0.127 mm tantalum foil was fabricated into an auxiliary electrode in a similar manner to that described for the platinum electrode.

Electrodepositions from 1.57×10^{-4} M Cu(II) solution were run at 850 rev min^{-1} for electrolysis periods ranging from 2.5 to 30 min and more than 99% of the Cu(II) was deposited in all cases except for the 2.5-min deposition (Table 3). The electrodeposition experiments were repeated in the same manner for a 1.57×10^{-5} M (1 ppm) solution of Cu(II) and the results are shown in Figure 9. For a 20-min deposition, essentially 100% of

Table 3. Comparison of percentage of Cu(II) removed with WIG, platinum and tantalum auxiliary electrodes from a 1.57×10^{-4} M solution at 850 rev min^{-1} (from Brooks and Mark, 1977)

Deposition time, min	Cu(II) removed, %		
	WIG	Pt	Ta
2.5	16.07	—	91.46
5.0	41.07	—	>99
10.0	85.27	—	>99
15.0	88.39	91.72	>99
20.0	92.75	95.54	>99
30.0	92.53	98.73	>99

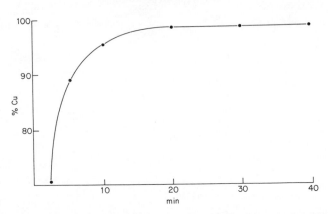

Figure 9. Plot of percentage of copper removed from solution *versus* deposition time in a cell with a rotating tantalum auxiliary electrode. Other conditions as in Figure 8. From Brooks and Mark (1977)

the Cu(II) was deposited from solution and similar behaviour was observed for a 1.57×10^{-7} M Cu(II) solution. The results obtained from this deposition series indicated that tantalum metal used as an auxiliary electrode solved the problems that hampered the WIGE and the platinum auxiliary electrodes. A possible mechanism for this behavior is discussed in the next section.

8.4.4. Kinetic effects of the auxiliary electrode substrates on the deposition process

The controlled potential electrolysis of a stirred solution is dictated by the Lingane equation (equation 2), which implies that the electrolysis should obey first-order kinetics. If the operational parameters of this equation are held constant, then the fraction of an electroactive species deposited, $(c_0 - c_t)/c_0$, becomes a function of time (t). It is desirable to collect the samples by exhaustive electrolysis in a reasonable length of time (about 20 min). The deposition rate constant, k, is dependent on the diffusion layer thickness, δ, which is in turn a function of the stirring rate of the solution (Lingane, 1948). As the value of k increases, obviously the time required to complete the exhaustive electrolysis decreases. The rate constant for first-order heterogeneous kinetic process should be independent of the concentration of the metal ion species, within limits. Also, as Lingane indicated, the rate constant should be independent of the type of species being reduced (Delahay, 1952; Reilley and Murray, 1963) for reversible systems, so the results of these studies should apply to any reversible system.

Thus, for well behaved deposition processes, a linear response curve should be obtained for a plot of $\ln (c_t/c_0)$ *versus* t, where the rate constant is given by the slope of the line. From the rate constant for a first-order kinetic process, the half-life may be determined by

$$t_{1/2} = \frac{2.303 \log 2.00}{k} \qquad (7)$$

$t_{1/2}$ may be used as a criterion for evaluating the efficiency of the deposition process. For example, if the process has a half-life of 4 min, then 99.4% of the electroactive species would be removed during a 30-min electrolysis. If, however, a secondary reaction became the controlling factor and the half-life was extended to 7 min, then only 95% of the electroactive species would be deposited in 30 min. A deposition time of about 53 min would be required to remove about 99.4% of the electroactive species under these conditions, if the secondary reaction did not create a steady-state or limiting condition in the cell. If a secondary reaction creates a steady-state condition in the electrolysis cell, a non-linear plot of the Lingane equation will result. Lingane plots of the rate curves shown in Figures 8 and 9 were found to have slopes of 0.0568 and 0.155 min^{-1} for the platinum and tantalum electrodes, respectively. Thus, the half-life for the platinum auxiliary electrode was found to be 20.8 min and a value of 4.47 min was obtained for the tantalum auxiliary electrode. The correlation coefficients obtained for the two deposition processes were 0.831 and 0.965, respectively.

If a competing side-reaction had not been produced at the platinum electrode, approximately 2.75 half-lives would have elapsed before approximately 85% of the copper was removed for a 30-min deposition time. In fact 1.44 half-lives had elapsed before 85% of the copper was deposited. The discrepancy between the observed and calculated values occurs because a competing side-reaction taking place at the auxiliary electrode causes the system to deviate from a simple first-order kinetic mechanism. In contrast, the competing side-reactions were essentially eliminated by using a tantalum auxiliary electrode and the deposition process agrees well with a first-order kinetic model.

The results of these studies suggest that the tantalum metal auxiliary electrode does shift the Cl_2 overpotential to sufficiently positive values for it not to interfere with the deposition process. This suggestion is supported by the work of Tomashov and Zalivalov (1970), who showed that the presence of Cl^- ions in solution had no effect on the anodizing of the tantalum anode and the oxidation process taking place at the anode is the decomposition of water to oxygen:

$$2H_2O \rightarrow 4e^- + 4H^+ + O_2$$

A mechanism for the decomposition reaction was proposed by Kokhanov *et al.* (1971) in which water is first decomposed to give atomic oxygen on the surface of the tantalum anode and this oxygen then leaves the surface as soluble molecular oxygen. Once in solution, oxygen is probably reduced back to water at the working electrode (the cathode) with apparently little effect on the copper deposition.

This proposed mechanism is supported by a phenomenon that was observed to take place on the surface of the tantalum. If the cell is potentiostated at a potential cathodic enough to reach the hydrogen overvoltage of the WIG working electrode, a film is formed on the surface of the tantalum auxiliary electrode that led to it becoming passive. The film was initially light green but became violet–red when the electrode reached its passive state. The film was at first thought to be a chloride complex, even though tantalum has been reported to be inert towards chloride attack, since tantalum chloride complexes usually range in colour from yellow to purple–red as solids (Cotton and Wilkinson, 1972). The film, however, was found to be insoluble in absolute ethanol and dilute sulphuric and hydrochloric acids, which is not characteristic of tantalum halide complexes. Also, X-ray fluorescence (XRF) spectrometry failed to detect the presence of chloride on the surface of the tantalum.

The other possibility was that the film was an oxide coating. This posibility would be consistent with the XRF spectrum, since oxygen is transparent to X-radiation, and with the fact that tantalum oxide complexes are chemically inert in most acids, except for concentrated hydrofluoric acid, in which the film readily dissolved. Also, Brunck (1912) stated that tantalum oxide is sometimes formed at an anode during the electrolysis of aqueous solutions. The only fact that was not consistent was the colour of the film, since tantalum oxide complexes are reported to be either clear, white, or grey (Weast and Selby, 1968; Cotton and Wilkinson, 1972). To resolve the question, strips of tantalum metal were electrolysed at 1.5 V *vs.* Ag|AgCl in 0.5 M NaCl/0.05 M MgSO$_4$, 0.5 M NaCl, and 0.5 M NaNO$_3$ solutions. A red-violet film was obtained on the surface of the tantalum strips in each case. These results show that the film formed was not a chloride or oxy-chloride complex and probably was an oxide complex.

Since the tantalum auxiliary electrode showed a tendency to form an oxide complex, the possibility of ionic or complexed tantalum species poisoning the surface of the working electrode existed. Therefore, a background electrolysis was run for 60 min and at −1.00 V *cs.* Ag|AgCl (the potential used for all of the depositions) with only electrolyte in the cell. XRF spectra were taken of the WIG working electrode and there was no evidence of tantalum being reduced or forming on the surface of the working electrode.

8.4.5. Laboratory investigation of two and three-component systems

Although Cu(II) was primarily used to evaluate the performance of the electrochemical system, mixed solutions containing 0.50 μg dm^{-3} of Cu(II) and Pb(II) and of Cu(II), Pb(II), and Cd(II) were also investigated. The simultaneous deposition of Cu(II) and Pb(II) posed no problems and both metal ions were quantitatively removed from solution. The results of the simultaneous deposition of Cu(II), Cd(II), and Pb(II), however, indicate that only about 81.0% of the Cd(II) is removed from solution when the cell is potentiostated at −1.00 or −1.13 V *vs.* Ag|AgCl. This phenomenon is not unusual for cadmium in chloride media. The effects of chloride on cadmium deposition have also been studied by Neeb and Kiehnost (1967) and Kadyrov and Golubev (1970), and these workers also report that Cl⁻ affects the electrochemistry of cadmium. Branica *et al.* (1977) observed such effects in their study of sea water and suggested that a soluble chloride complex (CdCl⁺) was formed in 0.5 M chloride solutions.

It is interesting that the addition of cadmium to the solution affected the quantitative removal of Cu(II) from solution but did not affect the quantitative removal of Pb(II). However, this is not expected to be a problem for the analysis of copper in sea water since copper is present at a considerably higher concentration than cadmium.

Other laboratory studies

The laboratory evaluation of the cell is by no means complete. In addition to the area of further study previously mentioned, studies are being initiated for deposition process with trace metals one and two orders of magnitude lower in concentration. The effects of different buffers and of complex buffer systems on the rate of deposition still need to be evaluated. Studies on repetitive cycling where the cell is filled with a fresh solution and the metal ions of this solution are deposited on the electrode together with those remaining from a previous deposition were initiated as a sample amplification technique. This method of pre-concentration amplification shows promise, but needs further work for complete evaluation of the technique. However, it has allowed semi-quantitative values of the amount of metal ions deposited to be determined by X-ray fluorescence spectrometry, a rather insensitive technique when very small quantities are to be determined. The effects of temperature changes on the deposition process from 5 to 40°C have been investigated for the Cu(II) system and complete removal of the copper was found for a 30-min deposition time over this range.

8.5. LIMITATIONS IMPOSED BY SEA WATER SYSTEMS

It is well realized by researchers studying trace metal characterization in sea water that to a great extent these metals exist not as free ions, but as complexed species (see Section 1.4). Because of the multitude of organic and inorganic complexing agents that exist in sea water, the number of possible complexed species that can be formed is large. Hence a major effort is required to identify the complexed species and to determine the stability constants and the kinetics of their formation. Where the species exist at sufficiently high concentrations, electrochemical methods are commonly employed for this work (Nürnberg and Valenta, 1975; Allen, 1974; Florence and Batley, 1976; Chau and Lum-Shue-Chan, 1974). The specific techniques used are discussed in Chapters 2 and 10. However, most of these methods require the removal of samples from the environmental system. Mark (1970) suggested that 'the metal ion valence and/or the degree of complexation can be expected to vary with pressure, oxygen content, biological condition, and light intensity and light energy'. It would be impossible to maintain these conditions if the sample were removed from the water system and stored for subsequent analysis.

The method described in this chapter should give direct information about the concentrations of unbound metal ions or those metal ions that exist in labile or unstable complexes (Turner and Whitfield, 1979a,b). Combining this information with information that could be obtained from a total analysis for a particular metal will give data on the degree that the metal is complexed. Also, this information, coupled with knowledge about the types and concentrations of complexing ligands present, may help lead to a better understanding about the *in situ* stability constants of the metal–ligand complexes.

8.6. FUTURE PLANS FOR THE CELL SYSTEM

At present, all studies on the cell system have been conducted in the laboratory. It is the contention of the authors that these studies must be completed prior to field evaluation studies in order to have a reference point to interpret correctly the results of field studies. Preparation of the cell system for field studies will require some minor modifications to the design used for laboratory studies. First the electrolysis cell case must be redesigned to add on another compartment to house the rotating auxiliary drive motor and coupling system. Since it is necessary to maintain a rotation velocity of only 850 rev min^{-1} to achieve exhaustive electrolysis in 30 min, a smaller drive motor may be substituted for field work. The controlling electronics, including the potentiostat, will be housed in a waterproof polyacrylate canister which was previously used for mass transport field studies (Mark, 1973, 1970).

The entire system can be powered by either a non-submerged automotive 12-V battery connected to the submerged electrolysis unit via a sturdy power cable or a submersible sea water battery. These changes will make the system portable and adaptable to field studies.

8.7. ACKNOWLEDGEMENTS

This research was supported in part by the National Science Foundation and the National Fellowship Fund.

REFERENCES

Adams, R. N. (1969). Mass transfer by forced convection. In *Electrochemistry at Solid Electrodes* (Ed. A. J. Bard). Marcel Dekker, New York, pp. 67–110.

Albery, W. J., and M. L. Hitchman (1971). The simple rotating disk system. In *Ring-disc Electrodes*. Clarendon Press, Oxford, pp. 9–16.

Allen, H. E., (1974). Characterization of aqueous trace metal species and measurement of trace metal stability constants by voltammetric techniques. *Thesis*, University of Michigan, Ann Arbor, Mich.

Andelman, J. B., and S. C. Caruso (1971). Concentration and separation techniques. In *Water and Water Pollution Handbook*, Vol. 2, (Ed. L. L. Ciaccio). Marcel Dekker, New York, pp. 483–591.

Anderson, J. E., and D. E. Tallman (1976). Graphite–epoxy mercury thin film working electrodes for anodic stripping voltammetry. *Anal. Chem.*, **48**, 209–212.

Branica, M., D. M. Novak, and S. Bubic (1977). Application of anodic stripping voltammetry to determination of the state of complexation of traces of metal ions at low concentration levels. *Croat. Chem. Acta*, **49**, 539–567.

Brewer, P. B., and D. W. Spencer (1970). Trace element intercalibration study. Woods Hole Oceanographic Inst., Woods Hole, Mass, Ref. No. 70–62.

Brooks, E. E., and H. B. Mark, Jr. (1975). Electroanalytical techniques in trace metal ion analysis. *Rev. Anal. Chem.*, **3**, 1–26.

Brooks, E. E., and H. B. Mark, Jr. (1977). Design of a new type of enclosed cell for rapid exhaustive electrodeposition of metal Ions. *J. Environ. Sci. Health*, **10**, 511–521.

Bruckenstein, S., and B. Miller (1977). Unraveling reactions with rotating electrodes. *Accounts Chem. Res.* **10**, 54–61.

Brunck, O. (1912). Tantalum electrodes. *Chem. Ztg.*, **36**, 1233–1234.

Chau, Y. K., and K. Lum-Shue-Chan (1974). Determination of labile and strongly bound metals in lake water. *Water Res.*, **8**, 383–388.

Cotton, F. A., and G. Wilkinson (1972). Chemistry of the transition elements. In *Advanced Inorganic Chemistry, A Comprehensive Text*. Interscience, New York, 3rd ed., pp. 934–942.

Delahay, P. (1952). Electrolysis at controlled potential and related methods. In *New Instrumental Methods in Electrochemistry*. Interscience, New York, pp. 282–297.

Farbenfabriken Bayer (1970). Anodes for electrolysis of sodium chloride. *Fr. Pat.*, 2 014 883.

Florence, T. M., and G. E. Batley (1976). Trace metal species in sea water. *Talanta*, **23**, 179–186.

Holcombe, J. A., and R. D. Sacks (1973). Exploding wire excitation for trace analysis of mercury, cadmium, lead, and nickel using electrodeposition for preconcentration. *Spectrochim. Acta*, **28B**, 451–467.

Horne, R. A. (1969). Instruments and techniques. In *Marine Chemistry*. Wiley-Interscience, New York, pp. 129–150.

Hume, D. H. (1967). Present capabilities and limitations. In *Analysis of Water for Trace Metals* (Ed. R. F. Gould). Advances in Chemistry Series, American Chemical Society, Washington, D.C., No. 67, pp. 30–44.

Hume, D. H. (1975). Equilibrium concepts in natural water systems. In *Analytical Methods in Oceanography* (Ed. T. R. P. Gibb, Jr.). Advances in Chemistry Series, American Chemical Society, Washington, D.C., No. 147, pp. 1–8.

Johnson, D. C., and R. E. Allen (1973). Stripping voltammetry with collection at a rotating disk electrode. *Talanta*, **20**, 305–313.

Kadyrov, M., and A. Golubev (1970). Behavior of cadmium in sodium chloride solutions. *Zasheh. Met.* **6,** 569–571.

Kester, D., K. Crocker, and G. Miller, Jr. (1973). Small-scale oxygen variation in the thermocline. *Deep Sea Res.,* **20,** 409–412.

Klein, A. (1908). Electrode materials for electrolytical processes. *U.S. Pat.,* 3 372 107.

Kokhanov, G. N., Y. V. Dobrov, and L. A. Khaneva (1971). Anodic evaluation of oxygen on tantalum. *Elektrokhimiya,* **7,** 928–931.

Laser, D., and M. Ariel (1974). Anodic stripping with collection using thin mercury films. *J. Electroanal. Chem.,* **49,** 123–132.

Levich, V. G. (1962). General theory of convective diffusion in liquids. In *Physico-chemical Hydrodynamics.* Prentice-Hall, Englewood Cliffs, N.J., pp. 60–72.

Lieberman S. H., and A. Zirino (1974). Anodic stripping voltammetry of zinc in sea water with a tubular mercury–graphite electrode. *Anal. Chem.,* **46,** 20–23.

Lingane, J. J. (1948). Controlled potential electrolysis. *Anal. Chim. Acta,* **2,** 584–601.

Mancy, K. H., D. A. Okun. and C. N. Reilley (1962). A galvanic cell oxygen analyzer. *J. Electroanal. Chem.,* **4,** 65–92.

Mark, H. B., Jr. (1970). Application of an electrochemical preconcentration technique to neutron activation analysis of trace metal ions. *J. Pharm. Belg.,* **25,** 367–399.

Mark, H. B., Jr. (1973). The *in situ* analysis of trace metal ions in seawater systems employing preconcentration sampling techniques. Unpublished research proposal.

Mark. H. B., Jr., R. Boczkowski, and K. E. Paulsen (1973). *In-situ* sampling techniques for trace analysis in water. *Proc. Joint Conf. Sensing Environ. Pollutants,* Instrument Society of America, Pittsburgh, Pa., pp. 295–307.

Means, D. K. (1970). Electrochemical photo effect. *Thesis,* University of Michigan, Ann Arbor, Mich.

Milsaps, K., and K. Pohlhausen (1952). Heat transfer by laminar flow from a rotating plate. *J. Aeronaut. Sci.,* **19,** 120–126.

Mitchell, R. L. (1957). Emission spectrochemical analysis. In *Trace Analysis* (Eds. J. H. Yoe and H. J. Koch). Wiley, New York, pp. 398–412.

Mizuike, A. (1965). Separation and Preconcentration. In *Trace Analysis.* (Eds. G. H. Morrison). Interscience, New York, pp. 103–159.

Neeb, R., and I. Kiehnost (1967). Anodic stripping voltammetry. VII. Effect of salts on the anodic peak height in inverse voltammetry. *Z. Anal. Chem.,* **226,** 153–159.

Nürnberg, H. W., and P. Valenta (1975). Polarography and voltammetry in marine chemistry. In *The Nature of Seawater* (Ed. E. D. Goldberg). Dahlem Konferenzen, Berlin, pp. 88–136.

Opekar, F., and P. Beran (1976). Rotating disc electrodes. *J. Electroanal. Chem.,* **69,** 1–105.

Reilley, C. N., and R. W. Murray (1963). Electrolytic separations. In *Treatise on Analytical Chemistry,* Part I, Vol. 4 (Eds. I. M. Kolthoff and P. J. Elving). Interscience, New York, pp. 2192–2195.

Riddiford, A. C. (1966). The rotating disk system. In *Advances in Electrochemistry and Electrochemical Engineering,* Vol. 4. (Eds. P. Delahay and C. W. Tobias). Interscience, New York, pp. 47–116.

Robertson, D. E. (1968). Role of contamination in trace element analysis of seawater. *Anal. Chem.,* **40,** 1067–1072.

Rottschafer, J. M., J. D. Jones, and H. B. Mark, Jr. (1971). Simple, rapid method for determining trace mercury in fish via neutron activation analysis. *Environ. Sci. Technol.,* **5,** 336–338.

Sacks, R. D., and J. A. Holcombe (1974). Radiative and electric properties of exploding silver wire. *Appl. Spectrosc.*, **28**, 518–535.

Schieffer, G. W., and W. J. Blaedel (1977). Study of anodic stripping voltammetry with collection at tubular electrodes. *Anal. Chem.*, **49**, 49–53.

Schimpff, W. K. (1971). Advances in anodic stripping voltammetry for the *in situ* analysis of trace metals in the presence of oxygen. *Thesis*, University of Michigan, Ann Arbor, Mich.

Thomas, P., and R. D. Sacks (1978). Parametric study of exploding wire continuum radiation. *Anal. Chem.*, **50**, 1084–1088.

Tomashov, N. D., and F. D. Zalivalov (1970). Anodic oxidation of zirconium, tantalum, titanium, and aluminum. *Zh. Prikl. Khim.*, **43**, 2474–2479.

Turner, D. R., and M. Whitfield (1979a). The reversible electrodeposition of trace metal ions from multi-ligand systems. Part I. Theory. *J. Electroanal. Chem.*, **103**, 43–60.

Turner, D. R., and M. Whitfield (1979b). The reversible electrodeposition of trace metal ions from multi-ligand systems. Part II. Calculations on the electrochemical availability of lead at trace levels in sea water. *J. Electroanal. Chem.*, **103**, 61–79.

Valenta, P., H. Rützel, H. W. Nürnberg, and M. Stoeppler (1977). Trace chemistry of toxic metals in biomatrices. *Z. Anal. Chem.*, **285**, 25–34.

Vassos, B. H., E. J. Berlandi, T. E. Neal, and H. B. Mark, Jr. (1965). Electrochemical preparation of thin metal films as standards on pyrolytic graphite. *Anal. Chem.*, **37**, 1653–1656.

Vassos, B. A., R. F. Hirsch, and H. Letterman (1973). X-ray microdetermination of chromium, cobalt, copper, mercury, nickel, and zinc in water using electrochemical preconcentration. *Anal. Chem.*, **45**, 792–794.

Weast, R. C., and S. M. Selby (1968). Physical constants of inorganic compounds. In *Handbook of Chemistry and Physics*, 48th ed. (Ed. R. C. Weast). The Chemical Rubber Co., Cleveland, Ohio, p. B-230.

Whitfield, M. (1971). Compact potentiometric sensor of novel design. *In situ* determination of pH, pS^{2-}, and E_H. *Limnol. Oceanogr.*, **16**, 829–837.

Whitfield, M. (1975). The electroanalytical chemistry of seawater. In *Chemical Oceanography*, Vol. 4, (Eds. J. P. Riley and G. Skirrow). Academic Press, New York, pp. 1–154.

Marine Electrochemistry
Edited by M. Whitfield and D. Jagner
© 1981 John Wiley & Sons Ltd.

K. GRASSHOFF†

Institut für Meereskunde an der Universität Kiel,
Düstenbrookerweg 20,
D-2300 Kiel,
Federal Republic of Germany

9

The Electrochemical Determination of Oxygen

† Deceased 11 March 1981

GLOSSARY OF SYMBOLS

a	Thickness of electrolyte film (Section 9.5.4)
A	Electrode surface area (equation 4)
b	Membrane thickness (Section 9.5.4)
c	Concentration (mol dm^{-3}) (equation 1)
c_E	Concentration of oxygen in electrolyte film (equation 22)
c_M	Concentration of oxygen in membrane (equation 22)
c_S	Concentration of oxygen in solution (equation 22)
ΔC_p	Heat capacity of solution of oxygen (Section 9.6.2)
D	Diffusion coefficient (equation 1)
D_E	Diffusion coefficient of oxygen in electrolyte film (equation 24)
D_M	Diffusion coefficient of oxygen in the membrane (Section 9.5.4)
E_D	Activation energy for membrane diffusion (equation 50)
E_p	Activation energy of permeation (equation 52)
f_c	Calibration factor (Section 9.3.6)
f_s	Salinity correction factor (Table 2)
f_t	Temperature correction factor (Table 1)
f_2	Gas fugacity (equation 59)
g	Acceleration due to gravity (equation 61)
h	Depth (m) (equation 61)
ΔH	Heat of solution of oxygen in water (equation 54)
ΔH_M	Heat of solution of oxygen in the membrane (equation 51)
i	Measured apparent diffusion current (Section 9.3.6)
i_b	Boundary current (equation 8)
i_d	Average diffusion current (equation 1)
i_c	Capacitance current (Section 9.3.6)
J_X	Flux of component X (equations 11–13)
k	Membrane constant ($= \pi^2 D_M/b^2$) (equation 45)
K	Henry's law constant (equation 59)
K_E	Distribution coefficient of oxygen between membrane and electrolyte (equation 22)
K_M	Distribution coefficient of oxygen between solution and membrane (equation 22)
m	Mercury flow-rate of DME (mg s^{-1}) (equation 1)
n	Number of Faradays of electricity per molar unit of electrode reaction (equation 1)
p	Partial pressure (Section 9.5.3)
P	Applied pressure (Section 9.6.3)
P_M	Permeability coefficient for oxygen in the membrane (Section 9.5.3)
r_e	Electrode radius (equation 18)
r	Oxygen radius (equation 64)

t	Time (s) (equation 19)
t_c	Time constant (equation 20)
v	Solution flow-rate (equation 64)
w	Amount of electroactive substance reaching the electrode surface (equation 4).
x_c	Degree of crystallinity (equation 49)
x_2	Molar fraction of dissolved gas (equation 59)
α	Distribution coefficient (Section 9.5.3)
α_T	Temperature coefficient (equation 47)
β	Bunsen coefficient (equation 57)
δ	Diffusion layer thickness (equation 6)
η	Viscosity (equation 64)
ν	Kinematic viscosity (equation 65)

9.1. INTRODUCTION AND HISTORICAL REVIEW

Apart from the classical hydrographic parameters salinity (chlorinity), temperature, and density, dissolved oxygen is the most important parameter for the description of water masses and for understanding mixing processes and oceanic circulation. In addition, oxygen is involved in all of the major biological reactions occurring in the sea and dissolved oxygen is, therefore, one of the non-conservative components in sea water (see Chapter 1). The depletion of oxygen indicates the occurrence of important processes (high rate of breakdown of organic material, slow renewal of enclosed water masses) and gives useful information about the properties of sea water (high age of a water body or a high load of degradable material). This is mainly due to the fact that there are only two sources of oxygen in sea water: gas exchange between the atmosphere and the sea surface and the assimilation process of phytoplankton and benthic algae. The difference between the theoretical saturation value (the oxygen concentration in sea water at equilibrium with gaseous oxygen in water vapour-saturated air) and the amount of oxygen actually found is called the apparent oxygen utilization (AOU). The AOU concept implies that sea water leaving the surface layer and contributing to abyssal water masses is 100% saturated with respect to the atmosphere. This concept can, however, only be accepted with some reservation since physical and biological processes may occur in the surface layers, leading to supersaturation or incomplete equilibration before the water breaks its contact with the atmosphere.

Even today the most widely used method for the determination of dissolved oxygen in sea water is the classical iodometric Winkler (1888) titration and its various modifications (Carritt and Carpenter, 1966; Carpenter, 1965; Grasshoff, 1962; Kalle, 1939; Montgomery *et al.*, 1964; Strickland and Parsons, 1960). The manual titrimetric methods used in field work usually allow a precision of $\pm 0.02 \text{ cm}^3$ of oxygen per dm^3 of sea water but the accuracy of the method may be considerably lower, especially at low oxygen contents, because of unavoidable sampling errors, e.g. outgassing of nitrogen and of oxygen, due to warming of the sample and pressure release.

One of the major drawbacks of the classical iodometric procedure is the visual determination of the end-point of the titration by means of the colour change of the starch–iodine complex. Therefore, many attempts have been made to increase the precision of the end-point detection by applying optical or electrochemical methods in which the reaction

$$I_3^- + 2e^- \rightleftharpoons 3I^-$$

is involved. Such methods will be dealt with in Section 9.2. Dissolved oxygen is a non-conservative component of sea water and the well established relationships (Richards, 1965) connecting dissolved oxygen content with the decomposition of biogenic organic substances and the release of nutrients have resulted in considerable interest in the application of *in situ* techniques to the determination of dissolved oxygen, especially if these techniques result in a continuous vertical (or horizontal) profile. The only possibility for *in situ* analysis is, with one exception, the application of electrochemical methods. It has taken a long time, however, before principal and technical difficulties could be overcome in the design of suitable field equipment. The major difference between the *in situ* recording techniques for physical and chemical parameters is the fact that in the latter case controlled chemical reactions must be converted into electrical signals in specific sensors, thus increasing the number of parameters which must be under strict control during the measurement. In addition, only techniques with a relatively fast response of the sensor not only to changes in the amount of oxygen but also to changes in temperature, salinity, and pressure can be applied to vertical profiling. If such techniques are available the advantages are, however, numerous. Besides the information which an oxygen profile represents in itself (revealing, for example, the fine structure of water masses which may not be evident from salinity–temperature records) the data enable water sampling for other non-conservative constituents (e.g. nutrients) to be carried out on a much more efficient basis (Grasshoff, 1976).

The general principles of electroanalytical techniques for the determination of oxygen in water have been known since the introduction of amperometry and polarography by Heyrovský nearly 50 years ago (Heyrovský, 1936, 1937), but the development and conversion of laboratory techniques into suitable instruments which could be used in oceanographic field work took more than 40 years. Even today, commercially available oxygen sensors for use in oceanography do not meet all the requirements which would make them comparable to temperature and salinity (conductivity) probes (see Chapter 5). Almost all available instruments exploit the same basic principle in their oxygen sensors: an amperometric procedure with a gas-permeable membrane separating the working cathode from the environment in which the oxygen is to be determined.

Therefore, a major part of this chapter will deal with membrane electrodes, but other principles will also be described, either for historical reasons or for the sake of completeness, or because of some interesting possibilities of application to special problems. These will include the use of the electrochemical determination of oxygen as an indicator for another process, such as primary production, bacterial activity, or, less obviously for the measurement of sea level changes, very low currents, and long seismic waves.

It seems to be very difficult to treat the electrochemical determination of oxygen in sea water in an exhaustive manner because many publications have appeared only as technical reports or reports from manufacturers, and many interesting developments remain confidential until they are protected by patents.

9.2. ELECTROCHEMICAL END-POINT DETERMINATION IN THE IODOMETRIC TITRATION OF OXYGEN (WINKLER TITRATION)

In all but exceptional cases, the chemical determination of oxygen dissolved in sea water is based on the method first proposed by Winkler (1888). Other methods, for example the microgasometric determination according to Scholander *et al.* (1955), the mass spectrometric method by Benson and Parker (1961), and the gas chromatographic procedure according to Swinnerton *et al.* (1962, 1964) and Weiss and Craig (1973) are used only for special purposes.

In the Winkler method, the oxygen dissolved in a measured amount of water is chemically bound by manganese (II) hydroxide in a strongly alkaline medium. The manganese (II) is oxidized to manganese (III), and not to manganese (IV) as is often stated, because of the large excess of manganese (II) hydroxide present. In this heterogeneous reaction all of the dissolved oxygen must be brought into contact with the precipitated manganese (II) hydroxide by vigorous shaking. Because of the instability of manganese (II) in an alkaline medium, the hydroxide is quantitatively oxidized with respect to the oxygen present in the reaction volume. It is self-evident that any contact of the sample with atmospheric oxygen must be carefully avoided during this step of the analysis.

After complete fixation of the oxygen and precipitation of the manganese hydroxides, the sample is acidified to a pH less than 2.5 but not less than 1. The precipitated hydroxides dissolve and the liberated manganese (III) ions oxidize iodide ions previously added to the water sample together with the potassium hydroxide reagent. The iodine formed is complexed with surplus iodide to minimize the loss of iodine through evaporation.

The third step of the analysis is titration of the iodine with thiosulphate, which is oxidized to tetrathionate. The thiosulphate is usually standardized with potassium iodate (or potassium hydrogen diiodate) as a primary standard. The stoichiometric equations for the reactions involved are

$$Mn^{2+} + 2OH^- \rightleftharpoons Mn(OH)_2$$

$$2Mn(OH)_2 + \tfrac{1}{2}O_2 + H_2O \rightleftharpoons 2Mn(OH)_3$$

$$2Mn(OH)_3 + 2I^- + 6H^+ \rightleftharpoons 2Mn^{2+} + I_2 + 6H_2O$$

$$I_2 + I^- \rightleftharpoons I_3^-$$

$$I_3^- + 2S_2O_3^{2-} \rightleftharpoons 3I^- + S_4O_6^{2-}$$

$$IO_3^- + 8I^- + 6H^+ \rightleftharpoons 3I_3^- + 3H_2O$$

Usually the end-point of the titration of iodine with thiosulphate is indicated visually with a starch indicator, giving a colour change from deep blue to colourless. The indication of the end-point incorporates both a subjective error from the visual end-point determination and a systematic error from the indicator itself. Bradburg and Hambly (1952) examined several iodine end-point techniques and stated the following sensitivities:

visual starch: 10 μequiv dm^{-3}
colorimetric starch: 2 μequiv dm^{-3}
amperometric: 0.02 μequiv dm^{-3}
UV absorption: 0.015 μequiv dm^{-3} (in 0.005 M iodine)

The authors concluded that the starch-indicated visual end-point is significantly different from the equivalence point in titrations of iodine solutions at concentrations corresponding to air-saturated water.

Theoretically three different principles can be applied for an electrometric determination of the end-point in the iodometric titration, *viz.* a potentiometric titration, an amperometric 'dead stop' titration, or a back-titration of surplus thiosulphate and indication by an amperometric end-point determination. As the reaction delivering electrons proceeds irreversibly in the potentiometric titration of iodine with thiosulphate, the equilibrium at the platinum indicator electrode is established rather slowly (Perley, 1939; Laitinen *et al.*, 1948; Lingane, 1949). Usually a significant number of samples must be analysed in a short period during practical work at sea. Therefore, a rapid procedure is advantageous and potentiometric end-point detection cannot be recommended for practical work. In the so-called 'dead stop' titration (Foulk and Bawden, 1926; see also Section 7.6.1) two bright platinum electrodes with small surface areas (3 mm platinum wire, 1 mm diameter, sealed in a glass rod) are used as indicator electrodes (Figure 1a). A

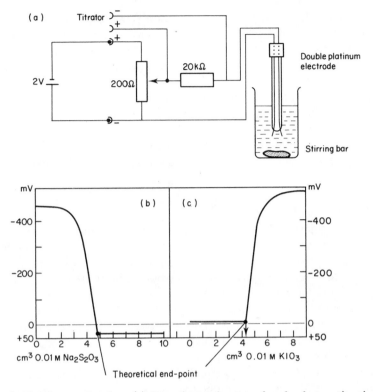

Figure 1. Dead stop titration. (a) Experimental set-up for dead stop titration with connection for an automatic titrator. (b) Dead stop titration curve of iodine with thiosulphate. (c) Dead stop titration curve of surplus thiosulphate with iodate (iodine). Grasshoff (1962)

d.c. voltage of about $100\,mV$ is applied to the electrodes. The cathode is depolarized by the reaction

$$I_3^- \rightleftharpoons I^- + I_2$$
$$I_2 + 2e^- \rightleftharpoons 2I^-$$

and the anode is depolarized by the reverse reaction. A current flows only when both electrodes are depolarized. This is, however, only possible if I_2 and I^- coexist in the solution. The reaction

$$2S_2O_3^{2-} \rightleftharpoons S_4O_6^{2-} + 2e^-$$

is irreversible and cannot depolarize the electrodes. At high iodine concentrations at the start of the titration, the current is controlled by diffusion. With decreasing iodine concentration during the titration the initial current drops until it is almost zero when all the iodine is reduced at the end-point of the titration. If the stirring of the solution is constant, the decrease in the current is proportional to the decrease in iodine concentration. The small remaining current after the end-point is caused by impurities in the solution. A typical current record of a 'dead stop' titration of iodine with thiosulphate is illustrated in Figure 1b.

The 'dead stop' method can be used as an automated titration procedure for the determination of oxygen in sea water. Experiments have shown that a better reproducibility at a faster titration speed can be reached if a reverse procedure is used. An excess of thiosulphate is added to the iodine sample and the excess is then titrated with iodate according to the reaction

$$IO_3 + 5I^- + 6H^+ \rightleftharpoons 3I_2 + 3H_2O$$

$$I_2 + 2S_2O_3^{2-} \rightleftharpoons 2I^- + S_4O_6^{2-}$$

If all surplus thiosulphate is consumed, the liberated iodine depolarizes the cathode and a rapid increase in the current can be observed. A method for the automatic titration of oxygen in sea water according to the latter procedure has been proposed by Grasshoff (1962). Figure 1c shows an example of the back-titration method with dead stop end-point determination. This method has the advantage that evaporation of iodine during the titration is further diminished. The relative standard deviation for the automated dead stop titration is 0.6%.

From this it follows that electrochemical end-point determination may be advantageous, especially if very precise measurements are required. It must be stated, however, that the cumulative error from other sources (reagents, sampling, and sub-sampling procedure) is usually much higher than the error introduced by the end-point determination in most procedures applied in field work.

9.3. POLAROGRAPHIC DETERMINATION OF DISSOLVED OXYGEN

9.3.1. Basic principles of polarographic oxygen determination

Since the introduction of polarographic methods for the qualitative and quantitative determination of dissolved chemical compounds by Heyrovsky (1926), the electrochemical behaviour of dissolved oxygen at the dropping mercury electrode (DME) has been studied by several authors. The main reason for these studies was the interference of dissolved oxygen in most polarographic determinations and not the determination of oxygen itself. The first publication on the application of polarography to the *measurement* of oxygen in sea water was by Giguère and Lauzin (1945). Ingols (1941), Moore *et al.* (1948), and Wood (1953) described methods for the polarographic determination of dissolved oxygen in fresh water, lake water, and waste water. Solid platinum electrodes have also been used for the determination of dissolved oxygen (Ingols, 1955; Lynn and Okun, 1955).

In view of the comprehensive books on polarography (Kolthoff and Lingane, 1952; Stackelberg, 1950; Meites, 1965; Galus, 1976) and periodic critical reviews (see, for example, Lingane, 1949; Hume, 1956; and Nürnberg, 1974) the discussion in the following section is restricted to basic principles.

The instrumental set-up for polarography (see Figure 2) consists of a narrow glass capillary tube which is connected to a mercury reservoir and from which mercury issues dropwise into a beaker containing the sample solution. The sample solution is connected by a low-resistance bridge to the reference electrode, which is usually a non-polarizable saturated calomel electrode or a silver/silver chloride electrode. As oxygen interferes in most polarographic determinations, the sample beaker is designed to allow for the stripping of the oxygen with an inert gas, usually pure nitrogen.

According to Lingane (1949) there are three unique virtues of the

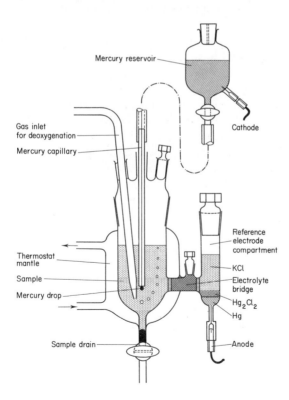

Figure 2. Experimental set-up for polarography with the dropping mercury electrode

dropping mercury electrode (DME). Firstly, the electrode surface is continuously renewed, causing a high degree of reproducibility of the current–potential curve regardless of the previous history of the electrolysis. Secondly, the DME produces ideal conditions for a diffusion-controlled limiting current. The continuous growth of the mercury drop more than compensates for the spreading of the diffusion layer from the drop surface into the solution and the actual diffusion layer remains thin, especially during the early life of the drop. This in turn makes the diffusion layer relatively insensitive to disturbances due to convection in the solution. Therefore, the quantitative interpretation of the polarographic current of the DME is far simpler than that of a solid electrode, where there is a mixture of diffusive and convective transport of the depolarizing compound towards the electrode. In addition, the voltage–current relationship is established without a time lag at the DME, whereas significant time elapses before the establishment of a steady state at a solid electrode surface after a voltage change. Thirdly, the DME has the advantage of a large overpotential with respect to hydrogen. For oceanographic work we may add a fourth virtue here, since the polarographic method enables oxygen analyses to be carried out on very small sample volumes.

The main disadvantage of the DME is the oxidation of mercury in the anodic range, giving rise to a large anodic current. This is, however, of no importance with respect to the application of the DME to the determination of dissolved oxygen.

When the cell (Figure 2) is used for analysis an increasing d.c. voltage is applied to the electrode couple with the DME acting as the cathode. As the voltage approaches the reduction potential of the compound in question the current increases exponentially with increasing voltage and then linearly according to Ohm's law. After this, the current continues to increase until all electroactive ions or molecules which reach the electrode surface are instantaneously reduced. From this moment on the diffusion controlled transport limits the current and this part of the voltage–current curve is known as the plateau. Theoretically, the current is independent of the applied voltage until the deposition potential of a further ion or compound is reached. The ultimate limitation in aqueous solutions is given by the decomposition potential of water. Each charge transfer at the DME surface is characterized by the so-called half-wave potential, i.e. the potential which corresponds to a current half the amount of the plateau current, or half the difference to the previous plateau if there are several successive reductions occurring. The height of the plateau current is a quantitative measure of the concentration of the electroactive species. This relationship may be disturbed, however, by complex equilibria. In special cases it is not necessary to record the whole polarogram and the current can be measured at a fixed voltage, normally that corresponding to the position of the middle of the plateau.

9.3.2. Limiting current

If the rate of the electrode reaction is governed by the flux of electroactive substance to the electrode surface, the resulting current is no longer dependent on the electrode potential. The electroactive substance can be transported to the electrode surface by diffusion, electrical migration, or convection, or a combination of these. In special cases the availability of the electroactive substance at the electrode surface can be controlled by a reaction in the solution. The limiting current is then a kinetic current. With the DME in an otherwise undisturbed solution, diffusion and electrical migration are the only processes contributing to the mass transfer. Normally the migration in the electrical field is limited to ions and is kept low through the addition of an excess of supporting salt to the solution. The decomposition potential of the supporting salt ions must, of course, be beyond the decomposition potential of the electroactive substance in question. The supporting electrolyte plays a double role: the transport number of the electroactive ion is decreased practically to zero and the ohmic potential drop due to the solution resistance (the 'IR drop') is kept small. When dissolved (molecular) oxygen is the electroactive species, migration in the electrical field is, however, of no importance.

The flux of the electroactive species to the electrode surface depends directly on the concentration gradient between the electrode surface and the bulk of the solution. As soon as the electrode potential is above the relevant decomposition potential, the concentration of the electroactive substance at the electrode surface decreases and the demand for it increases, as does the rate of diffusive transfer. Ultimately, the rate of the electrochemical reaction at the electrode surface is no longer controlled by the potential but by the diffusive transport of the electroactive species to the electrode. The difference between the concentration (activity) of the depolarizing substance at the electrode surface (zero in this case) and the bulk of the solution becomes constant and the result is the diffusion-controlled limiting current.

An equation for the limiting current was developed by Ilkovic in collaboration with Heyrovsky:

$$i_d = 607n D^{1/2} c m^{2/3} t^{1/6} \tag{1}$$

where i_d is the average diffusion current in μA during the drop life, n is the number of faradays of electricity per molar unit of the electrode reaction, D is the diffusion coefficient in $cm^2 s^{-1}$ ($\equiv 10^{-4} m^2 s^{-1}$) of the electroactive substance under the conditions prevailing in the solution, c is the concentration in $mmol\, dm^{-3}$, m is the rate of mercury flow from the dropping electrode in $mg\, s^{-1}$ and t is the drop time in seconds. The constant 607 is a combination of natural constants and is slightly temperature dependent.

An improved equation taking the curvature of the mercury drop into account was independently developed by Lingane and Loveridge (1950) and by Strehlow and Stackelberg (1950):

$$i_d = 607nD^{1/2}cm^{2/3}t^{1/6}\left(1 + \frac{AD^{1/2}t^{1/6}}{m^{1/3}}\right) \tag{2}$$

The value for the constant A is 39 according to Lingane and Loveridge (1950) and 17 according to Strehlow and Stackelberg (1950). In most cases, however, the correction term has a value very close to 1, so that the original Ilkovic equation can be used. (The thickness of the diffusion layer is only about one twentieth of the drop diameter so that the curvature can be neglected.)

The diffusion current is highly temperature dependent. The cumulative temperature coefficient, which combines the temperature dependence of the diffusion coefficient, the mercury flux, and the drop time, is usually between 1.3 and 1.6% $^{\circ}C^{-1}$ (Lingane, 1949). Therefore, polarographic determinations require temperature control, which can be achieved by a thermostatted polarographic cell (Figure 2). Of course, the diffusion coefficient is influenced by the composition and therefore the viscosity of the supporting electrolyte, and by the chemical nature and speciation of the electroactive substance, which may, in turn, be affected by the composition of the supporting electrolyte.

In addition to the diffusion current, a capacitance current is developed. This current is caused by the different potentials of the DME and the reference electrode and by the instant development of a new electrode surface. Usually the capacitance current can be compensated for by applying a current in the opposite direction so that the capacitance current is subtracted from the displayed signal.

In principle, the measured current should start from almost zero when a new drop is formed and rise to a maximum before the drop falls from the capillary tip. Suitable damping of the current oscillations results, however, in a curve which can easily be averaged to give the diffusion current.

9.3.3. Distortion of the polarographic waves by maxima

Convective movements of the supporting electrolyte solution may appear in the vicinity of the mercury drop. One of the reasons for the distortion of the diffusion layer is the development of different potentials at different points on the surface of the mercury drop (Stackelberg, 1950). The result is an increased transport of the electroactive substance to the electrode surface and thus an increased current. With increasing applied voltage the disturbance is usually minimized and the current drops back to the normal value. A detailed treatment of the development of maxima is given by Stackelberg (1950).

Since maxima considerably disturb the voltage–current curve, the addition of surface-active substances is beneficial in most cases. The surface-active substances, usually straight-chained molecules, stabilize the diffusion layer. According to their polarity they are arranged in the electrical field to form linear channels perpendicular to the electrode surface through which the electroactive substances diffuse. Currents parallel to the drop surface are thus prevented.

9.3.4. Reduction of oxygen at the dropping mercury electrode

Oxygen is dissolved in water in the molecular form. In sea water the concentration range varies between zero and slight oversaturation with respect to the partial pressure of oxygen in water vapour-saturated air. The saturation value is a function of temperature and salinity (Section 9.6). According to Kolthoff and Lingane (1952), molecular oxygen is reduced in two steps:

Step 1:

$$O_2 + 2H^+ + 2e^- \rightleftharpoons H_2O_2 \text{ (acidic medium)}$$

$$O_2 + 2H_2O + 2e^- \rightleftharpoons H_2O_2 + 2OH^- \text{ (neutral or alkaline medium)}$$

Step 2:

$$H_2O_2 + 2H^+ + 2e^- \rightleftharpoons 2H_2O \text{ (acidic medium)}$$

$$H_2O_2 + 2e^- \rightleftharpoons 2OH^- \text{ (alkaline medium)}$$

The half-wave potential of the first step is -0.05 V *vs.* SCE (saturated calomel electrode) and of the second step approximately -0.9 V *vs.* SCE. The first wave usually has no marked plateau and continues slowly into the second wave, which produces a relatively slight increase, indicating an irreversible reduction. Oxygen is not subject to a migration current in the electrical field. Several substances, e.g. ions of iron, copper, lead, and bismuth, are reduced in the same range, but their concentrations in natural sea water are usually so low that they do not interfere.

The polarographic reduction of dissolved oxygen in aqueous solution is very much disturbed by the development of maxima under certain conditions. In particular, the development of an interrupted maximum in the first reduction step must be avoided as the plateau of the first wave is usually used for the quantitative determination of dissolved oxygen with the DME. Figure 3 shows the interrupted maximum of dissolved oxygen reduction in air-saturated potassium chloride solutions. The formation of the maximum is much more pronounced at low electrolyte concentrations. A second factor

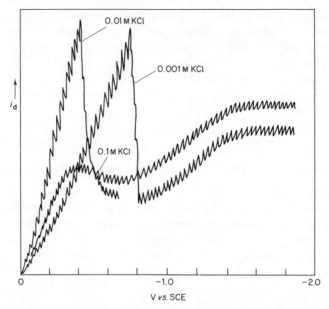

Figure 3. Interrupted maxima of dissolved oxygen reduction in air-saturated KCl solutions

which influences the formation of a maximum is the drop time. The shorter the drop time the smaller is the maximum (see Figure 4). The non-interrupted maximum disturbs the second wave and often leads to misinterpretation of the diffusion current. The current is biased to the high side, which can be overlooked in cases where the distortion of the wave is not too large (Stackelberg, 1950). Figure 5 shows a polarogram of dissolved oxygen

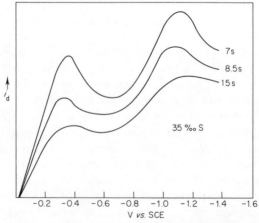

Figure 4. Dependence of the interrupted and non-interrupted maximum on the drop time in air-saturated sea water of 35‰ salinity. According to Rotthauwe (1958)

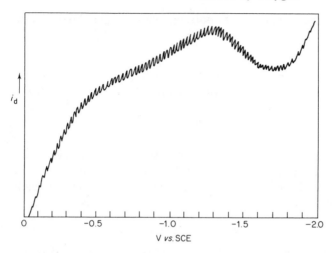

Figure 5. Non-interrupted maximum in air-saturated KCl solution

in 1 M potassium chloride solution with a non-interrupted maximum. Higher electrolyte concentrations appear to favour the development of the non-interrupted maximum. This is demonstrated in Figure 6 with sea water of 5, 15, 25, and 35‰ salinity. It is inconvenient to add a maximum suppressor to solutions in which dissolved oxygen is to be determined, since the depressing agent will contain oxygen and the sample solution must be stirred after the addition of the suppressor. Therefore, the experimental conditions must be selected so that maxima do not occur or interfere with the diffusion current.

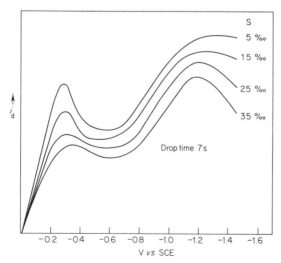

Figure 6. Salt effect on the maxima in air-saturated sea water of various salinities. According to Rotthauwe (1958)

9.3.5. Practical application of polarographic oxygen determination

Instrumentation

The polarographic determination of dissolved oxygen in sea water is limited to laboratory experiments, as one of the major disadvantages of the DME is its sensitivity to external vibrations and accelerations, which are unavoidable in shipboard use. Towards the end of the life of a hanging drop the slightest vibration causes disconnection of the drop from the capillary tip and therefore an irregularity in the polarographic curve. This is demonstrated in Figure 7. In solutions containing chloride another problem arises because mercury is oxidized by the dissolved oxygen and forms calomel, according to the equations

$$2Hg + \tfrac{1}{2}O_2 + 2Cl^- + 2H^+ \rightleftharpoons Hg_2Cl_2 + H_2O \text{ (acidic solutions)}$$

$$2Hg + \tfrac{1}{2}O_2 + 2Cl^- + H_2O \rightleftharpoons Hg_2Cl_2 + 2OH^- \text{ (neutral or alkaline solutions)}$$

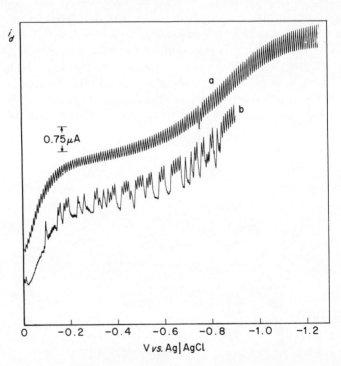

Figure 7. Curve b shows the influence of mechanical vibration (ship's engine) on the shape of a conventional DME polarogram (curve a) of oxygen in sea water (with addition of maximum suppressor). Grasshoff (1963)

This reaction is slow, however, and does not interfere with the determination of oxygen under normal conditions, especially if a negative potential is applied to the electrode.

Any conventional polarograph can be used for the determination of dissolved oxygen. If sea water with a salinity above 4‰ is analysed, a silver chloride-coated silver wire can be used as reference electrode and immersed directly in the polarographic cell (Grasshoff, 1963). A saturated calomel electrode which is connected to the polarographic cell with a KCl–agar-agar bridge can, of course, also be used.

Interference of maxima

For the usual drop time of about 3 s a maximum of the first kind (interrupted maximum) does not appear in sea water between 4 and 20‰ salinity. At salinities between 25 and 35‰ the maximum is flattened and rounded. Increased drop times favour the development of the first maximum, which then appears even at salinities less than 20‰. Shorter drop times also cause the appearance of a maximum of the second kind (non-interrupted maximum), which interferes with both oxygen waves.

The appearance of maxima causes non-reproducible distortion of the diffusion current and must therefore be avoided, especially if the determination of the dissolved oxygen is made at a pre-set voltage without recording the complete polarogram. In most practical cases the addition of a maximum suppressor (alkaloids, gelatin, organic dyes, or a surfactant) will cause interference with either the Winkler determination for calibration, or with the use of the water in a closed system for physiological experiments. Therefore, a careful study of the conditions under which maxima appear is necessary in the experimental set up in question.

Influence of the electrolyte concentration on the diffusion current

According to the Ilkovic equation (equation 1), the diffusion current may depend indirectly on the electrolyte concentration. In addition to influencing the diffusion coefficient of the electroactive substance, the composition of the supporting electrolyte may alter the surface tension of the mercury and thus affect the mercury flow from the capillary and hence the drop time. However, according to Rotthauwe (1958), the drop time is effectively independent of the salinity and of the applied potential as far as natural sea water is concerned.

The diffusion coefficient is a function of the viscosity of the supporting electrolyte. If all the other factors are known, the diffusion coefficient can be calculated from the Ilkovic equation. Rotthauwe (1958) quotes the following values for the diffusion coefficient at 25°C:

fresh water (tap water) $2.92 \times 10^{-9} \, m^2 \, s^{-1}$
sea water, 16‰ salinity $2.80 \times 10^{-9} \, m^2 \, s^{-1}$
sea water, 35‰ salinity $2.72 \times 10^{-9} \, m^2 \, s^{-1}$

The differences, which do not exceed $0.2 \times 10^{-9} \, m^2 \, s^{-1}$, have no practical importance, as the square root of the diffusion coefficient appears in the Ilkovic equation (equation 1). In other words, the calibration can be performed at one salinity and is then valid for the total range, provided that the other variables are known or kept constant.

If the concentration of the supporting electrolyte is so low that it results in a significant *IR* drop it will not only influence the diffusion current but can also cause a shift of the current–voltage curve. Therefore, the polarographic determination of dissolved oxygen in distilled water, tap water, and fresh water requires a tailored calibration. In sea water the resistance of the sample is relatively small, even at low salinities.

Influence of the temperature on the diffusion current

The temperature affects all of the terms in the Ilkovic equation except the number of electrons involved in the process at the DME and the Faraday constant. The most important influence of the temperature is on the diffusion coefficient and on the drop time. Rotthauwe (1958) quotes an overall temperature coefficient of 1.5% °C^{-1} based on the diffusion current at 20°C. This implies that the measuring temperature must be kept constant to within ±0.5°C if the diffusion current is not to be biased by more than 1–2%.

Working at constant temperature is possible only if the polarographic determination is used in connection with laboratory experiments. Usually the sample whose oxygen content is to be determined has a temperature different from room temperature. It is therefore more practical to determine the overall temperature coefficient with sufficient accuracy and apply a correction to the apparent diffusion current before using the calibration curve which is recorded at constant temperature (Grasshoff, 1963).

Influence of the hydrogen ion activity on the diffusion current

The pH of the supporting electrolyte has an influence on the development of the polarographic waves. This can be expected because hydrogen ions or hydroxyl ions are involved in both steps of the reduction of the dissolved oxygen (Section 9.4.4). At pH values around 5 the plateau of the first wave is considerably distorted, which results in an increased apparent diffusion current. No clear explanation of this phenomenon is to be found in the literature. At lower pH values the reduction of hydrogen ions at the DME is superimposed on the second oxygen wave. The effect of the pH on the diffusion current is of almost no practical importance in the polarographic determination of oxygen in natural sea water as the pH only varies between 8.4 and 7.0. In extreme cases at very high primary production values the pH may rise above 9, but this is exceptional.

9.3.6. Polarographic determination of dissolved oxygen with the rapid dropping mercury electrode

General principles

As pointed out earlier, conventional polarography with the DME is limited to laboratory determinations. The drop size and drop time are very sensitive to mechanical vibrations and no regular polarographic waves can be recorded unless the electrode assembly is installed on a vibration-free support. The power of adhesion of the mercury drop to the capillary tip is, however, relatively large during the first phase of its formation. If the fresh drop is separated from the capillary tip, shortly after its formation, by a short lateral acceleration of the capillary, the unstable phase is not reached and other 'normal' vibrations have no effect on the drop time.

Mechanical control of the drop time has been described by Heyrovsky (1949) and by Airey and Smales (1950) for differential polarography. Wolf (1960) described so-called 'rapid polarography' which allows a much faster scanning of a polarogram. Grasshoff (1963) used the rapid polarographic technique for the determination of dissolved oxygen in sea water in field work and on-board research vessels. Figure 8 shows a comparison between

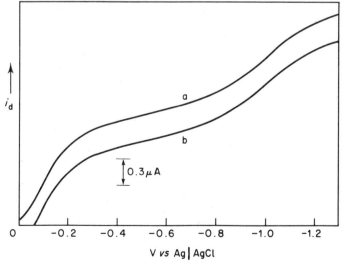

Figure 8. Rapid polarogram of oxygen in sea water (a) with and (b) without mechanical vibrations. Grasshoff (1963)

Marine Electrochemistry

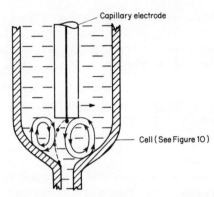

Figure 9. Schematic diagram of the convection in a polarographic cell with a rapid dropping mercury electrode. Grasshoff (1963)

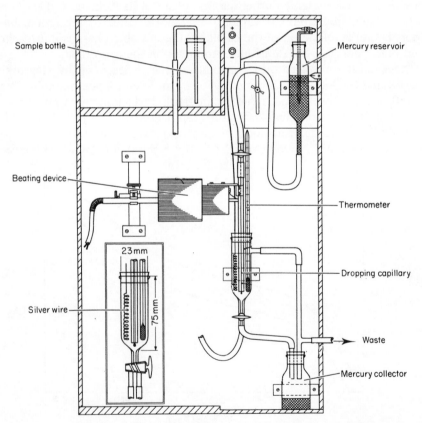

Figure 10. Cell assembly for rapid polarography for shipboard use. Grasshoff (1963)

conventional polarography of oxygen in sea water and rapid polarography with and without mechanical vibrations. The evaluation of the rapid polarogram is much easier because the current oscillations from the changing mercury drop surface are almost completely damped and mechanical vibrations of the system caused by pitching and rolling have no effect on the drop time.

The diffusion current of the rapid DME is smaller than that of the normal DME because of the smaller electrode surface area. This negative effect on the diffusion current is partly compensated for by the forced convection in the cell. The short, low-frequency vibration of the capillary tip (240 beats per minute) and the rather short drop time (0.25 s) compared with about 8 s in conventional polarography cause a convective movement of the solution in the cell, as is shown schematically in Figure 9. This convective transport of the electroactive substance diminishes the thickness of the diffusion layer, thereby increasing the concentration gradient between the electrode surface and the bulk of the solution with a proportional increase in the diffusion current. The mechanical acceleration of the capillary tip must be highly reproducible both in frequency and in strength. A suitable cell assembly for ship-board use is shown in Figure 10.

Temperature dependence of the diffusion current

As in conventional polarography, temperature has a cumulative effect on the diffusion current. Grasshoff (1963) found a linear correlation between

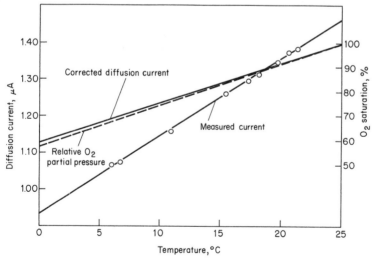

Figure 11. Relationship between corrected diffusion current and temperature for the determination of oxygen with the rapid dropping mercury electrode. Grasshoff (1963)

the diffusion current and the temperature. At 15°C the temperature coefficient is 1.92% °C^{-1}, i.e. higher than in conventional polarography
(1.5% °C^{-1}). If the corrected diffusion current (apparent diffusion current –
capacitance current) and the partial pressure of oxygen for the same sample
are plotted against temperature, both functions have the same coefficient,
which suggests that the real diffusion current for dissolved oxygen is a linear
function of the partial pressure and not of the concentration. These relationships are illustrated in Figure 11.

A correction table (Table 1) can be calculated from the experimental data,
which permits easy conversion of the diffusion current measured at an
arbitrary temperature to the diffusion current at a standard temperature
(15°C).

*Dependence of the diffusion current on the salinity at the rapid dropping
mercury electrode*

Experimental data (Grasshoff, 1963) show that the salinity of the sample
has a significant effect on the diffusion current. At 20‰ salinity the coeffi-

Table 1. Correction factors, f_t, for the reduction of diffusion currents measured at t_m to the
diffusion current at 15°C (Grasshoff, 1963)

t_m, °C	0	0.2	0.4	0.6	0.8
0	1.330	1.325	1.320	1.315	1.310
1	1.305	1.300	1.295	1.290	1.285
2	1.280	1.275	1.270	1.265	1.260
3	1.255	1.250	1.245	1.240	1.235
4	1.230	1.225	1.220	1.215	1.210
5	1.206	1.202	1.198	1.193	1.188
6	1.184	1.180	1.175	1.170	1.160
7	1.162	1.158	1.153	1.148	1.143
8	1.139	1.135	1.130	1.125	1.121
9	1.117	1.113	1.108	1.103	1.098
10	1.094	1.090	1.086	1.082	1.078
11	1.075	1.071	1.067	1.064	1.066
12	1.056	1.052	1.048	1.044	1.041
13	1.038	1.034	1.030	1.026	1.022
14	1.019	1.015	1.011	1.007	1.003
15	1.000	0.997	0.994	0.991	0.988
16	0.984	0.980	0.977	0.974	0.971
17	0.968	0.965	0.962	0.959	0.956
18	0.953	0.950	0.946	0.943	0.940
19	0.937	0.934	0.931	0.928	0.925
20	0.922	0.919	0.916	0.913	0.910
21	0.908	0.905	0.902	0.898	0.896
22	0.895	0.892	0.889	0.886	0.883
23	0.881	0.878	0.875	0.872	0.870
24	0.868	0.865	0.862	0.859	0.856
25	0.854	0.852	0.850	0.848	0.846

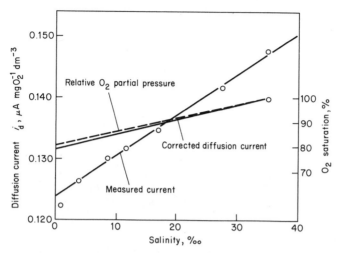

Figure 12. Relationship between corrected diffusion current and salinity for the determination of oxygen in sea water with the rapid dropping mercury electrode. Grasshoff (1963)

cient is about 0.5% for a 1‰ salinity change. There is an almost linear relationship between the relative diffusion current and the salinity between 4‰ and 35‰. As expected, this relationship is not applicable at salinities less than 4‰ because of the increasing resistance of the solution and its effect on the potential of the DME. Also here the experimental data suggest that the corrected diffusion current is proportional to the partial pressure of oxygen in the sample solution (see Figure 12). It is possible to normalize the measured diffusion current to zero salinity by applying a correction factor (Table 2). As is apparent from Table 2, the salinity must be known to within ±0.5‰ to make full use of the correction factors.

Practical application of the rapid dropping mercury electrode

As has been mentioned before, the rapid DME can be used for field work and at sea under moderate conditions. The polarogram must be recorded without delay to avoid any outgassing of the oxygen. The temperature of the sample is read during the scanning of the polarogram. One determination takes about 1.5 min. The oxygen content is calculated according to

$$c_{O_2} = (i - i_c)f_t f_s f_c$$

where

c_{O_2} = oxygen content in $cm^3 dm^{-3}$ or $mg dm^{-3}$;
i = measured apparent diffusion current in μA;
i_c = capacitance current in μA;

Table 2. Correction factors, f_s, for the reduction of the diffusion currents measured at a salinity, S_m, relative to zero salinity (Grasshoff, 1963)

S_m, ‰	f_s	S_m, ‰	f_s
0	1.000	21	0.901
1	0.995	22	0.897
2	0.990	23	0.892
3	0.985	24	0.888
4	0.980	25	0.883
5	0.975	26	0.879
6	0.970	27	0.874
7	0.965	28	0.870
8	0.961	29	0.866
9	0.956	30	0.861
10	0.951	31	0.857
11	0.946	32	0.852
12	0.942	33	0.848
13	0.937	34	0.844
14	0.933	35	0.840
15	0.928	36	0.836
16	0.924	37	0.834
17	0.919	38	0.832
18	0.915	39	0.830
19	0.910	40	0.827
20	0.906		

f_t = temperature correction factor from Table 1;
f_s = salinity correction factor from Table 2;
f_c = calibration factor.

A comparison between oxygen values from Baltic waters with different salinities determined by polarography and by the Winkler method is given in Table 3. The standard deviation (σ) of the polarographic determination was estimated from 10 parallel runs as $0.02 \, \text{mg dm}^{-3}$ and is therefore comparable to that of an accurate Winkler titration.

9.3.7. Determination of dissolved oxygen in sea water with a rapid dropping mercury electrode in a galvanic cell

General principles

The dropping mercury electrode can be applied to *in situ* measurements of dissolved oxygen in sea water. The method has been described by Føyn

Table 3. Comparison between oxygen values determined by rapid polarography and by Winkler titration (Grasshoff, 1963)

$S, ‰$	Oxygen (polarography), mg dm^{-3}	Oxygen (Winkler) mg dm^{-3}	Difference mg dm^{-3}
14.5	7.29	7.32	
	7.31	7.32	
	7.30	7.29	
	7.31	7.33	
	7.30	7.30	
Average:	7.30	7.31	−0.01
10.3	7.70	7.70	
	7.71	7.72	
	7.73	7.70	
	7.74	7.74	
	7.73	7.73	
Average:	7.72	7.72	0.00

(1955) and has been used in Norwegian fjords at depths down to 300 m. The general function of Føyn's oxygen probe is based on a principle proposed by Tödt (1929) in connection with his investigations on metal corrosion. He showed that a noble and a base metal, e.g. platinum and zinc, would generate an oxygen-dependent electric current in moving water (Section 9.4). In stagnant water or after longer periods of time the electrode reactions cause inactivation of the electrode surface. This can be overcome by using a DME in combination with a zinc anode. As the electrochemical reduction of oxygen at the DME starts at very low potentials, the e.m.f. of the cell is generated by the following processes:

$$\text{cathode: } \tfrac{1}{2}O_2 + 2e^- + H_2O \rightleftharpoons 2OH^-$$
$$\text{anode: } \qquad Zn \rightleftharpoons Zn^{2+} + 2e^-$$

The cathodic process is, of course, diffusion controlled and, therefore, highly dependent on the formation, maintenance, and thickness of the diffusion layer, as well as on temperature and oxygen concentration. Frequent calibration of the instrument against the Winkler titration seems to be necessary.

Description of the apparatus

The Føyn oxygen probe has a zinc tube as the anode with a large surface area, so that it can be assumed that concentration polarization does not

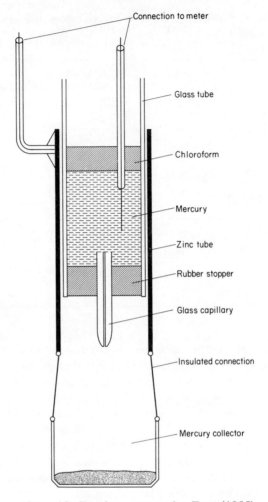

Figure 13. Føyn's oxygen probe. Føyn (1955)

occur under the conditions in which the instrument is used (Figure 13). The glass mercury reservoir is protected by the zinc tube. The lower end of the mercury reservoir is closed with a stopper which is penetrated by a short piece of a glass capillary drawn out at the tip. A layer of choloroform on top of the mercury serves as electrical insulation against the sea water and transmits the ambient hydrostatic pressure to the mercury. A cup for collection of the outflowing mercury is placed below the capillary tip. The upper end of the mercury reservoir is closed by a one-way stopcock not shown in the figure. No information is given on the drop time but from the construction it can be assumed that the drop time is much less than 1 s.

The zinc anode is connected to the mercury by a 500 m electric cable and a microammeter (100 μA full-scale) or a suitable recorder. The paper transport of the recorder is synchronized with the meter wheel of the cable winch. The probe was usually lowered at a rate of 10 m min^{-1}. According to Føyn (1965), the relationship between the recorded current and the oxygen content of the water was 11.4 μA cm^{-3} O$_2$ dm^{-3} at 22°C and 8.2 μA cm^{-3} O$_2$ dm^{-3} at 2°C, i.e. a variation of about 1.5% °C^{-1} at this temperature level. Føyn recommends the selection of a suitable standard temperature for each record, depending on the temperature profile in the region investigated (e.g. 6°C for the inner Oslo Fjord). The agreement between the oxygen values recorded with the probe and determined by the Winkler titration is within 0.1 cm^3 O$_2$ dm^{-3}, according to Føyn (1964, 1976).

9.4. AMPEROMETRIC DETERMINATION OF DISSOLVED OXYGEN WITH PLANE SOLID ELECTRODES

9.4.1. General principles and electrode processes

There are two principles which can be exploited to measure dissolved oxygen with solid electrodes: (i) the current density at suitable galvanic elements can be made proportional to the dissolved oxygen concentration in the electrolyte (Tödt, 1929) or (ii) a suitable potential can be applied externally to a working cathode and a non-polarizable anode giving a current which is proportional to the dissolved oxygen current (Kothoff and Laitinen, 1940). Platinum, gold, or amalgamated gold can be used as the material for the cathode, and zinc, iron, or lead as the material for the anode. If a large anode is externally connected to the cathode by a low-resistance connection and the resistance of the electrolyte is kept small, the cathode potential is governed by the non-polarizable anode. Usually the anode compartment is separated from the electrolyte compartment by a diaphragm. As the electrochemical determination of the dissolved oxygen in the electrolyte depends on the determination of a diffusion or boundary current, the electrolyte must be stirred with constant velocity or the cathode must be mounted in a flow-through system. According to Tödt (1958), there is direct correlation between the electrochemical reduction of oxygen and the measured current if the following conditions are fulfilled:

(i) The cathode potential must not be so negative that hydrogen ions are reduced to hydrogen. The potential *versus* the normal hydrogen electrode at which formation of hydrogen at the cathode commences can be obtained by multiplying the pH value of the electrolyte by -58 mV (at room temperature). If the cathode potential does not exceed

−400 mV at pH 7, formation of hydrogen will not occur. If the reduction of hydrogen ions at the cathode is associated with an overvoltage, the cathode potential can be more negative. A similar extension of the potential range is caused by the formation of hydroxyl ions at the cathode surface through the reduction of the dissolved oxygen.

(ii) The potential of the working cathode must be kept constant. This is the case if the anode remains non-polarized during the course of the measurement.

(iii) No ions with a deposition potential less than or equal to that for the reduction of oxygen should be present in the electrolyte. Ions such as copper will also be reduced and bias the depolarizing current of the oxygen.

(iv) The pH of the electrolyte must not change during the measurement and must be similar for different samples that are measured relative to a single calibration.

(v) The geometry of the electrode must be such as to guarantee a linear relationship between measured current and the concentration of the dissolved oxygen.

(vi) The measurement must be carried out at constant temperature or else the effect of the temperature on the current must be determined carefully, as the temperature coefficient is about 2–3% $°C^{-1}$.

(vii) The electrode system must be calibrated frequently by means of parallel Winkler titrations, by the addition of solutions with known amounts of oxygen to a closed system, or through addition of gas mixtures with known amounts of oxygen.

9.4.2. Theoretical background of the electrochemical determination of dissolved oxygen with a galvanic cell

The equilibrium established at the surface of an oxygen electrode is

$$O_2(g) + 2H_2O + 4e^- \rightleftharpoons 4OH^-$$

From thermodynamic data the normal potential at 25°C (for a partial pressure of oxygen of 1 atm and a pH of 14) is

$$E_{O_2}^{\ominus} = +402 \text{ V}$$

A labile, poorly reproducible potential is established at a platinum electrode submerged in an electrolyte containing oxygen. The potential is usually less positive than the thermodynamically reversible oxygen potential. The platinum electrode does not respond to changes in the concentration of oxygen and hydrogen ions in a predictable manner so that it does not provide a reversible working oxygen electrode. In addition, the redox buffer capacity in natural sea water is too small and the pH has too large an influence on the potential. This means in practice that with oxygen tensions varying between 100% and 10% of the oxygen partial pressure in air-saturated water almost no change of the potential can be observed.

If a cathodic current is applied to a platinum (or other noble metal) electrode, reduction of oxygen takes place even at very small current densities, the potential of the electrode is shifted to more negative values, and the electrode is polarized. With increasing current density the amount of oxygen reduced per unit time increases and the potential becomes still more negative (the reactions occurring at the anode are of no importance in this connection). It is not fully understood which reaction causes the shift of the potential (Schwarz, 1958). It seems unlikely that it is caused by coverage of the electrode with hydrogen, resulting from the reduction of hydrogen ions. This excludes the possibility that the hydrogen reduces the oxygen and acts as a depolarizer (Erlebach, 1954; Hässelbarth, 1954). Winkelmann (1956) has suggested a possible mechanism using the reaction scheme

$$O_2 + e^- \rightleftharpoons O_2^-$$
$$O_2^- + H^+ \rightleftharpoons HO_2^{\bullet}$$
$$HO_2^{\bullet} + e^- \rightleftharpoons HO_2^-$$
$$HO_2^- + H^+ \rightleftharpoons H_2O_2$$
$$H_2O_2 + e^- \rightleftharpoons OH^{\bullet} + OH^-$$
$$OH^{\bullet} + e^- \rightleftharpoons OH^-$$

Hydrogen peroxide is an intermediate in this scheme, which is in agreement with polarographic investigations and also with the potentials measured in natural sea water (Balzer, 1978). For the reaction

$$\tfrac{1}{2}O_2(g) + H^+ + e^- \rightleftharpoons \tfrac{1}{2}H_2O_2 \tag{3}$$

$$E_h = 0.681 + 0.030 \log p_{O_2} - 0.059 \, pH - 0.030 \log c_{H_2O_2}$$

At pH 8.2–7.0 and a calculated concentration of $c_{H_2O_2} = 10^{-11}$ M the thermodynamic potential is $560 \, mV < E_h < 631 \, mV$. If dissolved oxygen is reduced at a solid plane electrode surface in an undisturbed solution, the electrode reaction causes a decrease in the oxygen concentration in the vicinity of the electrode. A diffusion layer develops which increases in thickness during the electrolysis. The theoretical treatment is difficult and in practice the solution is never undisturbed. If the solution is stirred with constant velocity the bulk of the solution remains in turbulent motion and only a thin film at the electrode remains undisturbed. The thickness of this film depends on the stirring and is usually between 10 and 50 μm. The reduction of the dissolved oxygen at the electrode surface creates a concentration gradient across this undisturbed layer. The flux of oxygen here is only controlled by diffusion.

The diffusive transport is described by Fick's first law:

$$\frac{dw}{dt} = DA \cdot \frac{dc}{dx}$$
(4)

where

D = diffusion coefficient for oxygen (about $2.6 \times 10^{-9}\ \mathrm{m^2\,s^{-1}}$ at 25°C);
A = active electrode surface area;
w = amount of electroactive substance reaching the electrode surface;
c = concentration.

It is assumed that a linear concentration gradient exists between the outer boundary of the diffusion layer (concentration c_s) and the electrode surface (concentration c_e) therefore

$$dc = c_s - c_e$$
(5)

and

$$\frac{dw}{dt} = DA \cdot \frac{c_s - c_e}{\delta}$$
(6)

where δ is the thickness of the diffusion layer. On multiplication by nF we find that

$$i = \frac{dw}{dt} \cdot nF = DAnF \cdot \frac{c_s - c_e}{\delta}$$
(7)

A maximum current is obtained if c_e approaches zero. The boundary current is therefore

$$i_b = DAnFc_s/\delta$$
(8)

If the thickness of the diffusion layer is kept constant (constant stirring conditions), the boundary current is directly proportional to the concentration of oxygen in the solution. Of course, temperature has a marked effect as the diffusion coefficient is highly temperature dependent and so is the thickness of the diffusion layer. It is assumed that the active electrode surface area does not change, which is not always guaranteed, as will be discussed later. The sensitivity of the determination can be increased by increasing the turbulent mixing, and hence decreasing the thickness of the diffusion layer.

If the current is increased after the diffusion boundary current has been reached, the potential of the working electrode rises quickly. This continues until the reduction potential of another electroactive substance present in the solution is reached, e.g. that of hydrogen ions. If it is assumed that a linear correlation exists between the diffusion current and the oxygen concentration, it is possible to determine the oxygen concentration from the equation

$$c_{O_2} = c_{O_2}^* \cdot \frac{i_d}{i_d^*} \tag{9}$$

where $c_{O_2}^*$ is a known oxygen concentration and i_d^* the diffusion current for this concentration, provided that all other experimental conditions are kept constant.

The advantage of a solid electrode is the fact that there is no capacitance current. In practical applications the potential of the cathode is kept in the range of the diffusion boundary current. This can be done by connecting a suitable non-polarizable anode to the cathode. The potential of the cathode is then

$$E_{cathode} = E_{anode} + i\Sigma R \tag{10}$$

ΣR is the sum of all apparent resistances (from the electrolyte, from the diaphragm between the anode and the cathode, and from the external connection between the anode and the cathode). The product $i\Sigma R$ must be kept small and should not exceed 30 mV.

As the length of the plateau of the diffusion boundary current is dependent on the reduction current of hydrogen ions, which in turn increases rapidly shortly before the reversible hydrogen potential ($E_h = -0.059$ V pH), the optimal conditions for the reduction of oxygen are obtained in alkaline solutions. If the reduction has to be performed in acidic media, a cathode material with a high hydrogen overvoltage or an anode with a smaller potential must be used. A problem is the shift of the pH in the vicinity of the cathode due to the electrode reaction. Natural water usually has a relatively small buffer capacity.

In choosing the anode the following factors must be observed:

(i) it should be non-polarizable;
(ii) the potential must be in the range of the diffusion boundary current ($E_A \leq -0.059$ V pH + 0.050 V);
(iii) the anode potential should be constant over long periods;
(iv) no interfering substances should be released from the anode to the bulk of the solution.

9.4.3. Application of a galvanic cell with solid electrodes to the determination of dissolved oxygen

The possibility of using a galvanic cell for the *in situ* determination of dissolved oxygen was verified by Kolkwitz (1941). However, the effects of the composition of the electrolyte and the temperature were neglected. Ohle

(1953) thoroughly investigated the influence of external factors on the *in situ* determination with an oxygen probe due to Tödt (1942). He used a very simple probe consisting of a $70\,mm^2$ gold cathode, a $560\,mm^2$ zinc anode and a Plexiglas probe holder. The oxygen-dependent current was measured with a microammeter.

The instrument delivers a depolarizing current of $4.9\,\mu A\,mm^{-2}$ of electrode surface if applied in lake water with $8\,mg\,O_2\,dm^{-3}$, provided that the probe is moved rapidly. The current drops to one thirtieth to one fiftieth of this value if the electrode is still.

The electrolyte concentration of the water has a strong influence on the so-called residual current (Ohle, 1953). If the probe is used in brackish water with salinities between 6 and 18‰, the current value for identical oxygen concentrations changes by only about 5%. A change in the electrolyte concentration between 0.5 and 1.0‰, however, causes a change in the residual current of about 90%. Tödt (1942) has suggested that the apparent oxygen current be corrected with a nomograph. According to Ohle (1953), this is not possible in practice. He found a strong dependence of the current not only on the electrolyte concentration but also on the bicarbonate content of the natural water. The oxygen current was more than doubled in waters rich in bicarbonate under otherwise comparable conditions with the same absolute concentration of oxygen. This means that an individual calibration curve must be prepared for each type of natural water if the instrument is applied *in situ*. Apart from this type of correction, the effect of temperature must also be taken into consideration if the apparent current is to be converted into absolute values of oxygen.

Ohle (1953) has also reported that the bicarbonate content of the water has an undesirable effect on the stability of the oxygen current. This effect is also observed when using a platinum/cadmium amalgam electrode pair. A layer of calcium carbonate forms a crust on the cathode surface, which, therefore, becomes partially inactivated. Ohle (1953) observed a rapid formation of microscopically small nodules of calcium carbonate on the cathode when a platinum/zinc electrode pair was used in hard waters. The coating of the cathode through precipitation of calcium carbonate is significantly reduced if gold is used instead of platinum (Ohle, 1953). According to Ohle (1953), the concentration of other electrolytes in the water (e.g. sodium chloride) has no effect on the linear response of the current to the oxygen concentration. The undesirable effect of calcium carbonate formation remained the same, however, even at higher total electrolyte concentrations. It is recommended that the surface of the gold cathode be cleaned with acetic acid before the oxygen probe is used *in situ*. After the cleaning process the electrode must be immersed in the natural water for about 10 min to equilibrate. Ohle (1953) has also investigated the influence of the pH on the oxygen concentration–current relationship. No significant dependence of the current on the pH in natural waters could be observed between pH 5 and 9. This is in accordance with the findings of Tödt (1942), who found a negligible pH dependence of the current between pH 3 and 11.

As expected, temperature has a significant effect on the oxygen-dependent current. According to Ohle (1953) a linear relationship exists between the current and temperature in the range 0–25°C. The temperature coefficient of the electrode couple (gold/zinc) used by Ohle is 3.0–3.2% °C^{-1}. The influence of the temperature is independent of the oxygen concentration, as can be expected from the diffusion laws.

As the reduction of the oxygen at the cathode surface is a diffusion-controlled process, the velocity of flow in the vicinity of the electrode affects the apparent current. Depending on the construction of the probe the influence of flow, and therefore turbulent mixing, on the current is highest at low flow velocities and reaches a limiting value at higher velocities where the zone of linear diffusion is limited to the boundary layer at the cathode surface. Ohle (1953) reported that the necessary turbulent mixing near the probe can be obtained by manual agitation of the cable. Oxygen measurements with a gold/zinc probe and with Winkler titrations agreed to within 5% in the range between 0 and 10 mg O_2 dm^{-3}.

9.4.4. Odén's oxygen probe

General principles

Three different types of measurement may be mentioned, all of which are based on the cathodic reduction of dissolved oxygen (Odén, 1962):

(i) determination of the concentration or partial pressure of dissolved oxygen;
(ii) measurement of the diffusion of oxygen into the medium surrounding the electrode;
(iii) measurement of the flux of the solvent or hydrostatic pressure.

The first type of application requires the establishment of constant conditions with respect to the thickness of the diffusion layer, composition of the diffusion layer, and, of course, temperature. This can be accomplished by maintaining constant turbulent mixing by stirring of the medium or rotation or vibration of the electrode itself. Calibration and measurement must be carried out under identical conditions.

If the electrode and the medium remain completely undisturbed, a diffusion zone is established around the working cathode in which the oxygen content is determined by the construction of the electrode and by the transport conditions in the diffusion layer. If the diffusion is disturbed by particles or gas bubbles, this is reflected in the current–time relationship. The possibility of using an oxygen electrode for studying soil structure and for ecological studies was first investigated by Lemon and Erickson (1952) and Lemon (1955).

If the diffusion zone is established in an undisturbed solution the gradient of oxygen towards the cathode becomes small and, so does the electrical current. If the solution around the electrode is even slightly disturbed, the transport of oxygen molecules to the cathode surface and thereby the electrical current are markedly increased. Even very small flow velocities can be measured as the molecular diffusion coefficient of oxygen in water is as low as 2.6×10^{-9} m^2 s^{-1}. This fact is the basis for a most unusual application of the oxygen electrode (Odén, 1962). Details are described later in the section dealing with the 'total count electrode'. In connection with the development of suitable oxygen electrodes and the exploration of new fields of application, Odén (1962) carried out a number of investigations which are of fundamental interest for the understanding of oxygen electrodes and their function.

Measurement of the number of electrons involved in the reduction of dissolved oxygen

As mentioned earlier (Section 9.3.4), the reduction of dissolved oxygen at a metal cathode may take place in different ways, depending on the acidity or alkalinity of the solution. The number of electrons involved will be 2 if the reaction ends with the formation of hydrogen peroxide or 4 if there is complete reduction to hydroxyl ions. Microcoulometric measurements (Odén, 1962) show that in the cathodic voltage range (-0.4 to -1.1 V) the number of electrons involved is always 4 and that an equivalent amount of hydroxyl ion is formed. The coulometric measurements give no indication as to possible intermediate reaction steps but such steps cannot be excluded either. The reduction of oxygen is disturbed by the reduction of hydrogen ions at voltages below -1.1 V (Figure 14) and calculations of the number of electrons involved are no longer reliable.

The pH dependence of the electrode process

Data from previous literature on the pH dependence of the cathodic reduction process of dissolved oxygen are often inconsistent and, in the absence of sufficient experimental data, do not permit a firm conclusion to be drawn (Odén, 1962). The effect of the hydrogen ion concentration on the reduction process may stem partly from the reduction of the hydrogen ions themselves, resulting in an increase in the apparent 'oxygen' current (Figure

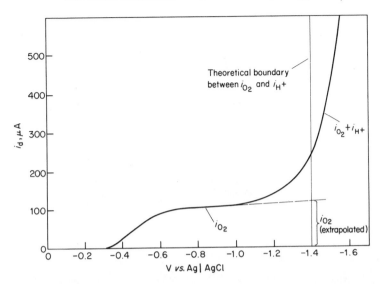

Figure 14. Relationship between current and voltage in a microcoulometric measurement of the oxygen and hydrogen ion reduction in water. According to Odén (1962)

14), or partly from direct interaction between intermediate reaction products and hydrogen ions (Kolthoff and Lingane 1952). The latter effect implies that the position of the wave plateau with respect to the applied voltage will vary continuously with the pH of the solution, according to the law of mass action. The experimental data obtained by Odén (1962) show no such relationship between pH 3.5 and 11 and it is therefore questionable whether the electrode processes formulated by Kolthoff and Lingane are applicable to the reduction of oxygen at a solid metal cathode (see Figure 15).

If an oxygen molecule is reduced at the surface of a platinum cathode, hydroxyl ions are formed even at -0.1 V. This can be easily demonstrated with an indicator such as phenolphthalein. This implies that the pH at the electrode surface must be equal to or higher than 9.3. If the bulk solution is acidic and the diffusion layer at a plane electrode has a thickness δ, the processes can be described by the following sequence:

(i) Oxygen molecules diffuse from the bulk of the solution towards the electrode surface with a flux, J_{O_2} (Figure 16), which is proportional to the oxygen concentration in the bulk of the solution, c_{O_2}, and the coefficient of diffusion, D_{O_2}, and inversely proportional to the diffusion layer thickness, δ:

$$J_{O_2} = \frac{c_{O_2} D_{O_2}}{\delta} \qquad (11)$$

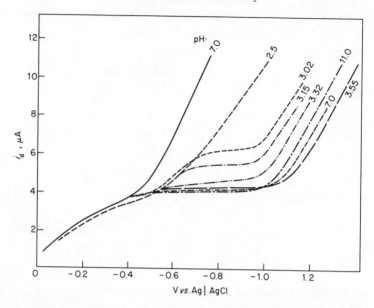

Figure 15. Current–voltage diagrams of the oxygen and hydrogen ion reduction at a stationary microelectrode for different pH values of the solution (KCl). According to Odén (1962)

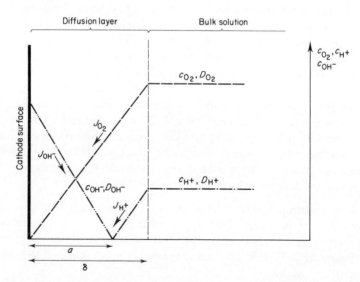

Figure 16. Schematic display of the diffusion processes occurring during the reduction of oxygen at a plane solid electrode. According to Odén (1962)

The overall equation for the reduction is (regardless of possible intermediate products)

$$O_2 + 4H^+ + 4e^- \rightleftharpoons 2H_2O$$

This equation cannot hold for the electrode reaction because of the formation of hydroxyl ions even when the bulk solution is acidic. Therefore, the electrode reaction is obviously better described by

$$O_2 + 2H_2O + 4e^- \rightleftharpoons 4OH^-$$

(ii) Hydroxyl ions formed at the electrode surface diffuse from the cathode towards the bulk of the solution to a point in the diffusion layer where the flux of approaching hydrogen ions (J_{H^+}, Figure 16) balances the flux of hydroxyl ions (J_{OH^-}, Figure 16) and neutralization takes place, giving a pH of 7. The flux of the hydroxyl ions is, therefore, described by

$$J_{OH^-} = \frac{c_{OH^-} D_{OH^-}}{a} \tag{12}$$

where a is the distance of the layer of neutrality from the electrode surface (Figure 16).

(iii) The flux of the hydrogen ions can be described by

$$J_{H^+} = \frac{c_{H^+} D_{H^+}}{\delta - a} \tag{13}$$

The heat of neutralization which may affect the thermal behaviour of an electrode couple is approximately $4.2 \times 10^{-1} \, J \, m^{-2} \, s^{-1}$.

As long as there is any reduction of oxygen, the formation of hydroxyl ions prevents the diffusion of hydrogen ions to the electrode surface until the pH of the bulk solution reaches a critical value (Odén, 1962). This critical value may be calculated from the equation

$$4C_{O_2} D_{O_2} \delta^{-1} = C_{H^+} D_{H^+} (\delta - a)^{-1} \tag{14}$$

when the flux of hydrogen ions towards the cathode balances the flux of oxygen since four hydrogen ions are required for the reduction of one oxygen molecule. Odén (1962) was able to show that a critical value of pH 3.48 was expected in air-saturated water. Only at lower pH values can the hydrogen ion concentration in the bulk solution be expected to affect the diffusion current. This can be seen from Figure 15. The interpretation of the curves in this figure is difficult at first glance. If, however, the experimental data are rearranged so as to show the change in the diffusion current as a function of pH at a given electrode potential, the effect of the pH becomes clear. Within the pH range 3.45–10 the pH has no significant influence on the cathodic reduction of oxygen (see Figure 17).

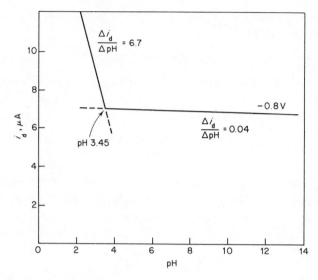

Figure 17. Graphical evaluation of the critical pH value for the reduction of oxygen at a solid electrode in an unbuffered medium. According to Odén (1962)

According to Odén (1962) the slight slope of the curve beyond pH 3.45 is caused by variations in the solubility of oxygen in water as a function of ionic strength. At lower pH values the effect of the pH on the diffusion current is rather strong and an oxygen measurement with bare electrodes cannot be carried out in this range. In the pH range 3.45–11 the reduction of oxygen at the cathode definitely prevents hydrogen ions from approaching the electrode surface and therefore no interference from hydrogen ion reduction is to be expected. This picture changes completely if the oxygen concentration approaches zero, e.g. if anaerobic conditions prevail. In this case the pH of the bulk solution and the cathode potential follow the Nernst equation when the reduction of hydrogen ions commences. This means that the reduction of hydrogen ions starts at pH 8 for an applied potential of −0.8 V.

Odén (1962) has calculated the pH at the electrode surface from the flux equations for oxygen, hydrogen ions, and hydroxyl ions:

$$c_{OH^-} \text{ (electrode surface)} = \frac{(4c_{O_2}D_{O_2}) - (c_{H^+}D_{H^+}) + (c_{OH^-}D_{OH^-})}{D_{OH^-}} \quad (15)$$

This implies that the pH at the electrode surface is almost constant at 10.75 when the pH of the bulk solution lies in the range 4.5–9.5. The point of neutrality in the diffusion layer shifts slowly towards the electrode surface with decreasing pH in the bulk solution, but half of the diffusion layer is still alkaline if the pH of the solution is 3.8 (Odén, 1962).

Influence of electrode geometry

As has been mentioned earlier, the amperometric measurement of oxygen with a solid metal electrode can be used for concentration, diffusion, or flux measurements. Different properties of the electrode and different geometries must be used for the different measurements. In order to select the most suitable geometry of the electrode, the variation of concentration or flux in time and space must be known. These relationships can be calculated using the Laplace differential equation and selected boundary conditions (Odén 1962):

$$\frac{\partial c}{\partial t} = D\left[\frac{\partial^2 c}{\partial x^2} + \frac{\partial^2 c}{\partial y^2} + \frac{\partial^2 c}{\partial z^2}\right] \tag{16}$$

where

c = concentration of the depolarizer;
t = time;
D = coefficient of diffusion;
x, y, z = coordinates.

The solution of this differential equation for a plane electrode is, according to Crank (1956)

$$\frac{c_x}{c_O} = \text{erf}(z) = \frac{2}{\sqrt{\pi}} \int_0^z e^{-u^2} \, du \tag{17}$$

where $z = x/2\sqrt{Dt}$, and for a spherical electrode

$$\frac{c_x}{c_O} = 1 - \frac{r_e}{x - r_e} \cdot \text{erfc}(z) \tag{18}$$

where $\text{erfc}(z) = \dfrac{2}{\sqrt{\pi}} \displaystyle\int_z^\infty e^{-u^2} \, du$

In these equations

x = distance from the electrode surface;
c_x = concentration at distance x;
r_e = radius of the electrode;
c_O = bulk concentration of oxygen in the medium;
u = a dummy variable.

Figure 18a and b shows the relative concentration, c_x/c_O, of the electroactive substance (O_2) as a function of the distance at different times. It can be seen that the establishment of the diffusion layer in its final form is a slow process. Further, it can be seen that the diffusion layer is significantly thinner at a spherical electrode. It is therefore obvious that the use of a

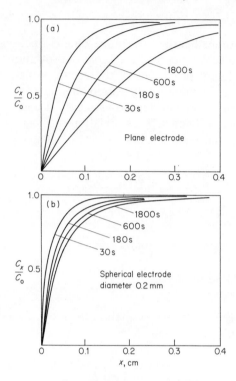

Figure 18. Relative concentration of oxygen as a function of the distance from the electrode surface and time after application of the polarization potential: (a) for a plane solid electrode; (b) for a spherical solid electrode. According to Odén (1962)

plane electrode is recommended if the porosity of the medium surrounding the electrode is to be studied (sediments). The difference in the current yield per unit electrode surface between an undisturbed diffusion layer and a diffusion layer containing solid particles is at least ten times higher with a plane electrode than with a spherical electrode.

On the other hand, the spherical electrode depletes the oxygen in its immediate surroundings much more effectively. This is to be expected because diffusion at a convex surface takes place in an expanding volume. (Equation 18 is only valid for a solid spherical electrode.) The behaviour of a cylindrical electrode is intermediate between the spherical and the plane electrode. The current yield is a function of the electrode radius and drops off rapidly with increasing values of r_e. The current yields for the two types of electrode (spherical and cylindrical) approach the current yield for a plane electrode with increasing radius (Figure 19). A spherical or cylindrical electrode with a radius of 1 mm behaves in a very similar manner to a plane electrode.

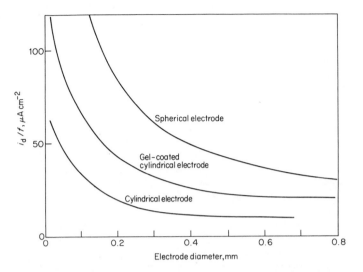

Figure 19. Current yield for different types of solid microelectrodes as a function of the electrode diameter. According to Odén (1962)

The time–current relationship for a spherical electrode can be calculated from the equation

$$i_t = nFAc_{O_2}D[r_e^{-1} + (\pi Dt)^{-1/2}] \tag{19}$$

where i_t = current (A) at time t, A = surface area of the electrode in cm^2, and c_{O_2} is expressed in $mol\ cm^{-3}$.

Differentiation of equation 19 with respect to time results in

$$\frac{di_t}{dt} = -0.5nFAc_{O_2}D^{1/2}t^{-3/2}\pi^{-1/2} \tag{20}$$

If one takes as the time constant, t_c, the point where the current–time curve is within 10% of the final slope, the relation between electrode radius and time constant becomes, for $(di_t/dt)/i_{t_c} = -0.1$ (from equations 19 and 20)

$$r_e = (60\pi D)^{1/2}t_c^{3/2}/(5 - t_c) \tag{21}$$

(Odén, 1962) (for t_c in minutes). If $r_e \to \infty$ then $t_c \to 5\ min$. The curves therefore approach asymptotically a value of $t_c = 5$, which is the theoretical time constant for a linear (plane) electrode. Experimental verification has resulted in $t_c = 4.5\ min$ for a plane electrode (Odén, 1962). Figure 20 depicts the relationship between the electrode diameter and the time constant for a spherical and a cylindrical electrode. It follows that a spherical electrode with a very small diameter (about 10 μm) gives the final diffusion current almost immediately. Such electrodes are, therefore, not significantly affected

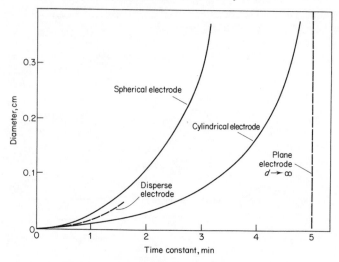

Figure 20. Time constant of different solid microelectrodes as a function of the electrode diameter. According to Odén (1962)

by turbulent processes in the solution near the electrode and would seem to be ideal for profiling instruments.

The disperse electrode

A high current yield, a small time constant, and insensitivity to flow in the vicinity of the electrode would seem to make the micro-electrode with diameters in the micrometre range the ideal electrode for *in situ* concentration measurements. On the other hand, any reduction of the active electrode surface area will reduce the diffusion current drastically. Further, the actual current is too small (below 1 nA) for practical applications. In consequence, Odén (1962) has developed a so-called 'disperse electrode' which consists of a large number (200–400) of individual microelectrodes, each 30 μm in diameter. The electrodes are embedded in epoxy resin with only the surface of the wires exposed to the medium. As the surface area of one electrode is 700 μm^2, 200 electrodes connected in parallel result in a current of 3 μA in oxygen-saturated water at room temperature. If the electrodes are exposed to a flow of 1 ms^{-1} the resulting current is only 1.08 times the current under quiescent conditions. Electrodes of this type have been used successfully for vertical oxygen profiling in lakes, rivers, and the Baltic (Odén, 1962). The probe was lowered at a speed of 1 ms^{-1} and successive profiles demonstrated good reproducibility of even the fine structures of the oxygen concentration. The disperse electrode was combined with a silver anode in a pencil-like probe. The resulting oxygen-dependent current must, of course, be corrected for the temperature effect.

The total count electrode

If the amperometric principle is used for the measurement of the oxygen concentration in a solution, the diffusion layer thickness must be kept constant. This can be achieved by controlled movement of the electrode, by exposing the electrode to a constant stream of the solution, or, as described above, by using spherical microelectrodes. In either case, only a very small fraction of the oxygen in the bulk solution is actually reduced.

Electrodes with large surface areas are very sensitive to any movement relative to the surrounding solution. This opens the possibility of using an oxygen electrode for the measurement of flow if the oxygen content of the solution is constant. If the electrode geometry is such that the surrounding medium is completely stripped with respect to oxygen, the electrode response mirrors the transport of oxygen into the electrode compartment by water currents.

Odén (1962) developed a 'total count' electrode consisting of a 20 mm silver cylinder with an axial channel of 1.5 mm diameter. A 1 mm platinum wire is mounted in the centre of the channel and isolated from the silver

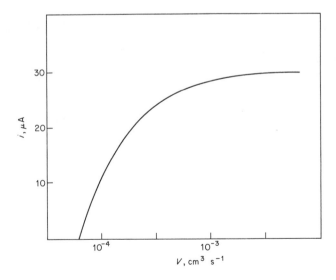

Figure 21. Relationship between flow and current in a total count electrode. According to Odén (1962)

cylinder which forms the anode. An inlet and an outlet are located at the ends of the anode cylinder. In a certain range, all oxygen entering the electrode is reduced before the solution has passed through the cell. In this range a linear relationship exists between flow (v) and current (i_t) (Figure 21), so that

$$i_t = nFvC$$

With a 10 mV recorder, an impedance of 20 kΩ and air-saturated water at 20°C the theoretical flow sensitivity of the total count electrode is 5×10^{-5} mm³ s⁻¹ through the electrode. In a suitable experimental set-up water velocities down to 10^{-8} m s⁻¹ can be measured.

The equilibration time between flow and amperometric current is of the order of 20 min at a flow-rate of 10^{-8} m s⁻¹ and 2 min at 10^{-6} m s⁻¹. Examples of current measurements in the range of 10^{-6}–10^{-2} m s⁻¹ at a depth of 1500 m in the Caribbean Sea were given by Odén (1962).

The total count electrode is, of course, sensitive to the oxygen content in the medium but is not affected by temperature because the coefficient of diffusion does not influence the actual reduction current. This electrode can therefore be used for concentration measurements if the flow through the electrode is kept constant. The reduction current is then proportional to the concentration of oxygen and is not influenced by temperature.

According to Odén (1962), the total count electrode can also be used to measure pressure differences. He was thus able to measure the change in sea level resulting from a passing rain shower with a total count electrode mounted in a 6 m tube and placed on the bottom (100 m) of the Gullmar Fjord (Sweden). The hydrostatic pressure difference from the slope of the sea level caused by the rain shower between the ends of the long tube created a flow through the tube which was sensed by the total count electrode.

9.5. AMPEROMETRIC DETERMINATION OF DISSOLVED OXYGEN WITH MEMBRANE-COVERED ELECTRODES

9.5.1. General principles

As has been pointed out earlier, the major disadvantage of an amperometric oxygen determination with bare electrodes in natural waters is that the current related to the oxygen concentration is a function of the active surface of the cathode. If the surface is partly deactivated by coating with calcium or magnesium carbonates or organic films as created by bacteria or by lipids, the response of the electrode is no longer a linear function of the oxygen concentration and the readings are not stable with time. This has led to the development of an electrode system in which the cathode and the anode are separated from the solution by a thin gas-permeable membrane. The first oxygen sensor of the membrane type was described by Clark *et al.* (1953), who used a membrane-covered system for the measurement of the partial pressure of oxygen in blood. Carritt and Kanwisher (1959) used a membrane-covered electrode to determine dissolved oxygen in water. Membrane-covered dissolved oxygen probes have been employed in a wide range of research and industrial applications (Hoare, 1969) and most, if not all, commercially available amperometric oxygen probes follow the concept of Clark.

The general construction of a membrane-covered electrode is illustrated in Figure 22. The inner compartment of the probe is separated by a thin, gas-permeable membrane, usually made of polyethylene, polypropylene, polytetrafluorethylene or silicone rubber. A noble metal cathode, usually gold or platinum, is located directly behind the membrane. In some devices

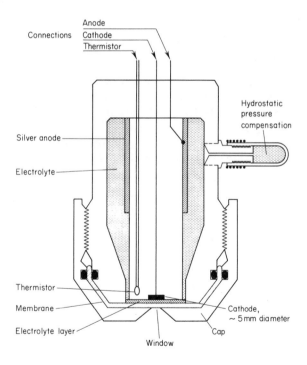

Figure 22. Schematic cross-section through a membrane-covered oxygen sensor with pressure compensation

(Gleichmann *et al.*, 1960; Grasshoff, 1962) a defined electrolyte layer between membrane and cathode is provided by a hydrophilic foil (Cellophane) which is soaked with electrolyte. The anode is usually silver.

9.5.2. Electrochemical process at the membrane electrode

The electrochemical reduction of oxygen at a membrane-covered electrode follows the same principle as for uncovered noble metal electrodes (Section 9.4). As oxygen is reduced at the working cathode (ideally to zero concentration), a gradient builds up and forces the oxygen dissolved in the bulk solution to diffuse through the membrane and the electrolyte layer covering the cathode.

$$\text{cathode: } O_2 + 2H_2O + 4e^- \rightleftharpoons 4OH^-$$
$$\text{anode: } \qquad\qquad 4Ag \rightleftharpoons 4Ag^+ + 4e^-$$

with KCl as electrolyte:

$$Ag^+ + Cl^- \rightleftharpoons AgCl$$

with KOH as electrolyte:

$$2Ag^+ + 2OH^- \rightleftharpoons Ag_2O + H_2O$$

It follows from the above reaction scheme that hydroxyl ions are produced at the cathode in proportion to the amount of oxygen reduced. Assuming a current of $10\ \mu A$, 0.37×10^{-6} mol of hydroxyl ions are generated at the cathode per hour or 8.9×10^{-6} mol per day. It is obvious that the pH of the electrolyte solution will change drastically if no hydroxyl ions leave the electrolyte layer in front of the cathode. According to Odén (1962), the pH of the solution has no effect on the reduction process up to a pH of 11 and probably up to 14. It can be assumed, however, that the ionic strength influences the coefficient of diffusion. If an alkaline electrolyte is used, the anodic process and the subsequent reaction consumes all of the hydroxyl ions generated. There are therefore two possible ways in which hydroxyl ions may be removed: either by using a relatively large chloride electrolyte volume into which the hydroxyl ions can diffuse, or by placing the cathode and anode relatively close together and using an alkaline electrolyte. A large electrolyte volume has the disadvantage that the oxygen dissolved in the electrolyte may diffuse to the cathode from behind and cause an erratic current which is then no longer proportional to the partial pressure of oxygen in the bulk solution (*cf.* Section 9.5.5).

It can also be demonstrated that the ohmic resistance of a thin electrolyte film at the cathode surface can cause a related drop in voltage which, in turn, results in complete inactivity of the centre of a plane electrode if the diameter is larger than 5 mm. In this case only the outer part of the electrode is active and almost no reduction of oxygen takes place in the centre.

Summing up, the reduction current for a given partial pressure of oxygen in the solution is affected by the distance between the anode and cathode, the surface area and shape of the cathode, and the composition and the concentration of the electrolyte. These relationships can be expressed as a function of the stability and sensitivity of the probe.

9.5.3. Diffusion process at the membrane-covered electrode

In principle, the diffusion of oxygen at a membrane-covered electrode can be divided into three steps: diffusion of the dissolved gas through a laminar boundary layer at the outer surface of the membrane, diffusion through the membrane, and diffusion through the electrolyte layer at the cathode surface. The boundary conditions for the diffusive process (thickness of the outer laminar layer, of the membrane, and of the electrolyte layer) must be strictly controlled if the sensor current is to be proportional to the amount of oxygen in the bulk solution. The dominating factor in the diffusion process at a membrane-covered electrode is the flux of oxygen through the membrane, if the outer laminar layer is kept thin, or if the oxygen consumption per unit area of the membrane is kept small, so that eddy diffusion within the solution is sufficient to replenish the oxygen consumed by the noble metal cathode.

Usually the membrane material used in oxygen-sensitive electrodes is hydrophobic. Only gas molecules may penetrate the membrane. This implies that the partial pressure of oxygen in the solution and in the membrane is in equilibrium if no potential is applied to the cathode behind the membrane. One can therefore assume that a membrane-covered electrode creates a current which is a function of the partial pressure of oxygen in the bulk solution. Migration of gases through a membrane is usually treated in terms of the permeability coefficient (P_M) rather than the diffusion coefficient (D_M). This simplifies calculations because the gas transport can be analysed as a function of partial pressure, which in turn maintains continuity through boundary layers. The permeability coefficient is related to the diffusion coefficient by

$$P_M = \alpha D_M$$

where α is the distribution coefficient of oxygen between the membrane and the inner electrolyte (Hitchman 1978; see Section 9.5.4). The diffusive flux (J) at any point in a cross-section through the membrane can be expressed as

$$J = P_M \left(\frac{\partial p}{\partial x} + \frac{\partial p}{\partial y} \right)$$

where p is the partial pressure of oxygen and the permeability, P_M, is assumed to have no directional dependence, which is valid for the type of membrane used in oxygen probes.

Assuming a steady state, the total number of oxygen molecules reaching any given point in the membrane must be equal to the total number of molecules leaving this point, i.e. the net flux must be zero. Ben-Yaakov *et al.* (1977) simulated the steady-state flux in a membrane when the electrode is working on the polarographic plateau, and the partial pressure of oxygen is zero at the cathode surface and is assumed to be unity in the bulk solution. A thin layer of electrolyte between the membrane and the cathode was taken into consideration (Mancy *et al.*, 1962). Assuming vertical diffusion through the membrane (the shortest path from the membrane surface to the electrode surface), the oxygen consumption per unit area of exposed membrane is equal to the oxygen consumption per unit area of the cathode, provided that the electrode surface is sufficiently and evenly covered with electrolyte solution. Under such conditions and with a relatively large electrode surface, the probe will require turbulent stirring of the solution at the membrane to give a correct current response to the partial pressure of oxygen in the bulk solution. All probes with a front 'window' of the same size as the cathode or smaller belong to this type.

If the cathode becomes smaller, oxygen is consumed from a larger area, owing to the spreading of the diffusion paths at the edge of the cathode. If the cathode is diminished to a point, it 'sees' the oxygen from a half sphere and maximum spreading of the diffusion streamlines is achieved. Since the resulting current is itself proportional to cathode surface area there is a limit to the extent to which the size of the electrode can be reduced. As has been pointed out in Section 9.4.4, the flow sensitivity of an electrode is also a function of the electrode size and can be reduced considerably if the cathode area is reduced to a diameter of 50 μm^2. Because of the spreading effect of the membrane, the flow sensitivity is further reduced if a small cathode is used in a membrane-covered electrode.

Various gas-permeable membranes have been proposed for use in oxygen sensors. Teflon (Du Pont Trade-Mark for polytetrafluoroethylene) is usually preferred because of its relatively high permeability to oxygen, its mechanical strength, its stability, and its inertness (the permeability of Teflon to oxygen is about 5 times larger than that of polyethylene or polypropylene). Ben-Yaakov *et al.* (1977) proposed that further spreading of the diffusive streamlines can be achieved by combining the Teflon membrane with a second outer membrane made of silicone rubber, a material which is about 60 times more permeable to oxygen than is Teflon.

In order to illustrate the function of such a combination, one may think of the analogy of a point heat source separated from a solution by a thermal insulator. It is evident that the heat from the source would spread over a

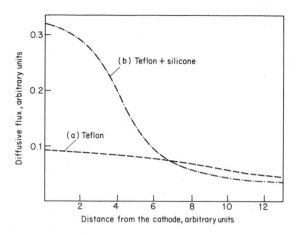

Figure 23. Model calculation of the vertical diffusion of oxygen in a plane perpendicular to the cathode surface: (a) for a 10-μm Teflon membrane; (b) for a 10-μm Teflon and a 25-μm silicone rubber membrane. According to Ben-Yaakov and Ruth (1980)

larger area if a thermally conducting layer was attached to the outer surface of the insulator. The model calculations of Ben-Yaakov *et al.* (Figure 23) illustrate the effectiveness of such a composite membrane.

9.5.4. Theory of the diffusion current at a membrane-covered amperometric oxygen probe

The general principles of the electrochemical processes that occur at a membrane-covered cathode and the diffusion of oxygen through a gas-permeable membrane have been discussed in Sections 9.5.1–9.5.3. Mancy *et al.* (1962) developed a mathematical derivation for the diffusion-controlled current at a membrane-covered cathode. Obviously the processes are complicated as the oxygen diffuses through different media, especially if non-steady-state conditions are involved.

Assuming the establishment of equilibrium in the distribution of oxygen between the solution, the membrane, and the electrolyte layer and before application of a potential, no gradient exists between the fugacity (partial pressure) of oxygen in the different media. If c_S is the oxygen concentration in the solution and K_M and K_E are the distribution coefficients of oxygen between the solution and the membrane, or the membrane and the electrolyte, the corresponding equilibrium concentrations within the membrane, c_M, and within the electrolyte film, c_E, are

$$c_M = K_M c_S \quad \text{and} \quad c_E = c_M/K_E \qquad (22)$$

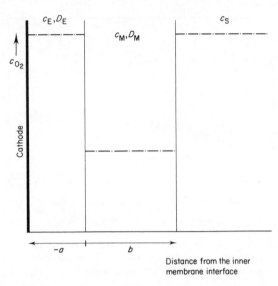

Figure 24. Distribution of oxygen in the electrolyte layer, in the membrane, and in the bulk solution in front of a plane solid electrode before a potential is applied to the electrode. According to Mancy *et al.* (1962)

K_E and K_M are almost identical if the ionic strength of the outer solution and the inner electrolyte are identical and if water is used as solvent in both cases. This implies that

$$c_S \approx c_E \tag{23}$$

Figure 24 illustrates schematically the conditions before application of a potential.

If the 'plateau' potential (usually *ca.* -0.8 V at a noble metal cathode) is applied, an immediate and complete reduction of oxygen occurs at the cathode surface. After this event, the current is diffusion controlled. The diffusion layer spreads from the cathode surface into the electrolyte and into the membrane. Assuming thorough turbulent mixing in the outer solution and neglecting a thin laminar boundary layer of solvent at the outer membrane surface, the oxygen concentration profile in a cross-section normal to the cathode varies until the outer surface of the membrane is reached. A steady-state profile is obtained as the electrolysis time, t, approaches t_∞.

Because it is postulated that the applied voltage is in the range of the polarographic plateau, the surface concentration of oxygen is reduced to zero. The resulting current is thus diffusion controlled and described by

$$i = nFD_E A \left(\frac{\partial c_E}{\partial x} \right)_{x=-a} \tag{24}$$

The following assumptions were made by Mancy *et al.* (1962):

(i) the oxygen penetrates into the cell only through the membrane in a direction perpendicular to the electrode surface;

(ii) the diffusion coefficients, D_M and D_E, for the oxygen in the membrane and in the electrolyte layer are constants independent of concentration and time;

(iii) concentrations are used instead of activities or chemical potentials.

The first assumption is realistic because the electrolyte layer is usually a capillary layer or a thin layer of hydrophilic material (e.g. Cellophane) and the thickness of the membrane is of the order of 5–20 μm. The contribution from non-perpendicular diffusion at the edges of the cathode is minimal and arises from a zone of about 0.02 mm around the edges (Hitchman, 1978). (As the current is directly proportional to the area of the diffusion zone, the error is approximately 0.1%, assuming an electrode diameter of 1 mm.)

The second assumption is valid if there are no structural changes in the membrane in the range of temperature in question and if the effect of pressure is not considered. The assumption is also valid for the electrolyte layer if the initial composition of the electrolyte is such that the generation of hydroxyl ions at the electrode surface during the electrode process does not alter the composition considerably. The effect of pressure on the diffusion coefficient is discussed later (*cf.* Section 9.6.3).

The third assumption was made by Mancy *et al.* (1962) in order to simplify the treatment of the derivation. The boundary conditions (Figure 24) are

$$c_E = 0 \qquad \text{at} \quad x = -a \text{ for } t > 0$$

$$c_M = K_E c_E \quad \text{at} \quad x = 0$$

$$c_M = K_M c_S \quad \text{at} \quad x = b \text{ for } t \geq 0$$

It is assumed that equilibrium prevails at the boundary between the outer solution and the membrane.

For comparison of the diffusion times we can define a time-dependent diffusion layer thickness using the equation

$$\delta_t = K D^{1/2} t^{1/2} \tag{25}$$

where K is a constant (Hitchman, 1978). In order to calculate the progress of the diffusion across the membrane and the electrolyte layer, $a/D_E^{1/2}$ must be compared with $b/D_M^{1/2}$. The solutions of the particular cases for $(\partial c_E / \partial x)_{x=-a}$ were given by Mancy *et al.* (1962) as

(1) $a/D_E^{1/2} \gg b/D_M^{1/2}$
(2) $a/D_E^{1/2} = b/D_M^{1/2}$
(3) $a/D_E^{1/2} \ll b/D_M^{1/2}$

Solution (1) corresponds to the situation when the diffusion of oxygen in the electrolyte layer controls the rate, whereas (3) describes the situation when the diffusion process through the membrane is the rate-controlling factor, and in (2) both diffusion processes contribute to the rate at which oxygen is transported to the electrode.

Immediately after application of the potential to the electrode, the diffusion layer will be confined to the immediate vicinity of the electrode (the electrolyte film). This implies that the zone with reduced oxygen content has not reached the membrane and that, therefore, the diffusion through the membrane has, as yet, no effect on the transport process. When a certain time has elapsed after the application of the potential to the electrode, the diffusion through both the electrolyte layer and the membrane control the rate of transport of oxygen to the electrode surface (condition 2). When the diffusion boundary has reached the membrane–solution interface, and if $a \ll 10b$ and $D_E \gg D_M$, then only condition (3) has to be considered. In recent designs of oxygen probes $a \approx 0.1b$ and $D_M \approx 10^{-2}D_E$. For Teflon, D_M is between 1×10^{-11} and 2×10^{-11} m^2 s^{-1} (Hitchman, 1978) and for the electrolyte D_E is of the order of 2×10^{-9} m^2 s^{-1} at 20°C (Horne, 1969). The expression for the steady-state current can therefore be based on assumption (3). The current a very short time after application of the potential is not usually of interest, although a special method of oxygen determination with membrane-covered probes may be based on this initial current.

Mancy *et al.* (1962) gave the equations for the three cases:

(i) For very short time intervals when the diffusion layer is confined to the electrolyte film:

$$(\partial c_E/\partial x)_{x=-a} = [c_{E,t=0}/(\pi D_E t)^{1/2}]\left[1 + 2\sum_{n=1}^{\infty} \exp\left(-n^2 a^2/D_E t\right)\right] \qquad (26)$$

(ii) For short times when the transport in the membrane is the controlling process:

$$(\partial c_E/\partial x)_{x=-a} = (c_{E,t=0}K_E/D_E)(D_M/\pi t)^{1/2}\left[1 + 2\sum_{n=1}^{\infty} \exp\left(-n^2 b^2/D_M t\right)\right] \qquad (27)$$

(iii) For longer times when a steady-state current is reached:

$$(\partial c_E/\partial x)_{x=-a} = (c_{E,t=0}D_M K_E/b D_E)\left[1 + 2\sum_{n=1}^{\infty} \exp\left(-n^2 \pi^2 D_M t/b^2\right)\right] \qquad (28)$$

Before a potential is applied to the electrode the concentration of oxygen in the electrolyte layer (c_E) is related to the oxygen concentration in the bulk solution by

$$c_E = \frac{K_E}{K_M} \cdot c_S \qquad (29)$$

The diffusive flux of the oxygen molecules through the membrane can be described by Fick's law:

$$J_M = D_M(c_{M,x=b} - c_{M,x=0})/b \qquad (30)$$

where $c_{M,x=b}$ and $c_{M,x=0}$ are the oxygen concentrations at the outer and inner membrane interfaces. According to Hitchman (1978), the permeability of the membrane is linked to the diffusion coefficient by

$$P_M = D_M \cdot \frac{K_M c_S - K_E c_{E,x=0}}{c_S - c_{E,x=0}} \qquad (31)$$

where P_M is the permeability coefficient of the membrane and D_M the diffusion coefficient. At longer times after the application of the potential $c_{E,x=0} \ll c_S$ and equation 31 becomes

$$P_M = K_M D_M \qquad (32)$$

Substituting equations 29 and 32 and the equation for $\partial c_E/\partial x$ at the different time intervals (equations 26 and 28) into equation 24, we can obtain the following relationships for the current generated by the cell:

(i) initial current:

$$i_{L,t} = (nFK_M c_S/K_E)(D_E/\pi t)^{1/2}\left[1 + 2\sum_{n=1}^{\infty} \exp(-n^2 a^2/D_E t)\right] \qquad (33)$$

(ii) current observed when the diffusion layer penetrates the membrane:

$$i_{L,t} = nFc_S(P_M/\pi t)^{1/2}\left[1 + 2\sum_{n=1}^{\infty} \exp(-n^2 b^2/D_M t)\right] \qquad (34)$$

(iii) for the steady-state current:

$$i_{L,t} = nFc_S(P_M/b)\left[1 + 2\sum_{n=1}^{\infty} \exp(-n^2 \pi^2 D_M t/b^2)\right] \qquad (35)$$

As the thickness of the electrolyte film is usually of the order of a few μm or less and, for $D_E \approx 10^{-9}\ m^2\ s^{-1}$, the diffusion layer progresses to the membrane–electrolyte interface within a few hundred milliseconds (Hitchman, 1978). Figure 25 illustrates the progression of the diffusion layer with increasing time of electrolysis according to Koryta et al. (1970). Therefore, equation 33 can only be valid for $t < 0.1$ s. Using practical values for a and

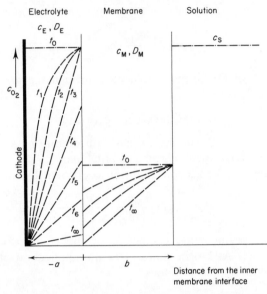

Figure 25. Schematic diagram of the formation of the diffusion layer in front of a plane membrane-covered electrode as a function of time after application of the plateau potential. According to Koryta *et al.* (1970)

D_E and a minimum value of n, equation 33 can be reduced to

$$i_{L,t} = nF \cdot \frac{K_M}{K_E} \cdot c_S \left(\frac{D_E}{\pi t}\right)^{1/2} \tag{36}$$

Equation 34 can also be reduced to a simplified expression assuming that after $t \geqslant 10$ s the diffusion layer has reached the interface between the membrane and the solution ($b \approx 20 \ \mu m$ and $D_M = 10^{-11} \ m^2 \ s^{-1}$). In this case the exponential term is small for all n:

$$i_{L,t} = nFc_S(P_M/\pi t)^{1/2} \tag{37}$$

If the current $i_{L,t}$ reaches its steady state ($t \geqslant 20$ s), equation 35 can be applied. All of these equations assume that the diffusion layer will not spread into the solution so that either a turbulent flow is maintained in the solution up to the outer surface of the membrane, or the depletion of oxygen at the interface of the solution and the membrane is balanced by molecular transport, which is the case if the electrode surface area is of the order of μm^2. The membrane layer thus acts as a diffusion layer with a defined thickness.

The exponential term in equation 35 becomes negligible when $n \geqslant 5$ (Hitchman, 1978). Assuming a membrane thickness, b, of 20 μm and with $D_M \approx 10^{-11} \ m^2 \ s^{-1}$, the time to reach a steady-state current is about 20 s, so that at the steady state

$$i_L = nFc_S P_M b^{-1} \tag{38}$$

The calculation of the steady-state current is valid only if diffusive transport in the membrane is the rate-determining process. If the electrolyte layer is large compared with the thickness of the membrane ($a \neq 0$), then

$$i_{L,a \neq 0} = i_L \left(1 + K_E \cdot \frac{D_M a}{D_E b}\right)^{-1} \tag{39}$$

(Hitchman, 1978).

Rearranging Equation 39, we find that

$$\frac{1}{i_{L,a \neq 0}} = \frac{1}{i_L} + \frac{1}{i_L} \cdot K_E \cdot \frac{D_M a}{D_E b} \tag{40}$$

The reciprocal of the diffusion current at $t \geqslant 20\,\text{s}$ is proportional to the thickness of the electrolyte layer, a. Of course, all probes must have a non-zero thickness of the electrolyte layer to establish electrolytic contact between cathode and anode and to support the electrochemical process in which water is consumed and hydroxyl ions are generated. Extrapolation of measurements with different but defined thicknesses of the electrolyte layer (which can be obtained by using increasing numbers of Cellophane spacers soaked with electrolyte between the electrode and the membrane) gives the intercept for zero thickness of the membrane. As the steady-state diffusion

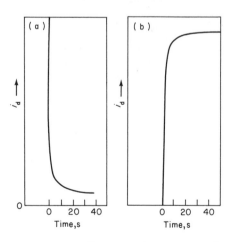

Figure 26. Current response curves for a membrane-covered polarographic oxygen sensor (Hydro-Bios, *cf.* Section 9.7) after the plateau potential has been applied and the sensor is immersed in water saturated to 10% with oxygen (a), and for a step change of the oxygen sensor from zero to saturation (b)

current i_L is related to the permeability of the membrane P_M by

$$i_L = nFc_S b^{-1} P_M \tag{41}$$

it is possible to determine P_M in a simple way from

$$P_M = i_L b (nFc_S)^{-1} \tag{42}$$

Since a is about 2–10 μm, the experimental determination of P_M from equation 42 is more reliable than from Equation 32, which assumes that $a = 0$. As reported by Hitchman (1978), P_M is $8.7 \times 10^{-11} \, m^2 \, s^{-1}$ for a 12.5 μm Teflon membrane and a cathode radius of 3.16 mm at 23°C. The calculation from equation 42 for the same conditions results in $P_M = 8.2 \times 10^{-11} \, m^2 \, s^{-1}$.

A practical example of the variation of current with time is given in Figure 26, which shows the validity of equations 35–37 for different lengths of time after the application of a potential to a membrane-covered oxygen probe.

9.5.5. Dependence of the diffusion current on time

In the practical application of a membrane-covered oxygen sensor it is important to ascertain how fast a steady-state current is reached after the polarization potential has been applied to the electrodes. As has been discussed in Section 9.5.4, the steady-state response to a constant partial pressure of oxygen in the medium should theoretically be reached within about 30 s with a 20 μm Teflon membrane. A thinner membrane reduces this time considerably. In practice the response time is longer. According to Hitchman (1978), the disagreement between theory and the real behaviour cannot be due to delays in the establishment of the steady-state concentration profile within the membrane because it is diffusion controlled and has a low energy of activation. The slower response may be due in part to the initial reduction of an oxide film on the cathode surface. Even a coverage with two or three molecular layers would prolong the establishment of the steady-state current significantly. Impurities in the electrolyte may be another source of sluggish response. A major source of the delay in the current response is certainly the change in pH in the electrolyte film at the cathode, causing either a change in the reduction path of oxygen ($n < 4$ for each molecule of oxygen) with H_2O_2 as an intermediate product (*cf.* Section 9.4.1) or a change in the viscosity close to the cathode. The effect of intermediate formation of H_2O_2 can be reduced by addition of a catalyst to the electrolyte.

A more interesting question from the practical point of view is how quickly an oxygen sensor responds to a step change in the oxygen concentration (partial pressure) of the medium. A rapid response is desirable if the probe is to be used for vertical profiling, especially if the measurements are used in conjunction with other environmental parameters (e.g. salinity or temperature) for which sensors are available which have time constants of the order of seconds, or even less. If the probe is to be used for the measurement of long-term changes or for monitoring purposes in rivers and estuaries or in controlled ecosystem experiments, however, a fast response is not as important as long-term stability and insensitivity towards mechanical stress and water movement in the vicinity of the electrode.

Heineken (1970) has developed a method which allows the calculation of the response of a membrane-covered oxygen sensor to a step change of the partial pressure of oxygen in the medium. Heineken (1971) also derived a simplified equation for the step-response of a membrane-covered oxygen detector if the oxygen partial pressure is changed from zero to saturation:

$$I_1 = 1 - 2\exp(-kt) + 2\sum_{n=2}^{\infty}(-1)^n \exp(-n^2 kt) \tag{44}$$

where I_1 is the amplified current, i_d, normalized to 1 at the steady state, and k is a material constant of the membrane:

$$k = \pi^2 D_M b^{-2} \tag{45}$$

where D_M is the diffusion coefficient of the membrane in $m^2\,min^{-1}$ and b is the thickness of the membrane (*cf.* equation 35). k can immediately be calculated from an experimental plot of $(1 - I_1/t)$ versus t by inserting the coordinates of two points from the curve in the region where $kt \geq 1.2$ and $(1 - I_1) < 0.6$ into the equation

$$k = \frac{\ln I_1(a) - \ln I_1(b)}{t_{a,b}} \tag{46}$$

(Grasshoff, 1978).

Figures 27a and b show the response of an oxygen probe to step changes in the oxygen partial pressure at different temperatures for a 25 μm Teflon membrane and for a 15 μm polyethylene membrane. A comparison of different membrane materials and thickness at 30°C is shown in Figure 28. As can immediately be seen from these diagrams, Teflon membranes show a significantly faster response. Of course, the response is delayed with decreasing temperature because of increasing k values which are, in turn, related to the diffusion coefficient (*cf.* equation 45). According to Heineken (1971), the experimentally determined curves show a significant deviation from the theoretical curve (*cf.* equation 44 and Figure 27) in the range where the steady-state response is approached (Grasshoff, 1978). This deviation is especially developed for polyethylene membranes, where the $(1 - I_1)$ curves show almost a discontinuity.

Grasshoff (1978) attributes the deviation of the practical $(1 - I_1)$ values from the theoretical curves to (i) the non-zero solubility of oxygen in the membrane material, which necessitates the re-establishment of the partition

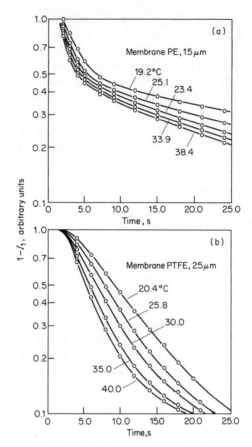

Figure 27. Current response of membrane-covered oxygen sensors to a step change in the partial pressure of oxygen in the solution for (a) a 15 μm and (b) a 25 μm Teflon membrane as a function of time and temperature

equilibrium with changing oxygen concentration (partial pressure) in the medium, and (ii) the non-linear diffusion in the marginal zones of the cathode where the approach of oxygen to the cathode is delayed by the longer diffusion path through the membrane. Further, the time required for the establishment of equilibrium conditions for the reduction of oxygen dissolved in the electrolyte film, especially in the vicinity of the cathode, and for the transport of oxygen from the bulk of the electrolyte to the diffusion zone may also prolong the response time, especially if the sensor is exposed periodically to low or zero oxygen concentrations.

Grasshoff (1978) proposed a method for correcting the response function for the margin effect. The current–time curve is recorded for several minutes after the step change has taken place. The response curve after about 30 s

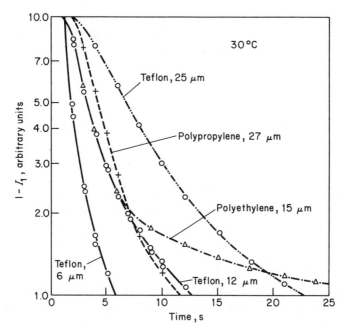

Figure 28. Comparison of the response of different membrane materials and thickness with respect to a step change in the oxygen partial pressure in the solution for a membrane-covered polarographic oxygen sensor. According to Grasshoff (1978)

can be described with good approximation by an exponential function which represents the non-ideal behaviour of the probe. This exponential function is then subtracted from the total response function. The results are exponential functions which give straight lines in a semi-logarithmic plot.

Table 4 gives the corrected k values for different Teflon membranes at various temperatures. The temperature dependence of k can be expressed by

$$k_T = k_R \exp \alpha_T(T - T_R) \tag{47}$$

Table 4. Corrected k values for Teflon membranes at various temperatures (according to Grasshoff, 1978)

Thickness, μm	Parameter	Values				
25	T, °C	20.4	25.8	30.0	35.0	40.0
	k, min^{-1}	8.02	10.09	11.92	14.06	16.17
12	T, °C	21.4	24.8	30.0	33.7	39.4
	k, min^{-1}	10.65	24.19	27.57	34.12	41.24
6	T, °C	20.2	25.6	29.7	34.5	39.4
	k, min^{-1}	35.36	43.00	49.5	60.3	70.4

where k_T is the value of k at temperature T, k_R is the value at a reference temperature, T_R, and α_T is the temperature coefficient (Grasshoff, 1978). From the data given in Table 4 a uniform temperature coefficient, $\alpha_T = 0.036$, was evaluated by Grasshoff (1978) for all Teflon membranes.

Hitchman (1978) gave another equation for the current–time response for a step change in the oxygen concentration (partial pressure) in the medium:

$$\frac{i_t - i_0}{i_\infty - i_0} = 1 + 2 \sum_{n=1}^{\infty} \left[(-1)^n \exp(-n^2 \pi^2 \tau) \right] \tag{48}$$

where i_t is the current at the time, t, after the step change, i_∞ is the steady-state current, i_0 is the background current, and $\tau = D_M t b^{-2}$. For Teflon membranes good agreement is found between calculated and experimental values up to about 14 s. After this initial period the experimental values depart from the theoretical curve for reasons discussed earlier.

9.5.6. Electrode material

In principle, the criteria discussed in Section 9.4 have to be applied in connection with membrane-covered polarographic oxygen sensors. The cathodic reaction in acid medium is

$$O_2 + 4H^+ + 4e^- \rightleftharpoons H_2O$$

$$E^{\ominus} = +1.229 \text{ V } vs. \text{ NHE}$$

(NHE = normal hydrogen electrode) and in alkaline or neutral medium

$$O_2 + 2H_2O + 4e^- \rightleftharpoons 4OH^-$$

$$E^{\ominus} = +0.401 \text{ V } vs. \text{ NHE}$$

The theoretical potential is difficult to obtain because of the low exchange current densities and the role of impurities (Bockris and Reddy, 1970). The net reactions are further complicated by the possible intermediate formation of H_2O_2 and the role of active sites at the electrode surface, forming intermediate metal oxides and metal hydroxides (Bockris and Reddy, 1970). The hydrogen peroxide may decompose at the metal surface according to

$$2H_2O_2 \rightleftharpoons 2H_2O + O_2$$

or it may diffuse away from the cathode before decomposition (Hahn et al., 1975). The electrode surface may be covered by oxides and the reduction may involve absorbed oxygen. In addition, few metals can withstand the highly positive electrode potentials without corrosion (Hitchman, 1978).

A typical experimental current–voltage diagram for a membrane-covered oxygen probe is shown in Figure 29. The plateau of the curve is at potentials more negative than ca. -0.5 V (vs. Ag|AgCl) and not at 0.9 V as expected theoretically (the Ag|AgCl electrode has a potential of $+0.22$ V vs. NHE).

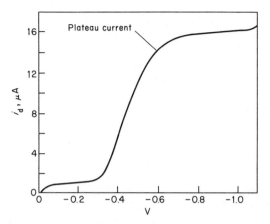

Figure 29. Current–voltage diagram for a working membrane-covered oxygen sensor

As discussed earlier (Section 9.4.4), the cathodic reaction produces hydroxyl ions and generates a high pH at the electrode surface (*ca.* pH 13). At such a high pH the potential should be shifted from +0.9 V (*vs.* NHE) to *ca.* +0.1 V (*vs.* NHE). With the Ag|AgCl couple as reference this potential will be *ca.* −0.1 V. In fact, the reduction starts at approximately −0.25 V (*vs.* Ag|AgCl) and the plateau is reached at about −0.65 V (*vs.* Ag|AgCl) if the electrode is performing well.

The rather anodic value of the equilibrium potential and the potential for the reduction of oxygen prevent the reduction of other gases which may be present in the environment, e.g. SO_2 (Hitchman, 1978). There are no gases present in sea water which are likely to interfere with the oxygen reduction (CO and the oxides of nitrogen are present in too small amounts).

Only the noble metals such as platinum, gold, palladium, rhodium, and iridium can be used as cathodes in the oxygen sensor. Other metals which have been proposed for oxygen probes (e.g. lead) form oxides which are reduced under the potential conditions required for the reduction of oxygen and are the source of a high background current. The only metal other than the noble metals which may be considered is silver for which

$$Ag_2O + H_2O + 2e^- \rightleftharpoons 2Ag + 2OH^-$$

$$E^\ominus = +0.35 \text{ V } (vs. \text{ NHE})$$

but the surface of a silver electrode is easily poisoned and the active sites for electron transfer blocked by sulphur or hydrogen sulphide. There are no advantages to be gained by using metals other than gold or platinum as cathode materials. Gold has no strong tendency to absorb oxygen on its surface and does not show ageing effects (Hitchman, 1978; Kessler, 1973). Therefore, gold is very often preferred to platinum (Hoare, 1969). On the other hand, platinum can be sealed directly to glass. Glass has a significant advantage as supporting material for the cathode and the probe head since it does not absorb oxygen in significant amounts and, therefore, oxygen leaking from the electrode carrier material cannot contribute to an undefined background current. Furthermore, the sealing between the gold cathode disc or wire may break, leak or be 'washed out', and form a pocket for the electrolyte containing an undefined amount of oxygen, which, in turn, contributes to the background current and causes a sluggish response of the electrode after application of the potential or after a step change of the oxygen pressure in the solution.

The requirements for the anode material are more simple. The anodic reaction should not be subject to concentration polarization, the reaction should produce a stable reference potential, and the oxidation products should not interfere with the cathodic reduction. The most suitable anode material is silver in combination with a chloride-containing electrolyte or with an alkaline electrolyte.

$$Ag + Cl^- \rightleftharpoons AgCl + e^-$$

$$2Ag + 2OH^- \rightleftharpoons Ag_2O + H_2O + 2e^-$$

As hydroxyl ions are formed at the cathode, a chloride electrolyte will slowly change into an alkaline electrolyte which will shift the reference potential of the anode by about $0.1\,V$ (Ag|AgCl, $+0.222\,V$ $vs.$ NHE; Ag|Ag$_2$O, $+0.35\,V$ $vs.$ NHE at 25°C). Therefore, an alkaline electrolyte and a silver–silver oxide anode is of advantage. The silver oxide does not adhere to the cathode surface as firmly as silver chloride, but this drawback can be overcome by a proper design of the anode compartment.

Concentration polarization can be avoided by using an anode with a larger surface area than that of the cathode. Assuming a steady-state current of $10^{-5}\,A$, a reference anode with $100\,mm^2$ of exposed surface is sufficient to avoid any concentration polarization. Assuming an average current of $10^{-5}\,A$ in the probe, a simple calculation shows that about $10^{-3}\,kg$ of silver will be converted into silver oxide with 3 years of continuous use.

9.5.7. Electrolyte

The electrolyte in a membrane-covered polarographic oxygen sensor should serve several purposes. It should form a medium with low ohmic resistance between the cathode and the anode. The components of the electrolyte should not interfere with the desired cathodic and anodic reactions. This is especially important for the unavoidable impurities which may be present in the salt or which may be introduced during the refilling process. Reaction products of the cathodic (OH$^-$) and the anodic reaction

(Ag_2O or AgCl) should be readily absorbed and should not cause considerable changes in the ionic composition of the electrolyte. The generation of the hydroxyl ions should not shift the pH in front of the cathode too much. The composition of the electrolyte should, furthermore, be such that the mechanism for the reduction of oxygen will not change to give a process requiring less than four electrons. This effect causes problems only if the pH is too low and changes rapidly to higher values through the generation of hydroxyl ions. The components of the electrolyte should not react with CO_2, forming precipitates.

The loss of water through the membrane must be kept to a minimum, especially if the sensor is used to measure oxygen in the gas phase. In fact, the loss of water from the electrolyte is possibly the most serious problem. It is therefore advisable always to keep the membrane moist or wet when not in use. If the electrolyte reservoir is hermetically sealed such a reduction of the electrolyte volume may cause a change in the thickness of the electrolyte layer between membrane and cathode, leading finally to partial or complete drying out of the film. This in turn increases the ohmic resistance and may result in inactivity for parts of the cathode surface, especially if relatively large electrodes are used.

If there is no hermetic sealing between the electrolyte compartment and the outside, a reduction in the electrolyte volume may cause air bubbles to enter the electrolyte. This will create serious problems because of the increased compressibility under hydrostatic pressure and the migration of micro-bubbles into the electrolyte layer in front of the cathode.

Loss of water vapour can be reduced by several methods. One possibility is to have a relatively large electrolyte reservoir with good connection to the layer in front of the cathode, where the loss occurs. The volume reduction can be compensated for by attaching a small thick-walled silicone rubber bulb (pipette bulb) which is filled with electrolyte and has a hydrostatic connection to the main electrolyte reservoir. Such a device also serves as a pressure compensator between the outside (hydrostatic) pressure and the inner part of the electrode, especially at the inside of the membrane. A disadvantage of a large electrolyte volume is the increased possibility of oxygen molecules dissolved in the electrolyte reaching the diffusion zone and the cathode to give a marked background current, which is generally neither constant nor reproducible. This problem can be overcome, however, by a proper design of the connection between the bulk of the electrolyte (the anode compartment) and the head of the probe where the cathode is located (for example, with several capillary channels at the outer edge of the top of the probe). Another way of reducing losses of water vapour is to use an electrolyte in the form of a gel or a paste. Such a gel electrolyte has the advantage of retaining the anodic oxidation products (Ag_2O, AgCl) near the anode. A disadvantage is that it is more difficult to change the electrolyte and to clean the electrolyte compartment. The most suitable method for the reduction of the loss of water from the electrolyte is the addition of a suitable deliquescent salt or glycerol. If a 2.3 M KCl solution is used the loss is 15% after 8 days and 40% after 190 days of exposure of the probe to air. If a small amount of KH_2PO_4 is added, the loss of water from the electrolyte is about the same during the first few days but is no greater than 10% after 190 days (Hitchman, 1978).

In order to provide a good conducting medium between the anode and cathode, the ionic strength of the electrolyte should be >0.5 M. Most frequently sodium chloride, potassium chloride, sodium or potassium hydroxide, or their mixtures are used as background electrolytes.

Niedrach and Stoddard (1972) have proposed the use of an ion-exchange resin as the electrolyte. Such a resin should contain ionic groups such as sulphonic acids for cation exchange and quaternary ammonium compounds for anion exchange. Both compounds are incorporated into a matrix such as polystyrene. Hydroxyl ions generated at the cathode are transported by 'jumping' from one molecule of the quaternary ammonium compound to the next in the direction of the electrical field until they encounter the silver ions which are dissolved from the anode by the oxidation process. Of course, the ion-exchange resin must be in intimate contact with the anode and the contact between the cathode and the resins must be kept moist.

9.5.8. Membrane characteristics

The membrane serves four purposes: (1) it separates the electrolyte compartment filled with an electrolyte of defined composition from the medium in which oxygen is to be measured; (2) it forms a defined diffusion layer in front of the working cathode; (3) it protects the cathode from being poisoned, because only gases diffuse through the membrane, and it protects the delicate cathode from mechanical damage; and (4) under certain circumstances it spreads the diffusion streamlines and reduces the sensitivity of the oxygen-dependent diffusion current to variations in the turbulence of the medium in front of the probe (*cf.* Section 9.5.3).

The commonly used materials are Mylar, Teflon, polyethylene, polypropylene, natural rubber, silicone rubber, and poly(vinyl chloride). A systematic study of the characteristics and suitability of membrane materials has been made by Sawyer *et al.* (1959). The permeability of plastic membranes to gaseous molecules depends on the solubility and diffusivity of the gas. Qualitatively it can be said that the thinner the membrane and the greater the permeability, the more sensitive is the polarographic oxygen sensor. Van Krevelen (1972) described a graphical method for calculating gas permeabilities from the glass transition temperature of the polymer and the molecular radius of the permeating gas. From these parameters the

Table 5. Calculated and experimental values of permeability data for oxygen in a Teflon membrane (from Hitchman, 1978)

	$D_M \times 10^{11}$, $m^2 s^{-1}$	$P_M \times 10^{11}$, $m^2 s^{-1}$	K_M
Calculated	1.0	13.7	13.7
Experimental	1.1	8.2	7.5

corresponding activation energy for diffusion and hence the diffusion coefficient can be estimated. For Teflon a value of 2.5×10^{-11} $m^2 s^{-1}$ at 20°C is found for the diffusion coefficient of oxygen (D_a) in the amorphous polymer. This value has to be corrected for the degree of crystallinity of the polymer as the evaluation is only valid for strictly amorphous materials, so that

$$D_c = D_a(1 - x_c) \qquad (49)$$

D_c is the corrected diffusion coefficient, and x_c is the degree of crystallinity. For Teflon, x_c is about 0.6, giving a value of 1.0×10^{-11} $m^2 s^{-1}$ for D_c at 20°C. The procedures were discussed in detail by Hitchman (1978, pp. 226–228).

The values of gas permeabilities calculated using such approximations and assumptions are helpful in assessing the suitability of a particular membrane material (Hitchman, 1978). There is fair agreement between the calculated and the experimentally determined values for the permeability, as can be seen from Table 5. These problems are treated in a more comprehensive form by Crank and Park (1968) and Hopfenberg (1974). Table 6 gives a comparison of different membrane materials with respect to their permeability to dissolved oxygen. The values are given for dissolved oxygen in water on both sides of a membrane with a gradient of 133 Pa partial pressure of oxygen (1 standard atmosphere = 101.325 Pa).

Theoretically, a silicone rubber membrane should be best suited for use in combination with a polarographic oxygen probe as its permeability is 40 times greater that of Teflon. In practice it is difficult to produce a very thin silicone rubber membrane with sufficient mechanical strength and which can

Table 6. Permeability of dissolved oxygen in polymers (from Yasuda and Stone, 1966)

Polymer	$P_M \times 10^{14}$, $m^2 s^{-1}$
Poly(dimethylsiloxane)	4000
Polyethylene, low density (Dow Corning, Medical Silastic)	50.0
Poly(fluorinated ethylene–propylene)(Teflon FEP)	105
Poly(tetrafluoroethylene) (Teflon)	91.0

be applied in a reproducible manner to the cathode surface. Therefore, fluorinated polymers have been most widely used because of their mechanical strength and elastic properties and because films of thickness 5–25 μm can be manufactured reproducibly. Polyethylene or polypropylene membranes have also been used but they have the disadvantage that they start to 'flow' when drawn over the top of the electrode and micropores often appear during the application of the membrane. Practical experience has shown that in three out of four cases such micropores are present after attachment. The test can be made by using an external anode, e.g. a silver wire. The apparent current should be close to zero when the potential is applied, corresponding to the insulating resistance of the membrane. If the current is significant, the membrane has a hole.

By using a thin membrane, the response to a step change in the partial pressure of oxygen in the solution can be considerably improved. Membranes of thickness less than 10 μm are very delicate, however. It is possible to manufacture Teflon membranes only 0.2–1 μm thick (Kessler, 1973). A combination of a thin Teflon membrane with an outer silicone rubber membrane has been proposed by Ben Yaakov and Ruth (1980) and has also been used in combination with a fine-mesh gauze imbedded in the silicone rubber, in order to maintain a reproducible silicone rubber layer and mechanical strength. (Inter Ocean Systems, 1977; *cf.* Section 9.7.) Another advantage of Teflon as a membrane material is its chemical inertness and purity (absence of additives or inhibitors). Furthermore, the permeability characteristics of Teflon do not alter significantly with time, if the membrane is not brought into contact with organic vapours (Izydorczyk *et al.*, 1976).

It follows from Sections 9.5.2 and 9.5.3 that the application of the membrane should guarantee the formation of an electrolyte layer of reproducible thickness. The electrolyte layer should be much thinner than the membrane (theoretically $a \ll 10b$) in order to ensure that the diffusion process is governed mainly by the membrane. In most practical cases $a \approx b$, which means that we are working at the limits of this condition. In some constructions a defined electrolyte layer is maintained by placing a Cellophane foil soaked with electrolyte on top of the cathode or by a fine-mesh synthetic fiber gauze as spacer between cathode and membrane. If the electrolyte layer is too thick the membrane may vibrate or change its distance relative to the cathode in other ways, e.g. through the action of hydrostatic pressure, or turbulence in front of the probe head. The change in the distance between membrane and cathode surface has, of course, an immediate impact on the current and leads to irreproducible readings or a sluggish response to changes in the oxygen concentration. Furthermore, the interference of oxygen diffusing from the bulk of the electrolyte to the front of the cathode is larger if the electrolyte layer is too thick. The most undesirable effect is, however, the 'pumping' of oxygen-containing electrolyte to the diffusion zone by a vibrating membrane. This can often be observed with *in situ* probes.

9.6. OPERATIONAL CHARACTERISTICS

9.6.1. Calibration

There are several procedures for the calibration of a membrane-covered oxygen sensor. The most suitable way of calibrating a sensor for application in oceanographic work is either a laboratory procedure, using a thermostatted water tank and simultaneous titration of the oxygen content with the Winkler titration, or comparison of the *in situ* recorded values with simultaneously taken water samples which are analysed by the Winkler titration. Both methods will be described here. It is not advisable to calibrate the probe against air or oxygen–nitrogen gas mixtures if the sensor is then used in water, because the oxygen saturation value and, therefore, the partial pressure, is defined on the basis of an atmosphere which is 100% saturated with water vapour and, more seriously, the diffusion layer of a membrane-covered oxygen sensor is not confined to the membrane. The diffusion layer in the medium is kept very thin either by design of the probe (e.g. by using a very small cathode or by spreading the diffusion streamlines through a combined Teflon–silicone rubber membrane) or by maintaining a sufficient turbulent flow in front of the membrane to guarantee that the transport of oxygen to the cathode is controlled by the membrane only. If the probe is calibrated in a gaseous medium, there is a considerable difference with respect to the diffusion process in the immediate vicinity of the outer membrane surface. Experiments have shown that at equal partial pressures of oxygen and equal temperatures the difference in reading between the measurement in water and air is about 8%! Wetting of the membrane surface is of no advantage. Firstly, the evaporation may cool down the membrane surface and thus change the diffusion coefficient, and secondly, the hydrophobic nature of the membrane prevents the formation of a defined film.

The experimental set-up for a calibration of a membrane-covered oxygen probe is illustrated in Figure 30. A suitable tank (aquarium type) containing about 30 dm^3 of water is insulated with Styrofoam. The surface of the water is also insulated with Styrofoam pellets. A stirrer, for convenience combined with a pump and a heater (bath thermostat) is mounted in such a way that the stream of water passes the electrode surface. A cooling device should also be installed to counterbalance the heater in the thermostat and to allow calibration at lower temperatures. The same apparatus can be used for the

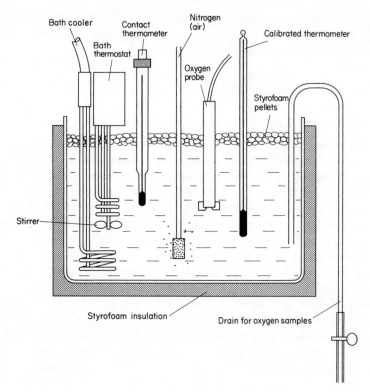

Figure 30. Calibration tank for membrane-covered oxygen probes. Grasshoff (1962)

determination of the overall temperature coefficient of the probe. A self-draining tube is attached for sampling the water for Winkler determinations. The oxygen content of the water is decreased by a stream of nitrogen introduced into the tank with an aquarium stone.

Several manufacturers recommend a two-point calibration, with one point either measured in air or in 100% air-saturated water and the second point (zero oxygen) in 35% w/v sodium sulphite solution. It must be stressed that it is difficult to prepare exactly 100% air-saturated water. Stirring, shaking, or bubbling of a stream of air through the water does not lead to complete equilibrium. Bubbling usually results in oversaturation. A device for the preparation of completely air-equilibrated water has been described by Grasshoff (1964).

The calibration procedure is started with nearly equilibrated water after a pre-polarization of the probe (usually 24 h is sufficient) and once the bath temperature is constant to within ±0.1°C. A reading is taken and then water is drained into three Winkler bottles. After the sampling, another reading of the meter of the oxygen detector is taken. Next, a gentle stream of nitrogen is passed through the 'bubble stone' until the reading is reduced by about 10%. It is convenient to connect a recorder to the meter to follow the decrease of oxygen during the outgassing procedure. The nitrogen is switched off and about 5 min is allowed to elapse before taking another reading, together with the next Winkler samples. This procedure is repeated until the reading is reduced to about 10% of the original value. The zero point is determined by inserting the probe into freshly prepared 35% w/v sodium sulphite solution. If the background current is very close to zero, which should be the case in a properly designed and prepared oxygen sensor, the temperature has, of course, no effect on the zero reading. The values obtained by the Winkler titration must be converted to saturation values, which can easily be done with the International Oceanographic Tables (Vol. II) (UNESCO, 1973) or with the equation given in Section 9.6.2. An abbreviated table for the 100% saturation values for temperatures between 0 and 30°C and salinities from 0 to 40‰ is given in Appendix XI.

A typical calibration curve is shown in Figure 31. The calibration is linear between zero and 120% saturation. Hitchman (1978) claims a linearity of the response of a membrane-covered oxygen probe over four orders of magnitude if gas mixtures are used for calibration. If the linearity of the response has been checked for a probe, the calibration can be checked by a

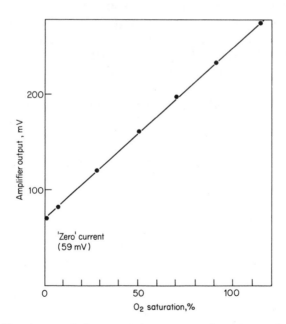

Figure 31. Calibration graph for a membrane-covered oxygen probe (Meerestechnik, *cf.* Section 9.7) at constant temperature (20°C)

single-point comparison at a controlled temperature with a chemical method and the measurement of the zero current with sodium sulphite solution.

A significant zero current can be attributed to (i) impurities in the electrolyte, (ii) an insufficient insulation resistance between the cathode and anode (e.g. humidity in the connectors, or a coating of silver oxide or silver chloride inside the electrolyte compartment, leading to a by-pass connecting the two electrodes), or (iii) a current generated by oxygen diffusing to the cathode from the bulk of the electrolyte or out of the material used for the construction of the probe (emphasizing the need to use glass as the material for the cathode support). It must be borne in mind that the resistance of a probe, assuming a saturation pressure of oxygen, is of the order of $500 \, k\Omega$ and increases to values $>10 \, M\Omega$ at zero oxygen pressure. A by-pass resistance of $1 \, M\Omega$ may therefore cause a 50% error! It is recommended that an impedance-matching amplifier be placed as close as possible to the sensor, especially if the very small cathode surfaces are used and the probes generate currents in the range of $50-100 \, nA$. Assuming strict linearity between the partial pressure of oxygen in water and the diffusion current, and assuming that the necessary amplification and treatment of the signal are also linear over the practical range (0–120% saturation), the precision can be obtained by comparing the chemical measurements with the readings from the probe. The precision of the Winkler titration under optimum conditions is $\pm0.02 \, cm^3 \, O_2 \, dm^{-3}$ (Grasshoff, 1962). The same precision is claimed for the conversion of oxygen concentrations into partial pressures with the Weiss equation (Table 7, Chapter 1) or the International Oceanographic Tables (UNESCO, 1973). If the temperature is kept constant during the calibration to within $\pm0.1°C$ the optimum precision which can be expected for the membrane-covered oxygen sensor is $\pm0.5\%$ at about 100% oxygen saturation or $\pm0.04 \, cm^3 \, O_2 \, dm^{-3}$. It must be borne in mind, however, that temperature has a considerable effect on the diffusion process and that it is difficult to measure the temperature right at the membrane surface with a temperature sensor that has the same time constant as the change in permeability of the membrane with temperature. A further complication is the effect of hydrostatic pressure on the partial pressure of oxygen and the permeability of the membrane (*cf.* Section 9.6.3). A careful measurement of the overall effect of pressure on the diffusion current must be made with each probe.

A very convenient method for the calibration of a membrane-covered oxygen sensor, especially during field work, is *in situ* calibration. Very often only a weak gradient of the partial pressure of oxygen can be observed in the upper mixed layer. In this case, a water sample from this homogeneous layer for a Winkler oxygen analysis and a salinity determination can be taken as one calibration point and the second point is obtained by immersing the probe in a sulphite solution. Of course, the temperature must be measured simultaneously with the combined oxygen–temperature probe to apply the necessary corrections.

If the oxygen sensor is used for monitoring relative changes in the partial pressure (e.g. in laboratory experiments for the measurement of respiration and aspiration in cultures) much smaller changes can be detected than is possible with chemical determinations (Schramm, 1978). It is obvious that only with oxygen sensors can a continuous record of the changes of the partial pressure of oxygen be obtained.

9.6.2. Effect of temperature on diffusion

Temperature has a considerable effect on the output of a membrane-covered oxygen detector. The overall effect is a combination of the effects of temperature on the diffusion process, the geometry of the probe and on the thermal expansion of the electrolyte. The effect on the anodic process and on the conductance of the electrolyte can be neglected. Usually the influence of the temperature on the probe geometry is also small compared with the influence on the diffusion process.

The thermal expansion of the electrolyte may have a marked effect especially if large electrolyte volumes are used. The contraction or expansion of the electrolyte is partly compensated for by the volume change of the compartment, owing to the thermal expansion of the material used for the probe. If the electrolyte compartment is hermetically sealed and no pressure compensation device is attached, expansion of the electrolyte will change the thickness of the electrolyte layer. This will, of course, change the geometry of the diffusion zone but, at the same time, oxygen dissolved in the electrolyte will be transported into the vicinity of the cathode and contribute to the current. This effect is most pronounced when the temperature changes frequently, for example when vertical profiles are recorded during field work at sea. The effects described so far cannot be treated theoretically and must be determined experimentally, using the apparatus described earlier (Figure 30).

If we assume that the major effect of temperature on the response of the membrane electrode results from the change in the transport characteristics across the membrane then we can derive a simple relationship between current and temperature. The influence of temperature on the diffusion coefficient for the diffusion of simple gases in membranes is described by

$$D_{M,T} = D_{M,0} \exp\left(-\frac{E_D}{RT}\right) \tag{50}$$

(Hitchmann, 1978), where $D_{M,T}$ is the coefficient of diffusion at temperature T, $D_{M,0}$ is the coefficient of diffusion at a reference temperature, and E_d is the activation energy in J mol^{-1} required to transfer a gas molecule from one site to the next in the membrane. This interpretation was treated theoretically by Crank and Park (1968). The temperature dependence of the solubility of a gas in a polymer can be described by an equation similar to that for the gas solubility in a liquid, but using the distribution coefficient (equation 22),

$$K_{M,T} = K_{M,0} \exp\left(-\frac{\Delta H_M}{RT}\right) \tag{51}$$

$K_{M,0}$ represents the distribution coefficient under the reference conditions, $K_{M,T}$ is the distribution coefficient at temperature T, and ΔH_M is the heat of the solution ($J\,mol^{-1}$) of oxygen in the membrane (van Krevelen, 1972). According to Hitchman (1978), the effect of temperature on the permeability coefficient (equation 32) can be expressed by

$$P_{M,T} = P_{M,0} \exp\left(-\frac{E_p}{RT}\right) \tag{52}$$

where E_p (the activation energy for permeation) $= E_D + \Delta H_M$. Using equation 52 and the equation for the steady-state current (equation 38) we obtain, according to Hitchman (1978),

$$i_T = \frac{nF}{b} P'_{M,0} \exp\left(-\frac{E_p^*}{RT}\right) \tag{53}$$

and

$$\ln i_T = -\frac{E_p^*}{RT} + \text{constant} \tag{54}$$

where $E_p^* = \Delta H - E_p$ and ΔH is the heat of solution of oxygen in water. Equation 54 is in reasonable agreement with experimental observations. This treatment assumes that the membrane material does not undergo structural change in the range of temperature in question. Such a change can be easily detected by a sudden break in the current–temperature curve. We have observed such inconsistencies at temperatures in excess of 40°C with polypropylene.

The separate determination of the temperature–current relationship for each type of probe and each batch of membranes is strongly recommended. The following procedure can be applied, using an experimental set-up as described for calibration (Figure 30). The water in the bath is stirred gently overnight with no Styrofoam pellets on the surface at the highest temperature which is to be considered. The water will then be approximately saturated with air. The surface is insulated with the Styrofoam pellets and the probe is inserted and polarized until a steady-state current is attained. The temperature should be controlled to within at least ±0.1°C. The readings are made and the Winkler samples are taken simultaneously. The cooling device is switched on and the thermostat is set to provide a stepwise decrease in temperature. When the pre-set temperature has been reached and the probe has equilibrated, temperature and current readings are taken and Winkler samples are withdrawn.

The degree of oxygen saturation at each temperature is calculated from the Winkler titrations and the current, i_T, measured at the respective temperatures, normalized to 100% saturation:

$$i_{T, 100\%} = i_T/\text{degree of saturation} \tag{55}$$

This procedure assumes that the response of the probe to the partial pressure is linear at all temperatures. The corrected current, $i_{T, 100\%}$, is plotted against temperature. A typical temperature curve is shown in Figure 32. For practical purposes it is convenient to select a current at a suitable reference temperature (i_0 at 0°C) and develop a correction equation of the general form

$$i_0 = i_t(1 + ft) \tag{56}$$

where f is a function of temperature and t is the temperature in °C.

Several manufacturers use an integrated temperature compensation network for the immediate correction of the overall temperature effect. Usually the dependence of the resistance of a thermistor on the temperature is not the inverse of the temperature dependence of the diffusion current. This means that a single thermistor cannot be shunted in such a way that its

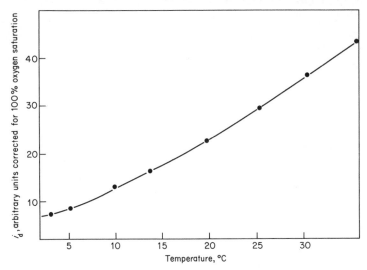

Figure 32. Dependence of the current normalized to the 100% O_2 saturation current on temperature for a Teflon membrane

characteristics mirror the characteristics of the diffusion current–temperature relationship. Two or more thermistors must be used. Furthermore, a direct temperature compensation implies that the temperature–time behaviour of the thermistor and that of the diffusion current are the same. In addition, the temperature coefficient of the permeability of different membranes may vary, even if the same polymer is used. It is better to measure the temperature separately, adjust the response characteristics of the temperature sensor as closely as possible to those of the membrane, and then use a numerical correction method.

Solubility of oxygen in water as a function of temperature and salinity

As the membrane-covered oxygen probe measures the partial pressure rather than the concentration of dissolved oxygen, it is necessary to know the dependence of the solubility of oxygen in sea water on temperature and salinity, in order to convert the readings of the probe into units of concentration. The relationships used are based on the work of Weiss (1970), since the equations derived by him (Table 7, Chapter 1) have been recommended for general application by the oceanographic community (6th meeting of the Joint UNESCO–ICES–SCOR–IAPSO Panel of Experts on Oceanographic Tables and Standards, Kiel, 1969) and are the basis for the conversion tables contained in Volume II of the International Oceanographic Tables (UNESCO, 1973). Oxygen solubility measurements covering a wide range of temperatures and salinity have been made independently with a high degree of precision by Green (1965), Douglas (1965), Green and Carritt (1967a,b), Carpenter (1966), and Murray and Riley (1969). The results are almost indistinguishable and agree within the precision of a well performed Winkler titration.

Solubilities are generally reported in terms of the Henry's law constant, K, the Bunsen solubility coefficient, β, or the air solubility, C^* ($cm^3 \, dm^{-3}$). The Bunsen coefficient is defined as the volume of gas (at STP) absorbed per unit volume of solution at the given temperature when the partial pressure of the gas is 1 atm (101.3 kPa). The air solubility is the volume of the gas (at STP) absorbed from water vapour saturated air at a total pressure of 1 atm (101.3 kPa) per unit of solution at the given temperature.

The temperature dependence of the Bunsen solubility coefficient, β, at constant pressure is derived from an integrated form of the Van't Hoff equation. Assuming that $\Delta C_p(T)$, the heat capacity of solution of the gas, is constant or proportional to T^{-1} the variation of β with temperature can be described by

$$\ln \beta = a_1 + a_2 T^{-1} + a_3 \ln T \qquad (57)$$

where a_1, a_2, and a_3 are constants and T is the absolute temperature. If

$\Delta C_p(T)$ is expressed as a power series in T, then equation 57 becomes

$$\ln \beta = a_1 + a_2 T^{-1} + a_3 \ln T + a_4 T + a_5 T^2 + \ldots \tag{58}$$

(Weiss, 1970).

Using the most recent data from different sources for the solubility of oxygen in water at different temperatures, the constants of equation 58 have been calculated according to the method of least-squares. It turned out that the use of the power series in T did not significantly improve the standard deviation (Weiss, 1970), so that the temperature dependence of the solubility can be described with sufficient precision by equation 57.

Measurements of the solubility of oxygen in sea water of varying salinities show that the solubility decreases with increasing salt concentration of the sea water. This behaviour is characteristic of the solubilities of many non-electrolytes when an electrolyte is added to the solution and is known as the 'salting-out' effect.

The empirical equation of Setschenow (1889) can be used to represent the salinity dependence of the solubility at constant temperature:

$$\ln \beta = b_1 + b_2 S\%_0$$

where b_1 and b_2 are constants (Weiss, 1970). Since none of the recent data for oxygen solubility in sea water showed a systematic error in the Setschenow treatment, it was adopted for the salinity dependence of the oxygen solubility. The final equation for β as a function of temperature and salinity is

$$\ln \beta = A_1 + A_2(100T^{-1}) + A_3(\ln 0.01T) + S\%_0(B_1 + 0.01B_2 T + B_3 T^2 \cdot 10^{-4})$$

where A_1, A_2, A_3, B_1, B_2, and B_3 are constants. In practice, it can be assumed that the gas exchange between the surface water and the atmosphere happens in a layer where the atmosphere is saturated with water vapour. The inclusion of the vapour pressure of water requires only one additional parameter for the temperature dependence to arrive at the same precision for the air solubility as was observed for β (Weiss, 1970). The final equation, which is recommended for general application by the Joint Panel of Experts for Oceanographic Tables and Standards, has the form

$$\ln C^* = A_1 + A_2(100T^{-1}) + A_3 \ln (0.01T)$$
$$+ A_4 \cdot 0.01T + S\%_0(B_1 + B_2 \cdot 0.01T + B_3 T^2 \cdot 10^{-4})$$

where C^* is the solubility in cm^3 at STP per dm^3 from water vapour saturated air at a total pressure of 1 atm (101.3 kPa), A_1–A_4 and B_1–B_3 are constants whose values are listed in Appendix X.

9.6.3. Effect of hydrostatic pressure

The effect of hydrostatic pressure on the current response of a membrane-covered electrode can be two-fold. The pressure influences (i) the solubility and therefore the partial pressure of oxygen in water and (ii) the density and the molecular structure of the membrane material and hence the permeability of the membrane.

Let us begin by considering the effect of pressure on the partial pressure of oxygen. A series of experiments by Enns *et al.* (1965) was evaluated by Fenn (1972). Extrapolation of the experimental data resulted in a four-fold increase of the partial pressure of oxygen at a depth of 10 000 m compared with that observed at the surface if the actual concentration remains the same. The variation of solubility and partial pressure of oxygen could be fitted by exponential equations, but Fenn (1972) was unable to give a thermodynamic derivation or a physical explanation of the equations. The matter was therefore taken up by Andrews (1972), who developed equilibrium criteria for solutions subject to variations in pressure and in the gravitational field, and further by Eckert (1973), who based his derivation on the definition of the reference state in Henry's law. Both authors arrived essentially at the same result.

The derivation according to Eckert starts from the logarithmic form of Henry's law:

$$\ln\left(\frac{f_2}{x_2}\right) = \ln K \tag{59}$$

were f_2 is the fugacity of the gas, x_2 is the molar fraction of the dissolved gas, and K is a proportionality constant which depends on the temperature and pressure of the standard state. The standard state pressure ($P^\ominus$) is taken as 1 atm (101.3 kPa) for practical reasons. If the pressure is increased from $P^\ominus$ to P at constant temperature, then equation 59 becomes (Krichevsky and Kasarnovsky 1935)

$$\ln\left(\frac{f_2}{x_2}\right) = \ln K + \bar{V}_2(P - P^\ominus)/RT \tag{60}$$

where $\bar{V}_2$ is the partial molal volume of the gas in the solvent. At the reference state $\bar{V}_2 = \bar{V}_2^\ominus$. According to Denbigh (1966), the isothermal variation of the fugacity of the dissolved substance in a gravitational field is

$$f_2 = f_2^\ominus \exp\left(M_2 gh/RT\right) \tag{61}$$

where $f_2^\ominus$ is the fugacity of the dissolved gas at the reference pressure $P^\ominus$, M_2 is the molecular weight of the gas, g is the acceleration due to gravity, and h is the depth. In the oceans the pressure varies with the depth, h, according to

$$P = P^\ominus + dgh \tag{62}$$

where d is the average density of sea water. Assuming that the ideal gas laws are obeyed at $P^\ominus$, the combination of equations 60, 61 and 62 gives

$$x_2 = (P^\ominus/K) \exp [(M_2 - d\bar{V}_2)gh/RT] \qquad (63)$$

According to this equation, at 10 000 m the partial pressure of oxygen will be 3.55 times that at the surface. Since the standard partial molal volume of oxygen ($\bar{V}_2^\ominus = 32 \text{ cm}^3 \text{ mol}^{-1}$) multiplied by the density of sea water (1.023 g cm^{-3}) is very close the molecular weight (32 for O_2), the *equilibrium* concentration of oxygen dissolved in sea water taken from the surface under isothermal conditions down to 10 000 m remains virtually unchanged. However, a significant correction must be applied to convert the reading of a membrane-covered amperometric oxygen probe into the absolute amount of oxygen per unit of volume or per kilogram of sea water (equation 61). Already at 400 m a correction of 5% is needed to compensate for the increase of the partial pressure (Hitchman, 1978).

Very little is known about the effect of pressure on the permeability of the various membrane materials and no firm guidelines for correction can be recommended. This means in practice that each probe should be tested empirically for the overall influence of pressure on the electrical output if it is to be used at depths greater than 200 m. Hitchman (1978) states that the compressibility of water is too low to cause any significant change in the thickness of the electrolyte layer. A pressure of 10^3 atm (1.01×10^6 kPa) will cause a volume change of about 1%. For the same reason, the effect of the pressure on the thickness of the membrane is negligible since the compressibility of plastics is in the region of $5 \times 10^{-5} \text{ atm}^{-1}$ (Hopfenberg, 1974). Of course, the external pressure acting on the membrane must be compensated for by an equal internal pressure in the electrolyte if membrane-covered oxygen probes are used at depths greater than 30–50 m. Pressure compensation is usually achieved through a non-permeable membrane between the electrolyte and the water, mounted on the shaft of the probe.

Another effect of the pressure may cause more serious problems. Under the normal working conditions the current is diffusion controlled. There seem to be no reliable data on the effect of pressure on the diffusion coefficient. Richardson *et al.* (1964) attempted to measure the pressure dependence of D_{NaCl} in artifical sea water. Although there appeared to be nearly a 7% increase in D_{NaCl} from 1 to 1000 bar, the experimental uncertainity was too great for firm conclusions to be drawn.

Nernst (1888) gave a molecular interpretation of the diffusion coefficient so that

$$D = RT/6\pi N\eta r \qquad (64)$$

where N is a constant, η is the viscosity of the solution and r is the radius of the diffusing particle. Assuming that r is not a function of pressure, the diffusion coefficient is inversely proportional to the viscosity of the solution. The pressure dependence of the viscosity of water is anomalous (Horne, 1969). Whereas the viscosity of a 'normal' fluid increases with increasing pressure, the viscosities of pure water and of many electrolyte solutions in water decrease relative to the 1 atm value, pass through a minimum (the position of which is strongly dependent on temperature) and finally increase. Increasing the temperature or adding an electrolyte reduces the depth of the minimum and shifts it to lower pressures. This implies that only a slight increase in the diffusion current should be observed down to a depth of about 6000 m, but as the diffusion process is to a greater extent controlled by the diffusion through the membrane, the influence of pressure on the diffusion process of oxygen through the electrolyte layer is probably negligible. Therefore, the effect of pressure on the diffusion process is reduced to its effect on the permeability of oxygen in the membrane. This effect remains largely unexplored.

9.6.4. Effect of flow in front of the membrane

All previous derivations of the diffusion current are based on the assumption that there is no gradient of oxygen partial pressure from the bulk of the solution to the outer interface of the membrane. If, however, the diffusion layer spreads from the membrane into the solution, the diffusive transport is not solely controlled by the membrane and the resulting diffusion current decreases. Hitchman (1978) used the equation

$$\delta_s \approx D_s^{1/3} \nu_s^{1/6} \left[\frac{x}{\nu_s} \right]^{1/2} \tag{65}$$

where δ_s is the thickness of the diffusion layer in the test solution, ν_s is the kinematic viscosity of the test solution (Chapter 1, Table 7), x is the diameter of the cathode, and ν_s is the flow velocity in the test solution parallel to the membrane surface. For diffusive transport to be controlled by the membrane only

$$\frac{\delta_s}{D_s} \ll \frac{b}{P_M} \tag{66}$$

or

$$\delta_s \ll 10b$$

where b is the membrane thickness.

Taking $\delta \approx 10^{-2}$ mm, $D_s \approx 10^{-3}$ mm^2 s^{-1}, $\nu_s \approx 10$ mm^2 s^{-1}, and $x = 6$ mm in equation 65, we find that the minimum flow velocity which allows diffusive transport to be controlled by the membrane alone is 0.25 m s^{-1}. This theoretical boundary value for v_s agrees well with experimental data (Hitchman, 1978). v_s may become smaller if there is turbulent flow in front of the membrane and if smaller cathode diameters are chosen. Sensitivity to turbulence may be further reduced if thick membranes or combination membranes of Teflon and silicone rubber are used (*cf.* Section 9.5.3). This can be so effective that practically no flow is required in the test solution to ensure that the diffusive transport is controlled by the membrane. Adequate minimum flow velocities must be determined for each type of membrane-covered oxygen sensor and can be obtained by using a mechanical stirring device (propellor type or vibrator) attached to the probe. Usually sufficient flow is maintained in profiling or towed instruments (Stack, 1967; Carritt and Kanwisher, 1959).

9.7. EXAMPLES OF MEMBRANE-COVERED AMPEROMETRIC OXYGEN SENSORS

Numerous membrane-covered oxygen sensors have recently been developed and are manufactured by several companies. In this section only a few of them can be considered, especially those which can be used in sea water for vertical and horizontal profiling. This section cannot be considered as an exhaustive review, and the instruments described are simply examples of the various types available for different applications. The specifications given below are those provided by the manufacturers.

One of the first commercially available oxygen probes for application in oceanographic work was manufactured by *Hydro-Bios* and was based on a prototype developed by Grasshoff (1969a). The instrument (see Figure 33a) consists of a deck unit with amplifiers, zero suppression, scale expansion, the stabilized voltage supply for the probe, and the meters. The instrument is equipped with rechargeable batteries. The temperature compensation is performed manually with a potentiometer. The instrument has outlets for analog recording and data logging. The probe consists of a streamlined, rather heavy body of plated brass and PVC (see G and L, Figure 33b) and contains a silver anode (I) and a platinum cathode (E) with a diameter of 5 mm. The electrolyte compartment is pressure equilibrated (F). Initially polypropylene was used as the membrane material but later a 16 μm Teflon membrane was found to be more convenient (*cf.* Section 9.5.5). A thin Cellophane foil (D) is placed between the membrane and the cathode to maintain a defined electrolyte layer. A 0.5 M potassium hydroxide solution is used as electrolyte. A thermistor (M) is integrated into the body of the probe for temperature measurement. The response is linear and the precision is ±2% full-scale. The probe can be used to a depth of 200 m (another version of the probe can be used to 1500 m). The design of the probe and the volume of the electrolyte allow continuous use for approximately 2–3

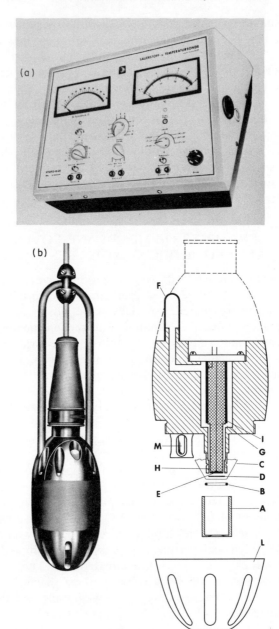

Figure 33. Hydro-Bios dissolved oxygen sensor: (a) reading unit; (b) probe. Reproduced by permission of Hydro-Bios, Am Jagersberg 7, D-2300 Kiel, Federal Republic of Germany

months. The probe is connected to the deck unit through a four–conductor plastic-mantled cable. The conductors are of cadmium–copper–tin alloy and have a breaking strength of 100 kg each. The probe can be lowered with a manual winch or with a cable winch with slip-ring connection for greater depths. The instrument has been used successfully in the Baltic (Grasshoff, 1969a), the Black Sea (Grasshoff, 1975), and the Red Sea (Grasshoff, 1969b) for vertical profiling. The flow in front of the membrane is kept sufficiently high by lowering the probe at a speed of $0.3 \, \text{m s}^{-1}$; 90% of the steady-state reading is reached within 10 s.

Another dissolved oxygen field probe is manufactured by *Yellow Springs International Inc.* (Figure 34). It also consists of a deck unit with batteries and power supply and a meter for either temperature or oxygen in a shock-resistant watertight case. The temperature effect is compensated for automatically with an accuracy of ±1% within a pre-set temperature range of ±5°C. The overall accuracy is ±3% full-scale reading. The instrument has a range selector for oxygen concentrations between 0 and 5, 0 and 10, and 0 and $20 \, \text{mg dm}^{-3}$. Corrections for ocean salinity can be made by directly dialing the salinity from 0 to 40‰ to cover the full range from fresh to ocean water. A strip-chart recorder can be connected. The probe consists of a plastic body with a gold cathode and silver anode, a pressure-compensated electrolyte compartment filled with potassium chloride solution, and a demountable plug for the cable. Thermistors for temperature measurement and temperature compensation are integrated in the probe. The probe can be used to a depth of 75 m. A submersible stirrer (with separate cable) can be attached to the probe and is strongly recommended.

WTW manufacture an oxygen meter for laboratory and field use consisting of a meter and power unit with a digital display of oxygen concentration, oxygen partial pressure, and temperature. The scales are from 0 to 50 mg $O_2 \, \text{dm}^{-3}$ or from 0 to 500% saturation. The precision is claimed to be ±2% full-scale reading with automatic temperature compensation within ±20°C of the calibration temperature. The instrument contains rechargeable batteries and a power supply (220/110 V a.c., 50/60 Hz). A 0–1 V recorder can be connected. The probe consists of a plastic body with a gold cathode and silver anode, and an electrolyte compartment containing potassium chloride solution. The electrolyte is not pressure compensated but the manufacturer states that the probe can be used down to 100 m (Nösel, 1976). A 12.5 μm Teflon foil is used as a membrane, mounted in an exchangeable plastic holder (Figure 35), 90% of the steady-state reading is reached within 15 s. A flow velocity of $0.5 \, \text{m s}^{-1}$ is recommended. One filling allows discontinuous use for 3–4 months. A submersible stirrer can be attached to the probe body, There is no possibility for salinity compensation.

Plessey Environmental Systems (Box 80845, San Diego, Calif. 92138, USA) produce a dissolved oxygen sensor for use with all Plessey signal

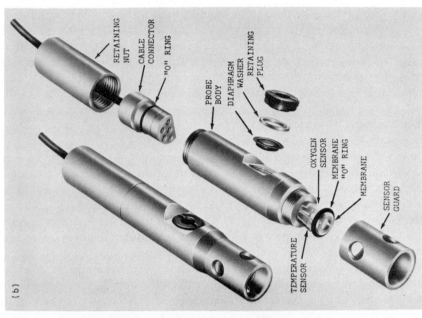

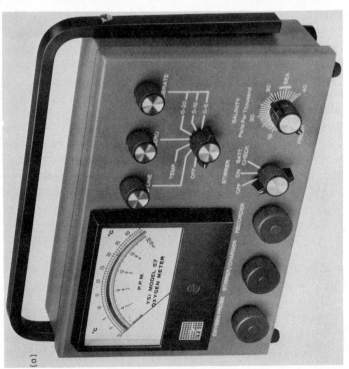

Figure 34. Yellow Springs dissolved oxygen sensor: (a) reading unit; (b) probe. Reproduced by permission of Yellow Springs International Inc., Box 297, Yellow Springs, Ohio 45387, USA

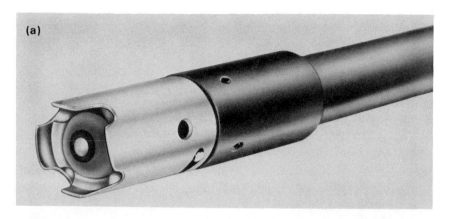

Figure 35. WTW dissolved oxygen probe. Reproduced by permission of WTW-Wissenschaftlich Technische Werkstätten GmbH, D-8120 Weilheim, Federal Republic of Germany

processors, especially in connection with an STD (salinity, temperature, depth) profiling system. The Model 5175 dissolved oxygen sensor provides continuous measurements of the concentration of dissolved oxygen in rivers, streams, and bodies of fresh or saline water to depths of 300 m. To obtain a frequency output, the transducer voltage is applied to a voltage-controlled oscillator (VCO) to provide a frequency-modulated (FM) analog of the measurement. This FM output provides good resolution, a data signal that can be readily adapted to d.c. or digital form for recording and analysis, and an economical and reliable means of transmitting signals over long distances without degradation. The manufacturer claims a precision of $\pm 2\%$ full-scale reading and a range from 0 to 15 mg O_2 dm^{-3}. The response time is 6 s for a 95% and 60 s for a 98% steady-state reading.

The *Beckman* Fieldlab Oxygen Analyzer (Figure 36) can be operated with line-current, mercury batteries, or rechargeable batteries. The meter allows the reading of temperature with an accuracy of $\pm 1\%$ full-scale

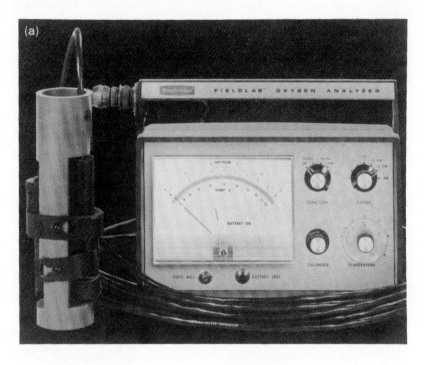

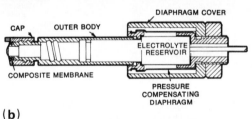

Figure 36. Beckman Instruments dissolved oxygen sensor: (a) reading unit with oxygen and temperature probe; (b) cross-section through the probe. Reproduced by permission of Beckman Instruments Inc., 2500 Harbor Boulevard, Fullerton, Calif. 192634, USA

reading (100°C) or oxygen in percentage saturation or concentration in the ranges 0–1, 0–10, and 0–25 mg dm^{-3}. The precision is ±0.5% of meter full-scale. A potentiometric recorder may be connected. The manufacturer claims that a single-point air calibration is sufficient (*cf.* Section 9.6.1). The *in situ* submersible dissolved oxygen sensor consists of two separate probes for temperature and oxygen. According to the manufactuere, no stirring is required, owing to the use of a very small cathode area, a special cathode geometry, and a composite membrane (Figure 36b). With a flow velocity of 0.05 m s^{-1} of the sample the readout is within 0.1% of the true value. Zero flow gives 97% of the true value. The response time for 90% of the steady-state reading is 30 s. The temperature effect on the diffusion current is compensated for manually. The probe has a pressure-compensated electrolyte reservoir and can be used down to 180 m. The instrument has no salinity correction (see Figure 36a).

The dissolved oxygen sensor of *Interocean Systems* (Figure 37) forms a part of an oceanographic sensor package and consists of a platinum cathode and a silver anode in a solution of potassium chloride. A potential difference of 0.55 V is established between the cathode and anode. The cell is

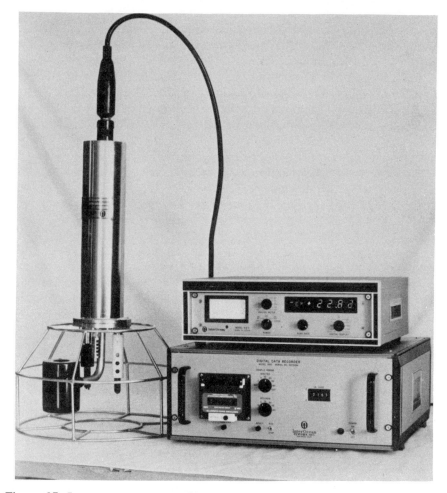

Figure 37. Interocean oceanographic sensor package with underwater unit and reading, and data logger unit. Reproduced by permission of Interocean Systems Inc., 3540 Aero Court, San Diego, Calif. 92123, USA

separated from the sample by a special membrane assembly which, according to the manufacturer, provides good dimensional stability, resulting in excellent long-term reliability without re-calibration. Dimensional stability is achieved by using stainless-steel construction in the sensor and by a special process of depositing the membranes on a fine-mesh stainless-steel screen. The screen is held securely in place by a stainless-steel screw-on assembly. The oxygen which enters through the membrane is reduced at the cathode and a very small current (0–50 nA) is produced. By designing the sensor to consume oxygen at a slow rate compared with the net mass transport of oxygen through the water, the sensor output is not affected by changes in water flow-rate past the membrane. No pump or stirrer mechanisms, which are subject to fouling and clogging, are required. Even in stagnant water the sensor will read 97% of the equilibrium value. The manufacturer claims a precision of ±1% and a time constant (90% steady state reading) of 5–10 s. A thermistor is used to measure the temperature and its output is used to adjust the gain of the signal processing amplifier. A salinity compensation network is used in combination with the *in situ* salinity probe. The long-term stability was tested during a period of 22 days without re-calibration. The average error was $0.05 \, \text{cm}^3 \, O_2 \, \text{dm}^{-3}$. During the test the probe was removed from the water three times and the membrane was allowed to dry out. No effects on calibration or accuracy were observed. The manufacturer claims a depth range down to 6000 m.

Another oxygen measurement system for application in oceanographic work has been designed by Orbisphere. This consists of an oxygen-indicating unit (Figure 38) with an optional analog computing salinity correction. The meter has ranges from 0 to 1, 0 to 3, 0 to 10, and 0 to $30 \, \text{cm}^3 \, O_2 \, \text{dm}^{-3}$ and a built-in temperature correction circuit. The salinity is set by two digital switches. The quoted accuracy is ±1% full-scale reading where the temperature of the unknown sample is within ±5°C of the calibration temperature and ±4% of full-scale reading if the temperature differences are greater than ±5°C in the range 0–50°C. The claimed precision at zero reading of the instrument is $\pm 0.03 \, \text{cm}^3 \, O_2 \, \text{dm}^{-3}$, and the linearity of the calibration graph of current output *versus* oxygen partial pressure is ±1% deviation. The response time for a true reading if the sensor is transferred from oxygen-free water to saturated water is 30 s. A flow velocity of $0.3 \, \text{m s}^{-1}$ is required to restrict the diffusion to the membrane only. The power is provided from a battery which allows continuous use for 40 days. The oxygen consumption of the sensor is $0.05 \, \mu\text{g h}^{-1}$ per cm^3 $O_2 \, \text{dm}^{-3}$. The sensor consists of a metal-encapsuled plastic body with an easily exchangeable sensor head. Teflon is used as the membrane material. The sensor is insensitive to hydrogen sulphide, sulphur dioxide, and ammonia. An alkaline electrolyte containing a deliquescent salt is used. The probe can be lowered to 200 m. A recorder can be connected to the

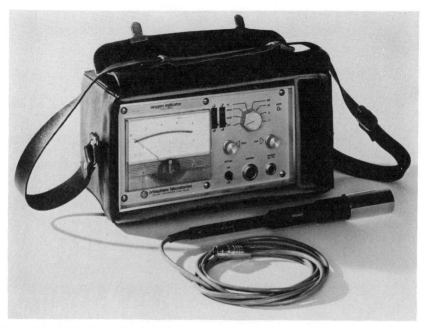

Figure 38. Orbisphere dissolved oxygen sensor with probe and reading unit. Reproduced by permission of Orbisphere Laboratories, 5 Rue Gustave-Moynier, CH-1202 Geneva, Switzerland

instrument. Comparison of the sensor with Winkler titration using 24 data points resulted in a correlation coefficient of 0.9994.

Meerestechnik has developed a membrane-covered oxygen sensor for use in a sensor package for oceanographic parameters (Figure 39a). This probe uses a membrane-covered disperse cathode, consisting of ten *ca.* 10 μm platinum wires moulded into a 6 mm glass rod. The surface of the disperse cathode is ground to form a network of small capillary electrolyte channels across the disperse cathode surface. Furthermore, this treatment guarantees a defined thickness of the electrolyte layer. A ring-shaped silver anode with a U-shaped cross-section is placed around the glass rod. The channel in the anode serves as the electrolyte reservoir. A pre-stretched 10–20 μm Teflon membrane mounted in a Teflon washer can be easily attached to the electrode top and is held in place by a screw-cap. Sealing of the electrolyte compartment is achieved with a double O-ring (Figure 39b). The steady-state current for a saturation pressure of oxygen is about 10–20 nA at 20°C. Each of the platinum wire surfaces acts as a point electrode. As has been discussed in Section 9.4.4, the flow velocity in the outer solution has no influence on the diffusive transport, as the molecular transport of oxygen in front of the membrane is sufficient to replace the oxygen consumed by the

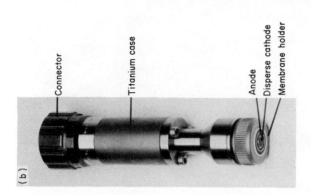

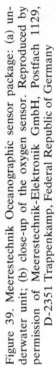

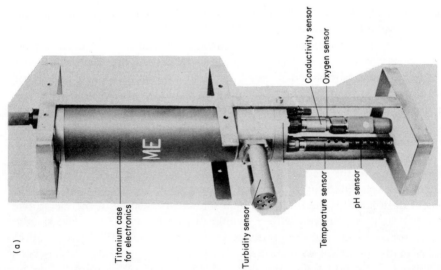

Figure 39. Meerestechnik Oceanographic sensor package: (a) underwater unit; (b) close-up of the oxygen sensor. Reproduced by permission of Meerestechnik-Elektronik GmbH, Postfach 1129, D-2351 Trappenkamp, Federal Republic of Germany

cathodic reaction. The accuracy of the sensor is ±1% and the sensitivity is 0.1%. Because of the special design of the probe head, a pre-polarization time of only a few minutes is required after switching on the potential. The probe can be used to a depth of 2000 m or deeper. It has an outer case of titanium and the electrolyte compartment head is pressure compensated with a silicone oil pressure transducer in order to prevent any deformation of the probe head by hydrostatic pressure. The probe has been used continuously in submerged monitoring units for 3 months without significant change of the calibration. Different types of membranes (6–30 μm Teflon or combined Teflon–silicone rubber) can be used for different applications, e.g. for profiling instruments, where small time constants are needed, or for an *in situ* monitoring instrument, where high mechanical stability and insensitivity to abrasion or fouling is of advantage, and longer time constants can be tolerated.

Oxygen probes of different design have frequently been used for monitoring long-term and short-term changes in field work, especially in connection with controlled ecosystems. An example is the determination of the oxygen consumption at the sea bed during large bell jar experiments (Balzer, 1978; Pamatmat, 1971; Pamatmat and Bause, 1969; Hallberg *et al.*, 1973; see also Chapter 2, Table 1). In the submerged bell jar system, enclosing more than 2 m³ of water at the bottom, the oxygen measurements were transmitted by radio together with 12 other parameters to a land station at a distance of 25 km and the raw data were corrected for temperature and salinity with an on-line computer. Hansen *et al.* (1980) have developed an independent system for the determination of the oxygen consumption rate at the sediment–sea water interface using an Interocean sensor. The partial pressure is measured within a 0.5 m³ bell jar together with temperature during a period of 3–8 days. Every 10–30 min readings are taken and the data are stored on magnetic tape. During the measurement the water within the bell jar is gently stirred to maintain homogeneity. The instrument can be used down to 250 m. After the measuring period the instrument is retrieved and the oxygen data are corrected for temperature and salinity and displayed on a plotter. The rate of oxygen consumption is evaluated from the oxygen–time curve (Sadjadi, 1979).

An oxygen sensor has also been used in a towed chemical profiling system, undulating from 3 to 100 m (Grasshoff and Hansen, 1978). Further examples of the use of oxygen sensors for vertical profiling are given in Chapter 2.

REFERENCES

Airey, L., and A. A. Smales (1950). Mercury drop control: application to derivative and differential polarography. *Analyst*, **75**, 287–304.
Andrews, F. C. (1972). Gravitational effects on concentrations and partial pressures in solutions: a thermodynamic analysis. *Science*, **178**, 1199–1201.

Balzer, E. (1978). *Thesis*, Kiel University.

Benson, B. B., and P. O. M. Parker (1961). Relations among the solubilities of nitrogen, argon and oxygen in distilled water and sea water. *J. Phys. Chem.*, **65,** 1489–1496.

Ben Yaakov, S., and E. Ruth (1980). A method of reducing the flow sensitivity of a polarographic dissolved oxygen sensor. *Talanta*, **27,** 391–395.

Bockris, J. O'M., and A. K. N. Reddy (1970). *Modern Electrochemistry*. Plenum Press, New York.

Bradburg, J. H., and A. N. Hambly (1952). Errors in the amperometric and starch-indicator methods for the titration of millinormal solutions of iodine and thiosulfate. *Aust. J. Sci. Res.*, **A5,** 541–554.

Carpenter, J. H. (1965). The accuracy of the Winkler method for oxygen analysis. *Limnol. Oceanogr.*, **10,** 135–140.

Carpenter, J. H. (1966). New measurements of oxygen solubility in pure and natural water. *Limnol. Oceanogr.*, **11,** 264–277.

Carritt, D. E., and J. W. Kanwisher (1959). An electrode system for measuring dissolved oxygen. *Anal. Chem.*, **31,** 5–9.

Carritt, D. E., and J. H. Carpenter (1966). Comparison and evaluation of currently employed modifications of the Winkler method for determining dissolved oxygen in sea water. *J. Mar. Res.*, **24,** 286–318.

Clark, L. C., R. Wolf, D. Granger, and Z. Taylor (1953). Continuous recording of blood oxygen tensions by polarography. *J. Appl. Physiol.*, **6,** 189–193.

Crank, J., and G. S. Park (1968). *Diffusion in Polymers*. Academic Press, New York.

Crank, L. E. (1956). *The Mathematics of Diffusion*. Clarendon Press, Oxford.

Denbigh, K. (1966). *The Principle of Chemical Equilibrium*. Cambridge University Press, London.

Douglas, E. (1965). Solubilities of argon and nitrogen in sea water. *J. Phys. Chem.*, **69,** 2608–2610.

Eckert, C. A. (1973). The thermodynamics of gases dissolved at great depths. *Science*, **180,** 426–427.

Enns, T., P. F. Scholander, and E. D. Bradstreet (1965). Effect of hydrostatic pressure on gases dissolved in water. *J. Phys. Chem.*, **69,** 389–391.

Erlebach, H. (1954). *Master Thesis*, Freie Universität, Berlin.

Fenn, W. O. (1972). Partial pressure of gases dissolved at great depth. *Science*, **176,** 1011–1012.

Foulk, C. W., and A. T. Bawden (1926). A new type of end-point in electrometric titration and its application to iodometry. *J. Am. Chem. Soc.*, **48,** 2045–2051.

Føyn, E. (1955). Continuous oxygen recording in sea water. *Fiskerdir. Skr. Havundersøk*, **11,** 1–8.

Føyn, E. (1965). *Progress in Oceanography*. Pergamon Press, Oxford.

Føyn, E. (1976). Oxygen recordings in seawater. In *Measurements of Oxygen* (Ed. H. Degn). Elsevier, Amsterdam.

Galus, Z. (1976). *Fundamentals of Electrochemical Analysis*. Ellis Horwood Ltd, Chichester.

Giguère, P. A., and L. Lauzin (1945). Analyse polarométrique de l'oxygène dissous avec une microélectrode de platine. *Can. J. Res.*, **23,** 223–233.

Gleichmann, U., D. W. Lübbers, W. Burger, and W. Eschweiler (1960). Die Messung des Sauerstoffdruckes in Gasen und Flüssigkeiten mit der Pt-Electrode unter besonderer Berücksichtigung der Messung im Blut. *Arch. Ges. Physiol. Pflügers*, **271,** 431–455.

Grasshoff, A. 1978. Die dynamische Sauerstoffmessung zur Ermittlung des volumetrischen sauerstoffsdurchgangskoeflizienten $K_L \cdot a$ mit der membranüberszogenen polarographischen Electrode. *Kieler milchwictschaftliché Forschungsberichte*, **30**, 67–88.

Grasshoff, K. (1962). Untersuchungen über die Sauerstoffbestimmung im Meerwasser. Teil I and II. *Kiel. Meeresforsch.*, **18**, 42–50 and 151–160.

Grasshoff, K. (1963). Determination of oxygen in sea water with the rapid dropping mercury electrode. *Kiel. Meeresforsch.*, **19**, 8–15.

Grasshoff, K. (1964). Apparatus to create absolute standards for oxygen determination by the Winkler method. *Kiel. Meeresforsch.*, **20**, 143–147.

Grasshoff, K. (1969a). Uber eine Sonde zur digitalen und analogen registrirung von Sauerstoffpartialdruck, temperatur und druck im Meerwasser. *Kiel. Meeresforsch.*, **25**, 133–142.

Grasshoff, K. (1969b). Chemistry of the Red Sea and inner gulf of Aden according to the observations made by the research vessel "Meteor" during the Indian Ocean expedition 1964–65. *"Meteor" Forschungsergeb.*, Reihe A6, 1–76.

Grasshoff, K. (1975). The hydrochemistry of landlocked basins and fjords. In *Chemical Oceanography*, Vol. 2 (Eds. J. P. Riley and G. Skirrow). Academic Press, London, New York and San Francisco, pp. 456–597.

Grasshoff, K. (1976). Sampling techniques. In *Methods of Seawater Analysis* (Ed. K. Grasshoff). Academic Press, London, pp. 1–20.

Grasshoff, K. and H. P. Hansen (1978). The towed chemical profiler, a versatile system for continuous measurement of environmental parameters in seawater. *ICES C. M. Pap. Rep.* no. C.M. 1978/C:4, 19 pp.

Green, E. J. (1965). *Thesis*, Massachusetts Institute of Technology.

Green, E. J. and D. E. Carritt (1967a). New tables for oxygen saturation of seawater. *J. Mar. Res.*, **25**, 140–147.

Green, E. J., and D. E. Carritt, (1967b). Oxygen solubility in sea water: thermodynamic influence of sea salt. *Science*, **157**, 191–193.

Hahn, C. E. W. (1975). Electrochemical improvement of the performance of the P_{O_2} electrode. *Respir. Physiol.*, **25**, 109–133.

Hallberg, R. O., L. E. Bågander, A. G. Engvall, M. Linström, S. Odén, and F. A. Schippel (1973). *Contrib. Askö Lab., Univ. Stockholm*, **2**, 1–117.

Hansen, H. P., K. Grasshoff, and J. Petersen, (1980) *In-situ* registration of oxygen utilisation at sediment–water interfaces. *Mar. Chem.*, **10**, 47–54.

Hässelbarth, U. (1954). *Masters Thesis*, Freie Universität, Berlin.

Heineken, F. G. (1970). Use of fast-response dissolved oxygen probes for oxygen transfer studies. *Biotechnol. Bioeng.*, **12**, 145–154.

Heineken, F. G. (1971). Oxygen mass transfer and oxygen respiration rate measurements utilizing fast response oxygen electrodes. *Biotechnol. Bioeng.*, **13**, 599–618.

Heyrovsky, J. (1926). Analysis by means of the dropping mercury cathode. *Chem. Listy*, **20**, 122–130.

Heyrovsky, J. (1936). Polarographie. In *Physikalische Methoden der chemischen Analyse* (Ed. W. Böttiger). Leipzig.

Heyrovsky, J. (1937). *Die Polarographische Methode, Theorie und Praktische Anwendung*, Leningrad.

Heyrovsky, J. (1949). The significance of derivative curves in polarography. *Chem. Listy*, **43**, 149–154.

Hitchman, M. L. (1978). *Measurement of Dissolved Oxygen*. Wiley, New York.

Hoare, J. P. (1969). *The Electrochemistry of Oxygen*. Wiley–Interscience, New York.

Hopfenberg, H. D. (1974). *Permeability of Plastic Films and Coatings to Gases, Vapours and Liquids*. Plenum Press, New York.

Horne, R. A. (1969). *Marine Chemistry*. Wiley–Interscience, New York.

Hume, D. N. (1956). Polarographic theory, instrumentation and methodology. *Anal. Chem.*, **28**, 625–637.

418   *Marine Electrochemistry*

Ingols, R. S. (1941). Determination of dissolved oxygen by the dropping mercury electrode. *Sewage Works J.*, **13**, 1097–1109.

Ingols, R. S. (1955). Experience with solid platinum electrodes in the determination of dissolved oxygen. *Sewage Ind. Wastes*, **27**, 4–7.

Interocean Systems Inc., (1977). Model 6660 Probe. Interocean Systems Inc., San Diego.

Izydorczyk, J., W. Misniakrewicz, and K. Raszka (1976). Working conditions of a Clark-type sensor with a poly-ethylene membrane in the presence of vapours of chlorine derivatives of methane. *J. Electroanal. Chem.*, **70**, 365–374.

Kalle, K. (1939). Oceanographic chemical investigations with the Zeiss Pulfrich photometer. *Ann. Hydrogr. Marit. Meteorol.*, **63**, 58–65.

Kessler, M. (1973). *Oxygen Supply.* Urban and Schwarzenberg, Munich.

Kolkwitz, R. (1941). Biological and electrometric determination of O. *Ber. Dtsch. Bot. Ges.*, **60**, 306–312.

Kolthoff, I. M., and J. J. Lingane (1952). *Polarography.* Interscience, New York.

Koryta, J., J. Dvaíak, and V. Boháckóv (1970). *Electrochemistry.* Methuen, London.

Krichevsky, I. R., and J. S. Kasarnovsky (1935). Thermodynamical calculations of solubilities of nitrogen and hydrogen in water at high pressure. *J. Am. Chem. Soc.*, **57**, 2168–2171.

Laitinen, H. A., T. Higuchi, and M. Czuha (1948). Potentiometric determination of oxygen using the dropping-mercury electrode. *J. Am. Chem. Soc.*, **70**, 561–565.

Lemon, E. R., and A. E. Erickson (1952). The measurement of oxygen diffusion in the soil with a platinum micro-electrode. *Soil. Sci. Soc. Am. Proc.*, **16**, 160–163.

Lemon, E. R. (1955). Principle of the platinum microelectrode as a method for characterizing soil aeration. *Soil. Sci.*, **79**, 383–392.

Lingane, J. J. (1949). Polarographic theory, instrumentation and methodology. *Anal. Chem.*, **21**, 45–60.

Lingane, J. J., and B. A. Loveridge (1950). A new polarographic diffusion current equation. *J. Am. Chem. Soc.*, **72**, 438–441.

Lynn, W. R., and D. A. Okun (1955). Experience with solid platinum electrodes in the determination of dissolved oxygen. *Sewage Ind. Wastes*, **27**, 4–7.

Mancy, K. H., D. A. Okun, and C. N. Reilley (1962). A galvanic cell oxygen analyzer. *J. Electroanal. Chem.*, **4**, 65–92.

Masterton, W. L. (1975). Salting coefficients for gases in sea water from scaled-particle theory. *J. Solut. Chem.*, **4**, 523–534.

Meites, L. (1965). *Polarographic Techniques.*, 2nd edition, Interscience, New York.

Montgomery, H. A. C., M. S. Thom, and A. Cockburn (1964). Determination of dissolved oxygen by the Winkler method and the solubility of oxygen in pure water and sea water. *J. Appl. Chem.*, **14**, 280–296.

Moore, E. W., J. Carrell, and D. A. Okun (1948). The polarographic determination of dissolved oxygen in water and sewage. *Sewage Works J.*, **20**, 1041–1053.

Murray, C. N., and J. P. Riley (1969). Solubility of gases in distilled water and sea water. II. Oxygen. *Deep-Sea Res.*, **16**, 311–323.

Niedrach, L. W., and W. H. Stoddard (1972). Sensor with anion exchange resin electrolyte. *U.S. Pat.* 3 703 457.

Nösel, H. (1976). Die Techologie der Sauerstoffmessung in der Siedlungswasser-wirtschaft. *Wasserwirtschaft*, **66**, 156–159.

Nürnberg, H. W. (Ed.) (1974). Electroanalytical chemistry. In *Advances in Analytical Chemistry and Instrumentation.* (Series Eds. C. N. Reilley and L. Gordon). Vol. 10. Wiley, New York.

Odén, S. (1962) Principilla problem rörende syre diffusans elektrode och deras ut formning. *Grundförbättring*, **3**, 150–178.

Ohle, W. (1952). Investigation and application of the electrochemical determination of dissolved oxygen for the examination of waters. *Vom Wasser,* **19,** 99–121.

Ohle, W. (1953). Die chemische und elektrochemische bestimmung des molekular gelösten Sauerstoffes der Binnengewässer. *Int. Ass. Theor. Appl. Limnol. Comm.,* **3,** 1–44.

Pamatmat, M. M., and K. Banse (1969). Oxygen consumption by the sea bed. II *In-situ* measurements to a depth of 180 m. *Limnol. Oceanogr.* **14,** 250–259.

Pamatmat, M. M. (1971). Oxygen consumption by the sea bed. IV. Shipboard and laboratory experiments. *Limnol. Oceanogr.* **16,** 536–550.

Perley, G. A. (1939). Determination of dissolved oxygen in aqueous solutions. *Ind. Eng. Chem., Anal. Ed.,* **11,** 240–242.

Richards, F. A. (1965). In *Chemical Oceanography,* Vol. 1 (Eds. J. P. Riley and G. Skirrow). Academic Press, London, pp. 197–226.

Richardson, J. L., P. Bergsteinsson, R. J. Getz, D. L. Peters, and R. W. Sprauge (1964). Philco Aerometric Div. Pub., Office of Naval Res.

Rotthauwe, H. W. (1958). Oxygen determination in sea and fresh water with the aid of dropping-mercury electrode and its application to physiological research. *Kiel. Meeresforsch.,* **14,** 48–63.

Sadjadi, S. (1979). *Thesis,* Kiel University.

Sawyer, D. T., S. G. Raymond, and R. C. Rhodes (1959). Polarography of gases. Quantitative studies of oxygen and sulfur dioxide. *Anal. Chem.,* **31,** 2–5.

Scholander, P. F., L. Van Dam, C. L. Claff, and J. W. Kanwisher (1955). Microgasometric determination of dissolved oxygen and nitrogen. *Biol. Bull.,* **109,** 328–334.

Schramm, K. (1977, 1978). Gesellschaft für Ökologie, 7. Jahresversammlung, Kiel.

Schwarz, W. (1958). *Elektrochemische Sauerstoffmessung.* Walter de Gruyter, Berlin.

Setschenow, J. (1889). Uber die konstitution der salzlösungen auf grund ihres verhaltens zu kohlensäure. *Z. Phys. Chem.,* **4,** 117–125.

Stack, V. T. (1967). Polarographic cell for dissolved oxygen. *U.S. Pat.,* 3 360 451.

Stackelberg, M. (1950). *Polarographische Arbeitsmethoden.* Walter de Gruyter, Berlin.

Strehlow, H., and M. v. Stackelberg (1950). Zur theorie der polarographischen Kurven. *Z. Elektrochem.,* **54,** 51–62.

Strickland, J. D. H., and T. R. Parsons (1960). A manual of sea water analysis with special reference to more common micronutrients and to particulate organic materials. *Bull. Fish. Res. Bd. Can.,* **125,** 1–185.

Swinnerton, J. W., V. J. Linnenbom, and C. H. Cheek (1962). Determination of dissolved gases in aqueous solutions by gas chromatography. *Anal. Chem.,* **34,** 483–485.

Swinnerton, J. W., V. J. Linnenbom and C. H. Cheek (1964). Determination of argon and oxygen by gas chromatography. *Anal. Chem.* **36,** 1669–1671.

Tiepel, E. W., and K. E. Gabbins (1973). Thermodynamic properties of gases dissolved in electrolyte solutions. *Ind. Eng. Chem. Fundam.,* **12,** 18–25.

Tödt, F. (1929). The continuous indication of dissolved oxygen content as well as of the rust-preventing action of sodium hydroxide by current density measurements. *Z. Ver. Dtsch. Zuckerind.,* **79,** 680–695.

Tödt, F., 1942. Chemical and electrochemical processes for combating corrosion of metals. *Wien. Chem. Ztg.,* **45,** 265–269.

Tödt, F. (1958). *Elektroschemische Sauerstoffbestimmung.* Walter de Gruyter, Berlin.

UNESCO (1969). Report of the 6th meeting of the UNESCO–ICES–SCOR–IAPSO panel of experts on oceanographic tables and standards. *Tech. Pap. Mar. Sci.,* 14, UNESCO, Paris.

UNESCO (1973). *International Oceanographic Tables,* Vol. II. National Institute of Oceanography, Wormley, England, and UNESCO, Paris.

Van Krevelen, D. W. (1972). *Properties of Polymers.* Elsevier, Amsterdam.

Weiss, R. F. (1970). The solubility of nitrogen, oxygen and argon in water and seawater. *Deep-Sea Res.,* **17,** 721–735.

Weiss, R. F., and H. Craig (1973). Precise shipboard determination of dissolved nitrogen, oxygen, argon and total inorganic carbon by gas chromatography. *Deep-sea Res.,* **20,** 291–303.

Winkelmann, D. (1956). Untersuchungen über das elektrochemische verhalten O_2–H_2O_2–H_2O am blanken und platinierten platin. *Z. Elektrochem.,* **60,** 731–740.

Winkler, L. W. (1888). Die bestimmung des im Wasser gelösten Sauerstoffes. *Ber. Dtsch. Chem. Ges.,* **21,** 2843–2855.

Wolf, S. (1960). Rapid-polarographie. *Angew. Chem.,* **72,** 449–454.

Wood, K. G. (1953). Polarograms of oxygen in lake water. *Science,* **117,** 560–561.

Yasuda, H., and W. Stone (1966). Permeability of polymer membranes to dissolved oxygen. *J. Polym. Sci.,* **4,** 1314–1316.

Marine Electrochemistry
Edited by M. Whitfield and D. Jagner
© 1981 John Wiley & Sons Ltd.

A. ZIRINO

Naval Ocean Systems Center,
San Diego,
California 92152, USA

10

Voltammetry of Natural Sea Water

GLOSSARY OF SYMBOLS

D_c Diffusion coefficient of complexed ion (equation 3)
$E_{1/2}^*$ Half-wave potential (equation 3)

F	Faraday (equation 1)
g	ln $10RT/nF$ (equation 3)
k	Cell constant (equation 1)
n	Number of equivalents per mole (equation 1)
r	Radius of mercury drop (equation 3)
R	Gas constant (equation 1)
t	Electrolysis time (equation 3)
v	Volume of solution (equation 1)
δ	Diffusion layer thickness

10.1. INTRODUCTION

In the last decade, the marine analytical community has increasingly adapted voltammetric techniques for collecting information about the natural marine environment (Davison and Whitfield, 1977). This interest was sparked by the technical advances in polarographic techniques and instrumentation which have made it possible to analyse for several of the trace constituents of sea water directly at mercury drop or mercury film electrodes without pre-concentration or extensive chemical treatment. Much of this interest has been in the field of anodic stripping voltammetry (ASV) or inverse polarography, which has the unique advantage of possessing extreme sensitivity for certain trace metals while requiring inexpensive instrumentation.

However, the popularity of voltammetric techniques cannot be accounted for solely on the basis of the analytical aspects and the inexpensive instrumentation. For the analysis of trace metals, flameless atomic-absorption spectrometry (AAS) has nearly the same sensitivity and is useful for the detection of a larger number of substances. One particular feature of voltammetric analysis is that, through judicious experiments, the possibility exists of determining the exact chemical nature of the trace constituents as they exist in the natural environment. Indeed, it is the great similarity between sea water and the usual polarographic media which makes available to the marine and natural media electrochemist all of the polarographic literature and experience of the last 50 years. In the recent literature efforts to deduce chemical speciation have gone hand in hand with and/or been part of larger analytical papers. This clearly indicates that when electrode measurements are made directly in sea water, analysis and speciation are inextricably related. The many difficulties encountered in the trace analysis of sea water have also emphasized the need for a better understanding of the nature of the analyte as well as the possible interferences occurring at the electrode.

Another attractive feature of polarographic measurements is that they are inherently suitable for continuous shipboard and (possibly) *in situ* analysis. This aspect leads to the possibility of eventually monitoring, from small vessels, coastal and oceanic waters for trace elements in order to gain a better understanding of natural marine processes. Indeed, combining polarographic and voltammetric technology with modern microcomputers to form 'intelligent' automated sampling and analysis systems promises to provide a way of following the spatial and temporal changes in trace constituents which must be understood before sound environmental judgements can be made.

This chapter attempts to demonstrate the application of voltammetric techniques to the analysis of natural marine samples by discussing certain examples from the literature. It is not intended to serve as a comprehensive survey, but rather as an illustrative one.

10.1.1. Polarography and anodic stripping voltammetry

The applications of voltammetric techniques to the analysis of natural samples may be placed into two categories: (a) those which measure the current arising from the direct reduction and/or oxidation of an electroactive species, and (b) those which employ a prolonged electrolysis to concentrate metal ions from solution into an amalgam. After a suitable time period, the amalgam-forming metal is oxidized back into solution and the oxidation current is interpreted as a measure of the original concentration of the electroactive component. The former methods are generally referred to as 'polarographic' techniques (they are primarily performed with polarized dropping mercury electrodes) and the latter are referred to as 'stripping' techniques (see Section 2.3 and Table 1 of Chapter 2 for further details).

In this section, only illustrative examples will be given to demonstrate the application of voltammetric techniques to the analysis of sea water. A thorough review of the application of these techniques in marine chemistry up to 1975 has been given by Whitfield (1975).

10.1.2. Natural sea water as an electrolytic medium

Natural sea water contains dissolved salts, dissolved gases, organic substances (mostly unidentified), colloidal matter, and particulates, both living and detrital. All of these have some bearing on voltammetric analysis and many may be analysable by voltammetry after suitable concentration and separation steps. Figure 1 presents many of the substances known or presumed to be in sea water as a function of their size and concentration. A first glance at the large number of entries makes one wonder how voltammetric analysis can be carried out in sea water with any degree of specificity at all! The answer, of course, is that most of the components of sea water are unreactive at the sensitivities and conditions commonly employed for *direct* analysis in sea water media. It is useful at this point to classify some of these components (see also Chapter 1).

INORGANIC AND ORGANIC SPECIES
IN "TRUE" SOLUTION

DIAMETER RANGE ← ———— 10 Å ———— 100 Å ———— 1000 Å

FREE IONS	INORGANIC AND ORGANIC SPECIES IN "TRUE" SOLUTION	HIGH MOL. WT. ORGANICS	HIGHLY DISPERSED COLLOIDS	LARGER COLLOIDS	PARTICULATES
Cl^-, Na^+, Mg^{2+}					
SO_4^{2-}, $NaSO_4^-$, $MgSO_4$ Ca^{2+}, K^+, HCO_3^-, $CaSO_4$ Sr^{2+}, $N_2(g)$, Br^-					
$MgHCO_3^+$, $O_2(g)$ $MgCO_3$, $NaCO_3^-$, $B(OH)_4^-$ CO_3^{2-} $B(OH)_3^0$, MgF^+ F^-, $CO_2(g)$, NO_3^- Li^+, Rb^+	Unidenified Substances 'Gelbstoffe', **Humic Acid types**				

10^2
10^0
10^{-2}
10^{-4}
(g/kg)

424

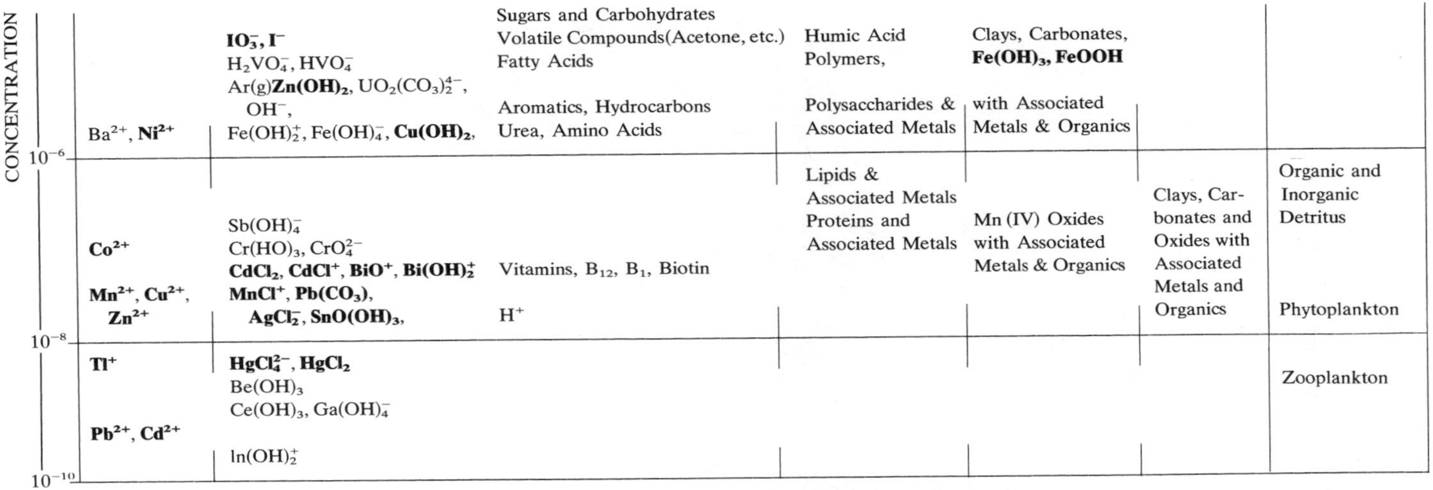

Figure 1.

CONCENTRATION

10^{-6}

10^{-8}

10^{-10}

Ba^{2+}, Ni^{2+}

IO_3^-, I^-
$H_2VO_4^-$, HVO_4^-
$Ar(g)/Zn(OH)_2$, $UO_2(CO_3)_2^{4-}$,
OH^-,
$Fe(OH)_2^+$, $Fe(OH)_4^-$, $Cu(OH)_2$,

Co^{2+}

$Sb(OH)_4^-$
$Cr(HO)_3$, CrO_4^{2-}
$CdCl_2$, $CdCl^+$, BiO^+, $Bi(OH)_2^+$

Mn^{2+}, Cu^{2+},
Zn^{2+}

$MnCl^+$, $Pb(CO_3)$,
$AgCl_2^-$, $SnO(OH)_3^-$,

Vitamins, B_{12}, B_1, Biotin

H^+

Tl^+

$HgCl_4^{2-}$, $HgCl_2$
$Be(OH)_3$
$Ce(OH)_3$, $Ga(OH)_4^-$

Pb^{2+}, Cd^{2+}

$In(OH)_2^+$

Sugars and Carbohydrates
Volatile Compounds(Acetone, etc.)
Fatty Acids

Aromatics, Hydrocarbons
Urea, Amino Acids

Humic Acid
Polymers,

Polysaccharides &
Associated Metals

Lipids &
Associated Metals
Proteins and
Associated Metals

Clays, Carbonates,
Fe(OH)₃, FeOOH

with Associated
Metals & Organics

Mn (IV) Oxides
with Associated
Metals & Organics

Organic and
Inorganic
Detritus

Phytoplankton

Clays, Car-
bonates and
Oxides with
Associated
Metals and
Organics

Zooplankton

Figure 1. Major and minor components of sea water as a function of concentration and estimated size. Constituents in bold type have been analysed by voltammetric methods. 1 Å = 0.1 nm

425

Major constituents

By definition, the major constituents of sea water are those present at concentrations greater than about 1 ppm. Cl^-, Na^+, Mg^{2+}, Ca^{2+}, SO_4^{2-}, K^+, Sr^{2+}, HCO_3^-, Br^-, and F^-, and the complexes formed by their association with one another, make up 99.99% of the total salt components in sea water and belong in this category. In general, the major constituents are not electroactive in sea water but serve to conduct the cell current. Polarographic analyses of the major cations and anions are, of course, possible (Kolthoff and Lingane, 1952; Meites, 1965; Heyrovsky and Kůta, 1966), and are reported in a subsequent section of this chapter.

Trace metals

The most extensive marine application of voltammetric techniques has been in the analysis of trace metals. To date, analyses have been reported for Cu, Zn, Ag, Cd, Sb, Bi, Hg, and Pb by anodic stripping techniques and for Co, Cu, Ni, Cd, Zn, and Pb by pulse polarography after pre-concentration on a chelating resin (see Chapter 2). By far the greatest effort has been directed toward the analysis of Cu, Zn, Cd, and Pb by direct and modulated ASV techniques. There are several reasons for this. Firstly, these metals are present in sea water at concentrations measurable by ASV. Secondly, their reduction in chloride media is reversible and thus they can be oxidized and recovered from the electrode almost quantitatively after pre-electrolysis, and finally, in sea water their oxidation potentials occur within the mercury 'window', i.e. between -1.5 and $+0.1$ V *vs.* SCE. Perhaps an additional reason for the amount of effort directed at Cu, Zn, Cd, and Pb is that through the study of these easily measured metals, important information may be obtained about general marine processes.

Dissolved gases and organic constituents

Oxygen is an extremely electroactive constituent of sea water and its concentration varies from about 300 μM at the surface of the ocean to below the detection limit of the micro-Winkler technique (Broenkow and Cline, 1969) in the 'O_2 minimum zone' of the eastern tropical Pacific Ocean. The sensitivity of the micro-Winkler method is approximately 1 μM and is primarily limited by the reagent blank. The electrochemical reduction of oxygen at the dropping mercury electrode in chloride solutions is well known, and proceeds in two steps, giving a wave at -0.05 V and a second wave at -0.09 V *vs.* SCE (Meites, 1965). Ordinarily, oxygen interferes with most polarographic determinations in sea water and is purged from solution with a high-purity inert gas before analysis. On the other hand, the polarographic determination of oxygen at the dropping mercury electrode and on membrane-covered noble metal electrodes has found wide application in oceanography (Whitfield, 1975) and is the subject of Chapter 9.

The concentrations of some of the organic constituents of sea water can be determined from the effects of adsorption at the mercury electrode. Ćosović and co-workers (Kozarac *et al.*, 1976; Ćosović *et al.*, 1977; Zutic *et al.*, 1977) have shown that Kalousek polarography (Chapter 2, Table 1) and the suppression of d.c. polarographic maxima can be used to estimate surfactant activity in a natural water sample. Cominoli, *et al.* (1980) have also assessed a number of polarographic techniques for the determination of humic and fulvic acids. In addition, exploratory experiments with San Diego Bay water (Zirino, unpublished work) have shown that with differential pulse polarography (DPP), unidentified peaks could be observed during cathodic scans. These peaks appeared to be from the reduction of organic components related to the high organic productivity in the bay. More recently, Lieberman (1979) studied the copper-complexing characteristics of humic acid-type compounds obtained from Lake Nitinat, an anoxic fjord in British Columbia. Initially he acidified the sea water samples to pH 2 with H_2SO_4 and concentrated the humic materials on a non-ionic polymeric adsorbent resin. The resin was then eluted with ammonia solution (pH 12) and the eluate was characterized by DPP in a sea water or phosphate buffer at pH 8. This author is of the opinion that the analysis of trace organic substances by DPP and by other high-sensitivity modulated polarographic techniques is a promising area of research in marine voltammetric analysis. The sea water constituents which have been analysed by voltammetric methods are indicated by bold type in Figure 1.

10.2. SAMPLING AND STORAGE OF WATER

The task of the marine analyst differs from that of the general analytical chemist in that he often has to describe phenomena of general scientific and sometimes public interest with data based on the analysis of unique and unrecoverable samples. This occurs because individual samples represent particular points in time and space. Given the variability associated with trace constituents in sea water, the data are already historical by the time the analysis is completed. There is a question of degree, of course, and the degree of variability of a particular characteristic depends on the phenomenon to be described. Considerable fluctuations in the concentrations of micro-constituents can be expected along coastal and inshore areas of great biological productivity and diversity. Smaller changes may be expected in central oceanic areas, away from the equator where the inherent stability of the water leads to a parallel stability in the biological and chemical features. On the other hand, such areas are more difficult to reach and elaborate expeditions must be launched to sample them. In summary, then, a great deal of effort should be spent to ensure that samples are as free from contamination and as unaltered by the sampling apparatus as possible. Although electroanalytical chemists are generally extremely careful about monitoring their own cells and instrumentation for contamination, they are often not aware of the many pitfalls that may be encountered when sampling the marine environment. What follows is a brief introduction to these problems. The books edited and co-authored by Zief (1976, 1972) are useful references on the subject of trace analysis, and the book *Strategies for Marine Pollution Monitoring* (Goldberg, 1976) discusses marine sampling and storage problems in particular.

10.2.1. Water sampling techniques

Commonly, sea water samples for trace metal analysis have been collected from ships by means of plastic sampling bottles. In the United States the most popular of these devices, called the Niskin bottle, after its designer, and marketed by General Oceanics, Hialeah, Florida, consists of a poly(vinyl chloride) (PVC) cylinder with circular lids at each end. When the bottle is in the closed position, the lids are held taut to each other and against the rim of the cylinder by a Teflon-covered steel spring. The heavy rubber tubing commonly sold with the bottles is unsuitable for trace metal analysis. The samplers can be opened and maintained open by pulling the lids back against the spring and fastening them to a release mechanism alongside the bottle. Nylon lanyards are provided for this purpose.

When new, or after prolonged storage (but generally not before every cast), the samplers are cleaned with dilute acid (approximately 1 M) and rinsed with doubly distilled water. Then, following traditional oceanographic practice, the samplers are clamped to a steel cable (hydrowire) and lowered into the sea open, that is free flushing, on the way to its sampling depth. After a brief equilibration period in which the sampler is allowed to bob through the same water parcel for a few minutes, the sample is collected. This is done by releasing a brass weight (messenger) at the surface, which then slides along the wire and eventually collides with the tripping mechanism, which in turn frees the lids to close snugly against the bottle rim. Teflon (not rubber) O-rings provide a water-tight seal. Once entrapped, the sample is lifted to the ship.

This procedure has been criticized on several points. The ship itself continually releases metals (Cu, Zn, and Pb) to the upper few metres of water from its brass screws, antifouling paints, and cathodic protection devices. Frequently, oil from the decks and bilges, sewage, and solid garbage are also released during or near the time of sampling. Oils in particular can cause serious interferences in polarographic analysis. Finally, soot from the stacks can contaminate the surface water. The steel wire used to lower the bottles is of necessity covered with grease and is usually rusty. Both grease and rust are released at the surface as the wire breaks the water, particularly after a few days of disuse. The messenger is commonly made of brass (plastic-covered messengers are now available). The bottle is lowered *open* through the interface, through water contaminated from the ship and where substances are already concentrated in naturally formed organic 'slicks' (Wallace *et al.*, 1977). This contamination is adsorbed on to the walls of the bottles and may not be desorbed during its descent to the sampling depth.

These and other sampling problems have caused investigators to re-examine their sampling procedures and a new sampling protocol is gaining acceptance. Surface or near-surface samples are not collected with bottles on a wire but with a rubber raft or small boat which is launched from the ship and rowed (at least for the last 100 m or so) to the sampling site. Sampling bottles, which in this case may be polyethylene or Teflon FEP storage bottles, are lowered by hand from the bow of the craft and dipped closed below the surface with one hand. They are then filled by removing the cap with the other hand. The sample is collected after several rinsings.

Deep samples cannot be taken in this manner and various attempts have been made to replace the hydrowire with nylon or plastic cables. Because these tend to stretch and break they are suitable only for shallow sampling. Recently, however, plastic-coated steel cable has been introduced and is in use in at least one laboratory. The most serious problem, that of contaminating the samples with surface water, has been solved by the introduction of a new bottle which is lowered closed below the surface. (Go-Flo, General Oceanics, Hialeah, Florida). A pressure-sensitive device then opens the bottle after it has been lowered to a pre-set depth, such as 5 m. The results of investigations on the concentrations of trace metals in sea water made by following the new procedure and by using over-all clean laboratory techniques have yielded lower surface and lower overall concentrations than had been observed previously (Moore and Burton, 1976; Bender and Gagner, 1976; Boyle *et al.*, 1976; Boyle and Edmond, 1975). Although these measurements were made by spectrophotometric techniques with pre-concentration, they indicate that careful sampling is even more crucial to the marine electrochemist, who by virtue of his methods tends to use small sample volumes.

10.2.2. Filtration

After the sample has been collected, it is often desirable to separate the particulate matter by filtration before a polarographic determination is made. This seemingly innocuous step is fraught with problems. Available filter-papers contain substantial quantities of contaminating elements (Table 1), which leach out of the material during filtration. The filtration apparatus

Table 1. Metal contents of materials used in trace analysis. [Reprinted with permission from Robertson (1968), *Anal. Chem.*, **40**, 1067–1072. Copyright © 1968 American Chemical Society]

	Metal content, $\mu g\,kg^{-1}$					
Material	Fe	Cu	Zn	Ag	Sb	Pb
Sampling materials						
Structural PVC	3×10^5	6×10^2	7.1×10^3	<5	2×10^3	—
Teflon	35	20	9	<0.3	0.01	—
Plexiglas	1×10^2	<10	<10	<0.03	0.01	—
Tygon	3×10^3	6.5	6×10^4	<10	5	$2\times10^{5*}$
Surgical rubber tube	—	<6	3×10^6	1×10^3	$<1\times10^2$	—
Steel hydrowire	—	2×10^4	—	—	5×10^4	—
Polyethylene hose	7.4	—	55	<300	9×10^3	—
Container materials						
Quartz tubing	4×10^2	0.03–2	1–30	<0.1	<0.01–2×10^3	—
Borosilicate glass	3×10^5	—	7×10^2	<0.001	3×10^3	—
Polyethylene	1×10^4	10	30	<0.1	1	—
Millipore filter apparatus	3×10^2	—	2×10^3	<0.1	40	—
Filters and papers						
Millipore HA	3×10^2	—	2×10^3	<0.05	40	3*
Nucleopore	3×10^4	2×10^3	2×10^3	—	<20	—
Whatman I	—	$<1\times10^2$	—	—	—	—
Glass Whatman GF/A	—	$<5\times10^4$	—	—	—	—
Kimwipe	1×10^3	—	5×10^4	<1	20	—
Selected reagents						
HCl (AR grade)	<2	—	20	<0.1	0.2	30*
HNO$_3$ (AR grade)	2×10^3	1	10	0.3	0.03	30*
NaOH (AR grade)	$<1\times10^3$	—	<20	<0.2	0.3	—
NH$_3$ solution (AR grade)	<0.1	6	2	<0.1	<0.01	—
KOH (AR grade)	3×10^3	—	1×10^3	1×10^2	2	—

— indicates not measured
* From Bertine *et al.* (1976).

itself may contaminate the sample or, conversely, after a recent acid cleaning, may adsorb trace elements from sea water. Zirino and Healy (1971) have shown that cellulose acetate membrane filters leach Zn. Burrell and Wood (1969) have shown that Zn can also be leached by doubly distilled water from the glass-frit support for membrane filters. Contamination of the samples was reduced by soaking the frit in *aqua regia* and rinsing with doubly distilled water. Brooks (quoted in Robertson, 1972) pre-cleaned filter membranes by soaking them for 24 h in a 1% solution of ammonium pyrrolidine dithiocarbamate. This treatment minimized the leaching of trace contaminants into the samples. Losses of trace elements by adsorption on the walls of the filtration apparatus are also possible. Robertson (1972) states that Pb, Cu, and Ra have been removed from solution by paper filters, while fritted-glass filters have been known to adsorb Be and Th. In our laboratory we have adopted the following filtration procedure. The 'clean' filtration apparatus (consisting of tower, sintered-glass filter holder, and modified Erlenmeyer flask) containing the filter membrane is assembled and two 100 cm³ aliquots of 5% HCl are filtered and discarded. Care is taken to wash all exposed surfaces with the 5% HCl, by gently rotating the filtration apparatus. This is followed by filtering two 100 cm³ aliquots of ultrapure water. Again the walls are carefully washed with the filtrate, which is then discarded. Finally, two 100 cm³ aliquots of the sample are filtered and discarded. This step is the most important because the sea water samples are used to bring the pH of the surface of the apparatus back to 8. Additionally, the sea water supplies major constituent ions to the walls of the vessel which occupy sites subject to adsorption and ion exchange. This reduces the loss of trace elements from the sample (Nürnberg *et al.*, 1976).

10.2.3. Materials for sampling and storage

Contamination

Because the main application of polarography and voltammetry to sea water has been in the determination of trace metals, a discussion of sampling and storage materials must first deal with these. Robertson (1968, 1972) constructed tables of the trace metal contents of many materials and reagents used in trace analysis. This table was augmented by Bertine *et al.* (1976) and portions of it are reproduced here (Table 1). Because Robertson's analyses were carried out by neutron activation, several elements (Cd, Pb, and Bi) of interest to polarographers were not included in his tables. Nevertheless, his work gives a good indication of the possible contamination which could occur from materials commonly used by marine electrochemists and oceanographers. The actual contamination, of course, depends on the rate of release of contaminants by the material, and this may vary with the physical structure of the material, the extent of pre-cleaning, and its general history. Indeed, we find from Robertson's tables that all materials contain possible trace contaminants although fluorocarbons (Teflon) and acrylics (Plexiglas) are by far the 'cleanest' structural materials, while quartz, borosilicate glass (Pyrex), Vycor glass, and polyethylene are good materials for containers.

Sorption

Robertson (1968) also determined the extent to which certain metals were adsorbed from sea water (pH approximately 8) on to polyethylene and Pyrex glass storage containers. He showed that serious losses of In, Se, Fe, Ag, U, and Co can occur after a few days of storage in both polyethylene

Table 2. Suggested method for cleaning plastic containers. [Reprinted with permission from Moody and Lindstrom (1977), *Anal. Chem.*, **49**, 2264–2267. Copyright © 1977 American Chemical Society]

Step No.	Procedure
1	Fill with 1 + 1 HCl (AR grade)
2	Allow to stand for 1 week at room temperature (80°C for Teflon)
3	Empty and rinse with distilled water
4	Fill with 1 + 1 HNO_3 (AR grade)
5	Allow to stand for 1 week at room temperature (80°C for Teflon)
6	Empty and rinse with distilled water
7	Fill with purest available distilled water
8	Empty and rinse two or three times with aliquots of the sample, carefully exposing all surfaces to the sample. If the sample has been acidified, rinse with the acidified sample

and Pyrex glass bottles. This adsorptive loss could be prevented by lowering the pH to about 1.5. Surprisingly, negligible adsorption of Zn, Cs, Sr, and Sb was observed for storage periods of up to 75 days, even at the normal pH of sea water. Robertson recommended that sea water samples be acidified to pH 1.5 immediately after collection and be stored in polyethylene bottles, which are low in trace metals (other than iron). These recommendations have been seconded by Moody and Lindstrom (1977) and Subramanian *et al.* (1978), who also studied the preservation and storage of trace elements in natural waters. Subramanian *et al.* (1978), however, make the additional recommendation that zinc samples be stored in Teflon vessels because polyethylene containers could leach out significant quantities of zinc.

Moody and Lindstrom (1977) have suggested a procedure for cleaning plastic containers which is in general agreement with that used by other workers (Patterson and Settle, 1975). A modified form of this procedure is outlined in Table 2.

Storage at the natural pH

Although storage at a pH of approximately 1.5 is suitable when the subsequent analyses are carried out by direct flameless AAS, neutron activation or isotope dilution mass spectrometry, polarographic determinations and atomic-adsorption extraction techniques require that the pH of the sample be raised again before analysis. This additional step leads to risks of contaminating the sample with the added base. Also, lowering the pH of the sample generally precludes the study of the trace element speciation at the natural pH. For these reasons it has become common practice to quick-freeze sea water samples and to store them in a frozen state until just before analysis. However, the effect of freezing is difficult to assess, and to date it is still unclear how freezing affects the trace element speciation or concentration. Sea water samples freeze from the outside and separate out fresh water ice and brine. Particulates and trace elements are concentrated at the centre, where they may undergo chemical and physical alteration. The degree of formation of fresh water ice depends on the rate of solidification and therefore rapid freezing should produce the smallest changes. Although Strickland and Parsons (1965) report that micronutrients (NO_3^-, PO_4^{3-}, and SiO_4^{4-}) can be preserved by freezing for several weeks, the effect of quick-freezing on the concentration and speciation of most trace elements remains unknown.

10.3. ANALYSIS BY POLAROGRAPHY

10.3.1. Analysis following chemical pre-concentration

Pre-concentration has often been employed to improve the sensitivity of polarographic analysis, and it also serves as a means of storing and preserving samples until an analysis can be made in a suitable laboratory environment. Joyner *et al.* (1967) have reviewed pre-concentration techniques for the analysis of sea waters up to about 1965. Of the various methods discussed, *viz.* lyophilization, co-crystallization, coprecipitation, and solvent extraction, only the last two were used in conjunction with polarographic techniques. Tikhonov and Zhavoronkina (1960) extracted copper from sea water into carbon tetrachloride after formation of the dithizone complex. The extract was then evaporated to dryness, re-dissolved, and analysed by d.c. polarography. Similarly, Tikhonov and Shalimov (1965) extracted nickel and manganese from sea water into chloroform as the diethyldithiocarba-mates. The extract was then evaporated to dryness, re-dissolved in HCl, buffered with NH_3-NH_4Cl, and finally analysed by d.c. polarography. Milner *et al.* (1961) extracted uranium from acidified sea water with di(2-ethylhexyl)phosphoric acid in CCl_4, followed by back-extraction into 11 M HCl. After several purification steps, two of which were wet oxidations to remove the organic chelates originally used to sequester the uranium from sea water, the uranium was analysed by derivative pulse polarography as the uranyl ion in a sodium tartrate supporting electrolyte.

Joyner *et al.* (1967) coprecipitated Mn, Fe, Ni, Co, Cu, Pb, and Zn from sea water by adding 50 cm^3 dm^{-3} of 0.2 M KOH. The resulting alkaline earth metal hydroxides and carbonates were then collected by filtration. The sea water samples were analysed for Cu, Pb, and Zn before and after coprecipi-tation by linear sweep ASV at the hanging mercury drop electrode

(HMDE). Approximately 90% of the metals originally present in the samples were removed by the precipitation. No attempt was made to re-dissolve the precipitate in acid and to analyse it directly, even though the resulting electrolyte would have been suitable for direct polarographic analysis. Muzzarelli and Sipos (1971) used chitosan (deacylated chitin) to pre-concentrate Zn, Cd, Pb, and Cu from sea water samples and measured the efficiency of the pre-concentration procedure by ASV at the mercury-coated graphite electrode (MCGE). When chitosan powder was added to natural sea water ($100\ mg\ dm^{-3}$) approximately 80% of the zinc and 100% of the copper were removed from solution. Cadmium and lead were not scavenged as effectively, however, and about 50% of these metals remained in the filtrate after the chitosan flakes had been collected on a $0.45\ \mu m$ filter. The effectiveness of chitosan packed in a chromatographic column was also tested. A $100\ cm^3$ sample of filtered sea water was percolated through a $50 \times 10\ mm$ chitosan column at a flow-rate of $10\ cm^3\ min^{-1}$. Zinc and copper were collected with 100% efficiency whereas only 65% of the lead and 38% of the cadmium were retained by the column. Muzzarelli and Sipos (1971) also mention that chitosan can be eluted with solutions which are suitable as electrolytic media in voltammetry: lead and zinc with $2\ M$ ammonium acetate, copper with $0.01\ M$ EDTA and cadmium with potassium cyanide. Because the most abundant component of zooplankton consists of small crustaceans with chitinous exoskeletons, it becomes interesting to speculate on the roles of chitin and chitosan, and other similar substances, in determining the speciation of copper and other trace metals in the open oceans.

Perhaps the most extensive use of a pre-concentration method in a polarographic analysis was that of Abdullah and Royle (1972) and Abdullah *et al.* (1972), who measured Cu, Pb, Cd, Ni, Zn, and Co in a single aliquot of sea water by derivative pulse polarography after passage through an iminodiacetic ion-exchange resin (Chelex 100, Biorad Laboratories, Sunnyvale, Calif., USA). A $10\ cm^3$ volume of sea water was percolated at $4\ cm^3\ min^{-1}$ through a $50 \times 10\ mm$ column in the calcium form. The column was then eluted with $70\ cm^3$ of $2\ M$ nitric acid and the eluate was subsequently irradiated using a 1 kW mercury vapour source in order to eliminate any residual organic matter. The solution, which now consisted primarily of $Ca(NO_3)_2$ and smaller amounts of sodium and potassium, was evaporated to dryness with HCl to remove the NO_3^- and the resulting $CaCl_2$ precipitate was then redissolved in dilute HCl. This solution was suitable for the direct polarographic analysis of copper, cadmium, and lead. Consequently, the same solution was made approximately $0.2\ M$ in NH_3–NH_4Cl by the addition of ammonia solution. This basic medium was now suitable for the determination of nickel and zinc. However, it was found that the cobalt wave interfered with that of zinc. Nevertheless, because the reduction of zinc is reversible in this medium whereas that of cobalt is not, zinc could be determined by using a reverse sweep. Cobalt was finally determined independently by adding dimethylglyoxime to the solution and recording the adsorption wave of the Co–dimethylglyoxime complex which is linear with respect to the cobalt concentration. All measurements were quantified by the standard addition method. The precision of the method was approximately 6% for the determination of Cu, Pb, and Co and about 3% for Cd, Ni, and Zn.

From the polarographic point of view, certain generalities may be stated about pre-concentration steps. (1) They are effective for concentrating and storing trace elements. However, the concentrating agent is generally incompatible with polarographic analysis and must be removed by oxidation, evaporation, and/or acidification. In this respect, polarographic determinations are not competitive with AAS techniques, which can handle a much larger variety of solvents. (2) At the high concentrations achieved after pre-concentration, it is possible to carry the analysis through many chemical steps without incurring excessive contamination and hence it is possible to analyse for many more elements than would have been possible without pre-concentration. (3) Pre-concentration may obscure information about the natural speciation of the elements in sea water.

10.3.2. Analysis without chemical pre-concentration

Whitnack (1961) determined Cl^- using cathode ray polarography with Ag|AgCl and DME electrodes. This procedure was a variant of cathodic stripping, a technique by which an insoluble chloride is formed by oxidizing a metallic electrode (mercury or silver) at a positive potential in chloride media, followed by reduction of the film to the metal plus chloride. The reduction current is then proportional to the concentration of chloride in solution. Similar techniques are discussed by Brainina (1971). Brainina and Sapozhnikova (1966) also analysed for I^- in sea water by cathodic stripping. Their value for I^- in sea water was considerably greater than the values reported by other workers, suggesting that IO_3^- was also being determined (Whitfield, 1975).

In an imaginative series of papers, Berge and Brügman described polarographic analyses for almost all of the major constituents. Initially, Berge and Brügman (1969) described a procedure for the determination of Na, K, Ca, and Mg. A 10-cm^3 volume of sea water was diluted to approximately 25 cm^3 and 1 cm^3 of 0.9 M tetramethylammonium hydroxide (TMAOH) was added to precipitate $Mg(OH)_2$. The white solid was then filtered off and redissolved in 10 cm^3 of 0.1 M HCl. To this solution were then added 25 cm^3 of 0.1 M CdEDTA, 32.5 cm^3 of concentrated ammonia solution and 25 cm^3 of 2 M KNO_3. The volume was made up to 100 cm^3 and direct polarographic analysis was carried out for magnesium.

Sodium, remaining in the filtrate after precipitation of magnesium, was measured as follows. The initial filtrate was first diluted to 50 cm^3 and then 2 cm^3 of the solution were mixed with 20 cm^3 of a magnesium uranyl acetate solution, prepared according to the method of Kahane (1933). The resulting Na–Mg uranyl precipitate was collected on a sintered-glass filter and washed with 1 cm^3 of Fehling's reagent and 2 cm^3 of ethanol Finally, the solid was dissolved in 10 cm^3 of 0.1 M HCl, diluted, and the uranyl wave was measured by scanning between −0.25 and −0.7 V *vs.* SCE.

Combined calcium and potassium could be measured by collecting the filtrate from the Na–Mg uranyl acetate precipitation, precipitating excess of magnesium with 0.9 M TMAOH and analysing the filtrate directly by applying a sweep between −1.75 and −2.25 V *vs.* SCE. Addition of a solution of 0.1 M CdEDTA in 1×10^{-2} M TMAOH then complexed the calcium and a repeat of the voltage scan showed only the potassium wave. All of the major constituents measured in this manner were measured with a precision of 1–2%.

Berge and Brügman (1970) later described a procedure by which SO_4^{2-} in sea water was determined by adding an acidified $BaCrO_4$ solution and measuring the resulting CrO_4^{2-} ion. The precision of the method was 0.5%. A procedure for the measurement of SO_4^{2-} in sea water which was more precise was developed by Luther *et al.* (1978), who titrated SO_4^{2-} with $Pb(NO_3)_2$ and then measured the excess of lead polarographically.

Procedures for the measurement of Br^- and F^- were also developed by Berge and Brügman. Br^- was oxidized to BrO_3^- with sodium hypochlorite and the BrO_3^- wave was measured directly in basic solution by scanning between −1.00 and −1.92 V *vs.* SCE. The precision of the method was approximately 1% (Berge and Brügman, 1971). Finally, these workers (Berge and Brügman, 1972, 1973) determined F^- in Baltic sea water by an indirect procedure in which F^- is reacted with Zr-alizarin S under acidic conditions to release the quinoid form of the alizarin S. The latter is then reduced at the DME and the reduction wave can be measured by either d.c. or a.c. methods.

Whitnack (1961, 1966, 1973) used single-sweep polarography to analyse for several trace elements in natural and spiked sea water samples. In single-sweep polarography (Rooney, 1966; Meites, 1965) a rapid linear voltage sweep is applied near the end of the drop life and the resulting potential–current plot is displayed on the screen of an oscilloscope or on an *x*–*y* recorder. While this is a very sensitive technique, signals appear against a sharply sloping background, thus it is often used in a subtractive manner, i.e., twin equivalent cells are used, one containing the substance of interest in a supporting electrolyte and the other the supporting electrolyte only. The voltage sweeps to both cells are synchronized and the output of the second cell (containing only the supporting electrolyte) is subtracted from the first, theoretically eliminating most of the capacitative component.

Whitnack (1966, 1973) used a Davis Differential Cathode-Ray Polaro-trace (Model A 1660, Southern Analytical Instruments, Ltd., England) and recorded the traces on a Moseley (now Hewlett-Packard) x–y recorder. The twin DME used had drop times of about 7 s and voltage sweeps of 500 mV were applied from a pre-set initial potential. A mercury pool electrode served as reference. A 2 cm^3 volume of sample was placed in 5 cm^3 quartz cells and purged of oxygen with nitrogen which had been purified by passage over copper turnings at 450°C and saturated with water. Voltage sweeps were applied for 2 s at the end of the lifetime of each drop. After a few sweeps, which could be seen on the oscilloscope, a permanent trace was made on an x–y recorder. The initial potentials for the sweeps were set according to the demands of each individual analysis.

Initially, Whitnack (1961) used Copenhagen sea water to show that Cu, Pb, Cd, Zn, Ni, Co, and Mn produced measurable peaks in sea water at concentrations above 50 μg dm^{-3}. The peak potentials for these elements in sea water are reported in Table 3. However, at lower concentrations the peaks were unclear, and sharper signals could be obtained by displaying the instrumental output in the derivative mode. Even under these conditions a maximum sensitivity of only 5 μg dm^{-3} could be obtained.

Table 3. Peak potentials of some trace elements in sea water. [From Florence (1972), *J. Electroanal. Chem.*, **35,** 237–245]

Element	Volts *vs.* Hg pool*	Volts *vs.* SCE†
Bi		−0.09
Sb(III)		−0.16
Cu	−0.25	−0.24
Pb	−0.50	−0.48
Sn		−0.43§
In		−0.58
Tl		−0.62
Cd	−0.67	−0.69
Zn	−1.09	−1.05
IO$_3^-$	−1.09‡	
Ni	−1.17	
Co	−1.46	
Mn	−1.58	

* Reduction at DME, mercury pool = +0.058 V *vs.* SCE (Whitnack, 1966).
† Oxidation at thin mercury film rotating disc electrode (RDGCE) (Florence, 1972).
‡ Reduction at DME *vs.* SCE.
§ Oxidation at hanging mercury drop electrode (HMDE–Pt).

Whitnack (1973) also measured copper directly in sea water while at sea, and reported that the equipment worked well in normal weather, while rough weather raised the detection limit another order of magnitude. In light of today's estimate of the copper concentration of sea water (Brewer, 1975; Boyle and Edmond, 1975; Bender and Gagner, 1976), it appears that Whitnack's samples were contaminated and that single-sweep polarography is not sufficiently sensitive for the measurement of trace metals directly in sea water. Nevertheless, single sweep-polarography is an elegantly simple, extremely rapid technique and may find use in estuarine work, where trace metal concentrations are higher.

In his pioneering work, Whitnack (1961, 1966, 1973) also pointed out that I⁻ (after oxidation to IO_3^-), Cr(III), Cr(VI), and As(III) could also be measured in sea water by single-sweep polarography with about the same detection limit as for Cu and Zn. As(III) could be determined directly after addition of $HClO_4$ while As(V) could be measured after addition of solid pyrogallol to the sea water as the As(V)–pyrogallol complex. Similarly, Cr(III) gave a well defined polarogram in the presence of CNS⁻ and acetic

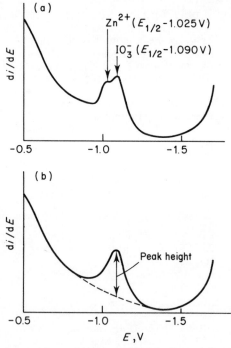

Figure 2. (a) Differential pulse polarogram of raw, filtered (0.45 μm) sea water. (b) Same as (a), except that an EDTA solution was added to a final concentration of 2.5×10^{-4} M. [Reprinted with permission from *Deep Sea Res.*, **21**, Herring and Liss (1974). Copyright © 1974 Pergamon Press Ltd.]

acid while Cr(VI) could be measured in NaOH–sea water media. Finally, Whitnack (1973) was able to detect IO_3^- at sea in freshly collected samples.

Herring and Liss (1974) utilized the enhanced sensitivity of differential pulse polarography and developed a new technique for the analysis of IO_3^- and I^- in sea water. Following Whitnack's (1966) suggestion, they verified that IO_3^- was directly reducible in sea water at the DME. However, the peak produced by the direct reduction of zinc interfered (Figure 2a). Addition of EDTA to a concentration of 2.5×10^{-4} M shifted the zinc peak beyond the hydrogen current, and removed the interference (Figure 2b). The procedure consisted simply of adding EDTA, bubbling a 15 or 20 cm^3 sample for 15 min with ordinary bottled nitrogen, and recording the polarogram. A standard addition of IO_3^- was made and the polarogram re-run. The increase in peak height resulting from the standard addition was used to calibrate the method. Replicate analyses gave a precision of 2.5% (one standard deviation).

Total iodine was determined in a separate aliquot of the sample by irradiating the water with a high-energy UV lamp (Armstrong *et al.*, 1966) to convert I^- and other reduced iodine compounds to IO_3^-, which was then measured in the manner previously described. If the sample was high in organic matter, a few drops of 30% H_2O_2 were added to complete the oxidation. The difference in the IO_3^- concentrations obtained before and after UV irradiation was assumed to be largely I^-. Herring and Liss (1974) carried out some of their determinations in the Santa Barbara Channel,

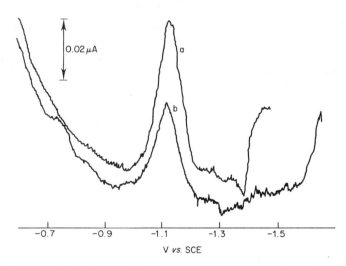

Figure 3. Differential pulse analysis of IO_3^- in sea water performed at sea: current 0.1 μA full-scale, sweep 2 mV s^{-1}, modulation amplitude 25 mV. (a) Raw sea water; (b) sea water plus IO_3^- standard. Concentration of IO_3^- in sea water 52.6 μg dm^{-3}.
Courtesy of J. R. Herring

California, aboard a small vessel, demonstrating that it is possible after all to use a DME on a ship, at least in a calm sea. Figure 3 shows such a polarographic determination carried out at sea. The lower plot gives the trace produced by the IO_3^- standard addition.

Herring and Liss also investigated the effect of pH on the peak potential and peak height of the IO_3^- wave. They observed that the peak potential did not change in the pH range 5–8 whereas the peak current decreased as the pH fell. Below pH 5, however, the peak potential shifted to about -0.4 V *vs.* SCE. Petek and Branica (1968) used the pH dependence of the IO_3^- peak to separate it from zinc. According to their technique, sea water is acidified to pH 3 with HCl and analysed for both IO_3^- and zinc by pulse polarography. Calibration is then effected by the standard addition technique. Adriatic Sea samples analysed in this manner gave values in the range 28–141 $\mu g\,dm^{-3}$ for Zn and 36–172 $\mu g\,dm^{-3}$ for IO_3^-.

10.4. ANALYSIS BY ANODIC STRIPPING VOLTAMMETRY

Because of its high sensitivity, ASV has been widely used in oceanography and today instrumentation for ASV and polarography is found in most marine laboratories. Over the last 15 years the literature dealing with stripping analysis has evolved from descriptions of single determinations illustrating some new facet of the analysis to full reports in which the methodology plays a secondary role to the elucidation of some natural phenomena. Predominantly, two types of electrodes have been used: mercury drops either extruded from capillaries or hung on platinum wires and thin mercury films plated on carbon substrates (Chapter 2, Table 3). Both kinds have inherent advantages and disadvantages which make them more or less suitable in specific applications. Both have been used at sea to a moderate extent. The modulation of the stripping wave form, designed to decrease the capacitance current, has brought improvements in sensitivity and has shortened analysis times. Finally, automated instrumentation has made it possible to carry out analyses on large numbers of samples with adequate precision.

10.4.1. Hanging mercury drop electrodes (HMDE)

The platinum contacted HMDE

The direct application of ASV to marine waters may originate with the work of Ariel and Eisner (1963), who used a hanging mercury drop electrode attached to a platinum wire (HMDE–Pt) to analyse for Zn and Cd in Dead Sea brine. The HMDE–Pt electrode (Shain, 1961; Ross *et al.*, 1956) is made by imbedding a platinum wire in flint-glass tubing, sanding the end flat to expose the disc-like platinum cross-section, and plating a mercury film on to the exposed metal. A mercury drop, formed with an ordinary polarographic capillary, is placed in a small spoon or ladle and attached to the amalgamated platinum disc. When clean, this electrode has excellent mechanical stability although it possesses a lower hydrogen over-voltage than a pure mercury electrode, possibly because some platinum is dissolved in the mercury drop. Also, when used in natural sea water, the disc tends to foul as new mercury drops are transferred on or off. Eventually, fresh drops will no longer stick on and the electrode must be re-sanded and freshly plated.

Zinc in Dead Sea brine was analysed using 3–8 min pre-electrolysis periods. In order to obtain the adequate sensitivity, fast voltage sweeps were used and the stripping curves were displayed on a dual-beam oscilloscope. Plating times of 30–90 min were required to produce enough signal current to be displayed on an ordinary strip-chart recorder. Copper in the brine could not be measured in the same way as the other metals because the copper oxidation peak coalesced with that of mercury.

Therefore, Ariel *et al.* (1964) devised the technique of medium exchange to separate the two metals. Copper was plated from Dead Sea brine at -0.48 V *vs.* SCE for 10–15 min, then, while the potential was changed to -0.6 V, a 0.5 M NH_3–0.5 M NH_4Cl solution was substituted for the brine, and the potential was swept. Copper in this new medium oxidized at -0.48 V and could be clearly resolved from mercury. The metals measured in Dead Sea brine were found to be present at the following levels: Zn, 5.7×10^{-7} M $\pm 2.5\%$; Cd, 4.4×10^{-8} M $\pm 0.8\%$; and Cu, 4.5×10^{-8} M $\pm 15\%$. To date, the technique of medium exchange has not been full utilized for the analysis of sea water where it may be useful to separate Ni from Zn, Sn from Pb, Tl from Cd, etc. Although Zieglerova *et al.* (1971) and Lieberman and Zirino (1974) have proposed cells for this purpose, routine analysis by medium exchange is still in the future.

Zirino and Healy (1972) improved the sensitivity of hanging drop strip-ping analysis by employing two HMDE–Pt electrodes in a subtractive mode: only one of the working electrodes receives the applied potential during the pre-concentration step. The second working electrode is activated only during the stripping step and its signal, representing essentially the capacity current, is subtracted from the first. This allows high amplification of the Faradaic component. The gain in sensitivity obtained by this technique allowed Zirino and Healy (1971) to analyse routinely for zinc in open ocean water. The gain is comparable to that obtained with differential pulse methods (Flato, 1972), although the nature of the signal is different. In differential pulse methods the electroactive metal is repeatedly oxidized from and reduced into the drop during the scan (see Chapter 2, Table 2). This makes it difficult to measure the actual charge transferred from the area under the peak. Areas obtained by differential linear sweep, on the other hand, are a true measure of the charges passed during the oxidation.

Zirino and Healy (1972) mounted their cell in a framework suitable for working at sea. Their objective was to achieve geometric stability, (i.e., maintaining a constant diffusion layer thickness), to facilitate sample handling and to reduce contamination. They mounted the electrodes on a moveable slide so that the HMDE–Pt electrode could be lifted out of solution (sometimes necessary to remove gas bubbles from the mercury drops) without permanently changing the geometry of the cell. Stirring was achieved with a paddle attached to a constant-speed motor mounted above the cell. This facilitated the transfer of samples and made possible the use of a 50 cm³ Teflon-FEP beaker as the sample holder. Because purging of the sample with an inert gas to remove oxygen caused natural samples to froth, only the lower part of the Teflon-FEP beaker was filled. This avoided the scavenging of dust from the cover, a source of contamination.

Zirino and Healy (1972) also controlled the pH of the analysis by using a carbon dioxide–nitrogen mixture as the sparging gas and equilibrating the sample at the new partial pressure of carbon dioxide. Zinc was routinely analysed at pH 5.6 (which yielded high peak currents without undue H^+ interference) by using a mixture of 20% carbon dioxide in nitrogen.

Abrupt pH changes in the cell, such as those occurring between the analysis of acidified and natural samples, were avoided because these led to sorption and desorption of metals from the cell walls. Before analysis all sample containers, (storage bottles, cell parts, etc.) were rinsed with dilute (1 M) HNO_3, followed by de-ionized water. Thereafter, all surfaces were equilibrated with sub-samples of the sample for several hours. These were then discarded. This procedure has also been adopted by Nürnberg *et al.* (1976). Although these steps were time consuming, it was necessary to perform them only once for a particular sample locality as long as the pH was not changed. Open ocean samples were found to have a narrow concentration range and it was unnecessary to re-equilibrate each sample.

The mercury contacted HMDE

Notwithstanding the relative success of Zirino and Healy's (1971, 1972) efforts to analyse a large number of samples at sea, the HMDE–Pt electrode was cumbersome to prepare and short-lived, at sea or in the laboratory. Therefore, efforts were made to develop an HMDE with the same properties as the classic DME. Macchi (1965) used narrow-bore capillaries to produce a DME with a very long drop life (up to 68 s) to analyse for zinc in Mediterranean Sea water. The pre-electrolysis potential (−1.30 V *vs.* SCE) was set for the 50 s of a drop life and then stirring was stopped. The solution was allowed to come to rest for the next 10 s. Finally, zinc was stripped from the amalgam with a rapid linear potential sweep ($200 \, mV \, s^{-1}$) and the peak was recorded on an oscilloscope. The instrument used (Model 451, AMEL, Milan) was essentially designed around this analysis and incorporated the oscilloscope and background-correcting circuitry (essential for fast sweeps), as well as the programming sequence to carry out sequential analyses. Zinc concentrations were measured by the standard addition method and a sensitivity of 3×10^{-9} M was claimed. The coefficient of variation was 4.5% at the 3 $\mu g \, dm^{-3}$ level. Even in the light of today's modulated techniques, Macchi's (1965) instrument was remarkable in that it combined the inherent reproducibility of the DME with high sensitivity (comparable to today's differential pulse methods) and an extremely rapid analysis time.

Prior to Macchi's work, Kemula and Kublick (1963) developed an extruded HMDE which made possible prolonged electrolysis times and allowed easy renewal of the mercury drop. In this electrode (Figure 4), forms of which are available commercially, mercury is forced from a reservoir through a capillary by a micrometer-controlled plunger. The electrical contact is made through the plunger to a screw mounted on the outside of the micrometer capillary. This electrode is easy to use and essentially requires only that a clean capillary and clean mercury be employed. Cleaning may be effected by filling a capillary with concentrated HNO_3, setting it in a Pyrex beaker containing enough concentrated HNO_3 to cover the tip

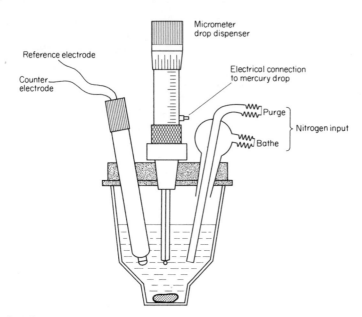

Figure 4. Micrometer controlled hanging mercury drop electrode (Kemula type) in a typical cell configuration

and heating on a hot-plate at a slow boil for 30 min. The electrode is then rinsed with pure water, air dried (a vacuum pump is very effective for this), and flushed with siliconizing fluid. Finally, the capillary is air dried again and filled with mercury. This is easily accomplished by filling the reservoir on the capillary with a plunger-type pipete until the mercury meniscus extends just above the reservoir. When the micrometer head and capillary are assembled, the plunger tip forces mercury through the capillary and the electrode is ready to use. Problems with this electrode include the changing of drop size with changes in room temperature which cause mercury in the reservoir to contract or expand, such as in a thermometer, air leaks between the mercury reservoir and the O-ring seal around the plunger, and fouling of the tip caused by the ingress of sea water into the capillary while changing drops. Organic substances in the sea water eventually cause the mercury thread to break in the capillary tip and the drop to fall into the solution. This problem is best cured by replacing the capillary with a clean one. Another consideration is the possible loss of metals in the capillary thread during electrolysis. However, Moorehead and Doub (1977) found that only a small loss of cadmium occurs during a typical analysis.

The advent of the Kemula electrode quickly led to some new instrumental applications and gave rise to extensive environmental studies in the Mediterranean, the Baltic, and the North Seas. Rojahn (1972) indicated that the sensitivity of voltammetric determinations in sea water could be enhanced if the stripping step was performed in the d.c. mode. He used a Tacussel PRG3 potentiostat and a Metrohm EH10 HMDE to analyse for Zn, Cd, Pb, and Cu in samples from the Oslofjord. Measurement of Cu, Pb, and Cd in acidified (pH 2) samples required only a 2 min pre-electrolysis time. Stripping was conducted at $150 \, mV \, min^{-1}$ with a superimposed a.c. signal of 20 mV at 15 Hz and a zero demodulation phase angle (see Chapter 2, Table 1). Zinc was measured in acetate buffer (pH 4.6) in a similar fashion. A feature of a.c. ASV is that it is able to resolve copper peaks from the mercury oxidation current more sharply than the linear sweep technique. This effect goes beyond the reduction of the capacitative component (Chapter 2). It probably occurs because the oxidation rate of copper from the bulk amalgam is slow and thus more responsive to the slow rate of sweep employed by the a.c. method. Also, copper at the surface of the electrode is repeatedly plated on and off, yielding larger cumulative currents.

A similar effect is observed when the stripping is carried out by pulse and differential pulse methods (Flato, 1972; Siegerman and O'Dom, 1972; Copeland et al., 1973). Donadey (1969) and Donadey et al. (1972) analysed for Zn, Cu, Pb, and Cd by pulse ASV using 3 min electrolysis times. Pulse modulation was effective in eliminating much of the capacity current and in sharply resolving copper from mercury.

Zirino (unpublished) used differential pulse ASV at the HMDE to analyse for copper at sea. Surface samples were collected while underway from a towed PVC hose, the inlet of which was kept just below the surface by a specially designed planing device. A peristaltic pump (with silicone rubber tubing in the pump head) was placed in the laboratory and was operated continuously for 24 h per day. Sea water from the inlet traveled through about 50 m of tubing to the ship's laboratory. Discrete samples were then obtained for analysis from this source. The instrumentation included a PAR 174A polarographic analyser and a Metrohm EH10 HMDE in a Metrohm polarographic vessel (Princeton Applied Research Corp.). The reference electrode was a double-junction Ag|AgCl electrode with sea water in the outer junction; platinum wire was used as the counter electrode. The solution was stirred with a Teflon-covered stirring bar coupled to a synchronous stirrer. Acid additions were made with concentrated HCl. Several types of measurements were made. The initial potential was set at -0.9 V and a current–voltage plot was recorded. Then the pH was lowered to 2 by adding HCl and another current–voltage plot was recorded. An addition of a standard copper solution was then made to the cell and the analysis was repeated. Finally, a subsample which had been made 3×10^{-5} M in ethylenediamine and allowed to stand for approximately 1 h was analysed (Figure 5). Peak areas were measured by photocopying the traces, cutting the peaks out of the copies and weighing them. Copper concentrations were measured by the standard addition method. The results of

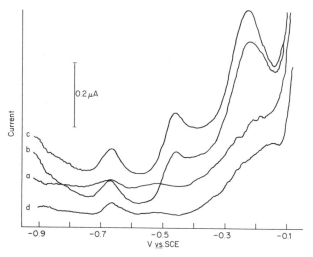

Figure 5. Shipboard analysis of sea water by differential pulse ASV at the HMDE: current 1 μA full-scale, initial potential -0.9 V, sweep 5 mV s^{-1}, modulation amplitude 50 mV. (a) Raw sea water, pH 8; (b) raw sea water, pH 2; (c) standard addition of 0.18 μg dm^{-3} of copper; (d) raw sea water made 3×10^{-5} M with ethylenediamine

Table 4. Shipboard measurements of Cu in the surface waters of the Eastern Tropical Pacific Ocean. The maximum estimated error is 42% using standard additions of 0.18 $\mu g\,dm^{-3}$ [equivalent to 2 mm^3 (2 μl)]. A dash indicates values below the limit of detection

	Copper concentration, $\mu g\,dm^{-3}$		
Position	pH = 8	pH = 2	With 3×10^{-5} M ethylenediamine
24°N, 111°W	0.15	0.26	0.16
22°N, 109°W	0.12	0.21	0.22
21°N, 109°W	0.10	0.64	0.11
19°N, 106°W	0.21	0.29	0.10
18°N, 105°W	0.06	0.16	0.08
16°N, 103°W	0.16	0.34	0.16
15°N, 103°W	0.23	0.41	0.22
13°N, 101°W	0.13	0.33	0.24
12°N, 100°W	—	0.17	—
10°N, 98°W	—	0.39	—
7°N, 95°W	0.19	0.14	—
6°N, 95°W	0.06	0.09	0.07

the analyses are summarized in Table 4. No particular precautions were required for the shipboard determinations other than ascertaining that the collection system was clean and making sure the cell walls were equilibrated with aliquots of each new sample before analysis (Zirino and Kounaves, 1980).

Kremling (1973) made shipboard measurements of Zn, Cd, Cu, and Pb by linear sweep ASV during two cruises in the Baltic Sea. Samples were collected with a 'metal-free' sampler (Hydrobios, Kiel) and transferred to 500 cm^3 polyethylene bottles which had been rinsed once with distilled HNO_3, twice with triply distilled water, and three times with portions of the sample. The time of storage in the bottles before analysis did not exceed 15 h. Samples were analysed with and without filtration (0.4 μm Nucleopore, General Electric Corp., USA). In order to compensate for the ship's motion, the electrolysis cell and HMDE (Metrohm 401) were mounted on a platform which floated on a high-viscosity oil. A Metrohm Polarecord E261 was used to apply the potential and the analyses were carried out in a 20 cm^3 thermostated quartz cell. At pH 8.2, no sorption-related changes in concentration were noted for a sea water sample after 30 min of contact with the cell. The analysis was initiated by purging a 15 cm^3 sample for 10 min with water-saturated nitrogen. This raised the pH to 8.2. For zinc, an electrolysis potential of -1.25 V *vs.* SCE was applied for 5 min, during which stirring was maintained with a Metrohm synchronous motor. The analysis for Cu, Cd, and Pb was carried out at -1.00 V *vs.* SCE and required 15–20 min of electrolysis. For all metals the voltage sweep was initiated at 33 mV s^{-1} after a rest phase of 30 s. Current–voltage curves were measured for the samples and for the samples plus standard additions on a strip-chart recorder. Concentrations were calculated in the usual manner. Analyses of ten replicates of a sample which contained 24.4 μg dm^{-3} of Zn, 0.99 μg dm^{-3} of Cd, 1.7 μg dm^{-3} of Pb, and 1.8 μg dm^{-3} of Cu produced relative standard deviations of 17, 23, 19, and 11%, respectively. This measure of precision is similar to that found by Zirino and Healy (1972), who attributed the large error to handling of the sample before its introduction into the electrolysis cell. Kremling (1973) remarked that copper peaks were poorly resolved from the mercury dissolution curve and that he was able to shift the peak cathodically by adding to his samples 2 cm^3 of a 0.3 M solution of triethanolamine in 3×10^{-2} M KOH.

Brügman (1974a,b, 1977) carried out detailed analyses of Zn, Cd, Pb, and Cu in waters from the Baltic Sea and northwest Africa. Determinations were made in the laboratory by linear sweep voltammetry at the HMDE (Radiometer, Copenhagen) with a GWP 563 Polarograph (Akademiewerk-statter, Berlin). Samples were collected with a plastic sampler (Hydro-bios, Kiel) stored in pre-cleaned polyethylene bottles, quick-frozen and analysed in the laboratory. Prior to the analyses, detailed examinations of the effect on the peak height of scan rate, stripping rate, mercury drop surface area, plating time, and initial potential were carried out (Brügman, 1974b). For the measurement of Zn, Cd, and Pb the electrolysis cell was rinsed with dilute HNO$_3$, triply distilled water and finally with the sample. A 50 cm^3 volume of sea water was then purged of oxygen for 8 min while stirring was maintained. An initial potential of -1.2 V was applied for 1–5 min followed by a 30 min rest period. Zinc was oxidized out of the amalgam using a potential sweep of 33 mV s^{-1}. The scan was then stopped for 2 min at -0.75 V *vs.* SCE and the current sensitivity of the instrument was increased. Cd, Pb, and Cu were then oxidized from the amalgam. Separate analyses for Cd, Pb, and Cu were carried out by plating at -0.8 V *vs.* SCE for 5–30 min. Concentrations were measured by means of the standard addition method. Most trace metal determinations were conducted at the natural pH of sea water, but tests for speciation were also conducted. Samples were measured before and after filtration and before and after oxidation with strong UV light following the procedure of Armstrong *et al.* (1966).

Brügman (1978) also developed a polarographic procedure for the analyses of particulate samples collected in northwest African waters. Plankton samples were collected with a 200 μm net, dried at 60°C to a constant weight, and homogenized in a plastic vessel. From 0.2 to 0.6 g of the dried substance was then ashed with concentrated HNO$_3$ in a Teflon bomb and the sample was heated in a muffle furnace at 450°C. After cooling, the sample was treated with dilute HNO_3. The samples were then analysed by linear sweep ASV in the manner described above.

Duinker and Kramer (1977) conducted an extensive study on the speciation of dissolved Zn, Cd, Pb, and Cu in surface Rhine river and North Sea water using differential pulse ASV. Samples were collected in a polypropylene bucket and filtered aboard ship through acid-washed 0.45 μm Millipore filters, with a specially constructed Teflon apparatus (Duinker and Nolting, 1977). The filtrates were then stored in pre-cleaned, high density polyethylene bottles and frozen for subsequent analysis in the laboratory. Subsamples were also measured by the APDC/MIBK extraction procedure (Brooks *et al.*, 1967) for the purpose of comparison with ASV.

Stripping determinations were made with a PAR 174 Polarographic Analyser, which for a portion of the work was equipped with a Model 315 Automated Electro-analysis Controller. The cell consisted of the Metrohm 410 electrode with Metrohm Ag|AgCl reference, and a platinum counter electrode placed in a Metrohm thermostatted cell (25.0 $\pm$ 0.1°C). Before an analysis, the cell was flushed once with 1 M HNO_3, twice with quartz-distilled water, and twice with the sample. Tank nitrogen was passed over BASF R3-11 catalyst and bubbled through the solution for 10 min with stirring. To eliminate the possible carryover of metal in the mercury thread from a previous analysis, two drops of mercury were discarded between determinations. It was also felt that the HMDE should be conditioned and therefore an initial 60 s electrolysis–stripping cycle was made but the data so obtained were not utilized. Samples were measured at the natural pH and at pH 2.7 (6.1 for zinc). Only a fraction of the electroactive concentration of Zn, Cu, and Pb detected at the low pH was measurable at pH 8.1. On the other hand, cadmium peak currents showed no pH dependence in accordance with the observations of Zirino and Healy (1972). The general precision (one standard deviation) of the method was 16% for cadmium, 17% for lead, and 10% for copper, which compared favourably with the precisions reported by others (Smith and Redmond, 1971; Kremling, 1973, Brügman, 1974b).

10.4.2. Thin-film mercury-coated graphite electrode (MCGE)

Matson *et al.* (1965) found that the sensitivity of the ASV analysis of sea water could be increased considerably if the hanging mercury drop could be replaced with an electrode made by electroplating a mercury film on an inert substrate. The inherent advantages of such a thin film electrode are: (1) The mercury surface area is increased over that of the drop; (2) the concentration of metal in the film is three to four orders of magnitude greater than that in bulk electrodes; (3) oxidation out of the film takes place virtually instantaneously; and (4) the peak current is directly proportional to the scan rate, rather than to the square root of the scan rate as with the mercury drop electrode (DeVries and Van Dalen, 1964, Roe and Toni, 1965; Perone and Bromfield, 1967). The thin-film electrode, when used in the linear sweep mode, gives sharp, well resolved peaks against a uniformly sloping (and easily correctable) baseline. A further advantage is that the oxidation peak of copper is easily separated from the dissolution current of mercury. This occurs because the higher concentration of copper in the thin film causes its peak potential to be cathodic over that for the drop.

Matson *et al.* (1965), Matson (1968), and Hume and Carter (1972) developed such a thin-film electrode by impregnating spectrographic grade graphite rods with paraffin wax (Baker, m.p. 56–58°C), under vacuum. Ceresine wax (Gilbert and Hume, 1973) and Sonneborn wax (Clem *et al.*, 1973) were used by later workers. Once the rod was fully impregnated, the end was scraped clear of wax with a knife and polished with 600-grade (very fine) carborundum paper. Finally, it was polished with paper tissue. The mercury film was applied in the same vessel as that used for ASV analysis. The cell was filled with 0.01 M KCl solution containing $HgCl_2$. A potential of -1.0 V was applied to the graphite electrode for a specified time which varied with the surface area of the electrode and concentration of mercury in solution, e.g., a $200 \, mm^2$ electrode was plated from a 1×10^{-4} M $HgCl_2$ solution for 15 min. Stirring was maintained by bubbling nitrogen for 15 min. The thickness of the mercury film was found to be critical and a mercury coverage of $2 \times 10^{-9} \, mol \, mm^{-2}$ was recommended. Microscopic analysis (Matson, 1968) revealed that the film consisted of separate droplets of mercury attached to the bare graphite and separated from each other by the wax which filled the pores of the graphite matrix. A coverage of less that $2 \times 10^{-9} \, mol \, mm^{-2}$ resulted in incomplete coverage of the graphite substrate and a lowering of the hydrogen overvoltage, as well as causing the peaks to broaden, approaching the characteristics of a pure graphite electrode. Similarly, excessive coverage caused the Hg to coalesce into large, unstable hemispheres and to leave bare graphite behind. The characteristics of the electrode and a critical review of its applicability to the detection of chemical speciation in natural waters has been given by Hume and Carter (1972).

The mercury-coated graphite electrode (MCGE) was found to have a short lifetime and later workers found that the replating of the mercury film on the graphite substrate became a daily necessity. Exposure of the MCGE to small quantities of oxygen, as would occur between the transference of sea water samples, could also result in a sudden loss of sensitivity. This was attributed to the oxidation of the mercury film to calomel.

The Matson (1968) cell has seen wide applicability even at sea and has been described often (Allen *et al.*, 1970; Hume and Carter 1972; Gilbert and Hume, 1973; Whitfield, 1975). Briefly, it consists of a quartz beaker equipped with a Teflon stopper. The MCGE, counter, and reference electrodes and the conduits for purging and blanketing the sample with an inert atmosphere are inserted through openings in the stopper. The counter electrode is a platinum wire sheathed in a quartz tube fitted with a Vycor plug. This prevents contamination of the sample with oxidation products, such as chlorine, which form at the platinum wire during electrolysis. Similarly, the reference electrode is simply a silver wire (coated with silver chloride) sheathed by another quartz tube fitted with a Vycor plug. The Vycor plug allows for the drainage and filling of sea water between samples. The Ag|AgCl wire reference electrode in sea water was found to have essentially the same potential as that of a saturated calomel electrode (SCE) and to be stable with time (Gilbert and Hume, 1973). Ag^+, which forms during the analysis, is prevented from entering the sample by the Vycor plug. The same approach was used by Zirino and Healy (1972).

Following Matson's work, the use of the MCGE in the analysis of marine waters was investigated in a series of postgraduate studies by Fitzgerald (1970), Gilbert (1971), Seitz (1970), and Huynh-Ngoc Lang, (1973). Fitzgerald (1970) conducted a study designed to establish the amounts of various copper species in waters near Falmouth, Mass. (USA). One dm^3 samples were collected in a pre-cleaned glass sampler, following the usual oceanographic protocol, and filtered. The filtrates were then stored in high-density polyethylene jerry cans until they could be analysed in a shore-based laboratory. The analysis was completed eleven days after collection. Several types of determination were made. Copper was measured at the natural pH, at pH between 2 and 3, and after acidification to pH 2–3 plus oxidation with a strong UV lamp (Armstrong et al., 1966). For the analyses, the plating potential was set at $-0.90\,V$ *vs.* Ag|AgCl (SCE), the electrolysis time was 10–40 min, and the sweep rate was $16.7\,mV\,s^{-1}$. The instrument was a Heathkit controlled-potential polarograph coupled to a Moseley x–y recorder. All analyses were carried out by the method of standard additions. Previous work had shown that the analysis could be carried out with a precision better than 10%. Studies on Zn, Cd, Pb, and Cu were also carried out in the Sargasso Sea. Sea water samples were analysed shortly after collection by the technique outlined above. The concentrations of Zn, Cu, and Cd measured were similar to those reported by other workers (Spencer and Brewer, 1969). However, values for lead were an order of magnitude higher than those reported by Chow and Patterson (1966) and Tatsumoto and Patterson (1963).

Seitz (1970) explored several new applications of the MCGE to the analysis of sea water and studied the interference of nickel and copper on zinc and that of silver on copper. The analyses of bismuth and thallium in sea water were shown to be feasible if bismuth was present at concentrations above $10^{-11}\,M$ and if the thallium peak could be resolved from the cadmium peak. Phase-selective a.c. ASV was shown to increase the sensitivity of lead and cadmium determinations. On the other hand, the determination of copper suffered from interfering a.c. currents believed to be caused by adsorption and desorption of Cl^- and Br^-.

Gilbert (1971) and Gilbert and Hume (1973) developed a procedure for the analysis of bismuth and antimony using the Matson *et al.* (1965) electrode. Graphite rods, 6 mm in diameter, were impregnated with molten ceresine wax under vacuum and prepared as previously described. The active graphite surface was then plated with mercury in sea water media to give a coverage of 5.0×10^{-9} mol mm^{-2}. The plating potential was maintained greater than -0.5 V *vs* the Ag|AgCl wire to ensure coverage of all of the graphite sites (Hume and Carter, 1973).

Bismuth was measured in sea water made 1 M in HCl by bubbling gaseous HCl through the sample until the appropriate weight increase was registered. After cooling, 20.0 cm^3 of the sample were transferred to the electrolysis cell and a potential of -0.40 V *vs.* Ag|AgCl was applied for 15 min. Peaks were registered on an x–y recorder by scanning to -0.10 V *vs.* Ag|AgCl at a rate of 17 mV s^{-1}. A specially built current-compensating circuit subtracted the large capacitative component present. The bismuth concentration was measured by the standard addition method. The sensitivity of the method was judged to be approximately 2×10^{-11} M, and the precision was 3% at the 0.1 μg dm^{-3} level.

Antimony was measured by acidifying another subsample to 4 M in HCl, again with gaseous HCl. The electrolysis was repeated but at -0.5 V *vs.* Ag|AgCl. Stripping and calibration were carried out similarly to those for bismuth. Antimony could be measured at the 0.1 μg dm^{-3} level with a precision of 20%. Both Sb(III) and Sb(V) are reduced to elemental antimony under these conditions, so that the original oxidation state cannot be established from this determination. Because antimony and bismuth have similar oxidation potentials in 4 M HCl, and because antimony is generally more abundant in sea water, they cannot be distinguished; thus plots of added antimony concentrations *vs.* peak area give the concentration of antimony plus bismuth. The previously measured concentration of bismuth must then be subtracted from the total to give the antimony concentration.

The analysis is made possible by the fact that antimony is electroinactive in 1 M HCl and does not interfere with bismuth. Gilbert and Hume (1973) tested this premise and concluded that at the concentrations present in sea water, 'there is no interference by a 9-fold excess of antimony' These workers also tested for interferences from copper and silver and found that these elements did not interfere at natural concentrations.

Samples were collected in Massachusetts coastal water, Gulf Stream water from the Florida Keys and Pacific Deep water from the Baja California Seamount Province in pre-conditioned 1 dm^3 polyethylene bottles containing sufficient HCl to make the samples 0.1 M in acid. Within 3 days additional acid was added to bring the acid content to 1 M. The samples were then stored until they could be analysed. As a check on their storage procedure, Gilbert and Hume (1973) collected six replicate samples of sea water and acidified four of them to 1 M with HCl shortly after collection, while the fifth sample was acidified only after 6 weeks and the sixth sample was stored at the natural pH for eight weeks and then transferred to a new bottle and acidified to 1 M. The results of this test indicate that essentially all of the bismuth in sea water is adsorbed on the walls of polyethylene containers during prolonged storage and that this loss can be recovered by acidification. Antimony, on the other hand, is adsorbed to a much smaller extent, in agreement with the observations of Robertson (1968).

The results of the analyses are presented in Table 5. The bismuth values obtained by ASV are in general agreement with values obtained by Portman and Riley (1966) for the Irish Sea and English Channel. Similarly the antimony values lie in the range of values reported for South African waters by Brooks (1960) and for oceanic averages reported by Shutz and Turekian (1965). A direct comparison between the ASV technique and neutron activation analysis for antimony was also carried out (P. G. Brewer and D. W. Spencer, 1970, unpublished work). The results were in agreement within experimental error.

Huynh-Ngoc Lang (1973) conducted an investigation of the spatial and temporal variability of zinc species in the Mediterranean Sea near Monaco (Baie des Anges) and in the Var River estuary using the MCGE and cell described by Matson *et al.* (1965) and Matson (1968). The study was preceded by a detailed investigation of the experimental variables which

Table 5. Bismuth and antimony contents of sea water. [From Gilbert and Hume (1973), *Anal. Chim. Acta.*, **65**, 451–459]

Sample	Bi, μg kg^{-1}	2σ ($n = 3$)	Sb, μg kg^{-1}	2σ ($n = 3$)
Boston Light Ship	0.015	±43%	0.18	±45%
Pacific Deep	0.040	±42%		
Caribbean Deep:				
Sample 1			0.48	±15%
Sample 2			0.39	±19%
Bahia Honda Key:				
Sample 1	0.090	±6%	0.40	±16%
Sample 2	0.086	±31%	0.48	±20%
Sample 3	0.094	±4%	0.43	±28%
Sample 4	0.080	±25%	0.44	±13%
Sample 5	0.075*	±9%	0.32*	±14%
Sample 6A	0.008†		0.27†	±8%
Sample 6B	0.04‡			

* Acidified after storage for 6 weeks at the natural pH, then analysed.
† Transferred to a new bottle after storage for 8 weeks at the natural pH, then acidified and analysed.
‡ Empty bottle from sample 6A extracted with 1 M HCl and analysed.

affected the analysis. Lang chose to plate the mercury film out of a 0.005 M H_2SO_4 solution 8×10^{-5} M in $Hg(NO_3)_2$ rather than sea water because he had noted that mercury precipitated in the natural medium. A similar observations was made later by Lieberman and Zirino (1974). A potential of -0.2 V *vs.* Ag|AgCl applied for 2 h produced a mercury film which maximized peak height and resolution. Apparently the problems caused by using a low overvoltage, discussed by Hume and Carter (1972), were not encountered. Huynh-Ngoc Lang (1973) critically examined the effect of purging time, plating potential, pre-electrolysis time, cell temperature, nitrogen sparging, and sweep rate on the zinc peak current and adopted the following experimental parameters: electroactive surface, 50 mm^2; pre-electrolysis potential, -1.25 V; pre-electrolysis time, 10 min; flow-rate of nitrogen, 40 cm^3 min^{-1}; quiescent period, 50 s, and sweep rate, 0.2 V min^{-1}. The effect of oxygen was also considered and it was determined that oxygen in solution raised the detection limit of the analysis (Figure 6). Similarly it was found that the pH of the solution increased as a function of the electrolysis time, presumably from the reduction of oxygen present as impurities in the purging gas, or diffusing into the cell during electrolysis. Huynh-Hgoc Lang also studied interferences among Zn and Cd, Cu, and Pb which plated simultaneously during analysis, and carried out several tests to determine the speciation of zinc.

Gardner and Stiff (1975) analysed Zn, Cd, Pb, and Cu in estuarine and sewage effluent water by using the MCGE in a slightly different manner than other investigators. Because many of these samples had a low ionic strength

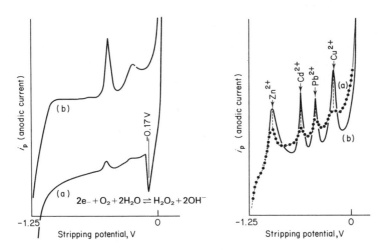

Figure 6. Diminution of sensitivity and deformation of peak currents caused by oxygen in sea water. Sample in right-hand figure contains additions of 1 μg dm^{-3} of zinc, cadmium, lead, and copper. (a) Natural sea water, without N_2 purge; (b) after removal of O_2. After Huynh–Ngoc Lang (1973)

and contained unknown organic materials which could interfere with the conventional partial electrolysis followed by the standard addition method, they chose to use exhaustive electrolysis and to concentrate essentially all of the platable metals into the film. Initially, they verified that the half-life of the plating session (the time required to plate half of all the electroactive ions out of the solution) was of the order of 12 min and that a complete electrolysis (greater than 98%) would require about 70 min. Then, from Faraday's law, the concentration could be calculated from the area under the resulting oxidation peak. Their electrolysis cell was similar to that described by other investigators (e.g. Matson, 1968; Allen, Matson, and Mancy, 1970).

Gardner and Stiff (1975) observed that although the charge yield was less than theoretical for all four metals, linearity between charge and concentration was maintained over two orders of magnitude for Cd, Pb, and Cu. The calibration graph for zinc was linear only when the copper concentration was negligible. An increase in the concentration of copper lowered the area of the zinc peak. At the same time, the peak in the copper position shifted anodically, eventually merging with the mercury peak. Interferences by Fe, Co, Ni, and complexing agents were also investigated. Finally, the results obtained by ASV were compared with those obtained by AAS after concentration by evaporation. Good agreement was obtained for all types of aqueous samples investigated.

10.4.3. Rotating disc glassy carbon electrode (RDGCE) with a mercury film plated *in situ*

Despite its wide use, the MCGE is cumbersome to prepare, difficult to cover adequately with a thin film (Hume and Carter, 1972), and difficult to maintain in an active state. The MCGE is sensitive to air (Perone and Davenport, 1966) and to undefined (presumably organic) substances in sea water (Zirino and Lieberman, 1974). Bradford (1972) reported that the sensitivity of the MCGE decreased rapidly with use and that it was necessary to replate the electrode with mercury daily. Zirino (unpublished work) noted that the lifetime of the electrode varied when used in different bodies of water, being longer in oligotrophic ocean waters and shortest in organic-rich estuarine environments. Hence there was a need to develop an equally sensitive but more reliable probe than the MCGE. An electrode which combines the sensitivity of the thin film with the convenience and reproducibility of the DME was first described by Florence (1970b). He suggested that glassy carbon would make a good substrate for mercury thin films because of its good conductivity, high hydrogen overvoltage, and chemical inertness. Furthermore, because of its high density, glassy carbon did not require impregnation with wax and could be polished metallographically to a high lustre. Also, since all of the metals commonly analysed by ASV are reduced at potentials more negative than mercury, a mercury thin film could be applied *in situ*, simply by adding a mercury (II) ion solution to the sample. Florence (1970b) constructed a rotating disc glassy carbon electrode (here called RDGCE) by sealing grade GC-20 carbon discs 5 mm high and 3 mm in diameter (Tokai Electrode Manufacturing Co., Ltd., Tokyo) into 250 mm lengths of 4 mm i.d. glass tubing using epoxy resins. The discs were then polished with diamond dust. Electrical contact was made by placing a small amount of mercury in contact with the disc and with a platinum wire which had been inserted into the tube. The electrode was rotated by a d.c. motor controlled by a tachogenerator. Rotational speeds from 50 to 4000 rev min^{-1} could be obtained with a precision of 0.2%.

When the electrode is rotated in a solution containing mercury (II) ions at 2000 rev min^{-1} and a suitable potential is applied, a thin mercury film is produced about one to two orders of magnitude thinner than that employed by Matson *et al.* (1965). Such a film inherently possesses a higher concentration factor (and higher sensitivity) while also maintaining higher resolution (Figure 7).

Initially Florence (1976) studied Pb^{2+} in 0.1 M KNO_3 in order to examine the relationship between peak currents produced at the RDGCE and various

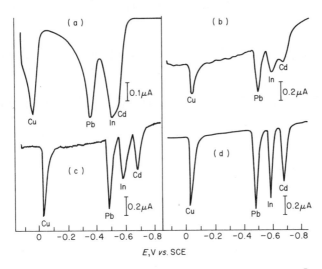

E,V *vs.* SCE

Figure 7. Comparison of several types of electrodes. Solution 2×10^{-7} M Cd^{2+}, In^{3+}, Pb^{2+}, and Cu^{2+} in 0.1 M KNO_3. Scan rate 0.3 V min^{-1}. Solutions b–d contain 2×10^{-5} M Hg^{2+}; rotation speed 2000 rev min^{-1}. (a) HMDE, 30-min deposition; (b) pyrolytic graphite electrode, 5-min deposition; (c) unpolished glassy carbon electrode, 5-min deposition; (d) polished glassy carbon electrode, 5-min deposition. [From Florence (1970), *J. Electroanal. Chem.*, **27**, 231–281]

experimental parameters. He observed that peak current was proportional to concentration over the range of interest, except that several plating–sweeping cycles were required before peaks of constant height were produced. Florence therefore used the peak obtained on the second or third sweep for his analytical work. This observation was not confirmed by Miguel and Jankowski (1974), who found that the peak height of the first few sweeps varied only randomly from the mean. Florence also found that a minimum concentration of 2×10^{-5} M Hg^{2+} in the sample was necessary for the production of sharp, reproducible peaks. Less mercury was relatively ineffective, while more produced only marginally bigger peaks. In agreement with theory (DeVries and Van Dalen, 1964), the peak current was found to be directly proportional to the deposition time and to the scan rate, although a relatively slow scan rate of 0.3 V min^{-1} (5 mV s^{-1}) was recommended. Similarly, the peak current, i_p, was found to be proportional to the square root of the rotation speed, but a practical rotational speed of 2000 rev min^{-1} was adopted. It is interesting that, contrary to previous practice with the MCGE, Florence continued the electrode rotation throughout the stripping cycle. Analyses were carried out by the standard addition method, and after completion of a determination the film was removed by wiping with a soft tissue. Thus, each determination was carried out with a fresh mercury film, an improvement over the MCGE. Some years after Florence's initial work, Miguel and Jankowski (1974) used a commercial RDGCE (Beckman Rotating Electrode, Catalogue No. 39084) to confirm most of Florence's initial observations.

Shortly after the development of the RDGCE, Florence (1972) described its applications to the analysis of sea water, seaweeds, fish, and other marine organisms taken from the Australian south coast. The metals analysed were Bi, Sb(III), Cu, In, Cd, Tl, and Zn in marine organisms and Zn, Cd, Pb, Cu, and Bi in sea water. Samples of surface sea water were collected in pre-cleaned, high-density polyethylene bottles and filtered through 0.45 μm Millipore filters on the day of collection. The filtrates were then stored in 1 dm^3 polyethylene bottles containing 2 cm^3 of 10 M HCl. The filters were wet-ashed with 2 cm^3 of 15 M HNO_3 and 1 cm^3 of 72% $HClO_4$, then diluted to 25 cm^3 after fuming off most of the acid. A 5.0 cm^3 volume of this solution was then diluted to 50 cm^3 with 0.2 M HCl for analysis.

The sea water samples were analysed for Bi, Cu, Pb, and Cd by adding 0.2 cm^3 of 0.01 M $Hg(NO_3)_2$ solution to 50 cm^3 of the filtered, acidified sea water. This solution was then purged of air, an electrolysis potential of -0.9 V *vs.* SCE was applied for 5 min while the electrode was rotated at 2000 rev min^{-1} and then the potential was swept at 3 V min^{-1}. The resulting peaks, however, were not recorded. A second electrolysis period of 30 min was then initiated and swept as before, this time recording the peak current. Cd, Pb, Cu, and Bi peaks occurred at -0.69, -0.48, -0.24, and -0.09 V, respectively. Concentrations were measured by the standard addition method. The precision of the determinations was between 5 and 10%.

In order to minimize the interference of copper on zinc, a separate determination for zinc was carried out on a diluted sample. A 5.0 cm^3 volume of acidified sea water was placed in a 50 cm^3 volumetric flask containing 0.1 cm^3 of 1 M HCl, 0.2 cm^3 of 0.01 M Hg(NO$_3$)$_2$, and 0.5 cm^3 of 2 M sodium acetate and analyzed as previously described, using a plating potential of -1.5 V. The zinc peak occurred at -1.05 V *vs.* SCE. A reagent blank was subtracted from the value obtained after standard addition. Solutions obtained from the acid digestion were analysed in the same manner as sea water.

Samples of marine organisms were dried in an oven at 105°C for three days and their dry to wet weight ratios were recorded prior to analysis by ASV. Florence compared several ashing procedures. Wet-ashing with HNO$_3$ followed by HClO$_4$ was compared with muffle ashing at 450°C for two days in silica dishes and with oxygen plasma ashing at about 200°C. However, the last method was found to be less efficient than the others and some samples were not oxidized even after exposure for 30 h in the plasma. After ashing, the ash was brought into solution with electronic grade HF and the remaining residue, consisting primarily of unburnt carbon, was filtered out on a filter-paper. The filtrate and acid washings were then evaporated to dryness in a platinum crucible. Next, the residue was dissolved in water and diluted to 25 cm^3. A 5 cm^3 volume of this solution was then made 0.03 M in HClO$_4$, to which 0.2 cm^3 of 0.01 M Hg(NO$_3$)$_2$ was added. For the analysis of Cd, Cu, and Pb, 5 cm^3 of this solution were added to 2 cm^3 of a second solution which was 0.5 M in trisodium citrate and 1.25 M in sodium nitrate. To this mixture 0.2 cm^3 of 0.01 M Hg(NO$_3$)$_2$ was added and the volume adjusted to 50 cm^3 with water in a volumetric flask. Zinc was then measured by the same procedure as used with sea water. Typical current–voltage plots for marine samples are shown in Figure 8.

Florence (1972) found that no significant differences were observed between any of the ashing methods, although muffle ashing at 450°C was found to be the most convenient. No appreciable losses due to volatility were observed.

In a later paper, Florence (1974) described a more extensive investigation of bismuth in marine samples using essentially the same procedure described earlier. However, it was found that when some marine organisms were ashed at 450°C, a preliminary separation was required in order to isolate the bismuth before ASV analysis could be performed. This separation consisted in dissolving the muffle ash in hot HCl and HNO$_3$, filtering off any solids (primarily silica) with Whatman No. 542 paper, re-heating the filtrate with concentrated HCl, and evaporating to 2 cm^3. This concentrated solution was then diluted to 20 cm^3 with water and passed through a 60 mm $\times$ 5 mm

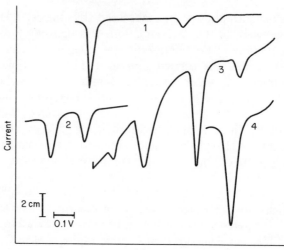

Figure 8. Typical voltammograms from marine samples. Sweep rate 3 V min^{-1}; rotation speed 2000 rev min^{-1}. (1) Fish skeleton, citrate buffer, sensitivity 0.79 μA cm^{-1}, peaks cadmium (−0.68 V), lead (−0.52 V), and copper (−0.06 V); (2) fish skeleton, citrate buffer, sensitivity 0.197 μA cm^{-1}, peaks cadmium (−0.68 V) and lead (−0.52 V); (3) sea water, pH 1.9, sensitivity 0.197 μA cm^{-1}, peaks cadmium (−0.69 V), lead (−0.48 V), copper (−0.24 V), and bismuth (−0.09 V); (4) sea water, pH 5.8, sensitivity 0.197 μA cm^{-1}, peak zinc (−1.05 V). [From Florence (1972), *J. Electroanal. Chem.*, **35**, 237–245]

diameter column of Bio-Rad AG1-X8 (100–200 mesh) anion-exchange resin (Cl$^-$ form) at a flow-rate of 0.7 cm^3 min^{-1}. The column was then washed with 10 cm^3 0.6 M HCl and the bismuth eluted by passing successively 5 cm^3 of water, 15 cm^3 of 0.5 M NaOH, 5 cm^3 of water and 10 cm^3 of HClO$_4$. The elements were then analysed by ASV after adding 5 cm^3 of 4 M HClO$_4$, 3 cm^3 of 0.6 M HCl, and 0.1 cm^3 of 0.01 M Hg(NO$_3$)$_2$ and adjusting the volume of the solution to 50 cm^3 with water in a volumetric flask.

In a separate study, Florence observed that only 90% of the natural bismuth present in sea water was retained on the column. Samples which had been passed through the column were therefore corrected for retention. Florence (1974) found that in a 0.4 M HClO$_4$–0.06 M HCl electrolyte, the bismuth peak current decreased when the electrolysis potential was increased cathodically from −0.2 to −0.5 V, probably owing to interference from copper. Thus, a plating potential of −0.2 V was adopted for the electrolysis, despite the fact that the peak occurs at −0.05 V, which allows only a small overvoltage for accumulation (Shain, 1963; Zirino and Kounaves, 1977). A plating potential of −0.25 V was adopted for sea water. Even at such low overvoltages, W(VI) and Mo(VI) were observed to interfere significantly with the determination, but neither was judged to be present in sufficient quantities in natural samples to be a problem. The determination of bismuth on the RDGCE with linear sweep was compared with its determination by differential pulse ASV on the hanging mercury drop, and it was found that both methods had similar sensitivities although the latter method was more affected by traces of surfactant in the sample.

Florence and Farrar (1974) extended the use of the *in situ* plated electrode by developing an analysis of tin in sea water and other marine samples. In most ASV media the $Sn \rightarrow Sn(II)$ oxidation peak is indistinguishable from lead and is tacitly ignored when lead determinations are discussed. The rationale for this is that in natural media tin is generally present as $Sn(IV)$, which hydrolyses and polymerizes to an electroinactive state. Therefore, in order to measure tin by ASV one must bring it to an electroactive state and be able to achieve a separation from lead. Florence and Farrar found that $Sn(IV)$ can be distilled as the bromide from H_2SO_4 and that the distillate proved to be an excellent medium for ASV. Two-dm^3 sea water samples were filtered through a 0.45 μm Millipore filter which had been washed with 5 M HCl and acidified to pH 1 with concentrated HCl. A 5 cm^3 volume of a 10 mg cm^{-3} Fe^{3+} solution were then added and then the iron was slowly precipitated (with stirring) by the addition of 15 M ammonia solution to pH 7–8. The precipitate was then collected on Whatman No. 541 paper (previously cleaned with 5 M HCl), washed with water, and dissolved in 40 cm^3 of hot 4.5 M H_2SO_4. The solution was then transferred into a distillation flask, 20 cm^3 of 9 M H_2SO_4 were added, and the mixture was heated gently to 150°C. A rapid stream of nitrogen passing through the solution via a fine glass tube prevented bumping. A 75 cm^3 volume of 48–50% HBr solution was next added to the flask and heated. The temperature of the sulphuric acid solution was then raised to 230 ± 20°C and the distillate was collected from the condenser at about 2 cm^3 h^{-1}. For analysis, 20 cm^3 of the distillate were transferred into a 100 cm^3 volumetric flask to which were added 30 cm^3 of ethanol (to prevent the precipitation of Hg_2Br_2 during stripping), 0.2 cm^3 of 2% hydrazinium hydroxide solution, and 0.8 cm^3 of 0.01 M $Hg(NO_3)_2$, and the mixture was diluted to 100 cm^3 with water. An aliquot of this solution was placed in the voltammetric cell and an electrolysis potential of -0.70 V was applied for at least 5 min. The tin peak occurs at -0.56 V *vs.* SCE. Tin concentrations were calculated by the standard additions method after subtraction of a reagent blank. Florence (1970a) also devised a procedure for the determination of iron in water based on the ligand-exchange method suggested earlier by Berge and Drescher (1967). Because of its insolubility in mercury, iron cannot be measured directly by ASV. However, if an excess of a Bi–EDTA complex is added to a sample containing Fe^{3+}, then the iron will quantitatively displace Bi^{3+} from the EDTA, which can then be measured directly with the RDGCE.

Initially, all of the iron in an aliquot (approximately 25 cm^3) of the sample containing $0.05-1\,\mu g$ of iron was oxidized to Fe^{3+} with $0.5\,cm^3$ of $5\,M\,HClO_4$ in a $50\,cm^3$ beaker heated to fuming. A fluted watch-glass was placed over the beaker to prevent loss by evaporation. After fuming, the contents of the beaker were washed into a $50\,cm^3$ volumetric flask, to which were added $0.2\,cm^3$ of $0.005\,M\,Bi\text{--}EDTA$ solution, $0.1\,cm^3$ of $0.01\,M\,Hg(NO_3)_2$, solution and $2.0\,cm^3$ of a solution which was $0.5\,M$ in sodium acetate and $1\,M$ in $NaCl$. The pH of the resulting mixture was adjusted to between 3.5 and 4.6 and the flask was immersed in a boiling water-bath for 15 min to facilitate the exchange of Fe^{3+} for Bi^{3+}. After completion of the exchange, the solution was placed in a voltammetric cell and analysed as previously described (Florence, 1970b). A potential of $-0.3\,V$ was applied to the RDGCE for 5 min while the electrode rotated at $2000\,rev\,min^{-1}$. The scan was conducted at $1\,V\,min^{-1}$, from $-0.3\,V$ to zero. In this medium the Bi^{3+} peak occurs at $-0.09\,V$. The calibration was effected by means of a plot of Fe^{3+} concentration *versus* peak height, which was linear up to $1\times10^{-7}\,M$. A reagent blank was also determined. Copper and antimony were found to interfere because their peaks are close to the stripping wave. However, in chloride solution a five-fold excess of copper over iron could be tolerated. Such an excess would be unusual for natural samples. The limit of detection for this method was found to be $0.15\,\mu g$ of iron (in the final $50\,cm^3$ volume) and was primarily determined by the reagent blank. Presumably lower levels could be measured if the reagents were purified.

Although Florence's method was not applied to the analysis of sea water samples, it suggests a general approach to the analysis of those elements in sea water which are not ordinarily measurable by ASV. Moreover, the rate of metal exchange may be useful in the elucidation of trace metal speciation.

10.4.4. Comparison between the HMDE and the RDGCE

The successful application of the thin-film electrode to the analysis of natural samples invited comparisons between it and the HMDE and raised the question as to which electrode was more suitable for environmental applications. The matter is complicated because each electrode has particular features which may make it more or less suitable for a particular analysis or a specific medium. Also, the suitability of an electrode for a particular application is dependent on the excitation mode, i.e. differential pulse, linear sweep, etc. (see Chapter 2).

Batley and Florence (1974) studied this problem by analysing several natural samples, e.g. sea water, orchard leaves, oyster ash, $0.1\,M\,HCl$ electrolyte, and $ZnSO_4$ electrolyte, using the HMDE and the RDGCE. The HMDE was used in three modes which were comparable to the RDGCE in sensitivity: differential pulse, and first and second harmonic a.c. ASV (Bond and Canterford, 1972). The RDGCE was used in the linear sweep mode. Initially, a pre-plated MCGE (Matson *et al.*, 1965) was compared directly with the *in situ* plated RDGCE and was found to give inferior precision and reliability in the determination of lead and cadmium in 2 M HCl. Therefore, all further tests were conducted with the RDGCE.

In the HCl medium it was observed that both the HMDE and RDGCE followed theoretical behaviour well, and that the sensitivity of the HMDE in the differential pulse mode was only slightly less than that of the RDGCE when experimental parameters such as pulse height and sweep rate were optimized (Copeland *et al.*, 1973).

Similarly, all methods were judged to be sensitive enough for the analysis of environmental samples where the limit of detection is primarily controlled by the blank. Surprisingly, only a small increase in sensitivity was obtained when the RDGCE was used in the differential pulse mode.

In general, the resolution of the RDGCE was found to be superior to that of the HMDE in the differential pulse mode, although some exceptions were noted. For instance, Tl(I) and Cd(II) in $ZnSO_4$ electrolyte are better resolved at the HMDE. Overall, the resolution and precision of a.c. stripping were found to be inferior to those of the other two methods (Figure 9). Further testing with the HMDE was then limited to the differential pulse mode. As was expected, interferences from inter-element compound formation (Section 10.5.2) were more pronounced with thin films, which have much larger element to mercury ratios. Adsorption of unoxidized organic matter from the oyster ash sample depressed the lead peak more when using the RDGCE than the HMDE. On the other hand, when the non-ionic detergent Triton X-100 was introduced into a sample, the cadmium peak was depressed more on the HMDE. It is obvious that the effect of adsorbed organics is variable from metal to metal and between the two substrates. This topic is discussed in the section on interferences.

Batley and Florence (1974) found that the RDGCE was considerably more sensitive to the solution chemistry than the HMDE. For instance, at the HMDE, Tl(I) could be determined in acetate buffer (pH 4.6) containing EDTA, the latter being necessary to mask the lead and cadmium peaks. On the other hand, well formed Tl(I) waves could be produced with the RDGCE only in strongly acidic solutions. Zinc peaks on the HMDE could be produced over a wide range of acidities (down to 0.05 M HCl), while at the RDGCE optimum peaks were produced at pH 5.5–6.0 in acetate buffer. Similarly, In(III) peaks were best obtained at the RDGCE in acetate–bromide electrolyte at pH 4–5 and only small peaks were observed in more acidic solutions. On the other hand, good peaks could be obtained in 1 M HCl when using the HMDE. Excellent peaks could be obtained at the HMDE for Sb(V) and Sb(III) in 4 M HCl and for Sn(IV) in 2 M HCl, while

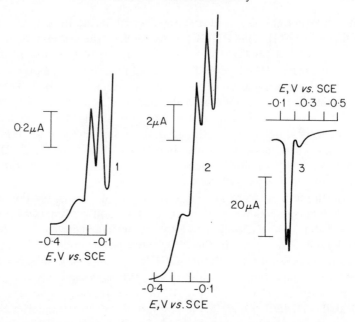

Figure 9. Resolution of Cu(II) and Bi(III) at HMDE and RDGCE. (1) 1×10^{-7} M Cu(II), 1×10^{-7} M Sb(III), 8×10^{-8} M Bi(III) in 1 M HCl using differential pulse anodic stripping at HMDE, scan rate 2 mV s^{-1}, 5 min deposition. (2) As (1), using fundamental a.c. anodic stripping at HMDE. (3) 1×10^{-7} M Cu(II), Sb(III), and Bi(III) in 1 M HCl + 0.02 M hydrazinium sulphate + 4×10^{-5} M Hg^{2+} at RDGCE, scan rate 83 mV s^{-1}, 5-min deposition. [From Batley and Florence (1974), *J. Electroanal. Chem.*, **55**, 23–43]

Sb(V) could not be reduced at the RDGCE and, in general, more complex media were required to prevent precipitations and interferences. These effects were attributed to general surface effects which could block electron transfer at a stationary film but not at the HMDE, whose liquid surface is constantly broken and renewed from the interior during an analysis.

Batley and Florence (1974) concluded that, although the greater precision and sensitivity obtained with the RDGCE may make this electrode more applicable for certain determinations, the HMDE in the differential pulse mode was the most versatile and reliable probe investigated.

The consequences of surface coatings which develop on the electrode when the solubility of the analyte in mercury is exceeded were mentioned by Copeland and Skogerboe (1974), who indicated that the resulting precipitation of metal on the mercury surface drastically changes the nature of the film, lowers the hydrogen overvoltage, and generally leads to irreproducibility and consequent loss of precision. Using a pre-plated, wax-impregnated thin-film electrode (Copeland *et al.*, 1973) as a model, they calculated the concentration of analyte required to cause precipitation during a typical

Table 6. Metal concentrations required to saturate a mercury film. [Reprinted with permission from Copeland and Skogerboe (1974), *Anal. Chem.*, **46,** 1275A–1268A. Copyright © 1974 American Chemical Society]

Metal	Concentration of metal in mercury, wt.-%	Analytic concentration in solution, M*
Cu	0.006	1×10^{-7}
Pb	1.2	2×10^{-5}
Cd	10.1	1.5×10^{-4}
Zn	6.4	3×10^{-5}
In	70.0	6.4×10^{-4}
Ga	3.6	2×10^{-5}
Tl	43.0	3×10^{-5}

* RDGCE, 6.5 mm diameter, 2500 Å (250 nm) thick, 2 min plate, 3600 rev min^{-1}, and a solution volume of 25 cm^3.

ASV experiment (Table 6). Their calculations were based on a 6.4 mm diameter electrode with a 250 nm film rotated at 3600 rev min^{-1} for 2 min. From Table 6 it can be seen that in the analysis of polluted waters, the concentration of copper may easily exceed the value required for solubility in the amalgam. It must also be concluded that the solubility of copper in mercury is often exceeded when sea water samples are analysed with the *in situ* plated thin-film electrode which has a film thickness of approximately 2.5 nm.

The effect of heterogeneous amalgam formation on ASV determinations was studied directly by Strojeck *et al.* (1976) who deliberately supersaturated mercury films less than 2 nm thick on the RDGCE with Cd, Pb, Zn, and Cu. They observed that the excess of solid metal formed on the surface leads to changes in peak width (distortions) and to systematic displacements of the peak currents along the potential axis. However, even though the thin-film electrode was grossly oversaturated, the dependence of the peak height on the concentration of analyte in solution was generally maintained. Studies with the HMDE at comparable analyte concentrations generally produced unproblematic, reversible behaviour.

It is clear that surface coatings on the thin-film electrode have only a minor effect on the electrode's ability to determine analytical concentrations by the standard addition method but have a much greater significance when the electrode is used for speciation studies which rely on peak potentials and absolute peak current measurements.

Lund and Onshus (1976) analysed a sample of sea water using the HMDE and the *in situ* plated RDGCE. Both electrodes were used in the linear sweep and the differential pulse modes. Initially, differential pulse parameters were optimized and the following values were recommended for both

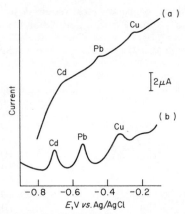

Figure 10. Comparison between linear sweep stripping (a) and differential pulse stripping (b) with an 'aged' TFE for 0.4 $\mu g\,dm^{-3}$ of cadmium, 1.3 $\mu g\,dm^{-3}$ of lead, and 2.4 $\mu g\,dm^{-3}$ of copper. Electrolysis time 30 min (a) and 5 min (b); scan rate 10 $mV\,s^{-1}$ (a). [From Lund and Onshus (1976), *Anal. Chim. Acta.*, **86**, 109–122]

types of electrodes: pulse amplitude, 25–50 mV; scan rate, 5 $mV\,s^{-1}$; and pulse repetition time, 0.5 s. The RDGCE was rotated at 2700 $rev\,min^{-1}$ and the film was plated out of a 4×10^{-5} M mercury solution. As in the work of Batley and Florence (1974), the sensitivity of the HMDE with differential pulse was comparable to that of the RDGCE, with linear sweep. However, in contrast to the previous observations, additional sensitivity could be gained when the RDGCE was used in the differential pulse mode. This was true especially for aged electrodes, which exhibited sharply sloping baselines in linear sweep (Figure 10).

10.4.5. Automated analyses

Substantial efforts have been made to automate polarographic and ASV analyses in order to reduce analysis times and to improve the sensitivity and resolution of the techniques. While most work has dealt with automating analyses in a laboratory environment, some have attempted to automate the sampling steps in order to develop sea water monitors suitable for use in the field. The computer control of conventional ASV cells has been described in Chapter 4.

Zirino and Lieberman (1975) and Zirino *et al.* (1978) developed an automated ASV analyser and employed it successfully in the field. In this system, ASV is performed with a tubular TFE constructed from hollowed-out, wax-impregnated graphite rods, through which mercury solution and sea water are passed alternately (Lieberman and Zirino, 1974). The tubular electrode and a tubular reference-counter electrode are part of a larger electrochemical cell which consists of electrodes, pumps, solenoid-operated

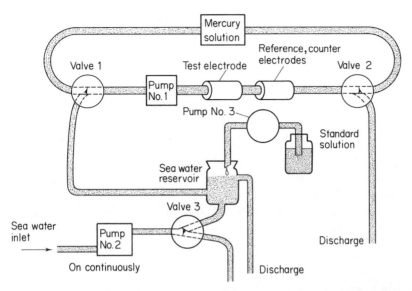

Figure 11. Schematic diagram of electrolysis cell for automated anodic stripping voltammetry. [Reprinted with permission from Zirino, Lieberman and Clavelli (1978), *Environ. Sci. & Tech.*, **12**, 73–79, Copyright © 1978 American Chemical Society]

valves, and solution reservoirs (Figure 11). The potential is applied with a PAR 174 Polarographic Analyser and the cell currents are recorded on a strip-chart recorder. A specially designed TTL programmer controls all cell and instrument functions. This device contains a power supply, timing and control logic circuits, operating controls, and program status displays. The sequence and duration of each program step are selectable using switches on the front panel. Samples can be analysed continuously and standard additions can be made when desired.

The electrolytic cell consists of two solution reservoirs, one for sea water and one for 2×10^{-5} M $Hg(NO_3)_2$. Each solution can be fed in turn into the electrode assemblage by activating valves 1 and 2 (Figure 11). These are all Teflon valves of the diaphragm type (Valcor Engineering Corp.). Three variable-speed peristaltic pumps are used in the instrument (Masterflex Pump No. 75-45-15, Cole Parmer Corp.). the first of these (No. 1 in Figure 11) circulates solutions through the electrodes. The second (No. 2) draws sea water continuously from the sampling source and feeds it to a discharge line after passage through a third three-way valve. The third pump (No. 3) is external to the cell and on command supplies standard to the sample for calibration. All tubing in the cell is made of Teflon and silicone rubber tubing is used in the pump head. The intake line from the sea, through which water flows continuously, is made of PVC. The system uses high flow-rates and makes no attempt to ensure laminarity (Lieberman and Zirino, 1974). A flow-rate of $250 \, cm^3 \, min^{-1}$ was found to be satisfactory.

An analytical cycle is performed as follows. Valve 3 is activated to divert sea water into the sample reservoir, which fills and discharges through an outlet at the top. The height of the outlet controls the volume of the sample, while the volume of flow controls the degree of rinsing. For survey work, $1 \, dm^3$ samples were used. As the sea water fills the reservoir, it is continually sparged with CO_2 to remove O_2 and to lower the solution pH to 4.9. While the sample is being collected and degassed, $Hg(NO_3)_2$ is circulated through the electrodes and the mercury film is deposited. In general, the potential used for depositing the film is the same as that used for the electrolysis of the sample: -1.4 to $-1.6 \, V$ *vs.* Ag|AgCl for the analysis of Zn, and $-1.0 \, V$ *vs.* Ag|AgCl for the analysis of Cd, Pb, and Cu. After the film has been deposited, a process which takes from 20 to $40 \, s$, the sample is introduced into the electrodes and is then discarded. Since a flow-rate of $250 \, cm^3 \, min^{-1}$ is maintained, electrolysis of a 1-dm^3 sample requires $4 \, min$. After the electrolysis period, the flow is stopped and the voltage is swept to $+0.5 \, V$ *vs.* Ag|AgCl. This potential is sufficient to oxidize the metals as well as the film itself. While the potential sweep can be made in the linear sweep mode, differential pulse ASV was found to be the most sensitive for this system (Zirino *et al.*, 1973). Calibration graphs were found to be linear for Zn, Cd, and Pb in the range 10^{-9}–$10^{-8} \, M$ and to show a small positive deviation from linearity for copper. The precision of consecutive determinations was better than 3%. (The analytical precision for replicate samples, which compares the values obtained for individual samples after standard addition, was of the order of 10–15%.)

When the system is in continuous use, i.e. when performing consecutive analytical cycles, it is observed that the electrode response increases for the first ten determinations and then decreases steadily to the point that after approximately 100 determinations the signal is insufficient for the analysis of natural sea water (Figure 12). The reasons for the non-linear response of the electrode with consecutive analytical cycles are complex. Lieberman and Zirino (1974) observed that the mercury coverage of the electrode increases for the first ten plating–stripping cycles, before reaching a steady state. Thereafter, however, the electrode response begins to decrease steadily, presumably because of adsorption of organic surfactants on mercury and because interfering trace substances accumulate in the film, changing its electrochemical characteristics.

It was found that frequent standard additions were necessary to compensate for the declining electrode performance. The efficacy of this correction was demonstrated when zinc was analysed at a stationary pier location in San Diego Bay over a five-day period; 380 consecutive analyses were made, one every 20 min, and every third sample contained an addition of zinc

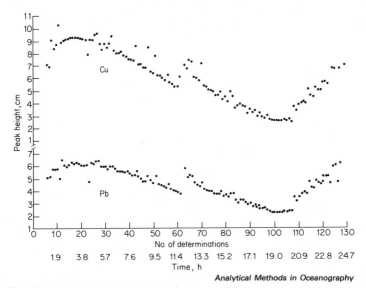

Analytical Methods in Oceanography

Figure 12. Time history of consecutive analyses of copper and lead performed with automated ASV analyser. LSASV, raw sea water, pH 4.9 (CO_2). Zirino and Lieberman (1975)

standard. Figure 13 (bottom) shows the peak current *vs.* time plots obtained over this period. These data can be plotted as two distinct curves: the locus of points of the peak currents obtained from the sea water and sea water plus standard. When zinc concentrations are calculated from the difference in the heights of the curves, and plotted against time (Figure 13, top) it is found that they follow the semidiurnal tidal cycles perfectly.

Wang and Ariel (1977a) proposed an ASV system with a flow-through cell similar in concept to the one originally proposed by Lieberman and Zirino (1974). However, they chose to carry out the voltammetry on glassy carbon TFEs which require essentially no preparation other than polishing. Additionally, by employing two TFEs, they were able to carry out differential ASV (Zirino and Healy, 1972) and improve the sensitivity of their technique. Their cell, which was not automated at the time of publication, also consisted of $Hg(NO_3)_2$ and sample reservoirs as well as a Masterflex peristaltic pump and three-way stopcocks. An important difference between the procedure of Wang and Ariel and that of Lieberman and Zirino is that the glassy carbon electrode is plated with mercury only once, at the beginning of a series of analytical determinations, and suffers no apparent deterioration even after 40 or more analytical cycles. In order to preserve the film on the electrode, the oxidations are carried out to only 0.1 V *vs.* Ag|AgCl. The flow and analytical characteristics were found to be very similar to those of the tubular, wax-impregnated TFE, but with the advantage of ease of preparation. When the twin TFEs were used in a differential

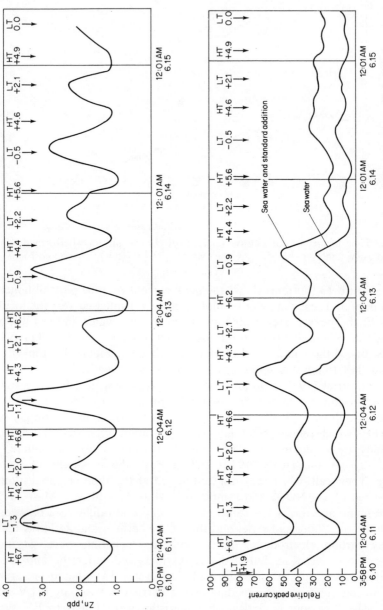

Figure 13. Above: tidal fluctuations of zinc in San Diego Bay during the period 10–15th June 1975. Below: peak currents for zinc in sea water and sea water plus standard addition. [Reprinted with permission from Zirino, Lieberman and Clavelli (1978), *Environ. Sci. & Tech.*, **12**, 73–79. Copyright © 1978 American Chemical Society]

system (Wang and Ariel, 1977b), the analytical sensitivity was judged to be superior to that obtained by differential pulse and to require considerably less analytical time.

10.5. INTERFERENCES

For the major constituents, the analysis of natural sea water provides few interferences that are not common to all saline media. At the trace level, however, two types of interferences are distinguishable: those produced by organic surfactants and those caused by the formation of intermetallic compounds. Few interferences have been found in the polarographic analysis of sea water samples (see Section 10.3).

On the other hand, a great deal has been written about interferences in ASV analyses. These will be discussed below.

10.5.1. Organic interferences

Natural sea water contains between 1 and 10 ppm of dissolved organic matter. Only 10% of this material, such as the fatty acid fraction and the algal-produced mucilagenous polysaccharides, is clearly surface active and likely to be adsorbed on mercury, hindering the electron transfer step during electrolysis and/or stripping. Some of this material may complex added standards and produce non-linear standard addition plots. In differential pulse ASV the desorption of organics from the mercury electrode produces peaks which are indistinguishable from those of the accumulated metal. The successful analysis of sea water by ASV clearly depends on understanding and minimizing the interference effects. Sometimes they are correctable by the standard addition technique, sometimes the analysis is possible only after removal of the interfering organic. Speciation studies which in general depend on *absolute* current measurements (rather than relative to a standard) are most prone to errors and misinterpretations. The problem is probably greatest for the analysis of inshore and near-shore waters which tend to be rich in dissolved organic matter.

Florence and Batley (1977) observed that the natural organic complexing agents present in a sea water sample complexed copper on standard addition and caused the calibration graph to have a positive deviation from linearity. This complexation occurred at pH 4.8 in 1.6×10^{-2} M acetate buffer. They interpreted the 'break' in the curve as representing the difference in plating rate between two species of copper in solution: the complexed copper, which plated out at a slower rate, and the 'free' copper, i.e. that present in excess of the organic ligand which was reduced at the maximum diffusion-controlled rate. When the sample was oxidized by exposure to a 500 W UV lamp for 6 h, in the presence of 0.5 cm^3 of 30% H_2O_2 per dm^3 of

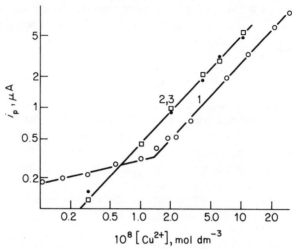

Figure 14. Stripping peak height–concentration plots for copper added to sea water and 0.5 M Suprapur sodium chloride. Deposition potential −0.9 V *vs.* SCE. (1) Sea water, 0.016 M acetate buffer pH 4.8; (2) sea water, 0.016 M HNO₃; (3) 0.5 M NaCl, 0.016 M acetate buffer, pH 4.8. [From Florence and Batley (1977), *J. Electroanal. Chem.*, **75**, 791–798]

sea water, the break point in the curve was lowered, but not eliminated, suggesting that inorganic complexing agents still existed in solution. This residual complexation was attributed to the adsorption of copper on inorganic colloids. Acidification of the sample to pH 2 eliminated even this form of complexation and produced a standard addition curve which possessed the same slope as that for copper in 0.5 M NaCl (Figure 14). However, organic interferences cannot always be removed by acidification. Zirino and Brann (unpublished) noted that when sea water from the coast of Maine was being analysed for copper by DPASV at pH 1.5, a broad current plateau occurred at the oxidation potential of copper. The area under the plateau which could have been easily interpreted as representing an irreversible oxidation of copper was not proportional to the electrolysis time and masked the true copper peak. The copper diffusion peak was noted only at long electrolysis times (Figure 15). The appearance of tensammetric peaks in pulse polarography and differential pulse polarography has already been noted (Anson *et al.*, 1976; Flanagan *et al.*, 1977; Jacobsen and Lindseth, 1976). The problem is just beginning to be appreciated for DPASV.

For example, Batley and Florence (1976b) also noted that when sea water samples were analysed by DPASV with the HMDE a broad wave occurred which interfered with the analysis of copper, lead, and cadmium. This wave was not seen when the analysis was carried out with linear scan ASV at the RDGCE (Figure 16). Lowering the pH of the sample merely shifted the

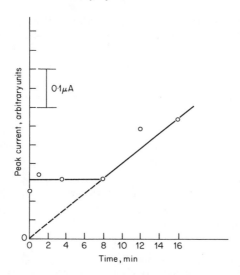

Figure 15. DPASV peak current *vs.* electrolysis time for a sample of Maine coastal water, pH 1.5

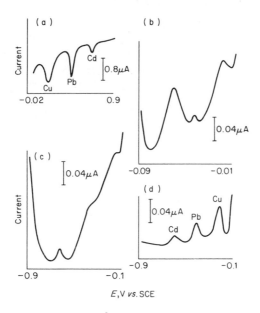

Figure 16. ASV of seawater (25 cm³) containing: (a) 1.6×10^{-2} M acetate buffer, (pH 4.8) $+8\times10^{-5}$ M Hg²⁺ at RDGCE; (b) 1.6×10^{-2} M acetate buffer (pH 4.8) at HMDE; (c) 1.6×10^{-2} M HNO₃ at HMDE; (d) (b) $+30$ μg dm⁻³ of Triton X-100, using 10 min deposition at -0.9 V. [From Batley and Florence (1976), *J. Electroanal. Chem.*, **72**, 121–126]

wave while irradiation with a strong UV lamp resulted in total elimination of this interference. From these characteristics the interference was attributed to the adsorption of natural organic surfactants on the HMDE. Competition from another surfactant, Triton X-100, added to the sample in concentrations as low as $30 \mu g\,dm^{-3}$ also succeeded in removing the interfering organic. Batley and Florence found surface-active organic compounds which interfered with ASV analysis on the HMDE in all samples collected from waters near Sydney, Australia, and the Southern Ocean south of Tasmania.

A comprehensive study of the possible effects of surfactants adsorbed on mercury on the analysis of natural waters was carried out by Brezonik *et al.* (1976). They selected a number of substances as models of the types of surface-active compounds to be found in the sea and they tested the effect of these compounds on the linear scan ASV analysis of cadmium and copper. An HMDE was used for all experiments.

Gelatin, commonly used in polarography as a maximum suppressor, was taken to represent colloidal proteinaceous matter. At neutral pH, up to $20\,mg\,dm^{-3}$ of gelatin caused no change in the peak current (i_p) or peak potential (E_p) when 2×10^{-6} M of cadmium was analysed. Reduction of the pH to 3, however, caused E_p to shift anodically by about $56\,mV$ and decreased i_p by 56%. Only $2\,mg\,dm^{-3}$ of the non-ionic surfactant Triton X-100 depressed the cadmium peak current up to 65% in neutral and acidic solution and caused the copper current peak to split into two when the pH was lowered towards pH 3 (Figure 17).

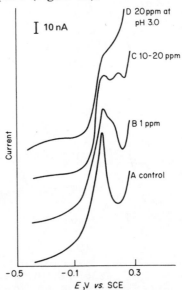

Figure 17. Distortion of the copper peak current after addition of Triton X-100 and acid. [Reprinted with permission from *Water Res.*, **10**, Brezonik, Brauner and Stumm (1976). Copyright © 1976 Pergamon Press, Ltd.]

Agar, starch, alginic acid, and polygalacturonic acid (PGUA) were tested as models for the polysaccharides found in sea water, notably around brown macroalgae. Agar did not shift E_p for either cadmium or copper but lowered i_p for both metals when the solution was kept neutral. Surprisingly, acidification to pH 3 produced peak currents which were higher than those at pH 7 but still not as high as for solutions of the metals without surfactants. Starch in concentrations of up to 10 mg dm^{-3} was found to have no effect on the E_p for cadmium and copper in either neutral or acidic media. On the other hand, the peak currents for both metals were significantly depressed in neutral solutions. These and other observations are summarized in Table 7. Brezonik *et al.* (1976) warn that many natural complexing agents are also surfactants and that *both* processes result in a lowering of i_p. They state that it may indeed be difficult to determine complexation simply from changes in i_p. As an example they cite humic and fulvic acids, which are known complexing agents for heavy metals and are strongly adsorbed on mercury. This problem has also been dealt with extensively by Cominoli *et al.* (1980).

From the analytical point of view, certain interferences can be tolerated if they can be corrected by the addition of a standard. Again, Brezonik *et al.* (1976) warn that any standard addition technique assumes that the added

Table 7. Effects of some sorbents on ASV diagnostic parameters. [Reprinted with permission from *Water Res.*, **10**, Brezonik, Brauner and Stumm (1976). Copyright © 1976 Pergamon Press, Ltd.]

Sorbent	Concentration range, mg dm^{-3}	Metal	Neutral conditions (pH 7)*		Acidic conditions (pH 3)*	
			i_p	E_p	i_p	E_p
Agar	1–10	Cd	D to 30%	NC	NC or I (but < control)	NC
	1–10	Cu	D to 30%	NC	NC or I (but < control)	NC
Alginic acid	4–40	Cd	D to 50%	NC	I (to control)	NC
	4–40	Cu	D to 50%	+	I (to control)	− (To control)
Alkaline phosphatase	1–36	Cd	D to 15%	NC	D to 90%	+
	1–36	Cu	D to 75%	Two peaks	I (to control)	Two peaks
Camphor	1–100	Cd	D to 17%	NC	D to 25%	+
	1–100	Cu	NC	NC		
Gelatin	0.1–20	Cd	NC	NC	D to 55%	+
Polygalacturonic acid	2–100	Cd	D to 23%	+	I (but < control)	− (Not to control)
	2–100	Cu	D to 60%	+	I (but < control)	NC
Starch	1–10	Cd	D to 15%	NC	NC	NC
	1–10	Cu	D to 50%	NC	NC	NC
Triton X-100	1–40	Cd	D to 40%	+	D to 65%	+
	1–20	Cu	D to 5%	Two peaks	Peak disappeared	

* D = decrease; I = increase; NC = no change, + = anodic shift, − = cathodic shift. Percentage changes are referred to controls for both neutral and acidic conditions. I and D under acidic conditions refer to changes beyond those occurring with a given level of sorbent under neutral conditions.

metal and the natural metal are in equilibrium at the time of analysis and that this assumption may not be valid when the rate of metal–ligand combination is slow. Nevertheless, in their example, standard addition fully compensates for the decrease in a Cd peak caused by the addition of a surfactant.

However, their point is well made. Unless the matrix is fully defined, the use of i_p in an absolute sense may lead to erroneous conclusions about the concentration and speciation of the metals in question. Standard addition is also important in the correction of the second type of interference, that caused by the formation of intermetallic compounds.

10.5.2. Metal to metal interferences

When differing metals are concentrated in mercury out of sea water at the same electrolysis potential, certain interferences with the analysis may occur. Primarily, the oxidation peaks of two or more metals may overlap when the metals are stripped out in the sea water medium. Examples of this problem include the simultaneous oxidation of nickel with zinc, silver with copper, and tin with lead. By choosing the electrolysis potential selectively, it is often possible to minimize these interferences. Alternatively, it may be possible to separate overlapping peaks by stripping in a different medium, in the manner of Ariel *et al.* (1964). However, the transfer must be performed carefully to avoid the inadvertant oxidation of metals from the amalgam and special cells should be constructed for this purpose. Zieglerova *et al.* (1971) and Lieberman and Zirino (1974) have described flow cells suitable for medium transfer.

A different problem arises when metals are concentrated in mercury to a level where they combine to form intermetallic compounds. In the stripping step, these intermetallic compounds may oxidize and give peaks at potential differing from that of the pure metal. Alternatively, the pure metal peaks are reduced by the amount contributed to the intermetallic peak. Often the intermetallic peak occurs at the same potential as one of the pure metals, thereby causing the concentration of the latter to be overestimated.

Clearly, the degree of metal to metal interference depends on the relative amounts of the metals which are plated out simultaneously from natural sea water and on their respective concentrations in the amalgam. Thus, the degree of intermetallic compound formation is found to be greatest for the thin-film RDE, less for the pre-plated TFE, and least for the HMDE.

Smith and Redmond (1971) systematically studied interferences on the HMDE in sea water media. They observed that Ni, Sb, and Zn all produce peaks at the same potential as zinc, Cd and Sn oxidized at the same potential as cadmium, while Cu, Ni, and V stripped at the same potential as copper. Only lead appeared to be free from interferences. However, Zirino and Healy (1972) pointed out that the oxidation of amalgamated tin to Sn(II) was indistinguishable from that of lead. Fortunately, these experiments were performed by the addition of metals to sea water and did not represent the natural equilibrium conditions. Because of inherent irreversibility and possible hydrolysis followed by colloid formation, at equal concentration, nickel peak currents were only one seventh those of zinc and one ninth those of cadmium. Vanadium produced currents only one eighth those of copper. Antimony produced peak currents one eighth those of zinc and a second peak at −0.78 V *vs.* SCE. Smith and Redmond noted that this peak was not visible in the stripping plots of natural sea water samples and concluded that antimony did not constitute a serious source of error, presumably because of hydrolysis followed by polymerization. For similar reasons, Zirino and Healy (1972) concluded that tin was not an interferent of lead under natural sea water conditions. Baier and Healy (1977) also showed that the tin did not interfere significantly with lead when the analysis was conducted at pH less than 5. Also, the low sensitivity of the HMDE to vanadium as well as its low concentration in sea water eliminated vanadium as a serious source of error in the analysis of copper. Finally, Zirino and Healy noted that nickel did not seriously interfere with zinc when the electrolysis was carried out at −1.25 V *vs.* SCE.

A detailed study of interferences at the pre-plated TFE (MCGE) was carried out by Seitz (1970). He found that nickel and zinc caused the copper peak current to broaden and that a Ni–Cu intermetallic compound did not strip out of the film during normal anodic scans. Similarly, Cu, Ni, and Co were found to reduce and to broaden the zinc peak. Silver and copper were found to oxidize at the same potential. Although the interference of Ni, Zn, and Co on copper can be easily avoided by plating at a reduction potential more anodic than that of the interferent metals, the reduction of silver with copper is unavoidable. Nevertheless, it is generally assumed that the concentration of silver in natural sea water samples is too low to interfere. This assumption has not been rigorously tested.

Because zinc requires the most negative plating voltage of all of the metals analysed by ASV, it is also the most subject to interferences. Bradford (1972) used the MCGE of Matson *et al.* (1965) to analyse for zinc in Chesapeake Bay water. The concentration of his samples was typically at the 1–10 μg dm^{-3} level. Bradford noted that Cu, Co, and Ni interfered with zinc, but that natural samples contained only copper in sufficient concentrations to interfere. The presence of copper resulted in lower, broader peaks. By titrating a sample of artificial sea water with copper at various concentrations, Bradford determined that copper and zinc formed a 1 : 1 intermetallic compound which dissociated during the stripping step to give broad peaks. Thus, plots of the zinc peak area were linear and analysis by the standard addition method was not affected. The interference of nickel on zinc was found to be similar to that of copper, but the stoichiometry of the Ni–Zn intermetallic compound was not determined.

In a similar study with the MCGE, Huynh-Ngoc Lang (1973) studied interferences among Cu, Cd, Pb, and Zn in sea water from the Gulf of Monaco and the Ligurian Coast. The four metals were added to sea water up to a concentration of 12 $\mu g\,dm^{-3}$. It was observed that a 12:1 ratio of copper to zinc resulted in a 48% diminution of the zinc peak current (notably, an 8:1 ratio did not cause any suppression). No significant interferences to the zinc peak current were found for copper and lead. The question of whether the diminution of the zinc peak current could have been compensated for by standard addition was not pursued. The effect of zinc on the copper peak current was found to be relatively small, with zinc causing reduction in the copper peak current of 15% when the Zn to Cu ratio was set at 12. Under the experimental conditions used, the relative interferences among Cu, Pb, and Cd were found to be negligible. The above results were verified by Crosmun et al. (1975), who conducted experiments similar to those of Huynh-Ngoc Lang but at considerably higher metal levels.

Also at higher concentrations (100–200 $\mu g\,dm^{-3}$) than those found in the analysis of sea water, Shuman and Woodward (1976) found that the zinc stripping peak diminished both in height and area in the presence of copper, and that the copper peak increased in the presence of zinc. The data suggested that intermetallic compounds were being formed which had stoichiometric ratios greater than 1:1. Using a thin film glassy carbon electrode and 0.05 M acetate buffer solutions (pH 4.5), Shuman and Woodward titrated zinc with copper and determined that $CuZn$, $CuZn_2$, and $CuZn_3$ intermetallic compounds could be formed at high metal loadings. An insoluble compound, $CuZn_3(s)$, with a solubility product of 3.1×10^{-5} M could also be formed. In a study at concentrations directly comparable to those found in sea water, Copeland et al. (1974) examined the interference of copper on zinc using a RDGCE in acetic acid–potassium acetate buffer. Titrations of zinc with copper produced values which were generally low for zinc and abnormally high for copper. These errors could not be compensated for by standard additions. When an excess of gallium was introduced into the sample, the zinc peak was restored to its height in the absence of copper and could be measured precisely by standard addition (Table 8). Copeland et al. suggest that gallium forms a Cu–Ga compound which has a larger formation constant than the Cu–Zn intermetallic compound, thus zinc is released and the copper peak is depressed. The technique of adding a third element to minimize undesirable intermetallic compound formation has been reviewed by Neiman et al. (1980).

The data of Copeland et al. (1974) strongly suggest that within the copper and zinc concentrations found in sea water, the effect of copper on zinc is within the 'normal' ASV experimental error (10–25%) and that zinc values obtained by the standard addition method are valid. This agrees with the

Table 8. Effect of copper on zinc determination using the method of standard additions. [Reprinted with permission from Copeland, Osteryoung and Skogerboe (1974), *Anal. Chem.*, **46**, 2093–2097. Copyright © 1974 American Chemical Society]

Copper concentration, μg dm^{-3}	Zinc concentration, μg dm^{-3}		
	Actual	Found	Error, %
2.4	6.3	6.3	0
	16.3	15.8	−3.1*
	26.3	25.1	−4.6*
	46.3	47.4	+2.4*
12.4	8.9	7.6	−14.6
	12.3	6.9	−44.0
	18.9	12.8	−32.4
	22.3	13.7	−37.4
	48.9	43.5	−11.1
42.4	8.9	2.4	−73.0
	18.9	12.6	−33.5
	42.4	11.6	−73.0
42.4 + 400 μg dm^{-3} Ga	16.4	16.4	0
	36.4	36.1	−0.8*
	46.4	45.8	−1.3*
	66.4	96	+44.6

* Errors within experimental error and therefore not significant.

results obtained in the direct analysis of sea water by Bradford (1972) and Huynh-Ngoc Lang (1973).

To summarize, the analysis of natural sea water samples by the standard addition method should give reasonably accurate results for Zn, Cd, Pb, and Cu. The effect of nickel on zinc can be lessened by judicious selection of the electrolysis potential. Similarly, plating below −1.0 V should eliminate the interference of zinc on copper, although it is questionable whether such a procedure is necessary for natural samples containing low concentrations of metals. The use of one electrolysis potential for zinc and another for copper essentially doubles the analysis time and may not be required if the effect of the interference can be compensated for by standard additions.

10.6. CALIBRATION

As for any voltammetric analysis, limiting currents and peak currents obtained with natural samples must ultimately be converted into the concentrations of the components of interest. Two possible complications make calibration difficult: (i) the chemical form (and electrochemical properties) of the element undergoing analysis, and (ii) extraneous interferences to the electroreduction arising, for example, from adsorption of organic compounds.

For natural sea water samples no assumption can be made that the concentration of the electroactive component represents the total concentration of that element in the sample. Indeed, various chemical steps (such as the lowering of the pH of the sample) have been recommended to turn electroinactive forms of the elements into species which are easily reduced at the electrodes. In turn, the processes with which inert moieties are turned into electroactive forms can yield information about chemical speciation (see next section). Progressively stronger chemical steps, such as UV irradiation, yield more and more of the total component. Oxidation with a hot, concentrated oxidizing acid is generally believed to yield electroactive forms of the natural elements commonly analysed by voltammetric methods. Thus, electrolysis following such an oxidation yields values which are representative of the total concentration. However, a final treatment with HF would probably yield additional amounts of the elements entrapped in colloidal silica. The adsorption of organic compounds can also change cell currents in an unpredictable manner (Brezonik *et al.*, 1976). Because of the above, the question of the concentration of trace elements in sea water as calculated from voltammetric data has received some discussion in the oceanographic literature (Zirino and Lieberman, 1975). This discussion has dealt largely with determinations by ASV rather than polarography, which deals primarily with the major constituents of sea water. However, some of the problems encountered are common to all electrometric measurements on sea water.

10.6.1. Exhaustive electrolysis

When dealing with a difficult medium, exhaustive electrolysis offers a simple and direct way of measuring concentration. The quantity of material (m) electroplated during a pre-electrolysis period is given by

$$m = vc_0[1 - \exp(-kT/vnF)] \tag{1}$$

where

v = volume of solution;
c_0 = concentration of the electroactive component at the beginning of the electrolysis period;
k = cell constant in amps per unit volume;
t = the electrolysis time in seconds;
n = number of equivalents per mole;
F = 96 500 coulombs per equivalent.

In theory, given enough time, all of the material can be recovered in the electrode even if k changes during the course of the electrolysis. Such a

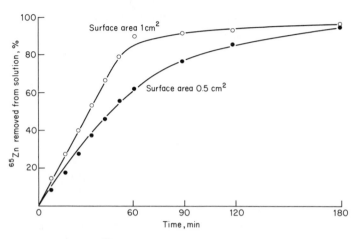

Figure 18. Percentage of ^{65}Zn removed from solution on MCGEs with 100 and 50 mm^2 surfaces. After Huynh-Ngoc Lang (1973)

technique was used by Gardiner and Stiff (1975) for the analysis of Cd, Pb, Cu, and Zn in organic-rich waters. The price, of course, is a very long analysis time. Huynh-Ngoc Lang (1973) added carrier-free ^{65}Zn to 50 cm^3 of sea water for electrolysis using the pre-plated TFE of Matson *et al.* (1965). Two experiments were conducted: one with an electrode area of 50 mm^2 and the second with an area of 100 mm^2. The progress of the electroreduction was followed by measuring the ^{65}Zn which remained in solution at various plating times. The results are presented in Figure 18. For the large electrode, essentially complete electrolysis occurred after 90 min, while 180 min were required for the smaller one. Such analysis times are generally deemed too long for routine analysis and most workers resort to the calibration of a partial electrolysis by the standard addition method.

10.6.2. Standard addition

If the cell constant k can be considered invariant with concentration, it is then possible to carry out only a partial electrolysis, calibrating the current or charge yield against a metal standard which is introduced to the sample and is assumed to take the same chemical form as the natural material. It is also assumed that the added standard reaches equilibrium with the natural material. Although these assumptions have never been directly tested, some idea about the correctness of these hypotheses is suggested by the literature.

Maljkovc and Branica (1971) observed that equimolar (10^{-6} M) mixtures of cadmium and EDTA in sea water reached equilibrium after approximately 40 min. A ten-fold excess of EDTA also required about 40 min to fully equilibrate although the initial rate of cadmium complexation was much greater than for the more dilute solution. Piro *et al.* (1973) studied the rate of exchange between ^{65}Zn and natural zinc in sea water. Zinc was measured by ASV at a DME with long drop times according to the method described by Macchi (1965). The ^{65}Zn fraction was measured with a gamma-ray spectrometer. The exchange experiments were conducted as follows. Natural sea water was acidified to pH 6 and placed in a cell containing a large mercury pool electrode. Exhaustive electrolysis at this pH removed all of the zinc available to the electrode *at that pH.* Inert forms of zinc not readily reducible at pH 6 but measurable at pH 2 remained in solution and were allowed to re-equilibrate. Equilibrium was re-established in about 16 h. If, after the initial electrolysis at pH 6, the pH of the sample was returned to 8, equilibrium was re-established in about 24 h. ^{65}Zn added to sea water at pH 6 did not significantly exchange with the natural zinc even 90 days after addition. Radioactive zinc added to sea water at pH 8 did not exchange with the natural zinc even after 1 year. These results were supported by experiments with marine algae, which showed that cultures of *Phaeodactylum tricornutum* (Bernhard and Zattera, 1969) contained a higher specific activity of ^{65}Zn than the culture medium of natural sea water. Duinker and Kramer (1977) made direct standard additions of Zn, Cd, Cu, and Pb to sea water at pH 8 and found that the electroactive forms of copper and lead (as measured by DPASV at the HMDE) decreased rapidly with time in the first 15 min after standard addition, reaching approximately 50% of the initial value after 18 min. On the other hand, zinc and cadmium showed no change in peak height over this interval. Duinker and Kramer (1977) conducted numerous DPASV determinations of Zn, Cd, Cu, and Pb in filtered water from the Rhine River estuary and the North Sea. All measurements were made by the standard addition method. They also compared the voltammetric results with the APDC–MIBK extraction followed by analysis by AAS (Brooks *et al.*, 1967). The results for sea water are presented in Table 9. In general, Zn, Cd, and Pb concentrations obtained at pH 8.1 were correlated with the values obtained by extraction at pH 8.1. Copper concentrations estimated by ASV were roughly one third of those obtained by extraction.

In a more comprehensive intercomparison, Fukai and Huynh–Ngoc Lang (1975) measured zinc by ASV on the pre-plated TFE and by spectrophotometry following a dithizone extraction. The spectrophotometric determinations were made on filtered sea water samples (0.45 μm) before and after oxidation with H_2SO_4 and $K_2S_2O_8$. Zinc measured by simple extraction was termed 'extractable zinc', while that determined after peroxodisulphate wet oxidation was termed 'total zinc'. Yields were checked with ^{65}Zn and it was established that the dithizone extraction alone fully complexed all added ionic ^{65}Zn. Several other experiments were also conducted. Some sea water samples were acidified to pH 1.5 and were left to stand for approximately

Table 9. Comparison of results from differential pulse anodic stripping voltammetry (DPASV) and atomic-absorbtion spectrometry (AAS) on North Sea samples. [From Duinker and Kramer (1977), *Marine Chemistry*, **5**, 207–228]

| | DPSAV | | | | AAS, pH 8.1 | |
| | pH 8.1 | | pH 2.7 | | | |
Element	Average concentration, $\mu g\,dm^{-3}$	S.D.	Average concentration, $\mu g\,dm^{-3}$	S.D.	Average concentration, $\mu g\,dm^{-3}$	S.D.
Zn	3.90	1.60	9.90*	4.90	3.90	1.90
Cd	0.23	0.07	0.20	0.07	0.11	0.07
Pb	0.30	0.10	2.50	0.50	0.50	0.20
Cu	0.60	0.20	1.70	0.50	1.60	0.70
	(0.30)†	(0.10)†				

* pH 6.1.
† Corrected for approximately 50% loss in Cu peak height with time.

half a day. Afterwards they were quickly neutralized and analysed by the dithizone methods. Other samples were allowed to stand in contact with a Chelex 100 ion-exchange resin for at least 24 h prior to filtration. They were then analysed by the spectrophotometric method. Similarly, samples were coprecipitated with $Fe(OH)_3$. The precipitate was then acidified and heated with $K_2S_2O_8$ and analysed. The solution was also analysed for zinc by the spectrophotometric method. The ASV determinations were conducted by the standard addition method at both pH 4 and pH 8.

The conclusions reached in this study were as follows: (1) extractable zinc determined spectrophotometrically gave approximately the same values as those measured by ASV at pH 8; (2) the chelating resin and hot acid treatment (but no oxidation) yielded values similar to ASV at pH 4; (3) hot acid oxidations were necessary to measure the total zinc in the samples; (4) the various chemical forms of zinc were probably *not* in equilibrium; and (5) the relative concentrations of the chemical forms of zinc varied over a wide range, depending on the time and place of collection. It was additionally surmised that the 'bound' forms of zinc existed as inorganic and organic complexes and colloids. The results obtained by Fukai and Huynh-Ngoc Lang (1975) are summarized in Figure 19.

A subsequent intercomparison for zinc and copper in San Diego Bay water was also carried out by Zirino *et al.* (1978). They compared zinc and copper concentrations obtained by ASV at pH 4.9 with those obtained by extraction on Chelex 100 columns with the method of Riley and Taylor (1968). Their results confirmed the earlier observations of Fukai and Huynh-Ngoc Lang (1975) (Figure 20).

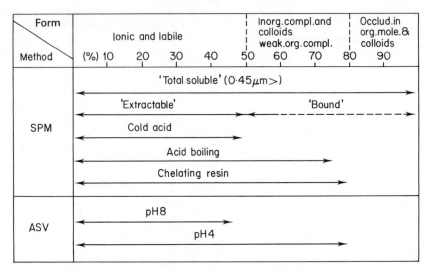

Figure 19. Relationships between the assumed chemical forms of zinc in sea water and measurable fractions obtained by different treatments in spectrophotometry with dithizone extraction (SPM) or in anodic stripping voltammetry (ASV). After Fukai and Huynh-Ngoc Lang (1975)

The good agreement between the ASV determinations for zinc and those obtained by complexation with dithizone or a chelating resin is notable because the analytical procedures differ markedly and because neither procedure measures the total concentration. Chelation or complexation followed by analysis by AAS is simply a process of collection and is not calibrated against an internal standard (although calibration of the spectrophotometer occurs during the final measurement). On the other hand, ASV measurements inherently *assume* equilibrium between the standard addition and the sample. The fact that the two sets of measurements agree implies that the added standard is *not* in equilibrium with the total quantity of metal but with only a portion of it. Similarly, it implies that the natural metal in the sample is relatively inert to exchange (at least over the period of measurement). This is in agreement with the findings of Piro *et al.* (1973). The observations by Fukai and Huynh-Ngoc Lang (1975), Duinker and Kramer (1977), and Piro *et al.* (1973) that the analysis of zinc by ASV at pH 8 yields lower values than a similar analysis at pH 4–6 implies that acidification of the sample cleaves off a portion of a relatively inert form of natural zinc and renders it electroactive. The added standard then equilibrates with the newly electroactive component and provides a measure of its concentration. A similar conclusion may be drawn for copper (Duinker and Kramer, 1977; Bubic *et al.*, 1973). A simple test of whether or not an inert fraction exists consists simply in making an ASV determination by the

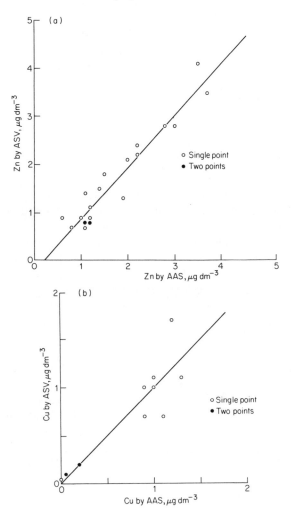

Figure 20. Results of an intercomparison of concentrations obtained by ASV and Chelex 100 extraction followed by atomic-absorption spectrometry: (a) zinc; (b) copper. [Reprinted with permission from Zirino, Lieberman and Clavelli (1978), *Environ. Sci. & Tech.*, **12**, 73–79. Copyright © 1978 American Chemical Society.

standard addition method at pH 8 and under acidic conditions. If the same concentration value is obtained in each case, then the implication is that the natural material has exchanged with the standard, i.e. it is labile under the experimental conditions. Similarly, a smaller value at pH 8 implies the presence of an inert form of the metal.

Of the other two metals commonly measurable by ASV, cadimum appears to be present in a labile form at least at one location. Filtered samples

(0.45 μm) from the North Sea gave identical values at both pH 8 and 2.7 (Duinker and Kramer, 1978). On the other hand, lead appears to be present in sea water in both forms. Zirino and Lieberman (1974) found that San Diego pier water yielded identical values at both pH 8 and 4.9, implying lability and perhaps pollution; similar observations were made by Jagner and Kryger (1975). On the other hand, Duinker and Kramer (1977) found that a major portion of lead in the North Sea was present in an inert form.

10.7. THE FORMS OF TRACE METALS IN NATURAL SAMPLES

Oceanographers have long recognized that trace substances in sea water existed in a variety of forms which had to be defined experimentally. This gave rise to terms such as 'reactive' silicate (that which reacted with the test), 'organic' phosphate (measurable only after a strong acid oxidation), and 'dissolved' iron, i.e. that which passed through a 0.45 μm filter even though colloidal in nature. The tendency to classify the various chemical forms of the elements by analytical procedures has also applied to those elements which are measured voltammetrically.

For trace metals, marine chemists have begun to recognize that the various chemical forms or species must be part of a continuum which eventually come to equilibrium, even though, at present, very little can be said about the time scales required to reach equilibrium. By analogy with the trace nutrients, it is implicit that in the oceans, bacterial processes re-mineralize detrital materials (and metals) which eventually become useful to the near-surface planktonic algae (Boyle and Edmond, 1975). However, at the same time, it is also recognized that the net transport of pelagic trace metals is downwards, to the sediments (Krauskopf, 1956), and that the re-mineralized portion can only be a small fraction of the total metal present in oceanic water columns. Hence any sea water sample collection for trace metal analysis must then be expected to contain both re-mineralized (reactive, labile) and inert fractions, the percentages of each depending on location and history. The manner in which these fractions interact with the phytoplankton is currently the subject of numerous investigations. An understanding of this interdependence is subject to a clear definition of the nature of the chemical speciation of the trace elements.

The measurement of chemical species perhaps has been the most important application of polarographic techniques to the analysis of sea water. The thrust of this work has been carried out by ASV and has progressed mainly in two directions: (i) the study of the shift in trace metal peak potentials and (ii) the study of the changes in metal peak height or peak area under different experimental conditions. The techniques based on the shift in peak potentials depend upon the degree of reactivity of the oxidized metal with ligands in the reaction layer. They can describe the species undergoing reduction (i.e. the natural forms) only indirectly and are therefore more suitable for model studies (Zirino and Healy, 1970; Zirino and Yamamoto, 1972; see Chapter 2) and for the determination of stability constants in known media (Bradford, 1973; Stumm and Bilinski, 1973; Bilinski *et al.*, 1976) than for the determination of natural speciation. This section will deal with speciation as obtained from peak currents. Peak current techniques include pH and complexometric titrations in which natural ligands are titrated with metal ions or alternatively ions are titrated with ligands. All of these depend on being able to discriminate between the 'free' uncomplexed metal and the 'bound' form. Such definitions are functions of the applied overvoltage and require some clarification.

10.7.1. Characterizing the overvoltage

Kounaves and Zirino (1979) have developed an equation which predicts the plating potential required for the reduction of a reversibly complexed metal ion. The equation comes from a treatment of stripping polarography and is analogous to textbook treatments of d.c. polarography (Zirino and Kounaves, 1977; Meites, 1965; Heyrovsky and Kůta, 1966).

For the reduction at the HMDE of a free metal ion (M^{2+}), complexed in solution in the presence of an excess of a ligand L^-

$$ML_j^{(2-j)+} \underset{\quad}{\overset{K_d^{\ominus}}{\rightleftharpoons}} M^{2+} + jL^- \xrightarrow{2e^-} M(Hg) \tag{2}$$

Kounaves and Zirino (1979) obtained the equation

$$E_{1/2,c}^* = E^{\ominus} + (g/2) \log \left[\frac{2r\delta\gamma_{ML_j^{(2-j)+}}}{3D_c\gamma_{M(Hg)}} \right] + (g/2) \log K_d^{\ominus}$$
$$- (jg/2) \log a_L - (g/2) \log t \tag{3}$$

where t is the deposition time and $K_d^{\ominus}$ is the thermodynamic dissociation constant for the complexation reaction expressed in terms of solute activities relative to an infinite dilution reference state. $E_{1/2,c}^*$ is the half-wave potential of the pseudo-polarographic curve obtained from a plot of peak current *versus* plating potential. $E^{\ominus}$ is the standard potential of the amalgam electrode. The other terms have their usual meanings and are defined in the glossary of symbols. The general forms of equation 3 for reduction from a multicomponent system at a variety of different electrodes have been derived by Turner and Whitfield (1979).

In the absence of ligands it can be shown (Zirino and Kounaves, 1977) that

$$E_{1/2,s}^* = E^{\ominus} - (g/2) \log \left[\frac{2r\delta\gamma_{ML_j^{(2-j)+}}}{3D_c\gamma_{M(Hg)}} \right] - (g/2) \log t \tag{4}$$

Ignoring small differences, the excess voltage required for the reduction of the complex relative to the simple ion is therefore given by

$$E^*_{1/2,c} - E^*_{1/2,s} = (g/2) \log K^{\ominus}_d - (jg/2) \log a_L \tag{5}$$

This relationship is analogous to the Lingane equation in d.c. polarography and it also predicts the minimum concentration of ligand required to achieve full complexation ($E^*_{1/2,c} - E^*_{1/2,s} = 0$). For example, equation 4 predicts that a minimum of 10^{-4} M ethylenediamine is required to complex a 1×10^{-8} M cadmium solution. In sea water it is found that the concentration of ethylenediamine required is closer to 10^{-3} M. This occurs because of competition by Ca^{2+} and Mg^{2+} for the ethylenediamine as well as by the Cl^- for Cd^{2+} (Kounaves and Zirino, 1979). Similarly, if we assume that sea water contains approximately 1 mg dm^{-3} of dissolved organic matter which has an average molecular weight of approximately 100, we can then estimate that under any circumstances the free organic ligand concentration in natural sea water cannot really exceed 10^{-5} M. We can then use equation 5 to estimate the possibility of metal–ligand interaction in the natural system. It is easy to see that only very strong chelators will be significant under equilibrium conditions (Davison, 1978).

Equation 5 can also be used to note that even in the presence of a strong chelator an overvoltage of approximately 200 mV is sufficient to reduce the complex fully at the same rate as the simple ion. It can also be shown by analogy with d.c. polarography (Delahay, 1954) that even for a highly irreversible reaction an additional 200 mV are sufficient to achieve full reduction of the complexed metal ion. Thus, it may be concluded that even when allowing for complexation, irreversibility and prolonged electrolysis, the usual overvoltages employed for the reduction of Cd, Pb, and Cu are sufficient to reduce most complexes. When the electrolysis is carried out at -1.0 V, the overvoltage is overwhelming for copper ($E^*_{1/2,s} = 0.3$–0.4 V *vs.* SCE), somewhat less for lead ($E^*_{1/2,s} = 0.4$–0.5 V *vs.* SCE) and marginal for cadmium complexes ($E^*_{1/2,s} = 0.6$–0.7 V *vs.* SCE). We find then that for the above metals the popular concept that ASV measurements in sea water measure only the free ion, is for the most part, incorrect since it does not take into account the overvoltage.

Only for zinc can it be said that the measurement favours the free ion, since under most circumstances the analyst is restricted to only a small overvoltage by the evolution of hydrogen at the electrode. Commonly, an electrolysis potential of -1.25 V *vs.* SCE is applied for the analysis of zinc even though $E^*_{1/2,s}$ (equation 4) for zinc in sea water is approximately -1.1 to -1.2 V *vs.* SCE. The low current densities obtained for zinc under these conditions have prompted workers to acidify their samples before analysis.

10.7.2. pH titrations

Zinc has been studied extensively by pH titrations by Piro *et al.* (1969), Zirino and Healy (1972), and Bernhard *et al.* (1976). When Bernhard *et al.* analysed sea water from the Gulf of La Spezia and the Gulf of Taranto for zinc at pH values ranging from 1.4 to 8.1, they found that the peak currents increased with the addition of acid (Figure 21). The characteristic curve obtained when the zinc peak height was plotted as a function of pH was interpreted as suggesting that the increase in peak current was due to the release of ionic zinc from particulate and organically complexed forms naturally present in sea water. The plateau occurring between pH 4 and 6 was attributed to the presence of organic ligands which competed for the zinc ion as it was released from the particulate substance. In fact, when the organic matter was destroyed with a strongly oxidizing UV lamp, the plateau was eliminated. (It should be noted that after several years of sampling the plateau no longer occurred.) The hypothesis that the increased acid content of the sample liberated zinc from an organic ligand system was well demonstrated by Bernhard *et al.* (Figure 22), and is compatible with the overvoltage considerations mentioned previously.

Similar titrations, which yielded increased peak currents with increased acid content of sea water, have also been shown for copper and lead (Fitzgerald, 1970; Baier, 1971; Florence and Batley, 1976; Duinker and Kramer, 1977). An example of such a pH titration for copper in North Atlantic sea water is given in Figure 23. However, for copper, unlike zinc, the decrease in peak height with pH is not attributable to complex formation because at the applied potential (-0.9 V *vs.* SCE) it is probable that any complex would be reduced. Indeed, it is easily verifiable that CuEDTA is

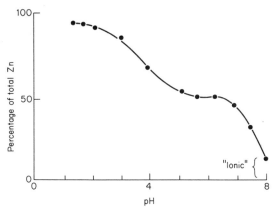

Figure 21. Influence of a pH titration on the electrochemically available fraction of zinc in a sea water sample from the Ligurian Sea. After Bernhard *et al.* (1976)

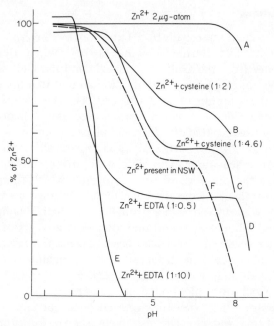

Figure 22. pH titration of 150 μg dm^{-3} of zinc in natural sea water (NSW) with and without various complexing agents. After Bernhard *et al.* (1976)

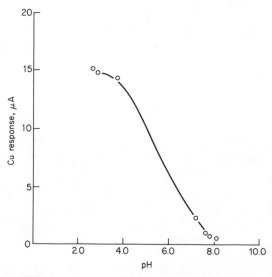

Figure 23. Copper response as a function of pH in a sea water sample (acidified with 6 N HCl) from the North West Atlantic. The peak is at -0.30 V *vs.* Ag/AgCl. After Fitzgerald (1970)

easily reduced at -0.3 V *vs.* SCE at the HMDE (Meites, 1965). Zirino and Kounaves (1980) have indicated that copper at pH 8 forms hydroxo or hydroxocarbonato complexes, as is postulated by current speciation models (Zirino and Yamamoto, 1972; Dyrssen and Wedborg, 1974; Bilinski *et al.*, 1976). Such complexes could agglomerate into particles which would effectively insulate the metal species from the applied potential. A pH titration would then be effective in releasing the metal.

Such a mechanism would also apply to lead, which appears to follow the same chemical patterns as copper and zinc (Petri and Baier, 1978; Petri and Baier, 1976). The peak current of cadmium, on the other hand, appears to show no pH dependence when the ionic metal is *added* to sea water (Zirino and Healy, 1970, 1972). According to the model of Zirino and Yamamoto (1972) and Dyrssen and Wedborg (1974), cadmium in sea water forms strong chloro complexes and the peak current is independent of pH in the range 2–8. Nevertheless, this should not be taken to mean that cadmium could not be absorbed on pH-sensitive colloids (such as iron and manganese oxides) and freed when titrated with acid. Such an explanation could account for the different observations of Duinker and Kramer (1977), who found no pH dependence for natural cadmium in North Sea water, whereas Florence and Batley (1976) observed at least some increase in cadmium peak height with increasing acidity in Australian coastal water.

10.7.3. Photolysis and chelating resins

Photo-oxidation of sea water has already been mentioned as a means of removing dissolved organic components while risking a minimum of contamination. Indeed, voltammetry combined with photolysis offers an elegant method of studying natural speciation in sea water (Moore and Burton, 1976). Photo-oxidations performed by the method of Armstrong *et al.* (1966) are generally considered to oxidize natural organics quantitatively to carbon dioxide (Williams, 1969). In voltammetric terms, the freeing of metals for analysis implies going from a very slow electroplating rate to a fast one identical with that of a solution of the uncomplexed ion. It is generally not appreciated that photo-oxidation will also free the metal into a form which is likely to be adsorbed on iron and manganese oxides and other particulate surfaces which have formed as a result of the oxidation. This can result in an actual lowering of the peak current, as can be seen in Figure 24, which shows the results of a photo-oxidation of a fjord sample being analysed for zinc by ASV at the HMDE before photolysis. Zinc gives a clear sharp peak even at pH 8.1 (curve A). Curve B shows the peak produced by the sample after destruction of the organic matter by UV oxidation. The zinc peak has diminished and has become noisy, and a secondary peak, tentatively identified as $Zn(OH)_2^0$ (Zirino, 1970; Wopshall and Shain, 1967),

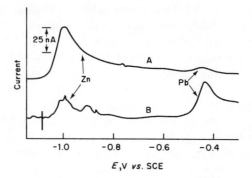

Figure 24. Differential ASV of sea water from a fjord: (A) before irradiation; (B) after irradiation. Zirino *et al.* (1973). The figure was originally presented at the Fall 1972 meeting of the Electrochemical Society Inc., held in Miami Beach, Florida

can be seen at −0.9 V. The increase in the lead peak with oxidation, also appearing in Figure 24, is probably due to contamination. The problem of adsorption and inorganic colloid formation is easily circumvented by acidifying the sample before or after oxidation. However, it is possible that acidification of the sample before photolysis may lead to incomplete oxidation of the organic matter (P. M. Williams, personal communication). Since much of the dissolved (<0.45 μm) organic matter in sea water is colloidal in nature, it appears that photolysis proceeds equally effectively for truly dissolved and colloidal organic species. Chelating resins may offer a means of distinguishing between these two fractions.

Florence and Batley (1975, 1976) observed that when filtered (0.45 μm) natural sea water was passed through a chelating resin of the iminodiacetic type (Chelex 100), a significant fraction of the total metal was not retained by the column. Chelex 100 columns, 60×120 or 100×8 mm in size, were packed with 50–100 mesh resin in the H^+ form. Filtered sea water was then passed through the columns at a flow-rate of $3\ cm^3\ min^{-1}$ in the manner described by Riley and Taylor (1968). Elution was performed with $2\ M\ HNO_3$. Retention of cadmium, lead, and copper was measured by LSASV with the RDGCE while zinc was measured by DPASV at the HMDE. Several types of measurements were made. The labile fractions in the sea water and column effluent were analysed after acidifying the sea water to pH 2 with HNO_3. The sample was acidified to pH 5 for the measurement of labile zinc. Total metal was measured after the sea water had been acidified to pH 0.7 with HNO_3 and boiled gently in a covered beaker for 10–15 min. Total zinc was measured after the boiled, acidified sea water had been brought to pH 4.8 with sodium acetate. Volume corrections were made for the losses incurred in boiling. Bound metal was estimated as the fraction obtained by subtracting labile from total metal.

Initially it was observed that column retention was poor for all trace metal fractions until the sea water had neutralized the hydrogen ion bound to the resin and brought the pH to about 7. Thereafter the column quantitatively removed all of the labile copper and zinc, 77% of the labile cadmium and approximately 40% of the labile lead. The column *did not* effectively remove any of the bound metal fraction of zinc, cadmium, and lead but removed approximately 40% of the bound copper. When the ammonium form of the resin was used, the column was considerably more effective, retaining 91, 74, and 78% of labile cadmium, lead, and copper, respectively, and 15, 20, and 42% of the bound fractions of these metals.

Florence and Batley (1976) surmised that the bound fraction consisted either of metal bound to organic chelates or as metal adsorbed on or occluded in organic or inorganic particles. Photolysis did not fully eliminate the difference between labile and total metals, suggesting that not all of the metal was combined with a simple organic ligand. By analogy, these workers concluded that the bound fraction must indeed be colloidal since the resin pore size (about 9 nm) was too small to allow colloids to enter the resin network. Cation-exchange resins have been used in this manner to remove ionic impurities from colloidal zirconia and other salts (Samuelson, 1963).

The work summarized above suggests that zinc, cadmium, copper, and lead exist in at least three forms: (i) colloidal, either pure or adsorbed on a

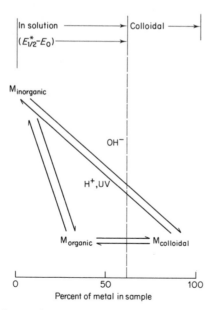

Figure 25. Chemical form of trace metals in filtered sea water as seen from the voltammetric point of view

colloidal matrix; (ii) organically complexed but in true solution (electron transfer is possible within the available overvoltage range); and (iii) inorganically complexed. The last fraction includes the simple aqueous ion, $Me_{(aq)}$, chloro and sulphato complexes, and uncharged species such as $Me(OH)_2^0$ and $Me(CO_3)^0$. The latter are indistinguishable even at small overvoltages but show reduced plating currents when they agglomerate or are adsorbed on particles in the water. They then become colloidal and are classified with the particulate fraction. This model is shown in Figure 25.

Needless to say, not all fractions are expected to be present in the same amounts in every sample, nor is equilibrium between the various fractions easily reached (Bernhard *et al.*, 1976). Coastal water samples can be expected to have a higher portion of the labile reactive component (Fukai *et al.*, 1975), since many anthropogenic inputs are in inorganic forms. The open ocean, on the other hand, probably contains the less reactive forms which have resisted bacterial degradation, adsorption, and precipitation.

10.7.4. The Batley–Florence scheme

Batley and Florence (1976a) made use of the separations afforded by photolysis and the chelating resin to develop a speciation scheme for identifying the forms of copper, lead, and cadmium in sea water. Their experimental design, which makes use of the diagnostic properties of ASV, distinguishes among nine forms, which are either directly measurable or whose concentration may be estimated by difference. These forms are separable into four classes as shown in Table 10. The letter symbols represent the following:

FM = free metal or aqueous complex;
  L = labile metal complex subject to measurement by ASV (pH 4.8);
  R = metal complex not subject to direct measurement by ASV at pH 4.8;
  C = metal bound to a colloidal form not retained by an iminodiacetic resin (Chelex 100);
O, I = indicate whether the complexing or colloidal moiety is organic (O) or inorganic (I).

These classifications are based on behaviour rather than actual chemical or physical properties, but have the virtue of being differentiable and measurable at the μg to $ng\,dm^{-3}$ level in the laboratory.

A simplified flow scheme based on the paper of Batley and Florence (1976) is presented in Figure 26. A filtered sea water sample, containing all of the named species, may be fractioned as follows:

1. The labile forms may be measured by ASV at pH 4.8 directly. This separates all of the labile metal species (FM, LO, LI, LCO, LCI) from the inert forms (RO, RI, RCO, RCI).

Table 10. Chemical forms of trace metals in sea water as defined by Batley and Florence (1976). The notation is defined in the text

Class 1	
FM+LO+LI	Free aqueous metal ions, ASV labile (pH 4.8) organic and inorganic complexes. Example: $Cu(H_2O)_n^{2+}$, Cu citrate, $CuCl_2^0$
Class 2	
LOC	Metals adsorbed on colloidal organics (these become measurable by ASV because they are desorbed from the colloid at pH 4.8) and metals bound to Chelex-resistant organics in 'true' solution
LIC	Metals adsorbed on colloidal Fe and Mn. They become measurable by ASV because of desorption at pH 4.8 and by solubilization of the colloid
Class 3	
RO	Organically complexed metals dissociated by Chelex 100 but resistant to analysis by ASV at pH 4.8. Example: humic acid-type complexes
RI	Inorganically complexed metals which follow the above criteria (no example of this category is offered by Batley and Florence, 1976)
Class 4	
ROC	Organically complexed metals resistant to both ASV (at pH 4.8) and Chelex 100 and measured only after a vigorous oxidation procedure. Example: metals adsorbed on organo-colloids not dissociated at pH 4.8
RIC	Inorganically complexed metals resistant to both ASV (at pH 4.8) and Chelex 100 and measured only after a vigorous oxidation procedure. Example: metals adsorbed on colloidal clays, not released at pH 4.8

2. Colloidal forms of the metals (or metals adsorbed on colloids) may be distinguished by passing the sample through a Chelex 100. An initial direct analysis of the sample by ASV will yield FM, LO, LI, LOC, and LIC, while analysis of the eluate will give FM, LO, and LI. Thus, LOC and LIC can be obtained by difference. UV oxidation followed by a second Chelex 100 separation will differentiate between these organic and inorganic colloidal forms.
3. The total concentrations of the metals in the sample can be obtained only after treatment with hot oxidizing acid (Florence and Batley, 1976; Fukai and Huynh-Ngoc Lang, 1975). Additional separations between organic, inorganic, colloidal, and truly soluble moieties are possible by retracing through the flow diagram. For instance, ROC and RIC are separable if a second pass through Chelex 100 is made.

Batley and Florence (1976a) applied their speciation scheme to the analysis of cadmium, lead, and copper in a surface sea water sample obtained near Walla Molla Beach (near Sydney) in an area relatively free from pollution. The results of their measurements appear in Table 11.

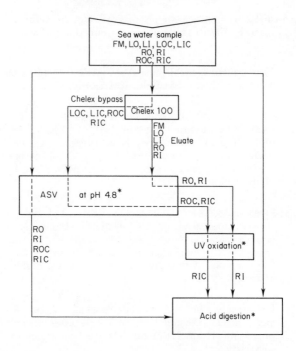

Figure 26. Flow chart of a scheme to determine the chemical forms of electroactive metals proposed by Batley and Florence (1976a). ---, Indicates those metals which are not rendered measurable or are not directly measurable by the indicated procedure. * The species made measurable by the indicated operation can be obtained by subtracting the contents of the outgoing arrow from the contents of the incoming arrow

Table 11. Partitioning of electroactive trace metals in a sea water sample according to the scheme of Batley and Florence

Metal species	Cd, %	Pb, %	Cu, %
Labile forms (FM, LO, LI)	11	12	22
Labile organic colloids (LOC) and labile complexes not dissociated by Chelex 100	75	66	42
Labile inorganic colloids (LIC)	0	0	0
Resistant organic complexes (RO)	0	0	25
Resistant inorganic complexes (RI)	0	0	0
Resistant organic colloids (ROC)	11	0	0
Resistant inorganic colloids (RIC)	7	24	25
Totals*	104	102	114
Total concentration of the elements ($\mu g\,dm^{-3}$)	0.24	0.39	0.26

* The total concentrations exceed 100% because of experimental error.

For all three metals, the predominant form is the labile organic colloid or complex, measurable directly by ASV at pH 4.8. The remaining copper is almost equally distributed between a labile component of organic and inorganic composition, a resistant component that is nevertheless dissociable by Chelex 100 and a totally resistant fraction that can be measured only after vigorous acid oxidation. Approximately one quarter (24%) of the lead is also present in the resistant fraction and a small portion of this metal (12%) is present in labile forms. Cadmium is overwhelmingly present in labile forms (86%), with a small portion (18%) bound to resistant colloids, both organic and inorganic.

Although the results of one beach sample cannot be extrapolated to the oceans, these findings are in general agreement with the speciation results of Fukai and Huynh-Ngoc Lang (1975), Bernhard *et al.* (1976), and others. A small fraction of the metal is labile and presumably also measurable at pH 8. A significant portion is measurable at pH 5; a great deal of this fraction must be colloidal. A third portion is present in highly resistant organic and inorganic colloids.

REFERENCES

Abdullah, M. I., and L. G. Royle (1972). The determination of copper, lead, cadmium, nickel, zinc and cobalt in natural waters by pulse polarography. *Anal. Chim. Acta*, **58**, 283–288.

Abdullah, M. I., L. G. Royle, and A. W. Morris (1972). Heavy metal concentrations in coastal waters. *Nature*, **235**, 158–160.

Allen, H. E., W. R. Matson, and K. H. Mancy (1970). Trace metal characterization in aquatic environments by anodic stripping voltammetry, *J. Water Pollut. Control Fed.*, **42**, 573–581.

Anson, F. C., J. B. Flanagan, K. Takahashi, and A. Yamada (1976). Some virtues of differential pulse polarography in examining adsorbed reactants. *J. Electroanal. Chem.*, **67**, 253–259.

Ariel, M. and U. Eisner (1963). Tracemetal analysis by anodic stripping voltammetry. I Trace metals in Dead Sea brine. 1. Zinc and cadmium. *J. Electroanal. Chem.* **5**, 362–374.

Ariel, M., U. Eisner, and S. Gottesfeld (1964). Trace analysis by anodic stripping voltammetry. II. The method of medium exchange. *J. Electroanal. Chem.*, **7**, 307–314.

Armstrong, F. A. J., P. M. Williams, and J. D. H. Strickland (1966). Photo-oxidation of organic matter in sea water by ultra-violet radiation, analytical and other applications. *Nature*, **211**, 481–483.

Baier, R. (1971). *Thesis*, University of Washington, Seattle, Wash.

Baier, R., and M. L. Healy (1977). Partitioning and transport of lead in Lake Washington. *J. Environ. Qual.*, **6**, 291–296.

Batley, G. E., and T. M. Florence (1974). An evaluation and comparison of some techniques of anodic stripping voltammetry. *J. Electroanal. Chem.*, **55**, 23–43.

Batley, G. E., and T. M. Florence (1976a). A novel scheme for the classification of heavy metal species in natural waters *Anal. Lett.*, **9**, 379–388.

Batley, G. E. and T. M. Florence (1976b). The effect of dissolved organics on the stripping voltammetry of sea water. *J. Electroanal. Chem.* **72**, 121–126.

Bender, M. L., and C. Gagner (1976). Dissolved copper, nickel and cadmium in the Sargasso Sea. *J. Mar. Res.*, **34**, 327–339.

Berge, H., and L. Brügman (1969). Possibilities for the polarographic determination of some main components in sea water. *Beitr. Meereskd.*, **26**, 47–57.

Berge, H., and L. Brügman (1970). Indirect polarographic determination of sulfate ions in sea water. *Beitr. Meereskd.*, **27**, 5–13.

Berge, H., and L. Brügman (1971). Polarographic methods for the determination of bromide ions in sea water. *Beitr. Meereskd.*, **28**, 19–32.

Berge, H., and L. Brügman (1972). Indirect determination of fluoride ions in sea water by a.c. polarography. *Beitr. Meereskd.*, **29**, 115–127.

Berge, H., and L. Brügman (1973). Die Bestimmung von Alizarin S nach der Anreichung am stationären Quecksilbertropfen und ihre Anwendung zur Fluoridanalyse. *Anal. Chim. Acta*, **63**, 175–183.

Berge, H., and A. Drescher (1967). Indirect, inverse voltammetric determination of elements by application of displacement reactions. Determination of iron and alkaline earth metals. *Fresenius Z. Anal. Chem.*, **231**, 11–17.

Bernhard, M., and A. Zattera (1969). A comparison between the uptake of radioactivity and stable Zn by a marine unicellular alga. In *Symposium of Radioecology* (Eds. D. J. Nelson and F. C. Evans). Fed. Sci. Tech. Information, U.S. Dept. Commerce. Springfield, Virg.

Bernhard, M. E., E. D. Goldberg, and A. Piro (1976). Zinc in seawater, an overview. In *The Nature of Seawater* (Ed. E. D. Goldberg). Dahlem Konferenzen, Berlin, pp. 43–68.

Bertine, K. K., J. H. Martin, and J. M. Teal (1976). Aids to analysis of seawater. In *Strategies for Marine Pollution Monitoring* (Ed. E. D. Goldberg). Wiley-Interscience, New York, pp. 217–235.

Bilinski, H., R. Huston, and W. Stumm (1976). Determination of the stability constants of some hydroxy and carbonate complexes of Pb(II), Cu(II), Cd(II) and Zn(II) in dilute solutions by anodic stripping voltammetry and differential pulse polarography. *Anal. Chim. Acta*, **84**, 157–164.

Bond, A. M., and D. R. Canterford (1972). Comparative study of a wide variety of polarographic techniques with multifunctional instrumentation. *Anal. Chem.*, **44**, 721–731.

Boyle, E. A., and J. M. Edmond (1975). Copper in surface waters south of New Zealand. *Nature*, **253**, 107–109.

Boyle, E. A., F. Sclater, and J. M. Edmond (1976). On the marine geochemistry of cadmium. *Nature*, **263**, 42–44.

Bradford, W. L. (1973). Determination of a stability constant for the aqueous complex $Zn(OH)_2^0$ using anodic stripping voltammetry. *Limnol. Oceanogr.*, **28**, 757–762.

Bradford, W. L. (1972). A study of the chemical behaviour of Zn in Chesapeake Bay water using anodic stripping voltametry. *Chesapeake Bay Inst. Tech. Rep.*, No. 76.

Brainina, Zh. K., and E. Ya. Sapozhnikova (1966). Concentration of substances in polarographic analysis. IX. Determination of iodide ions. *Zh. Anal. Khim.*, **22**, 1342–1347.

Brainina, Zh. K. (1971). Film stripping voltammetry. *Talanta*, **18**, 513–539.

Brewer, P. G. (1975). Minor elements in sea water. In *Chemical Oceanography*, Vol. 1 (Eds. J. P. Riley and G. Skirrow). Academic Press, New York, pp. 415–496.

Brezonik, P. L., P. A. Brauner and W. Stumm (1976). Trace metal analysis by anodic stripping voltammetry: effect of sorption by natural and model organic compounds. *Water Res.*, **10**, 605–612.

Broenkow, W. W., and J. D. Cline (1969). Colorimetric determination of dissolved

oxygen at low concentrations. *Limnol. Oceanogr.*, **14**, 450–454.

Brooks, A. R. (1960). The use of ion-exchange enrichment in the determination of trace elements in sea water. *Analyst*, **85**, 745–748.

Brooks, R. R., B. J. Presley, and I. R. Kaplan (1967). APDC–MIBK extraction system for the determination of trace elements in saline waters by atomic-absorption spectrophotometry. *Talanta*, **14**, 809–816.

Brügman, L. (1974a). Determination of trace elements in sea water by means of a stationary mercury electrode. *Acta Hydrochim. Hydrobiol.*, **2**, 123–138.

Brügman, L. (1974b). Determination of zinc, cadmium and lead in the Baltic Sea by inverse voltammetry. *Beitr. Meereskd.*, **34**, 9–21.

Brügman, L. (1977). Occurrence and analysis of trace metal forms in water. *Acta Hydrochim. Hydrobiol.*, **5**, 421–429.

Brügman, L. (1978). Zum Gehalten einiger Schwermetalle in Marinen Organismen aus dem Auftriebsgeliebt vor NW-Afrika. *Fischereiforschung*, **16**, 53–58.

Bubic, S., L. Sipos, and M. Branica (1973). Comparison of different analytical techniques for the determination of heavy metals in sea water. *Thalassia Jugosl.*, **9**, 55–63.

Burrell, D. C., and G. G. Wood (1969). Direct determination of zinc in seawater by atomic absorption spectrophotometry. *Anal. Chim. Acta*, **48**, 45–49.

Chow, T. J., and C. C. Patterson (1966). Concentration profiles of barium and lead in Atlantic waters off Bermuda. *Earth. Planet. Sci. Lett.*, **1**, 397–400.

Clem, R. G., G. Litton, and L. D. Orneales (1973). New cell for rapid anodic stripping voltammetry. *Anal. Chem.*, **45**, 1306–1317.

Cominoli, A., J. Buffle, and W. Haerdi (1980). Voltammetric study of humic and fulvic substances. Part III. Comparison of the capabilities of the various polarographic techniques for the analysis of humic and fulvic substances. *J. Electroanal. Chem.*, **110**, 259–297.

Copeland, T. R., and R. K. Skogerbae (1974). Anodic stripping voltammetry. *Anal. Chem.*, **46**, 1257A–1268A.

Copeland, T. R., J. H. Christie, R. A. Osteryoung, and R. K. Skogerboe (1973). Analytical applications of pulsed voltammetric stripping at thin film mercury electrodes. *Anal. Chem.*, **45**, 2171–2174.

Copeland, T. R., R. A. Osteryoung, and R. K. Skogerboe (1974). Elimination of copper–zinc intermetallic interferences in anodic stripping voltammetry. *Anal. Chem.*, **46**, 2093–2097.

Ćosović, B., V. Zutic, and Z. Kozarac (1977). Surface active substances in the sea surface microlayer by electrochemical methods. *Croat. Chem. Acta*, **50**, 229–241.

Crosmun, S. T., J. A. Dean, and J. R. Stokely (1975). Pulsed anodic stripping voltammetry of zinc, cadmium and lead with a mercury-coated wax-impregnated graphite electrode. *Anal. Chim. Acta*, **75**, 421–430.

Davison, W. (1978). Defining the electroanalytically measured species in a natural water sample. *J. Electroanal. Chem.*, **87**, 395–404.

Davison, W., and M. Whitfield (1977). Modulated polarographic and voltammetric techniques in the study of natural water chemistry. *J. Electroanal. Chem.*, **75**, 763–789.

Delahay, P. (1954). *New Instrumental Methods in Electrochemistry*. Interscience, New York.

De Vries, W. T., and E. van Dalen (1964). Theory of anodic stripping voltammetry with a plane, thin mercury-film electrode. *J. Electroanal. Chem.*, **8**, 366–377.

Donadey, G. (1969). Redissolution anodique par polarographie à impulsions. Application au dosage direct des traces de métaux lourds en solution dans les eaux de mer. Thesis, Faculty of Science, University of Paris.

Donadey, G., R. Rosset, and G. Charlot (1972). Anodic dissolution in pulse polarography. *Chem. Anal.* (*Warsaw*), **17**, 575–602.

Duinker, J. C., and C. J. M. Kramer (1977). An experimental study on the speciation of dissolved zinc, lead and copper in river Rhine and North Sea water by differential pulsed anodic stripping voltammetry. *Mar. Chem.*, **5**, 207–228.

Duinker, J. C., and R. F. Nolting (1977). Dissolved and particulate trace metals in the Rhine estuary and the southern Bight. *Mar. Pollut. Bull.*, **8**, 65–71.

Dyrssen, D., and M. Wedborg (1974). Equilibrium calculations of the speciation of elements in seawater. In *The Sea*, Vol. 5 (Ed. E. D. Goldberg). Wiley–Interscience, New York, pp. 181–195.

Fitzgerald, W. F. (1970). *Thesis*, Massachusetts Institute of Technology, Cambridge, Mass.

Flanagan, J. B., K. Takahashi, and F. C. Anson (1977). Reactant adsorption in differential pulse polarography. *J. Electroanal. Chem.*, **81**, 261–273.

Flato, J. B. (1972). Renaissance in polarographic and voltammetric analysis. *Anal. Chem.* **44**, 75A–87A.

Florence, T. M. (1970a). Determination of iron by anodic stripping voltammetry. *J. Electroanal. Chem.*, **26**, 293–298.

Florence, T. M. (1970b). Anodic stripping voltammetry with a glassy carbon electrode mercury-plated *in situ*. *J. Electroanal. Chem.*, **27**, 273–281.

Florence, T. M. (1972). Determination of trace metals in marine samples by means of anodic stripping voltammetry. *J. Electroanal. Chem.*, **35**, 237–245.

Florence, T. M. (1974). Determination of bismuth in marine samples by anodic stripping voltammetry. *J. Electroanal. Chem.*, **49**, 255–264.

Florence, T. M., and G. E. Batley (1975). Removal of trace metals from seawater by a chelating resin. *Talanta*, **22**, 201–204.

Florence, T. M., and G. E. Batley (1976). Trace metal species in sea-water. I. *Talanta*, **23**, 179–186.

Florence, T. M., and G. E. Batley (1977). Determination of copper in seawater by anodic stripping voltammetry. *J. Electroanal. Chem.*, **75**, 791–798.

Florence, T. M., and Y. J. Farrar (1974). Determination of tin by thin film anodic stripping voltammetry. Application to marine samples. *J. Electroanal. Chem.*, **51**, 191–200.

Fukai, R., and Huynh-Ngoc-Lang (1975). Chemical forms of zinc in sea water. Problems and experimental methods. *Nippon Kaiyo Gakkai-Shi*, **31**, 179–191.

Fukai, R., C. N. Murray and Huynh-Ngoc-Lang (1975). Variations of soluble zinc in the Var River and its estuary. *Estuarine Coastal Mar. Sci.*, **3**, 177–188.

Gardiner, J., and M. J. Stiff (1975). Determination of cadmium, lead, copper and zinc in ground water, estuarine water, sewage and sewage effluent by anodic stripping voltammetry. *Water Res.*, **9**, 517–523.

Gilbert, T. R. (1971). Electrochemical studies of environmental trace metals. *Thesis*, Massachusetts Institute of Technology, Cambridge, Mass.

Gilbert, T. R., and D. N. Hume (1973). Direct determination of bismuth and antimony in sea water by anodic stripping voltammetry. *Anal. Chim. Acta*, **65**, 451–459.

Goldberg, E. D. (1976). *Strategies for Marine Pollution Monitoring*. Wiley–Interscience, New York.

Herring, J. R., and P. S. Liss (1974). New method for the determination of iodine species in sea water. *Deep-Sea Res.* **21**, 777–783.

Heyrovsky, J., and J. Kůta (1966). *Principles of Polarography*. Academic Press, New York.

Hume, D. N., and J. M. Carter (1972). Characteristics of the mercury-coated graphite electrode in anodic stripping voltammetry. Application to the study of trace metals in environmental systems. *Chem. Anal.* (*Warsaw*), **17,** 747–759.

Huynh-Ngoc-Lang (1973). L'Application d l'électrode indicatrice 'graphite-mercure' en redissolution anodique par voltammétrie pour la détermination simultanée du zinc, du cadmium, du plomb et du cuivre dans l'environnement aquatique. *Thesis,* Institute of Mathematics and Science, University of Nice, Nice.

Jacobsen, E., and H. Lindseth (1976). Effects of surfactants in differential pulse polarography. *Anal. Chim. Acta,* **86,** 123–127.

Jagner, D., and L. Kryger (1975). Computerized electroanalysis. Part III. Multiple scanning anodic stripping and its application to sea water. *Anal. Chim. Acta,* **80,** 255–266.

Joyner, T. M., M. L. Healy, D. Chakravarti, and T. Kenyanagi, (1967) Preconcentration for trace analysis of seawaters. *Environ. Sci. Technol.,* **1,** 417–424.

Kahane, E. (1933). Detection of sodium as the triple acetate of uranyl, magnesium and sodium. *Bull. Soc. Chim. Fr.,* **53,** 555–563.

Kemula, W., and Z. Kublik (1971). Applications of hanging mercury drop electrodes in analytical chemistry. *Adv. Anal. Chem. Instrum.,* **2,** 342.

Kolthoff, I. M., and J. J. Lingane (1952). *Polarography,* 2nd ed. Interscience. New York.

Kounaves, S. P., and A. Zirino (1979). Studies of cadmium–ethylenediamine complex formation in seawater by computer-assisted stripping polarography. *Anal. Chim. Acta,* **109,** 327–339.

Kozarac, Z., B. Ćosović, and M. Branica (1976). Estimation of surfactant activity of polluted sea water by Kalousek commutator technique. *J. Electroanal. Chem.,* **68,** 75–83.

Krauskopf, K. B. (1956). Factors controlling the concentrations of thirteen rare metals in sea water, *Geochim. Cosmochim. Acta,* **9,** 1–32.

Kremling, K. (1973). Voltammetrische Messungen über die Vertailung von Zink, Cadmium, Blei und Kupfer in der Ostsee. *Kiel. Meeresforsch.,* **29,** 77–84.

Kryger, L., and D. Jagner (1975). Computerized electroanalysis. Part II. Multiple scanning and background subtraction. A new technique for stripping analysis. *Anal. Chim. Acta,* **78,** 251–260.

Lauer, G., and R. A. Osteryoung (1968). A general purpose laboratory data acquisition and control system. *Anal. Chem.,* **40,** 30A–43A.

Lieberman, S. H., and A. Zirino (1974). Anodic stripping voltammetry of zinc in seawater with a tubular mercury–graphite electrode. *Anal. Chem.,* **46,** 20–23.

Lieberman, S. H. (1979). *Thesis,* University of Washington, Seattle, Wash.

Lund, W., and D. Onshus (1976). The determination of copper, lead and cadmium in sea water by differential pulse anodic stripping voltammetry. *Anal. Chim. Acta,* **86,** 109–122.

Luther, G. W., III, A. L. Meyerson, and A. D'Addio (1978). Voltammetric methods of sulfate ion in analysis in natural waters. *Mar. Chem.,* **6,** 117–124.

Macchi, G. (1965). The determination of ionic zinc in sea-water by anodic stripping voltammetry using ordinary capillary electrodes. *J. Electroanal. Chem.,* **9,** 290–298.

Maljókovic, D., and M. Branica (1971). Polarography of sea water. II. Complex formation of cadmium with EDTA. *Limnol. Oceanogr.,* **16,** 779–785.

Matson, W. R., D. K. Roe, and D. E. Carritt (1965). Composite graphite–mercury electrode for anodic stripping voltammetry. *Anal. Chem.,* **37,** 1594–1595.

Matson, W. R. (1968). Trace metals. Equilibrium and kinetics of trace metal complexes in natural media. *Thesis,* Massachusetts Institute of Technology, Cambridge, Mass.

Meites, L. (1965). *Polarographic Techniques,* 2nd ed. Interscience, New York.

Miguel, A. H., and C. M. Jankowski (1974). Determination of trace metals in aqueous environments by anodic stripping voltammetry with a vitreous carbon rotating electrode. *Anal. Chem.*, **46**, 1832–1834.

Milner, G. W. C., J. D. Wilson, G. A. Barnett, and A. A. Smales (1961). The determination of uranium in sea water by pulse polarography. *J. Electroanal. Chem.*, **2**, 25–38.

Moody, J. R., and R. M. Lindstrom (1977). Selection and cleaning of plastic containers for storage of trace element samples. *Anal. Chem.*, **49**, 2264–2267.

Moore, R. M., and J. D. Burton (1976). Concentrations of dissolved copper in the eastern Atlantic Ocean 23°N to 47°N. *Nature*, **264**, 241–243.

Moorhead, E. D., and W. H. Doub, Jr. (1977). Digital microcoulometric measurements of cadmium anodic stripping at the micrometer hanging mercury drop electrode. *Anal. Chem.*, **49**, 199–205.

Muzzarelli, R. A. A. (1971). Chitosan for the collection from seawater of naturally occurring zinc, cadmium, lead and copper. *Talanta*, **18**, 853–858.

Neiman, E. Ya, L. G. Petrova, V. I. Ignatov, and G. M. Dolgopolova (1980). The third element effect in anodic stripping voltammetry. *Anal. Chim. Acta*, **113**, 277–285.

Nürnberg, H. W., P. Valenta, L. Mart, B. Raspor, and L. Sipos (1976). Applications of polarography and voltammetry to marine and aquatic chemistry. II. The polarographic approach to the determination and speciation of toxic trace metals in the marine environment. *Fresenius Z. Anal. Chem.*, **282**, 357–367.

Patterson, C. C., and D. Settle (1975). The reduction of orders of magnitude errors in lead analysis in biological materials and natural waters. In *Accuracy in Trace Analysis: Sampling, Sample Handling, Analysis*. (Ed. P. Lafleur), U.S. National Bureau of Standards Special Publication 422, pp. 321–351.

Perone, S. P., and K. K. Davenport (1966). Application of mercury-plated graphite electrodes to voltammetry and chronopotentiometry. *J. Electroanal. Chem.*, **12**, 269–276.

Perone, S. P., and A. Brumfield (1967). Theoretical and experimental study of anodic stripping voltammetry with mercury-plated graphite electrodes. *J. Electroanal. Chem.*, **13**, 124–131.

Petek, M., and M. Branica (1968). Hydrographical and biotical conditions in North Adriatic. III. Distribution of zinc and iodate. *Thalassia Jugosl.*, **5**, 257–261.

Petrie, L. M., and R. W. Baier (1976). Lead (II) transport processes in anodic stripping voltammetric analysis. *Anal. Chim. Acta*, **82**, 255–264.

Petrie, L. M., and R. W. Baier (1978). Thin mercury-film voltammetry of inorganic lead (II) complexes in seawater. *Anal. Chem.*, **50**, 351–357.

Piro, A., M. Verzi, and C. Papucci (1969). The importance of the physico-chemical state of elements for their accumulation in marine organisms. I. The physico-chemical state of zinc in sea water. *Publ. Staz. Zool. Napoli*, **37**, 298–310.

Piro, A., M. Bernhard, and M. Branica (1973). Incomplete exchange reaction between radioactive ionic Zn and stable natural Zn in seawater. In *Radioactive Contamination of the Marine Environment, Proc. Symp. Seattle, 10–14 July*, IAEA, Vienna, pp. 29–45.

Portmann, J. E., and J. P. Riley (1966). The determination of bismuth in sea and natural waters. *Anal. Chim. Acta*, **34**, 201–210.

Riley, J. P., and D. Taylor (1968). Chelating resins for the concentration of trace elements from sea water and their analytical use in conjunction with atomic absorption spectrophotometry. *Anal. Chim. Acta*, **40**, 479–485.

Riley, J. P., and D. Taylor (1968). The use of chelating ion exchange in the determination of molybdenum and vanadium in sea water. *Anal. Chim. Acta*, **41**,

175–178.

Robertson, D. E. (1968). Role of contamination in trace element analysis of sea water. *Anal. Chem.*, **40**, 1067–1072.

Robertson, D. E. (1972). Contamination problems in trace element analysis and ultra purification. In *Ultra-purity: Methods and Techniques* (Eds. M. Zief and R. M. Speights). Marcel Dekker, New York.

Roe, D. K., and J. E. A. Toni (1965). An equation for anodic stripping curves of thin mercury film electrodes. *Anal. Chem.*, **37**, 1503–1506.

Rojahn, T. (1972). Determination of copper, lead, cadmium and zinc in estuarine water by anodic-stripping alternating-current voltammetry on the hanging mercury drop electrode. *Anal. Chim. Acta*, **62**, 438–441.

Rooney, R. C. (1966). Derivative cathode ray polarography. *J. Polarogr. Soc.*, **9**, 45–48.

Ross, J. W., R. D. DeMars, and I. Shain (1956). Analytical applications of the hanging drop mercury electrode. *Anal. Chem.*, **28**, 1768–1771.

Samuelson, O. (1963). *Ion Exchange Separations*. Wiley, New York.

Seitz, W. R. (1970). Trace metal analysis in seawater by anodic stripping voltammetry. *Thesis*, Massachusetts Institute of Technology, Cambridge, Mass.

Shain, I. (1963). Stripping analysis. In *Treatise on Analytical Chemistry*, Vol. 4, Part I, Sect. D2 (Eds. I. M. Kolthoff and P. D. Elving). Interscience, New York, pp. 2533–2568.

Shuman, M. S., and G. P. Woodward, Jr. (1976). Intermetallic compound formation between copper and zinc in mercury and its effects on anodic stripping voltammetry. *Anal. Chem.*, **48**, 1979–1983.

Shutz, D. F., and K. K. Turekian (1965). The investigation of the geographical and vertical distribution of several trace elements in sea water using neutron activation analysis. *Geochim. Cosmochim. Acta*, **29**, 259–313.

Siegerman, H., and G. O'Dom (1972). Differential pulse anodic stripping of trace metals. *Am. Lab.*, **4**, 59–60, 62, and 64–68.

Smith, J. D., and J. D. Redmond (1971). Anodic stripping voltammetry applied to trace metals in sea water. *J. Electroanal. chem.*, **33**, 169–175.

Spencer, D. W., and P. G. Brewer (1969). The distribution of copper, zinc and nickel in sea water of the Gulf of Maine and the Sargasso Sea. *Geochim. Cosmochim. Acta*, **35**, 325–339.

Strickland, J. D. H., and T. R. Parsons (1965). A manual of seawater analysis, 2nd ed. *Fish. Res. Bd. Can. Bulletin.* 125.

Strojek, Z., B. Stepnik, and Z. Kublik (1976). Cyclic and stripping voltammetry with graphite based thin mercury film electrodes prepared *in situ*. *J. Electroanal. Chem.*, **74**, 277–295.

Stumm, W., and H. Bilinski (1973). *Advances in Water Pollution Research*. Pergamon Press, Oxford and New York.

Subramanian, K. S., C. L. Chakrabarti, J. E. Sueiras, and I. S. Marnes (1978). Preservation of some trace metals in samples of natural waters. *Anal. Chem.*, **50**, 444–448.

Tatsumoto, M., and C. C. Patterson (1963). The concentration of common lead in seawater. In *Earth Science and Meteorites*. North-Holland, Amsterdam, pp. 74–89.

Tikhonov, M. K., and V. Z. Zhavoronkina (1960). *Sov. Oceanogr.*, **3**, 22.

Tikhonov, M. K. and G. A. Shalimov (1965). *Gidrofiz. Gidrokhim. Issled.* Acad. Nauk Ukr. SSR, p. 33.

Turner, D. R., and M. Whitfield (1979). The reversible electrodeposition of trace metal ions from multi-ligand systems. Part 1. Theory. *J. Electroanal. Chem.*, **103**, 43–60.

Wallace, G. T., Jr., G. L. Hoffman, and R. A. Deuce (1977). The influence of organic matter and atmospheric deposition on the particulate trace metal concentration of northwest Atlantic surface seawater. *Mar. Chem.*, **5,** 143–170.

Wang, J., and M. Ariel (1977a). Anodic stripping voltammetry in a flow-through cell with a fixed mercury film glassy carbon disc electrode. Part I. *J. Electroanal. Chem.*, **83,** 217–224.

Wang, J., and M. Ariel (1977b). Anodic stripping voltammetry in a flow-through cell with fixed mercury film glassy carbon disc electrodes. Part II. The differential mode. *J. Electroanal. Chem.*, **85,** 289–297.

Whitfield, M. (1975). The electroanalytical chemistry of seawater. In *Chemical Oceanography*, 2nd ed. (Eds. J. P. Riley and G. Skirrow). Academic Press, London, pp. 1–154.

Whitnack, G. C. (1961). Applications of cathode-ray polarography in the field of oceanography. *J. Electroanal. Chem.*, **2,** 110–115.

Whitnack, G. C. (1966). Application of single-sweep polarography to the analysis of trace elements in seawater. In *Polarography 1964* (Ed. G. J. Hills). Interscience, New York, pp. 641–651.

Whitnack, G. C. (1973). Recent applications of single-sweep polarography, aboard ship and in the laboratory to the analysis of trace elements in seawater. In *Marine Electrochemistry* (Eds. J. B. Berkowitz, R. A. Horne, M. Banus, P. L. Howard, M. J. Pryar, G. C. Whitnack and H. V. Weiss). Electrochemical Society, Princeton, N.J., pp. 342–351.

Williams, P. M. (1969). The determination of dissolved organic carbon in seawater. A comparison of two methods. *Limnol. Oceanogr.*, **14,** 297–298.

Wopshall, R. H. and I. Shain (1967). Effect of adsorption of electroactive species in stationary electrode polarography. *Anal. Chem.*, **39,** 1514–1527.

Zief, M., and J. W. Mitchell (1976). *Contamination Control in Trace Element Analysis.* Wiley–Interscience, New York.

Zief, M., and R. M. Speights, Eds. (1972). *Ultrapurity: Methods and Techniques.* Marcel Dekker, New York.

Zieglerova, L., K. Stulik, and J. Dolezal (1971). Use of chelating agents in anodic stripping voltammetry. *Talanta*, **18,** 603–613.

Zirino, A. (1970). Voltammetric measurements, speciation and distribution of zinc in ocean water. *Diss. Abstr. Int.*, **31,** 2155.

Zirino, A., and M. L. Healy (1970). Inorganic zinc complexes in sea water. *Limnol. Oceanogr.*, **15,** 956–958.

Zirino, A., and M. L. Healy (1971). Voltammetric measurement of zinc in the north-eastern tropical Pacific Ocean. *Limnol. Oceanogr.*, **16,** 773–778.

Zirino, A., and M. L. Healy (1972). pH-controlled differential voltammetry of certain trace transition elements in natural waters. *Environ. Sci. Technol.*, **6,** 243–249.

Zirino, A., and S. P. Kounaves (1977). Anodic stripping peak currents: electrolysis potential relationships for reversible systems. *Anal. Chem.*, **49,** 56–59. (Correction, *idem*, **51,** 592, 1979).

Zirino, A. and S. P. Kounaves (1980). Stripping polarography and the reduction of Cu(II) in sea water at the hanging mercury drop electrode. *Anal. Chim. Acta*, **113,** 79–90.

Zirino, A., and S. H. Lieberman (1975). *Analytical Methods in Oceanography.* American Chemical Society, Washington, D.C., p. 88.

Zirino, A., S. H. Lieberman, and M. L. Healy (1973). Anodic stripping voltammetry of trace metals in sea water. In *Marine Electrochemistry* (Eds. J. B. Berkowitz,

R. A. Horne, M. Banus, P. L. Howard, M. J. Pryar, G. C. Whitnack and H. V. Weiss). Electrochemical Society, Princeton, N.J., pp. 319–332.

Zirino, A., S. H. Lieberman, and C. Clavell (1978). Measurement of Cu and Zn in San Diego Bay by automated anodic stripping voltammetry. *Environ. Sci. Technol.*, **12**, 73–79.

Zirino, A. and Yamamoto, S. (1972). pH-dependent model for the chemical speciation of copper, zinc, cadmium and lead in seawater. *Limnol. Oceanogr.*, **17**, 661–671.

Zutic, V., B. Ćosović, and Z. Kozarac (1977). Electrochemical determination of surface active substances in natural waters. On the adsorption of petroleum fractions at mercury electrode/seawater interface. *J. Electroanal. Chem.*, **78**, 113–121.

Appendixes: Tabulations of properties of sea water of direct relevance to electrochemical analysis

For detailed information on the physical chemistry of sea water see the reviews by Leyendekkers (1976), Millero (1974a,b, 1979), Millero and Leung (1976), Pytkowicz and Kester (1971), Walton Smith (1974), and Whitfield (1975b, 1979b).

References cited in the Appendixes are given in the bibliography to Chapter 1.

Units. The units used are those quoted in the original references. For consistency, conversion factors to the appropriate SI units are given in footnotes

I. DENSITY ($g\,cm^{-3}$)

Temperature, °C	S, ‰							
	0	5	10	15	20	25	30	35
0	0.9999	1.0039	1.0080	1.0120	1.0160	1.0201	1.0241	1.0281
5	1.0000	1.0040	1.0079	1.0119	1.0158	1.0198	1.0237	1.0277
10	0.9997	1.0036	1.0075	1.0114	1.0153	1.0191	1.0231	1.0270
15	0.9991	1.0030	1.0068	1.0106	1.0145	1.0183	1.0222	1.0260
20	0.9982	1.0020	1.0058	1.0096	1.0134	1:0172	1.0210	1.0248
25	0.9971	1.0008	1.0046	1.0083	1.0120	1.0158	1.0196	1.0234
30	0.9957	0.9994	1.0031	1.0068	1.0106	1.0143	1.0180	1.0218
35	0.9941	0.9978	1.0015	1.0052	1.0088	1.0125	1.0163	1.0200

* $1\,g\,cm^{-3} \equiv 10^3\,kg\,m^{-3}$.

II. DIELECTRIC CONSTANT*

S, ‰	t, °C							
	0.00		5.00		10.00		15.00	
0.00	87.85	24.49	85.87	21.21	83.94	18.06	82.05	15.04
5.00	86.40	26.87	84.45	24.35	82.54	21.93	80.67	19.60
10.00	84.99	29.17	83.07	27.39	81.19	25.67	79.34	24.01
15.00	83.63	31.40	81.73	30.33	79.88	29.29	78.06	28.28
20.00	82.31	33.55	80.44	33.17	78.61	32.79	76.81	32.40
25.00	81.04	35.64	79.19	35.93	77.38	36.18	75.60	36.40
30.00	79.80	37.66	77.97	38.60	76.18	39.46	74.43	40.27
35.00	78.59	39.62	76.79	41.18	75.02	42.65	73.30	44.02
40.00	77.43	41.52	75.65	43.69	73.90	45.73	72.19	47.65

S‰	t, °C							
	20.00		25.00		30.00		35.00	
0.00	80.19	12.15	78.38	9.38	76.60	6.74	74.86	4.21
5.00	78.84	17.37	77.05	15.23	75.30	13.18	73.58	11.21
10.00	77.54	22.42	75.77	20.88	74.04	19.41	72.35	17.98
15.00	76.27	27.30	74.53	26.35	72.82	25.42	71.15	24.53
20.00	75.05	32.02	73.33	31.63	71.64	31.24	70.00	30.86
25.00	73.87	36.59	72.17	36.75	70.50	63.88	68.88	36.98
30.00	72.72	41.01	71.04	41.70	69.40	42.33	67.79	42.90
35.00	71.60	45.30	69.95	46.50	68.33	47.61	66.74	48.64
40.00	70.52	49.46	68.89	51.14	67.29	52.73	65.72	54.20

* Dielectric constant ≡ relative permittivity (SI). Calculated from equations of Ho and Hall (1973, sea water data) and Bradley and Pitzer (1979, pure water data). The first column gives the dielectric constant (ε_s', Table 7, Chapter 1) and the second column the dielectric loss (ε_s'', Table 7, Chapter 1). Leyendekkers (1976, p. 82) gives data on the individual salt contributions.

III. VISCOSITY, $\eta(g\,cm^{-1}\,s^{-1} \times 10^2)$*†

$t,°C$	0	5	10	15	20	25	30	35
					$S,‰$			
0	1.791	1.804	1.817	1.831	1.844	1.857	1.870	1.884
5	1.519	1.531	1.543	1.555	1.567	1.579	1.592	1.604
10	1.307	1.318	1.329	1.341	1.352	1.363	1.374	1.385
15	1.138	1.149	1.160	1.170	1.180	1.190	1.201	1.211
20	1.002	1.012	1.022	1.032	1.041	1.051	1.061	1.070
25	0.890	0.900	0.909	0.918	0.927	0.936	0.946	0.955
30	0.797	0.807	0.816	0.824	0.833	0.841	0.850	0.858
35	0.719	0.729	0.737	0.745	0.753	0.761	0.769	0.777

* Calculated from equation 2, Table 7, Chapter 1, using a linear interpolation for A and B and assuming that for pure water

$$\log(\eta_t/\eta_{20}) = [1.1709(20-t) - 0.001827(t-20)^2]/(t+89.93)$$

References to measurements of η as a function of pressure are given in the Appendices to Riley and Skirrow (1975).

† $1\,g\,cm^{-1}\,s^{-1} \equiv 0.1\,kg\,m^{-1}\,s^{-1}$.

IV. KINEMATIC VISCOSITY, $\nu(cm^2\,s^{-1} \times 10^2)$*

$t,°C$	0	5	10	15	20	25	30	35
					$S,‰$			
0	1.791	1.797	1.803	1.809	1.815	1.821	1.826	1.832
5	1.519	1.525	1.531	1.537	1.543	1.549	1.555	1.561
10	1.307	1.314	1.320	1.325	1.331	1.337	1.349	1.359
15	1.139	1.146	1.152	1.157	1.163	1.169	1.175	1.181
20	1.004	1.010	1.016	1.022	1.028	1.033	1.039	1.044
25	0.893	0.899	0.905	0.911	0.916	0.922	0.927	0.933
30	0.801	0.807	0.813	0.819	0.824	0.829	0.835	0.840
35	0.724	0.730	0.736	0.741	0.746	0.752	0.757	0.762

* $1\,cm^2\,s^{-1} \equiv 10^{-4}\,m^2\,s^{-1}$.

V. OSMOTIC COEFFICIENT, ϕ*

	S, ‰						
t, °C	5	10	15	20	25	30	35
0	0.9149	0.9013	0.8950	0.8919	0.8907	0.8909	0.8925
5	0.9151	0.9020	0.8963	0.8937	0.8929	0.8936	0.8954
10	0.9151	0.9024	0.8972	0.8950	0.8946	0.8956	0.8978
15	0.9149	0.9026	0.8977	0.8959	0.8959	0.8971	0.8996
20	0.9146	0.9026	0.8980	0.8965	0.8967	0.8982	0.9009
25	0.9143	0.9025	0.8981	0.8967	0.8972	0.8989	0.9017
30	0.9138	0.9021	0.8979	0.8968	0.8974	0.8993	0.9023
35	0.9132	0.9016	0.8976	0.8966	0.8973	0.8994	0.9025

* Calculated from equation 4, Table 7, Chapter 1.

VI. OSMOTIC PRESSURE, π(bars)*†

	S, ‰						
t, °C	5	10	15	20	25	30	35
0	3.34	6.62	9.91	13.23	16.60	20.03	23.54
5	3.40	6.74	10.10	13.50	16.95	20.40	24.05
10	3.46	6.87	10.29	13.76	17.28	20.88	24.54
15	3.52	6.99	10.47	14.01	17.60	21.27	25.02
20	3.58	7.10	10.65	14.25	17.91	21.64	25.46
25	3.63	7.21	10.82	14.48	18.21	22.01	25.90
30	3.69	7.32	10.98	14.70	18.49	22.35	26.31

* Calculated from equation 5, table 7, Chapter 1.
† 1 bar = 10^5 N m^{-2}.

VII. VAPOUR PRESSURE, p(mmHg)*†

	S, ‰						
t, °C	0	5	10	15	20	25	30	35
0	4.58	4.57	4.56	4.55	4.53	4.52	4.50	4.50
5	6.54	6.52	6.50	6.49	6.47	6.46	6.44	6.42
10	9.21	9.18	9.16	9.14	9.11	9.09	9.06	9.04
15	12.79	12.76	12.72	12.69	12.66	12.62	12.59	12.55
20	17.54	17.49	17.45	17.40	17.36	17.31	17.26	17.21
25	23.77	23.70	23.64	23.58	23.52	23.45	23.39	23.32
30	31.84	31.76	31.67	31.59	31.51	31.42	31.33	31.25
35	42.02	42.09	41.98	41.87	41.76	41.65	41.58	41.41

* Calculated from equation 6, Table 7, Chapter 1, using values of p_0 taken from Ambrose and Lawrenson (1972).
† 1 mmHg ≡ 133.322 N m^{-2}.

VIII. FREEZING POINT DEPRESSIONS, $t_f(°C)$, AND BOILING POINT ELEVATIONS, $t_b(°C)$, at 1 atm (1.01 kPa)

$S, ‰$	5	10	15	20	25	30	35	40
t_f^*	−0.275	−0.541	−0.810	−1.082	−1.359	−1.638	−1.922	−2.209
$t_b\dagger$	0.08	0.15	0.23	0.31	0.39	0.49	0.54	0.64

* Calculated from equation 7, Table 7, Chapter 1.
† Interpolated from values given by Dietrich (1963).

IX. SURFACE TENSION, $\tau(N\ m^{-1} \times 10^3)^*$

$S, ‰$				$t, °C$				
	0	5	10	15	20	25	30	35
0	75.64	74.92	74.20	73.48	72.76	72.04	71.32	70.60
10	75.86	75.14	74.42	73.70	72.98	72.26	71.54	70.82
20	76.08	75.36	74.64	73.92	73.20	72.48	71.76	71.04
30	76.30	75.58	74.86	74.14	73.42	72.70	71.98	71.26
35	76.41	75.80	74.97	74.36	73.53	72.92	72.09	71.48
40	76.52	76.03	75.08	74.95	73.64	73.15	72.20	71.71

* Calculated from equation 8, Table 7, Chapter 1, for clean sea water uncontaminated by surfactants.

X. GAS SOLUBILITY $(cm^3\ l^{-1}\ atm^{-1})^*$

Gas	A_1	A_2	A_3	A_4	B_1	B_2	B_3
N_2	−172.4965	248.4262	143.0738	−21.7120	−0.049781	0.025018	−0.0034861
O_2	−173.4292	249.6339	143.3483	−21.8492	−0.033096	0.014259	−0.0017000
Ar	−173.5146	245.4510	141.8222	−21.8020	−0.034474	0.014934	−0.0017729
Ne	−160.2630	211.0969	132.1657	−21.3165	−0.122883	−0.077055	−0.0125568
He	−152.9405	196.8840	126.8015	−20.6767	−0.040543	0.021315	−0.0030732
H_2	−49.641	67.460	21.028	—	−0.077314	0.046580	−0.0074291
$CO_2\dagger$	−58.0931	90.5069	22.2940	—	0.027766	−0.025888	0.0050578

* Parameters for equation 9, Table 7, Chapter 1 $(1\ cm^3\ dm^{-3}\ atm^{-1} \equiv 9.8692 \times 10^{-9}\ N^{-1}\ m^2)$.
† Solubilities in mol dm^{-3} sea water atm^{-1} corrected for non-ideal gas behaviour.

XI. OXYGEN SATURATION VALUES ($cm^3 dm^{-3}$) FOR WATER AND SEA WATER RELATED TO WATER VAPOUR-SATURATED AIR AT A TOTAL PRESSURE OF 101.3 kPa (1 atm)

Calculated according to the equation of Weiss (1970) (see Table 7, Chapter 1).

$t, °C$	0.0	1.0	2.0	3.0	4.0	5.0	6.0	7.0	8.0	9.0	10.0	11.0	12.0	13.0
0.0	10.22	10.15	10.08	10.01	9.94	9.87	9.81	9.74	9.67	9.61	9.54	9.48	9.41	9.35
1.0	9.94	9.87	9.80	9.74	9.67	9.60	9.54	9.48	9.41	9.35	9.28	9.22	9.16	9.10
2.0	9.67	9.60	9.54	9.47	9.41	9.35	9.28	9.22	9.16	9.10	9.04	8.98	8.92	8.86
3.0	9.41	9.35	9.28	9.22	9.16	9.10	9.04	8.98	8.92	8.86	8.80	8.74	8.68	8.63
4.0	9.16	9.10	9.04	8.98	8.92	8.86	8.81	8.75	8.69	8.63	8.57	8.52	8.46	8.41
5.0	8.93	8.87	8.81	8.75	8.70	8.64	8.58	8.53	8.47	8.41	8.36	8.30	8.27	8.19
6.0	8.70	8.65	8.59	8.53	8.48	8.42	8.37	8.31	8.26	8.20	8.15	8.10	8.05	7.99
7.0	8.49	8.43	8.38	8.32	8.27	8.22	8.16	8.11	8.06	8.00	7.95	7.90	7.85	7.80
8.0	8.28	8.23	8.17	8.12	8.07	8.02	7.97	7.91	7.86	7.81	7.76	7.71	7.66	7.61
9.0	8.08	8.03	7.98	7.93	7.88	7.83	7.78	7.73	7.68	7.63	7.58	7.53	7.48	7.44
10.0	7.89	7.84	7.79	7.74	7.69	7.64	7.60	7.55	7.50	7.45	7.41	7.36	7.31	7.27
11.0	7.71	7.66	7.61	7.56	7.52	7.47	.7.42	7.38	7.33	7.28	7.24	7.19	7.15	7.10
12.0	7.53	7.49	7.44	7.39	7.35	7.30	7.26	7.21	7.17	7.12	7.08	7.03	6.99	6.95
13.0	7.37	7.32	7.27	7.23	7.18	7.14	7.10	7.05	7.01	6.96	6.92	6.88	6.84	6.79
14.0	7.20	7.16	7.12	7.07	7.03	6.98	6.94	6.90	6.86	6.81	6.77	6.73	6.69	6.65
15.0	7.05	7.00	6.96	6.92	6.88	6.84	6.79	6.75	6.71	6.67	6.63	6.59	6.55	6.51
16.0	6.90	6.86	6.81	6.77	6.73	6.69	6.65	6.61	6.57	6.53	6.49	6.45	6.41	6.37
17.0	6.75	6.71	6.67	6.63	6.59	6.55	6.51	6.47	6.44	6.40	6.36	6.32	6.28	6.24
18.0	6.61	6.58	6.54	6.50	6.46	6.42	6.38	6.34	6.31	6.27	6.23	6.19	6.16	6.12
19.0	6.48	6.44	6.40	6.37	6.33	6.29	6.25	6.22	6.18	6.14	6.11	6.07	6.03	6.00
20.0	6.35	6.31	6.28	6.24	6.20	6.17	6.13	6.09	6.06	6.02	5.99	5.95	5.92	5.88
21.0	6.23	6.19	6.15	6.12	6.08	6.05	6.01	5.98	5.94	5.91	5.87	5.84	5.80	5.77
22.0	6.11	6.07	6.04	6.00	5.97	5.93	5.90	5.86	5.83	5.79	5.76	5.73	5.69	5.66
23.0	5.99	5.96	5.92	5.89	5.85	5.82	5.79	5.75	5.72	5.69	5.65	5.62	5.59	5.56
24.0	5.88	5.84	5.81	5.78	5.74	5.71	5.68	5.65	5.61	5.58	5.55	5.52	5.49	5.46
25.0	5.77	5.74	5.70	5.67	5.64	5.61	5.58	5.54	5.51	5.48	5.45	5.42	5.39	5.36
26.0	5.66	5.63	5.60	5.57	5.54	5.51	5.48	5.44	5.41	5.38	5.35	5.32	5.29	5.26
27.0	5.56	5.53	5.50	5.47	5.44	5.41	5.38	5.35	5.32	5.29	5.26	5.23	5.20	5.17
28.0	5.46	5.43	5.40	5.37	5.34	5.31	5.28	5.25	5.23	5.20	5.17	5.14	5.11	5.08
29.0	5.37	5.34	5.31	5.28	5.25	5.22	5.19	5.16	5.14	5.11	5.08	5.05	5.02	5.00
30.0	5.28	5.25	5.22	5.19	5.16	5.13	5.10	5.08	5.05	5.02	4.99	4.97	4.94	4.91
31.0	5.19	5.16	5.13	5.10	5.07	5.05	5.02	4.99	4.96	4.94	4.91	4.88	4.86	4.83
32.0	5.10	5.07	5.04	5.02	4.99	4.96	4.94	4.91	4.88	4.86	4.83	4.80	4.78	4.75
33.0	5.01	4.99	4.96	4.93	4.91	4.88	4.86	4.83	4.80	4.78	4.75	4.73	4.70	4.68
34.0	4.93	4.91	4.88	4.85	4.83	4.80	4.78	4.75	4.73	4.70	4.68	4.65	4.63	4.60
35.0	4.85	4.83	4.80	4.78	4.75	4.73	4.70	4.68	4.65	4.63	4.60	4.58	4.55	4.53

XI (*Continued*)

t, °C	\multicolumn{14}{c}{S, ‰}

t, °C	14.0	15.0	16.0	17.0	18.0	19.0	20.0	21.0	22.0	23.0	24.0	25.0	26.0	27.0
0.0	9.29	9.22	9.16	9.10	9.04	8.97	8.91	8.85	8.79	8.73	8.67	8.61	8.56	8.50
1.0	9.04	8.97	8.91	8.85	8.79	8.73	8.68	8.62	8.56	8.50	8.44	8.39	8.33	8.27
2.0	8.80	8.74	8.68	8.62	8.56	8.51	8.45	8.39	8.34	8.28	8.22	8.17	8.11	8.06
3.0	8.57	8.51	8.45	8.40	8.34	8.29	8.23	8.18	8.12	8.07	8.01	7.96	7.91	7.86
4.0	8.35	8.29	8.24	8.19	8.13	8.08	8.02	7.97	7.92	7.87	7.81	7.76	7.71	7.66
5.0	8.14	8.09	8.03	7.98	7.93	7.88	7.83	7.77	7.72	7.67	7.62	7.57	7.52	7.47
6.0	7.94	7.89	7.84	7.79	7.74	7.69	7.64	7.59	7.54	7.49	7.44	7.39	7.34	7.29
7.0	7.75	7.70	7.65	7.60	7.55	7.50	7.45	7.40	7.36	7.31	7.26	7.22	7.17	7.12
8.0	7.57	7.52	7.47	7.42	7.37	7.33	7.28	7.23	7.19	7.14	7.09	7.05	7.00	6.96
9.0	7.39	7.34	7.30	7.25	7.20	7.16	7.11	7.07	7.02	6.98	6.93	6.89	6.84	6.80
10.0	7.22	7.17	7.13	7.08	7.04	6.99	6.95	6.91	6.86	6.82	6.78	6.73	6.69	6.65
11.0	7.06	7.01	6.97	6.93	6.88	6.84	6.80	6.75	6.71	6.67	6.63	6.58	6.54	6.50
12.0	6.90	6.86	6.82	6.77	6.73	6.69	6.65	6.61	6.56	6.52	6.48	6.44	6.40	6.36
13.0	6.75	6.71	6.67	6.63	6.59	6.55	6.50	6.46	6.42	6.38	6.34	6.31	6.27	6.23
14.0	6.61	6.57	6.53	6.49	6.45	6.41	6.37	6.33	6.29	6.25	6.21	6.17	6.14	6.10
15.0	6.47	6.43	6.39	6.35	6.31	6.27	6.24	6.20	6.16	6.12	6.08	6.05	6.01	5.97
16.0	6.34	6.30	6.26	6.22	6.18	6.15	6.11	6.07	6.03	6.00	5.96	5.93	5.89	5.85
17.0	6.21	6.17	6.13	6.10	6.06	6.02	5.99	5.95	5.91	5.88	5.84	5.81	5.77	5.74
18.0	6.08	6.05	6.01	5.97	5.94	5.90	5.87	5.83	5.80	5.76	5.73	5.69	5.66	5.63
19.0	5.96	5.93	5.89	5.86	5.82	5.79	5.75	5.72	5.69	5.65	5.62	5.59	5.55	5.52
20.0	5.85	5.81	5.78	5.74	5.71	5.68	5.64	5.61	5.58	5.54	5.51	5.48	5.45	5.42
21.0	5.74	5.70	5.67	5.64	5.60	5.57	5.54	5.51	5.47	5.44	5.41	5.38	5.35	5.32
22.0	5.63	5.60	5.56	5.53	5.50	5.47	5.44	5.40	5.37	5.34	5.31	5.28	5.25	5.22
23.0	5.52	5.49	5.46	5.43	5.40	5.37	5.34	5.31	5.28	5.24	5.21	5.18	5.15	5.12
24.0	5.42	5.39	5.36	5.33	5.30	5.27	5.24	5.21	5.18	5.15	5.12	5.09	5.06	5.03
25.0	5.33	5.30	5.27	5.24	5.21	5.18	5.15	5.12	5.09	5.06	5.03	5.00	4.98	4.95
26.0	5.23	5.20	5.17	5.14	5.12	5.09	5.06	5.03	5.00	4.97	4.95	4.92	4.89	4.86
27.0	5.14	5.11	5.08	5.06	5.03	5.00	4.97	4.94	4.92	4.89	4.86	4.83	4.81	4.78
28.0	5.05	5.03	5.00	4.97	4.94	4.91	4.89	4.86	4.83	4.81	4.78	4.75	4.73	4.70
29.0	4.97	4.94	4.91	4.89	4.86	4.83	4.81	4.78	4.75	4.73	4.70	4.67	4.65	4.62
30.0	4.89	4.86	4.83	4.81	4.78	4.75	4.73	4.70	4.68	4.65	4.62	4.60	4.57	4.55
31.0	4.80	4.78	4.75	4.73	4.70	4.68	4.65	4.62	4.60	4.57	4.55	4.53	4.50	4.48
32.0	4.73	4.70	4.68	4.65	4.63	4.60	4.58	4.55	4.53	4.50	4.48	4.45	4.43	4.41
33.0	4.65	4.63	4.60	4.58	4.55	4.53	4.50	4.48	4.46	4.43	4.41	4.38	4.36	4.34
34.0	4.58	4.55	4.53	4.50	4.48	4.46	4.43	4.41	4.39	4.36	4.34	4.32	4.29	4.27
35.0	4.51	4.48	4.46	4.43	4.41	4.39	4.36	4.34	4.32	4.30	4.27	4.25	4.23	4.21

XI (*Continued*)

t, °C	28.0	29.0	30.0	31.0	32.0	33.0	34.0	35.0	36.0	37.0	38.0	39.0	40.0	41.0
								S, ‰						
0.0	8.44	8.38	8.32	8.27	8.21	8.16	8.10	8.05	7.99	7.94	7.88	7.83	7.77	7.72
1.0	8.22	8.16	8.11	8.05	8.00	7.94	7.89	7.84	7.78	7.73	7.68	7.63	7.58	7.52
2.0	8.01	7.95	7.90	7.85	7.79	7.74	7.69	7.64	7.59	7.53	7.48	7.43	7.38	7.33
3.0	7.80	7.75	7.70	7.65	7.60	7.55	7.50	7.45	7.40	7.35	7.30	7.25	7.20	7.15
4.0	7.61	7.56	7.51	7.46	7.41	7.36	7.31	7.26	7.22	7.17	7.12	7.07	7.03	6.98
5.0	7.42	7.37	7.33	7.28	7.23	7.18	7.14	7.09	7.04	7.00	6.95	6.90	6.86	6.81
6.0	7.25	7.20	7.15	7.11	7.06	7.01	6.97	6.92	6.88	6.83	6.79	6.74	6.70	6.66
7.0	7.08	7.03	6.98	6.94	6.89	6.85	6.81	6.75	6.72	6.67	6.63	6.59	6.55	6.50
8.0	6.91	6.87	6.82	6.78	6.74	6.69	6.65	6.61	6.57	6.52	6.48	6.44	6.40	6.36
9.0	6.76	6.71	6.67	6.63	6.59	6.54	6.50	6.46	6.42	6.38	6.34	6.30	6.26	6.22
10.0	6.61	6.56	6.52	6.48	6.44	6.40	6.36	6.32	6.28	6.24	6.20	6.16	6.12	6.08
11.0	6.46	6.42	6.38	6.34	6.30	6.26	6.22	6.18	6.14	6.10	6.07	6.03	5.99	5.95
12.0	6.32	6.28	6.24	6.21	6.17	6.13	6.09	6.05	6.01	5.98	5.94	5.90	5.87	5.83
13.0	6.19	6.15	6.11	6.07	6.04	6.00	5.96	5.93	5.89	5.85	5.82	5.78	5.74	5.71
14.0	6.06	6.02	5.99	5.95	5.91	5.88	5.84	5.80	5.77	5.73	5.70	5.66	5.63	5.59
15.0	5.94	5.90	5.87	5.83	5.79	5.76	5.72	5.69	5.65	5.62	5.58	5.55	5.52	5.48
16.0	5.82	5.78	5.75	5.71	5.68	5.64	5.61	5.58	5.54	5.51	5.48	5.44	5.41	5.38
17.0	5.70	5.67	5.64	5.60	5.57	5.53	5.50	5.47	5.43	5.40	5.37	5.34	5.31	5.27
18.0	5.59	5.56	5.53	5.49	5.46	5.43	5.40	5.36	5.33	5.30	5.27	5.24	5.21	5.17
19.0	5.49	5.45	5.42	5.39	5.36	5.33	5.29	5.26	5.23	5.20	5.17	5.14	5.11	5.08
20.0	5.38	5.35	5.32	5.29	5.26	5.23	5.20	5.17	5.14	5.10	5.07	5.05	5.02	4.99
21.0	5.28	5.25	5.22	5.19	5.16	5.13	5.10	5.07	5.04	5.01	4.98	4.95	4.93	4.90
22.0	5.19	5.16	5.13	5.10	5.07	5.04	5.01	4.98	4.95	4.92	4.89	4.87	4.84	4.81
23.0	5.10	5.07	5.04	5.01	4.98	4.95	4.92	4.89	4.87	4.84	4.81	4.78	4.75	4.73
24.0	5.01	4.98	4.95	4.92	4.89	4.86	4.84	4.81	4.78	4.75	4.73	4.70	4.67	4.65
25.0	4.92	4.89	4.86	4.84	4.81	4.78	4.75	4.73	4.70	4.67	4.65	4.62	4.59	4.57
26.0	4.83	4.81	4.78	4.75	4.73	4.70	4.67	4.65	4.62	4.59	4.57	4.54	4.52	4.49
27.0	4.75	4.73	4.70	4.67	4.65	4.62	4.60	4.57	4.54	4.52	4.49	4.47	4.44	4.42
28.0	4.67	4.65	4.62	4.60	4.57	4.55	4.52	4.50	4.47	4.45	4.42	4.40	4.37	4.35
29.0	4.60	4.57	4.55	4.52	4.50	4.47	4.45	4.42	4.40	4.37	4.35	4.33	4.30	4.28
30.0	4.52	4.50	4.47	4.45	4.43	4.40	4.38	4.35	4.33	4.31	4.28	4.26	4.24	4.21
31.0	4.45	4.43	4.40	4.38	4.36	4.33	4.31	4.28	4.26	4.24	4.22	4.19	4.17	4.15
23.0	4.38	4.36	4.33	4.31	4.29	4.26	4.24	4.22	4.20	4.17	4.15	4.13	4.11	4.08
33.0	4.31	4.29	4.27	4.24	4.22	4.20	4.18	4.15	4.13	4.11	4.09	4.07	4.04	4.02
34.0	4.25	4.23	4.20	4.18	4.16	4.14	4.11	4.09	4.07	4.05	4.03	4.01	3.98	3.96
35.0	4.18	4.16	4.14	4.12	4.10	4.07	4.05	4.03	4.01	3.99	3.97	3.95	3.93	3.91

XII. SPECIFIC CONDUCTIVITY (mmho cm^{-1})*†

$S, ‰$	$t, °C$					
	0	5	10	15	20	25
10	9.341	10.816	12.361	13.967	15.628	17.345
20	17.456	20.166	23.010	25.967	29.027	32.188
30	25.238	29.090	33.137	37.351	41.713	46.213
31	26.005	29.968	34.131	38.467	42.954	47.584
32	26.771	30.843	35.122	39.579	44.192	48.951
33	27.535	31.716	36.110	40.688	45.426	50.314
34	28.298	32.588	37.096	41.794	46.656	51.671
35	29.060	33.457	38.080	42.896	47.882	53.025

* Adapted from Cox (1965). For information on the effect of pressure on the conductivity of sea water, see Walton Smith (1974, p. 69).
† 1 mmho cm$^{-1} \equiv 10^{-1} \, \Omega^{-1}$ m.

XIII. LIMITING EQUIVALENT IONIC CONDUCTIVITIES IN AQUEOUS SOLUTIONS ($\Omega^{-1} m^2$ equiv^1)*

Ion	$t, °C$			
	0	18	25	100
H$^+$	0.0225	0.0315	0.03497	0.0637
Li$^+$	0.00191	0.00334	0.003868	0.0120
Na$^+$	0.002585	0.00435	0.005010.	0.0150
K$^+$	0.00403	0.00646	0.007350	0.0200
Rb$^+$	0.00435	0.00675	0.00775	—
Cs$^+$	0.0044	0.0068	0.00768	0.0200
NH$_4^+$	0.00403	0.0064	0.00737	0.01843
$\frac{1}{2}$Be^{2+}	—	—	0.0045	
$\frac{1}{2}$Mg^{2+}	0.00285	0.0046	0.005306	0.0170
$\frac{1}{2}$Ca^{2+}	0.00308	0.0051	0.00595	0.0187
$\frac{1}{2}$Sr^{2+}	0.0031	0.0051	0.00595	—
$\frac{1}{2}$Ba^{2+}	0.00336	0.00543	0.00637	0.0200
OH$^-$	0.0105	0.0174	0.01976	0.0446
Br$^-$	0.00431	0.00676	0.00784	—
CH$_3$CO$_2^-$	0.0020	0.0034	0.0041	0.0130
$\frac{1}{2}$CO$_3^{2-}$	0.0036	0.00605	0.00693	—
HCO$_3^-$	—	—	0.00445	—
Cl$^-$	0.00414	0.00655	0.00763	0.0207
ClO$_4^-$	0.00373	0.00591	0.00673	0.0179
F$^-$	—	0.00466	0.00554	—
I$^-$	0.00420	0.00665	0.00769	—
H$_2$PO$_4^-$	—	0.0028	0.0036	—
$\frac{1}{2}$HPO$_4^{2-}$	—	—	0.0057	—
$\frac{1}{2}$SO$_4^{2-}$	0.0041	0.00683	0.00798	0.0256

* Data selected from Dobos (1975).

Marine Electrochemistry

XIV. DIFFUSION COEFFICIENTS AT INFINITE DILUTION (10^6 cm^2 s^{-1})*†

Cation	t, °C			Anion	t, °C		
	0	18	25		0	18	25
H^+	56.1	81.7	93.1	OH^-	25.6	44.9	52.7
Li^+	4.72	8.69	10.3	F^-	—	12.1	14.6
Na^+	6.27	11.3	13.3	Cl^-	10.1	17.1	20.3
K^+	9.86	16.7	19.6	Br^-	10.5	17.6	20.1
Rb^+	10.6	17.6	20.6	I^-	10.3	17.2	20.0
Cs^+	10.6	17.7	20.7	IO_3^-	5.05	8.79	10.6
NH_4^+	9.80	16.8	19.8	HS^-	9.75	14.8	17.3
Ag^+	8.50	14.0	16.6	S^{2-}	—	6.95	—
Tl^+	10.6	17.0	20.1	HSO_4^-	—	—	13.3
$Cu(OH)^+$	—	—	8.30	SO_4^{2-}	5.00	8.90	10.7
$Zn(OH)^+$	—	—	8.54	SeO_4^{2-}	4.14	8.45	9.46
Be^{2+}	—	3.64	5.85	NO_2^-	—	15.3	19.1
Mg^{2+}	3.56	5.94	7.05	NO_3^-	9.78	16.1	19.0
Ca^{2+}	3.73	6.73	7.93	HCO_3^-	—	—	11.8
Sr^{2+}	3.72	6.70	7.94	CO_3^{2-}	4.39	7.80	9.55
Ba^{2+}	4.04	7.13	8.48	$H_2PO_4^-$	—	7.15	8.46
Ra^{2+}	4.02	7.45	8.89	HPO_4^{2-}	—	—	7.34
Mn^{2+}	3.05	5.75	6.88	PO_4^{3-}	—	—	6.12
Fe^{2+}	3.41	5.82	7.19	$H_2AsO_4^-$	—	—	9.05
Co^{2+}	3.41	5.72	6.99	$H_2SbO_4^-$	—	—	8.25
Ni^{2+}	3.11	5.81	6.79	CrO_4^{2-}	5.12	9.36	11.2
Cu^{2+}	3.41	5.88	7.33	MoO_4^{2-}	—	—	9.91
Zn^{2+}	3.35	6.13	7.15	WO_4^{2-}	4.27	7.67	9.23
Cd^{2+}	3.41	6.03	7.17				
Pb^{2+}	4.56	7.95	9.45				
UO_2^{2+}	—	—	4.26				
Sc^{3+}	—	—	5.74				
Y^{3+}	2.60	—	5.50				
La^{3+}	2.76	5.14	6.17				
Yb^{3+}	—	—	5.82				
Cr^{3+}	—	3.90	5.94				
Fe^{3+}	—	5.28	6.07				
Al^{3+}	2.36	3.46	5.59				
Th^{4+}	—	1.53	—				

* Reproduced with permission from Li, Y.H., and S. Gregory (1974), Diffusion coefficients of ions in sea water and deep-sea sediments, *Geochim. Cosmochim. Acta*, **38**, 703–714.
† 1 cm^2 s$^{-1} \equiv 10^{-4}$ m^2 s^{-1}.

Index